Geophysical Monograph Series

Including

IUGG Volumes

Maurice Ewing Volumes

Mineral Physics Volumes

Geophysical Monograph Series

158 The Nordic Seas: An Integrated Perspective *Helge Drange, Trond Dokken, Tore Furevik, Rüdiger Gerdes, and Wolfgang Berger (Eds.)*

159 Inner Magnetosphere Interactions: New Perspectives From Imaging *James Burch, Michael Schulz, and Harlan Spence (Eds.)*

160 Earth's Deep Mantle: Structure, Composition, and Evolution *Robert D. van der Hilst, Jay D. Bass, Jan Matas, and Jeannot Trampert (Eds.)*

161 Circulation in the Gulf of Mexico: Observations and Models *Wilton Sturges and Alexis Lugo-Fernandez (Eds.)*

162 Dynamics of Fluids and Transport Through Fractured Rock *Boris Faybishenko, Paul A. Witherspoon, and John Gale (Eds.)*

163 Remote Sensing of Northern Hydrology: Measuring Environmental Change *Claude R. Duguay and Alain Pietroniro (Eds.)*

164 Archean Geodynamics and Environments *Keith Benn, Jean-Claude Mareschal, and Kent C. Condie (Eds.)*

165 Solar Eruptions and Energetic Particles *Natchimuthukonar Gopalswamy, Richard Mewaldt, and Jarmo Torsti (Eds.)*

166 Back-Arc Spreading Systems: Geological, Biological, Chemical, and Physical Interactions *David M. Christie, Charles Fisher, Sang-Mook Lee, and Sharon Givens (Eds.)*

167 Recurrent Magnetic Storms: Corotating Solar Wind Streams *Bruce Tsurutani, Robert McPherron, Walter Gonzalez, Gang Lu, José H. A. Sobral, and Natchimuthukonar Gopalswamy (Eds.)*

168 Earth's Deep Water Cycle *Steven D. Jacobsen and Suzan van der Lee (Eds.)*

169 Magnetospheric ULF Waves: Synthesis and New Directions *Kazue Takahashi, Peter J. Chi, Richard E. Denton, and Robert L. Lysal (Eds.)*

170 Earthquakes: Radiated Energy and the Physics of Faulting *Rachel Abercrombie, Art McGarr, Hiroo Kanamori, and Giulio Di Toro (Eds.)*

171 Subsurface Hydrology: Data Integration for Properties and Processes *David W. Hyndman, Frederick D. Day-Lewis, and Kamini Singha (Eds.)*

172 Volcanism and Subduction: The Kamchatka Region *John Eichelberger, Evgenii Gordeev, Minoru Kasahara, Pavel Izbekov, and Johnathan Lees (Eds.)*

173 Ocean Circulation: Mechanisms and Impacts—Past and Future Changes of Meridional Overturning *Andreas Schmittner, John C. H. Chiang, and Sidney R. Hemming (Eds.)*

174 Post-Perovskite: The Last Mantle Phase Transition *Kei Hirose, John Brodholt, Thorne Lay, and David Yuen (Eds.)*

175 A Continental Plate Boundary: Tectonics at South Island, New Zealand *David Okaya, Tim Stem, and Fred Davey (Eds.)*

176 Exploring Venus as a Terrestrial Planet *Larry W. Esposito, Ellen R. Stofan, and Thomas E. Cravens (Eds.)*

177 Ocean Modeling in an Eddying Regime *Matthew Hecht and Hiroyasu Hasumi (Eds.)*

178 Magma to Microbe: Modeling Hydrothermal Processes at Oceanic Spreading Centers *Robert P. Lowell, Jeffrey S. Seewald, Anna Metaxas, and Michael R. Perfit (Eds.)*

179 Active Tectonics and Seismic Potential of Alaska *Jeffrey T. Freymueller, Peter J. Haeussler, Robert L. Wesson, and Göran Ekström (Eds.)*

180 Arctic Sea Ice Decline: Observations, Projections, Mechanisms, and Implications *Eric T. DeWeaver, Cecilia M. Bitz, and L.-Bruno Tremblay (Eds.)*

181 Midlatitude Ionospheric Dynamics and Disturbances *Paul M. Kintner, Jr., Anthea J. Coster, Tim Fuller-Rowell, Anthony J. Mannucci, Michael Mendillo, and Roderick Heelis (Eds.)*

182 The Stromboli Volcano: An Integrated Study of the 2002–2003 Eruption *Sonia Calvari, Salvatore Inguaggiato, Giuseppe Puglisi, Maurizio Ripepe, and Mauro Rosi (Eds.)*

183 Carbon Sequestration and Its Role in the Global Carbon Cycle *Brian J. McPherson and Eric T. Sundquist (Eds.)*

184 Carbon Cycling in Northern Peatlands *Andrew J. Baird, Lisa R. Belyea, Xavier Comas, A. S. Reeve, and Lee D. Slater (Eds.)*

185 Indian Ocean Biogeochemical Processes and Ecological Variability *Jerry D. Wiggert, Raleigh R. Hood, S. Wajih A. Naqvi, Kenneth H. Brink, and Sharon L. Smith (Eds.)*

186 Amazonia and Global Change *Michael Keller, Mercedes Bustamante, John Gash, and Pedro Silva Dias (Eds.)*

187 Surface Ocean–Lower Atmosphere Processes *Corinne Le Quèrè and Eric S. Saltzman (Eds.)*

188 Diversity of Hydrothermal Systems on Slow Spreading Ocean Ridges *Peter A. Rona, Colin W. Devey, Jérôme Dyment, and Bramley J. Murton (Eds.)*

189 Climate Dynamics: Why Does Climate Vary? *De-Zheng Sun and Frank Bryan (Eds.)*

190 The Stratosphere: Dynamics, Transport, and Chemistry *L. M. Polvani, A. H. Sobel, and D. W. Waugh (Eds.)*

191 Rainfall: State of the Science *Firat Y. Testik and Mekonnen Gebremichael (Eds.)*

192 Antarctic Subglacial Aquatic Environments *Martin J. Siegert, Mahlon C. Kennicutt II, and Robert A. Bindschadler (Eds.)*

Geophysical Monograph 193

Abrupt Climate Change: Mechanisms, Patterns, and Impacts

Harunur Rashid
Leonid Polyak
Ellen Mosley-Thompson
Editors

American Geophysical Union
Washington, DC

Library of Congress Cataloging-in-Publication Data

Abrupt climate change: mechanisms, patterns, and impacts / Harunur Rashid, Leonid Polyak, Ellen Mosley-Thompson, editors.
p. cm. — (Geophysical monograph, ISSN 0065-8448; 193)
Includes bibliographical references and index.
ISBN 978-0-87590-484-9
1. Climate changes. I. Rashid, Harunur, 1969- II. Polyak, Leonid. III. Mosley-Thompson, Ellen
QC902.9.A27 2011
551.6—dc23

2011039755

ISBN: 978-0-87590-484-9
ISSN: 0065-8448

Cover Image: (left) Plotted oxygen isotope data from the North Greenland Ice Core Project (black) and Mount Kilimanjaro (blue) and seawater oxygen isotope data (red) from the planktonic foraminifera *Globigerinoides ruber* (white) from the Andaman Sea of the northern Indian Ocean. (top middle) X-radiograph of centimeter thick sediment slice of Heinrich layer 2 from the northwest Labrador Sea (courtesy of H. Rashid). (top right) Photograph of ice cores collected from the Guliya ice field of China (courtesy of L. G. Thompson). (bottom) For the atmosphere, vectors depicting the winds in the lowest model layer and the sea level pressure. Ocean model surface temperatures are shown in blue to red colors; sea ice concentration is shown in white gray. This coupled model consists of the National Center for Atmospheric Research Community Climate Model for the atmospheric component, the Los Alamos National Laboratory Parallel Ocean Program for the ocean component, and the Naval Postgraduate School sea ice model (courtesy of G. Strand of NCAR). (backdrop) Sea ice in the Southern Ocean during the early autumn (courtesy of NOAA).

Printed in the United States of America.

CONTENTS

PREFACE

When the first book on abrupt climate change was published in 1987, the idea of millennial climatic oscillations now known as the Dansgaard-Oeschger events, named after seminal researchers Willi Dansgaard and Hans Oeschger, was slowly taking shape, while periodic collapse and rebuilding of the Northern Hemisphere ice sheets at millennial timescales, recognized as the Heinrich (named for Hartmut Heinrich) iceberg-rafting events, were not yet known. Using marine sediment data from the northeast Atlantic deep-sea cores, Gerard Bond and Wallace Broecker, and colleagues in 1992 and 1993 placed Heinrich's work into a wider climate context; climatologists began to consider mechanisms of reorganization of the global ocean-atmospheric system at millennial and finer timescales. Several collections of papers and reports published since that time reflect an impressive development in our understanding of the history and mechanisms of abrupt climate events: *Abrupt Climatic Change: Evidence and Implications* (NATO-ASI series published in 1987); the 1999 AGU Geophysical Monograph *Mechanisms of Global Climate Change at Millennial Scale Time Scales*; *Abrupt Climate Change: Inevitable Surprises*, published by the National Academy Press in 2002; and "Abrupt Climate Change," a 2008 report by the U.S. Climate Change Science Program and Subcommittee on Global Change Research.

This volume arises from contributions to the 2009 American Geophysical Union Chapman Conference on Abrupt Climate Change that addressed the progress made in understanding the mechanisms of abrupt climate events in the decade following the previous Chapman Conference on this topic (*Mechanisms of Global Climate Change at Millennial Scale Time Scales* and *Abrupt Climate Change: Inevitable Surprises*). The 2009 conference was held at the Byrd Polar Research Center of the Ohio State University, Columbus, Ohio, on 15–19 June 2009. A total of 105 scientists from 24 countries working in disciplines ranging from paleoclimatology, paleoceanography, and atmospheric and marine chemistry to paleoclimate model-data comparison to archaeology attended and presented their cutting-edge research at the weeklong conference. The basic purpose of the 2009 conference was to understand the spatiotemporal extent of abrupt climate change and the relevant forcings. This monograph covers a breadth of global paleoclimate research discussed at the conference and provides a list of critical topics that need to be resolved to better understand abrupt climate changes and thus advance our knowledge and the tools required to project future climate.

We would like to express our deep appreciation to participants and presenters for their cutting-edge research both at the conference and in this monograph. The conference was sponsored by the Office of Research and the Climate, Water and Carbon Program (CWC) of the Ohio State University, the Consortium for Ocean Leadership, and the National Science Foundation (OCE-0928601). As a result, we were able to provide travel support to all of the participating graduate students, postdoctoral researchers, and a few young investigators and most of the invited speakers. We acknowledge numerous reviewers for their critical assessment of the papers and help in streamlining the manuscripts. We would also like to acknowledge invaluable assistance provided by the staff of the American Geophysical Union.

Harunur Rashid
Byrd Polar Research Center, Ohio State University

Leonid Polyak
Byrd Polar Research Center, Ohio State University

Ellen Mosley-Thompson
Byrd Polar Research Center, Ohio State University
Department of Geography, Ohio State University

Abrupt Climate Change: Mechanisms, Patterns, and Impacts
Geophysical Monograph Series 193

10.1029/2011GM001138

Abrupt Climate Change Revisited

Harunur Rashid and Leonid Polyak

Byrd Polar Research Center, Ohio State University, Columbus, Ohio, USA

Ellen Mosley-Thompson

Byrd Polar Research Center, Ohio State University, Columbus, Ohio, USA
Department of Geography, Ohio State University, Columbus, Ohio, USA

This geophysical monograph volume contains a collection of papers presented at the 2009 AGU Chapman Conference on Abrupt Climate Change. Paleoclimate records derived from ice cores, lakes, and marine sedimentary archives illustrate rapid changes in the atmosphere-cryosphere-ocean system. Several proxy records and two data-model comparison studies simulate Atlantic meridional overturning circulation during the last deglaciation and millennial-scale temperature oscillations during the last glacial cycle and thereby provide new perspectives on the mechanisms controlling abrupt climate changes. Two hypotheses are presented to explain deep Southern Ocean carbon storage, the rapid increase of the atmospheric carbon dioxide, and retreat of sea ice in the Antarctic Ocean during the last deglaciation. A synthesis of two Holocene climate events at approximately 5.2 ka and 4.2 ka highlights the potential of a rapid response to climate forcing in tropical systems through the hydrological cycle. Both events appear contemporaneous with the collapse of ancient civilizations in low-latitude regions where roughly half of Earth's population lives today.

1. INTRODUCTION

Remarkable, well-dated evidence of high-magnitude, abrupt climate changes occurring during the last glacial period between ~80 and 10 ka have been documented in Greenland ice cores and marine and terrestrial records in the Northern Hemisphere. These climate perturbations are known as the Dansgaard-Oeschger (D-O) warming and cooling cycles of 2–3 kyr duration. These are bundled into Bond cycles of 5–10 kyr and are terminated by Heinrich events consisting of massive iceberg rafting from the late Pleistocene Laurentide Ice Sheet (LIS). Since the discovery of the D-O cycles in ice cores and their counterparts in marine sediments of the North Atlantic, the search for abrupt, millennial- or finer-scale events has intensified across the globe (see *Voelker et al.* [2002] and *Clement and Peterson* [2008] for an overview). In recent years, an increasing number of paleoclimatic records, mostly in the Northern Hemisphere show teleconnections with the D-O cycles recorded in Greenland. The most commonly inferred link between these rapid climate events is related to the release of cold fresh water by iceberg melting. The introduction of this fresh water is assumed to slow or shut down the formation of the North Atlantic Deep Water (NADW), thus preventing the penetration of the North Atlantic Drift (the northern branch of the Gulf Stream) into high latitudes. The climatic importance of the Gulf Stream stems from the enormous quantity of heat it transports to northwestern Europe and its facilitation of the exchange of moisture between the ocean and atmosphere.

Abrupt Climate Change: Mechanisms, Patterns, and Impacts
Geophysical Monograph Series 193

10.1029/2011GM001139

Marine and terrestrial paleoclimate records from the Southern Hemisphere (SH) are sparse and lack sufficient temporal resolution to characterize the timescales relevant for high-frequency climate variability. Although the evidence for abrupt climate change in the SH is not clear, a one-to-one correlation has been established between the new Eastern Droning Maud Land (EDML) Antarctic ice core record and that from Greenland [*EPICA Community Members et al.*, 2006]. The EDML ice core was recovered from a location facing the South Atlantic and is assumed to record climate changes in the Atlantic. Paleoclimatic records from the northern tropics and subtropics mainly show changes concordant with those in the North Atlantic, while asynchronous or even anticorrelated events are exhibited in records from the southern tropics and high latitudes of the SH. For example, the Indian and East Asian monsoon histories seem to correlate with the North Atlantic climate, whereas South American monsoon records are anticorrelated with the Greenland records [*Wang et al.*, 2006]. Furthermore, paleoclimatic proxy records from the equatorial Pacific are characterized by a complex pattern of abrupt climate change that borrows elements from both the Northern Hemisphere and Southern Hemisphere end members, suggesting that the tropical Pacific may have played a significant role in mediating climate teleconnections between the hemispheres [*Clement et al.*, 2004; *Li et al.*, 2010].

The Chapman Conference on Abrupt Climate Change (ACC) held at the Ohio State University in June 2009 brought together a diverse group of researchers dealing with various paleoclimatic proxy records (such as ice cores, corals, marine sediments, lakes, and speleothems) and coupled ocean-atmosphere climate models to discuss advances in understanding the ACC history and mechanisms. Special attention was given to the three most commonly invoked ACC mechanisms: (1) freshwater forcing, in which meltwater from the circum–North Atlantic ice sheets may have disrupted the meridional overturning circulation (MOC) by preventing or slowing down the formation of NADW in the Labrador and Nordic Seas; (2) changes in sea ice extent, which affect the ocean-atmosphere heat exchange, moisture supply, and salt content; and (3) tropical forcing that calls for a combination of Earth's orbital configuration, El Niño–Southern Oscillation, and sea surface temperature conditions. Several contributions dealing with various aspects of these topics are presented in this volume.

2. KEY QUESTIONS FOR THE ACC CONFERENCE

Without providing an exhaustive list of topics raised at the conference, we list a number of pressing scientific questions that were discussed:

1. Do paleoproxies suggest a one-to-one proxy-based relationship between circum–North Atlantic freshwater pulses and the strength of the MOC, as well as the determination of meltwater sources?
2. Are kinematic and nutrient proxies for the strength of the MOC (Pa/Th, Cd/Ca, and $\delta^{13}C$) congruent across abrupt climate changes that occurred during the Younger Dryas (YD) or Heinrich events? How do these proxies differ between periods of stable and transient climate?
3. Do current general circulation models capture abrupt strengthening and gradual weakening of thermohaline circulation, consistent with the rapid warming and gradual cooling of D-O events? If not, what other factors must be considered?
4. Is there robust evidence for sea ice in the North Atlantic during the last glacial cycle? How much has sea ice extent fluctuated on millennial timescales? How have the fluctuations influenced the surface salinity and thus water mass stratification?
5. With the exception of the tropical Atlantic, most tropical paleorecords show a clear lack of D-O cooling events. Does this indicate that various parts of the tropics respond differently to the North Atlantic freshwater forcing because of local hydrological and temperature variability?
6. Given the dramatic changes in Arctic sea ice and circulation, how did the Arctic freshwater budget affect the MOC in the North Atlantic?
7. Why do Antarctic temperatures show more gradual and less pronounced warmings and coolings compared to the D-O events in Greenland? Does this suggest that the deep ocean circulation is modulating the abrupt climate change?
8. In light of the current concern about instabilities of the West Antarctic and Greenland ice sheets, how can paleoceanographic records be used to decipher past ice sheet dynamics?
9. What is the link between sea ice extent and ice sheet dynamics? How does ocean heat transport influence the ice sheet margin? Might coastal ice shelves be slaves to ocean currents?
10. Was there a relationship between the demise of past civilizations and climatic deterioration? What are the climate tipping points that have driven past civilizations to collapse?

One of the important points raised at the conference is that a close examination of paleoclimatic data and modeling results does not show adequate support for many of the widely accepted explanations for abrupt climate change. For example, it is almost taken for granted that fresh water released from circum–North Atlantic ice sheets during Heinrich events perturbed the Atlantic meridional overturning circulation (AMOC), which caused abrupt changes recorded around the globe. However, with the exception of the Younger Dryas, there is no paleoproxy evidence from deep waters that would indicate a shutdown or slowdown of the AMOC during Heinrich events [*Lynch-Stieglitz et al.*, 2007]. This situation does not necessarily mean that changes in paleo-bottom water composition did not occur but simply shows that adequate supporting data are lacking. Even less certainty can be applied to paleorecords older than the last glacial cycle, when these abrupt climate changes were a recurrent phenomenon. Whether such recurrent events occurred during previous glacial cycles is not well documented because of the scarcity of long paleoclimatic records with the requisite spatial and temporal resolution.

Several new areas of inquiry were discussed during the meeting including (1) development of a new chronostratigraphy for Antarctic ice cores based on local insolation and independent from bias-prone orbital tuning [*Kawamura et al.*, 2007; *Laepple et al.*, 2011], (2) phasing between the deep ocean and surface water warming during terminations as derived from oxygen isotope records on benthic and planktonic foraminifers [*Rashid et al.*, 2009], (3) indication of monsoon failure from atmospheric oxygen isotopes and deep ocean temperature change from inert gases [*Severinghaus et al.*, 2009], (4) timing of elevated subantarctic opal fluxes and deep ocean carbon dioxide release to the atmosphere and phasing between these features and the position of the westerlies [*Anderson et al.*, 2009], (5) use of dynamic circulation proxies (Pa/Th, Nd isotopes, etc.) and models of freshwater forcing in assessing the strength of MOC, and (6) role of the Antarctic Intermediate Water in distributing heat and transporting old carbon around the ocean [*Marchitto et al.*, 2007; *Basak et al.*, 2010].

Members of the paleoclimate modeling community stressed the need to improve the temporal resolution and age constraints on paleoclimatic records. In addition, it was suggested to further explore "the meaning of proxies," resolve the leads and lags in important paleorecords, and provide benchmark tests for climate sensitivity analysis. For example, oxygen isotopes in speleothems could indicate the amount of precipitation, changes in seasonality, or changes in source region of moisture. More research is needed to clarify the relationship of oxygen isotopes with these variables. Modelers also asked questions regarding the sources of carbon dioxide increase during the Antarctic isotope maximum (AIM) warming events and factors that could be attributed to the deep ocean warming during Heinrich events.

3. DISCUSSION OF MAJOR FINDINGS

3.1. Last Glacial-Interglacial Climate Cycle

After four decades of intense research, there are not that many high-resolution (millennial to submillennial scale) paleoceanographic records available from the North Atlantic, the critical area for understanding changes in the MOC. *Voelker and de Abreu* [this volume] provide an overview of the last four glacial cycles from high-accumulation-rate sites on the Iberian margin based mostly on data from *Martarat et al.* [2007], *Voelker et al.* [2009], and *Salgueiro et al.* [2010]. Latitudinal (43.20° to 35.89°N) and longitudinal (10.39° to 7.53°W) gradients in sea surface temperature and density are reconstructed in this chapter using a foraminiferal assemblage–based transfer function (SIMMAX [*Pflaumann et al.*, 1996]) and $\delta^{18}O$ from 11 sediment cores from the northeastern Atlantic. In addition to sea surface conditions, $\delta^{18}O$ and $\delta^{13}C$ compositions in various planktonic foraminifers covering the depth range of 50 to 400 m indicate changes in calcification depth during glacial intervals. Voelker and de Abreu also evaluated changes in seasonality based on the difference in $\delta^{18}O$ between *Globigerina bulloides* and *Globorotalia inflata* [*Ganssen and Kroon*, 2000]. As a result, migration of subpolar and subtropical boundaries and hydrographic fronts during abrupt climate events were identified. Voelker and de Abreu suggest that nutrient levels and thus ventilation of the upper 400 m of the water column were not driven by millennial-scale events. In contrast, millennial-scale oscillations in ventilation were recorded in the intermediate to bottom waters, indicating the status of the overturning circulation either in the North Atlantic or in the Mediterranean Sea that admixed deep flowing Mediterranean Overflow Water to the Glacial North Atlantic Intermediate Water. One of the important aspects of Voelker and de Abreu's contribution is the detailed documentation of the upper water structure during the penultimate glacial (marine isotope stage (MIS) 6) that has been characterized by only scarce data thus far.

Flower et al. [this volume] review planktonic Mg/Ca-$\delta^{18}O$ evidence from the Gulf of Mexico (GOM) to investigate the role of meltwater input from the LIS in abrupt climate change during MIS 3 and the last deglaciation. The chapter provides an important summary of the current understanding of the ACC in the GOM by synthesizing data mostly presented by *Flower et al.* [2004], *Hill et al.* [2006], and *Williams et al.* [2010]. These data show that the ice volume–corrected seawater $\delta^{18}O$ in the GOM matches the East Antarctic ice core $\delta^{18}O$ [*EPICA Community Members et al.*, 2006]. *Flower et al.* [this volume] conclude that (1) LIS meltwater pulses started during Heinrich stadials and lasted through the subsequent D-O events; (2) LIS meltwater pulses appear to coincide with the major AIM events; and (3) LIS meltwater discharge is associated with distinct changes in deep ocean circulation in the North Atlantic during H events. These observations lead the authors to propose a direct link between GOM meltwater events and the weakening of the

AMOC as modulated by the Antarctic climate. Furthermore, Flower et al. hypothesize that LIS melting is linked to the Antarctic climate, so that it was the AMOC reduction (via the bipolar seesaw and Antarctic warming) that drove increased LIS meltwater input to the GOM and not vice versa. This causality, however, may have been limited as LIS meltwater input to the GOM continued throughout D-O 8 and Bølling/Allerød events despite a lack of evidence for AMOC reduction during these intervals [*Rahmstorf*, 2002].

Since the D-O cycles were first discovered in Greenland ice cores and then later confirmed in deep-sea sediments of the Atlantic, a large number of modeling efforts were directed toward understanding the origin and significance of these paleoclimatic events. *Rial and Saha* [this volume] simulate the D-O cycles using a conceptual model, based on simple stochastic differential equations, termed the sea ice oscillator (SIO), which borrowed elements from the *Saltzman et al.* [1981] concept. Simulation results show that sea ice extent and mean ocean temperature can be driven by changes in orbital insolation, which play a dominant role in controlling atmospheric temperature variability. The SIO model has two internal parameters controlling the abruptness of temperature change: the free frequency of the oscillator and the intensity of positive feedbacks. The model best reproduces the D-O cycles if the free oscillation period is set around 1.5 kyr, as originally proposed by *Bond et al.* [1997]. The performance of the SIO model is tested against two of the best resolved paleoclimate records: the Greenland ice core $\delta^{18}O$ [*North Greenland Ice Core Project members*, 2004] and planktonic and benthic foraminiferal $\delta^{18}O$ records of sea surface and deep-sea temperature from the Portuguese margin (core MD95-2042 [*Shackleton et al.*, 2000]). *Shackleton et al.* [2000] have shown that the $\delta^{18}O$ of planktonic foraminifer exhibits changes similar to those in the Greenland ice core, whereas the $\delta^{18}O$ record of benthic foraminifer (water depth below 2200 m) varies in a manner similar to the Antarctic air temperature. At any rate, Rial and Saha conclude that the time integral of the surface ocean proxy history is proportional to that of the deep ocean temperature, and the latter is shifted by $\pi/2$ with respect to sea ice extent. Surprisingly, a similar relationship was found between methane-synchronized temperature proxies from Greenland and East Antarctic ice core records [*Blunier and Brook*, 2001]. Rial and Saha's model output is reproduced using ECBilt-Clio, another well regarded model [*Goosse et al.*, 2002].

To summarize up-to-date results on abrupt climate events of the last glacial cycle, Figures 1 and 2 show records representative of the global climate during this period. These records were selected based on the geographic location and temporal resolution adequate to assess millennial- or finer-scale climate events. Regardless of the nature of climate archives, the climatic expression of the D-O cycles is evident from the southwest Pacific Ocean to northeastern Siberia.

3.2. Deglaciation Period

One of the most studied periods of the last glacial cycle is the time between 10 and 25 ka, commonly known as the last deglaciation. This period contains four very well studied abrupt climate events: the Younger Dryas (also known as H0 event, 11.6–12.9 ka); Bølling-Allerød (B-A) (14.6 ka); the so-called "Mystery Interval" (14.5–17.5 ka); and the last glacial maximum (LGM) (19–26.5 ka). The rationales behind focusing on the last deglaciation interval include the availability of paleoclimate records with high temporal resolution and a robust age control constrained by ^{14}C–accelerator mass spectrometry and U-Th dating. It is not surprising that 5 out of 14 chapters in this volume focus on climate events from this interval. In addition, we outline several new hypotheses (relative to the deglaciation events) that have emerged since the 2009 conference.

Simulating the impact of freshwater discharge into the North Atlantic during the Heinrich and other meltwater events has gained a significant momentum after the work of *Stouffer et al.* [2006] and *Liu et al.* [2009]. Numerous modeling studies, where freshwater discharge was artificially added to the North Atlantic, have shown large climate impacts associated with abrupt AMOC changes. These results highlight the need to better understand the mechanisms of the AMOC variability under the past, present, and future climate conditions. The mechanisms controlling the recovery of the AMOC are, however, difficult to investigate as they require long simulations with models that can capture very different internal variabilities. *Cheng et al.* [this volume] describe the AMOC recovery stages from a model simulation of the last deglaciation built on the work of *Liu et al.* [2009]. These authors perform a remarkably long simulation using an atmosphere-ocean general circulation model [*Liu et al.*, 2009], which allows for a clear identification of transient states of the recovery process including the AMOC overshoot phenomenon [*Renold et al.*, 2010]. This 5000 year simulation time covers the last deglaciation period using insolation and fresh water estimated from paleoproxies as forcings. In particular, the simulation offers an explanation for the B-A warm period by a recovery of the AMOC in less than 400 years. An in-depth analysis of the AMOC recovery suggests that the two convection sites in the North Atlantic simulated in the model do not recover at the same time: the first stage of the recovery occurred in the Labrador Sea and was then followed by convection recovery in the Greenland-Iceland-Norwegian (GIN) Seas. The study suggests that reinitiation of convection in the Labrador Sea is related to the reduction

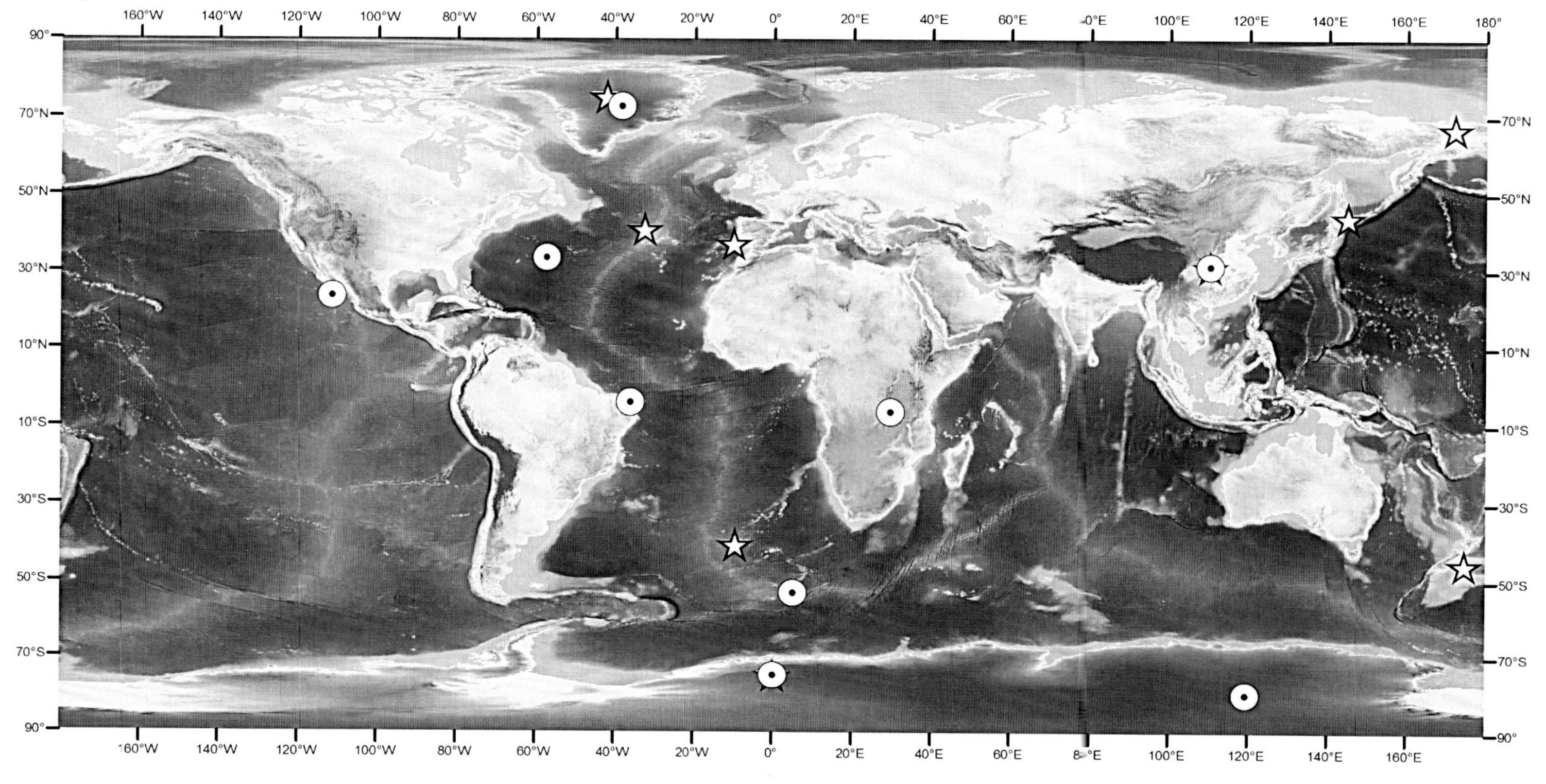

Figure 1. Geographic distribution of paleoclimate proxy records for the last glacial cycle (stars) and the last deglaciation (circles) as shown on Figures 2 and 3.

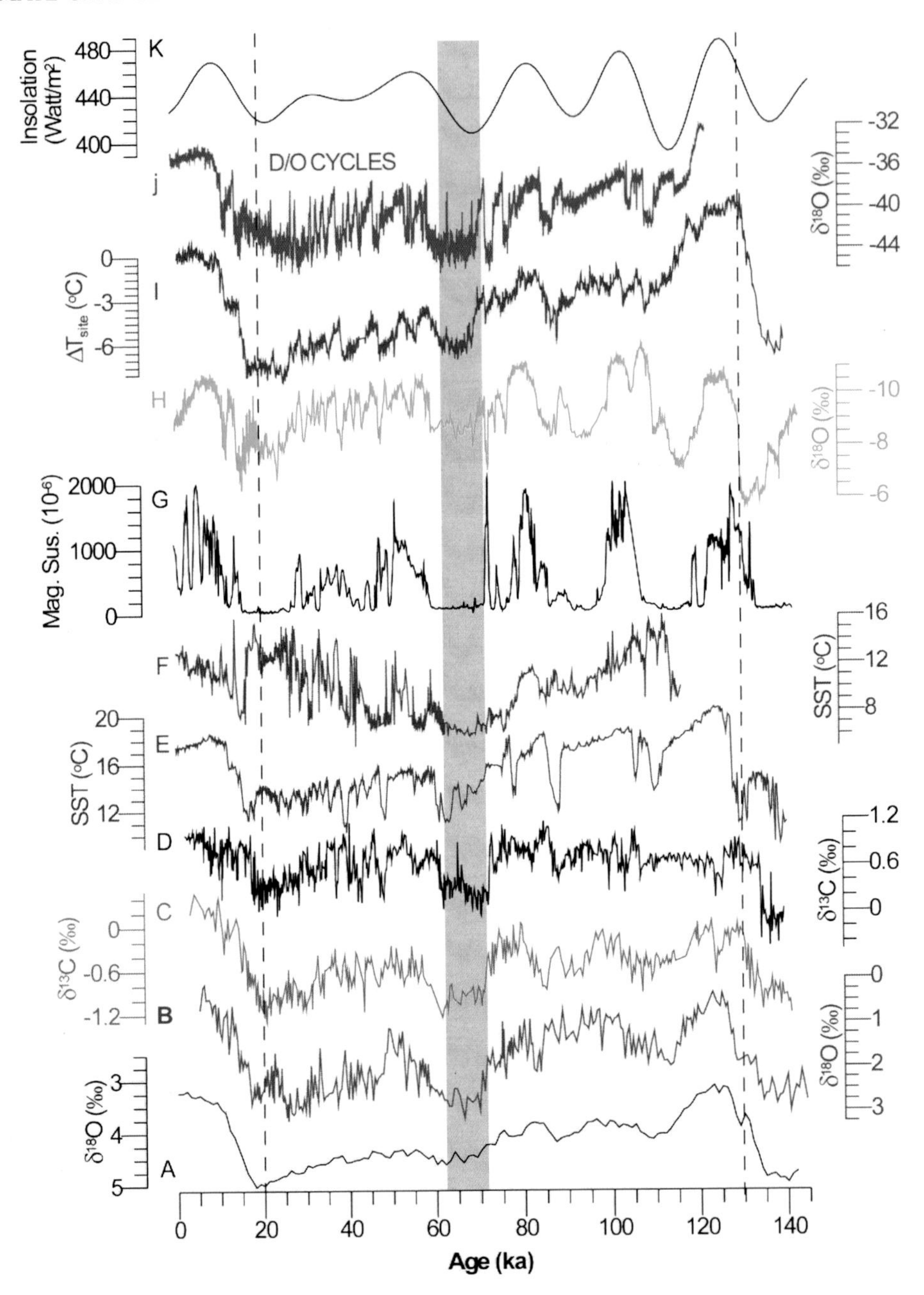

Figure 2. Comparison of various proxy records that demonstrate abrupt climate changes during the last glacial cycle. (a) Global stack of benthic foraminiferal oxygen isotopes ($\delta^{18}O$) [*Lisiecki and Raymo*, 2005]. (b) The $\delta^{18}O$ in planktonic foraminifera *Globigerina bulloides* from the North Atlantic Integrated Ocean Drilling Program Site 1313 (H. Rashid, unpublished data, 2011). (c and d) Carbon isotopes ($\delta^{13}C$) in benthic foraminifera *Cibicidoides wuellerstorfi* from the South Atlantic Ocean Drilling Program Site 1089 [*Charles et al.*, 2010] and core MD97-2120 from the SW Pacific Ocean [*Pahnke and Zahn*, 2005], respectively. (e and f) Alkenone (U^{K}_{37}) derived sea surface temperatures from the NE Atlantic core MD01-2443 [*Martarat et al.*, 2007; *Voelker and de Abreu*, this volume, Figure 5] and core MD01-2412 from the NW Pacific Ocean [*Harada et al.*, 2008], respectively. (g) Magnetic susceptibility record from Lake El'gygytgyn, NE Siberia [*Nowaczyk et al.*, 2007]. (h) The $\delta^{18}O$ in the Shanbao speleothems, NE China [*Wang et al.*, 2008]. (i) Antarctic temperature reconstructed from the deuterium content ((ΔT) elevation corrected) in the Eastern Droning Maud Land facing the South Atlantic [*Stenni et al.*, 2010]. (j) The $\delta^{18}O$ in the North Greenland Ice Core Project ice core with a revised chronology [*Svensson et al.*, 2008]. (k) June insolation at 65°N [*Berger and Loutre*, 1991]. All records are plotted according to their independent age models. Note that the vertical grey bar indicates the duration of marine isotope stage 4 and the two vertical discontinuous lines indicate the 20 and 130 ka time horizons.

of freshwater forcing, leading to an increase in salinity. Convection in the GIN Seas is then forced by a (partial) recovery of the AMOC and associated salinity transport. Overall, Cheng et al. provide a convincing mechanism, in which both local and remote salinity feedbacks play a role in the AMOC recovery.

New data from four sediment cores in addition to the synthesis of earlier results from the Labrador margin [*Rashid et al.*, this volume] shed light on the development of the YD event. There has been a long-standing debate about the origin of YD event [e.g., *Mercer*, 1969; *Johnson and McClure*, 1976; *Ruddiman and McIntyre*, 1981; *Teller and Thorleifdson*, 1983; *Teller et al.*, 2005; *Broecker et al.*, 1989; *Andrews et al.*, 1995; *Lowell et al.*, 2005]. Most of these studies propose freshwater flooding at sites of deepwater formation in the North Atlantic; however, a freshwater signature has not been found in proxy records near these sites thus far. Rashid et al. show that a $\delta^{18}O$ depletion in planktonic foraminifers indicates a YD freshwater signature in the Hudson Strait but not in the more distal cores despite the presence of a H0 high-carbonate bed in these cores. Rashid et al. hypothesize that if the fresh water discharged through Hudson Strait was admixed with fine-grained detrital carbonates, it would form a hyperpycnal flow transported through the deep Labrador Sea Northwest Atlantic Mid-Ocean Channel to distal sites of the NW Atlantic Ocean such as the Sohm Abyssal Plain. With the release of entrained sediment, fresh water from the hyperpycnal flow would buoyantly rise and lose its signature by mixing with the ambient sea water. This mechanism explains a lack of YD $\delta^{18}O$ depletion in sediment cores retrieved from the NW Atlantic. Thus, the long-sought smoking gun for the signature of YD (H0) freshwater flood [e.g., *Broecker et al.*, 1989] remains elusive.

Applegate and Alley [this volume] evaluate the potential of using cosmogenic radionuclides (CRN) to date glacial moraine boulders to trace the geographic expression of abrupt climate change. The chapter provides a synthesis of the current state of the knowledge using CRN to determine the extent and retreat of glaciers in a terrestrial setting. Problems highlighted deal with selecting samples for exposure dating and with the calculation of exposure dates from nuclide concentrations. Failure to address these issues will yield too-young exposure dates on moraines that have lost material from their crests over time. In addition, geomorphic processes are likely to introduce errors into the calibration of nuclide production rates. The authors point to a conclusion from a recent study by *Vacco et al.* [2009] that the ages of true YD moraines should cluster around the end of YD, however the recent modeling study complicated this simplistic assumption. Applegate and Alley demonstrate their points using beryllium-10 exposure dates from two "not so primed" moraines of inner Titcomb Lakes (Wind River Range, Wyoming, United States) and Waiho Loop of New Zealand. The sampling strategies prescribed may help in minimizing problems with defining true age of the moraines. The chapter includes a guide for determining the minimum number of samples that must be collected to answer a particular paleoclimate question.

During the termination of the last ice age, atmospheric CO_2 rapidly increased in two steps [*Monnin et al.*, 2001], and atmospheric $\Delta^{14}C$ ($\Delta^{14}C_{atm}$) decreased by ~190% between 17.5 and 14.5 ka [*Reimer et al.*, 2009] (Figure 3), coined the "Mystery Interval" by *Denton et al.* [2006]. From two eastern Pacific sediment cores off Baja California, *Marchitto et al.* [2007] have documented two strong negative excursions of $\Delta^{14}C$, which were corroborated by another eastern Pacific record of *Stott et al.* [2009, this volume] and a record from the western Arabian Sea [*Bryan et al.*, 2010]. One of those negative excursions of $\Delta^{14}C$ corresponds to the Mystery Interval and the other to the YD. The emerging understanding is that during the LGM, there was a hydrographic divide between the upper 2 km of the water column and the deep Southern Ocean, where dense and salty deep waters hosted the depleted $\Delta^{14}C$ [*Adkins et al.*, 2002]. Accordingly, these depleted $\Delta^{14}C$ waters were termed the "Mystery Reservoir." The origin of the "Mystery Reservoir" has been linked to freshwater release into the North Atlantic during H1 and YD (H0) [*Toggweiler*, 2009]. It has been further inferred that the resulting AMOC weakening initiated a chain of events reaching to the tropics and Southern Hemisphere [*Anderson et al.*, 2009]. According to this interpretation, as the northward heat flow slowed, the heat accumulated in the tropics and warmed the Southern Ocean, resulting in the reduction of sea ice extent around Antarctica. This sea ice retreat shifted the Southern Hemisphere westerlies poleward through a mechanism that remains unknown, allowing the ventilation of the deep ocean that stored CO_2 and some other chemical components such as silica. As a result, a rise in atmospheric CO_2, depleted $\Delta^{14}C_{atm}$, and a dramatic increase in the accumulation of siliceous sediments were observed in the Southern Ocean [*Anderson et al.*, 2009].

Broecker and Barker [2007], *Broecker* [2009], and *Broecker and Clark* [2010] launched an extensive search for the "Mystery Reservoir" in the Pacific Ocean and did not find any evidence for it. *Stott and Timmermann* [this volume] put forward a provocative "CO_2 capacitor" hypothesis that resembles the "clathrate gun" hypothesis of *Kennett et al.* [2003] and offers a way to account for a decrease in the depleted $\Delta^{14}C_{atm}$ and increase in the atmospheric CO_2. The main point of the "CO_2 capacitor" hypothesis is the presumed existence of an unspecified source of CO_2 that can explain both the glacial to interglacial CO_2 change and the signal of old radiocarbon in the atmosphere during the deglaciation. This source is inferred to be a combination of liquid and hydrate CO_2 in subduction zones and volcanic centers other than on mid-ocean ridges. During glacial times, CO_2 hydrates in the cold intermediate ocean are stable and able to trap liquid CO_2 bubbling up from below the seafloor. During deglaciation, a swath of these hydrates becomes unstable and releases radiocarbon-dead CO_2 to the ocean-atmosphere system, thus increasing atmospheric CO_2 and decreasing $\Delta^{14}C_{atm}$. This hypothesis is likely to stimulate wide interest in the scientific community; however, it requires vigorous testing through experiments, observations, and modeling.

In a related study that takes into account the relationship between the rise in atmospheric CO_2 and temperature in low to high latitudes, *Russell et al.* [2009] reconstruct precipitation and temperature histories from southeast African Lake Tanganyika based on compound-specific hydrogen isotopes and paleotemperature biomarker index TEX_{86} in terrestrial leaf waxes. It has been shown that Lake Tanganyika paleotemperature followed the temperature rise in Antarctica at 20 ka [*Monnin et al.*, 2001] as well as the Northern Hemisphere summer insolation at 30°N [*Tierney et al.*, 2008]. Furthermore, temperatures in Lake Tanganyika began to increase ~3000 years before the rise of atmospheric CO_2 concentrations during the last ice age termination (Figure 3). This is a significant discovery that needs to be replicated from similar climate settings in other regions of the world. An extensive statistical testing is also needed to confirm whether the temperature record from Lake Tanganyika can be applied more generally throughout the tropics or whether it represents only a regional warming. If the reconstructed temperature history from Lake Tanganyika survives the scrutiny of proxy validation, it would strongly suggest that the primary driver for glacial-interglacial termination, at least for the last ice age, lies in the tropics rather than in high latitudes.

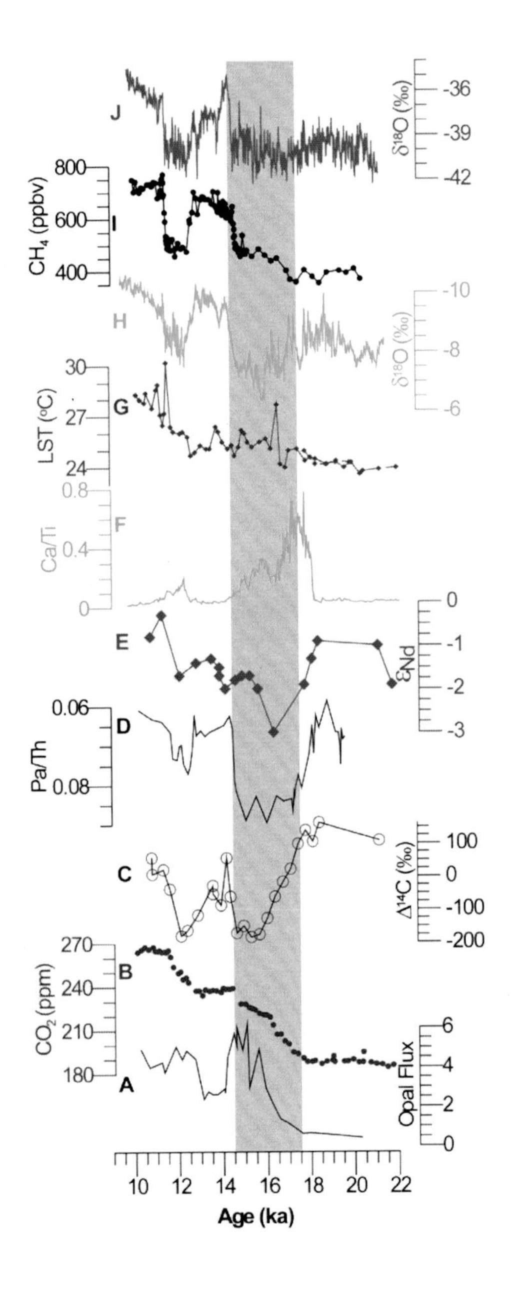

Figure 3. Paleoclimatic proxy records of the last deglaciation. (a) Biogenic opal flux in the Southern Ocean, interpreted as a proxy for changes in upwelling south of the Antarctic Polar Front [*Anderson et al.*, 2009]. (b) Atmospheric CO_2 from Antarctic Dome C [*Monnin et al.*, 2001] placed on the Greenland Ice Sheet Project 2 (GISP2) timescale. (c) Baja California intermediate water $\Delta^{14}C$ [*Marchitto et al.*, 2007]. (d) The $^{231}Th/^{230}Th$ ratios from the Bermuda Rise (increasing values reflect reduced Atlantic overturning circulation) [*McManus et al.*, 2004]. (e) The ε_{Nd} of fossil fish tooth/debris record suggesting variations of water mass at Baja California [*Basak et al.*, 2010]. (f) Ti/Ca ratios in bulk sediment from western equatorial Atlantic, interpreted as a proxy for high Amazon River runoff [*Jaeschke et al.*, 2007]. (g) TEX_{86}-derived surface temperature of Lake Tanganyika [*Tierney et al.*, 2008]. (h) The $\delta^{18}O$ in Shanbao speleothems [*Wang et al.*, 2008]. (i and j) GISP2 methane and oxygen isotope ratios ($\delta^{18}O$) [*Stuiver and Grootes*, 2000; *Blunier and Brook*, 2001], respectively. All records are plotted according to their independent age models.

3.3. Holocene Climate

The conventional view based primarily on Greenland ice core $\delta^{18}O$ records suggests that climate in the Holocene (current interglacial) was rather uniform and remarkably stable in comparison to the preceding glacial and interstadial periods (MIS 2–3). However, as more new paleoclimatic records emerge, such as palynological and paleolimnological data from North America [*Viau et al.*, 2006; *Gajewski and Viau*, this volume] and tropical climate records reconstructed from high-elevation ice cores [*Thompson*, this volume], it appears that Holocene climate variability may have been greater than has been assumed based on prior data sets.

In the early Holocene, large climatic fluctuations were related to freshwater pulses from the disintegrating LIS into the western North Atlantic [e.g., *de Vernal et al.*, 2000]. Using palynological data (terrestrial palynomorphs and marine dinocysts) from the Newfoundland and northern Scotian Shelf sediment cores, *Levac et al.* [this volume] assess the impact of fresh water on the circulation over the Labrador margin and related climate change during the final demise of the proglacial Lake Agassiz. The authors identify a circa 8.7 ka detrital carbonate bed accompanied by two meltwater pulses that lowered sea surface salinity (SSS) in the coastal water of Newfoundland. In contrast, data from the Scotian Shelf do not show changes in the SSS at this time. The divergent impact of freshwater pulses at these two sites is explained by the absence of a diluting effect of the North Atlantic Current at the Newfoundland margin as opposed to the warmer Scotian Shelf. *Keigwin et al.* [2005] have documented a cooling event around 8.5 ka, likely correlative to that identified by Levac et al., as far south as Cape Hatteras, suggesting a large geographic impact of the Lake Agassiz drainage. Surprisingly, such a pronounced cooling event has not been found in the northern Labrador Sea and Baffin Bay, possibly because of a lack of high-resolution records due to the stronger winnowing conditions of the Labrador coastal current [*Chapman*, 2000; *Rashid et al.*, 2011].

Ruzmaikin and Feynman [this volume] investigate a quasiperiodic 1500 year climate oscillation during the Holocene and its relation to the AMOC. They employed a simple conceptual model to simulate the forcing mechanism for this oscillation. The analysis of six paleoclimatic records from the Atlantic Ocean and sunspot numbers using wavelet and empirical mode decomposition methods allowed the authors to overcome the leaks from one mode to another, a common problem in conventional band-pass filtering. Ruzmaikin and Feynman conclude that solar forcing does not drive the 1500 year climate cycle "directly," contrary to the inference by *Bond et al.* [2001]. Instead, they suggest a simple model of excitation of this oscillation in a nonlinear dynamical system with two equilibrium states. They reason that the transitions between the two states are caused by the noise, ocean, and solar variability, and the implication is that the beats between the centennial ocean variability and ~90 year solar cycles produce the 1500 year oscillation in the noisy system. This conceptual framework for the 1500 year oscillation requires more rigorous testing by more sophisticated, coupled ocean-atmosphere models.

Gajewski and Viau [this volume] suggest that the Holocene in North America can be divided into four general periods with abrupt transitions at ~8, 6, and 3 ka, generally consistent with the hypothesis of *Ruzmaikin and Feynman* [this volume]. Gajewski and Viau further infer that the late Holocene 5.2 ka, 4.2 ka, and Little Ice Age (circa 1300–1870 A.D.) cooling/drying events were part of a continual series of millennial-scale climatic fluctuations.

Paliwal [this volume] documents abrupt climate events during the Holocene from a series of lake records and groundwater data from the western Indian peninsula. Some of these abrupt climate events, in combination with neotectonic activity, altered the drainage pattern that caused a disappearance of the Vedic Sarasvati River and other tributaries of the Indus River. Paliwal infers that the onset of the arid climate in the Thar Desert around 3.5 ka may have caused the decline of the Indus Valley Harappan and Mohenjo-Daro civilizations. Interestingly, fossils of elephants and bamboo curtains discovered in the Quaternary gypsum deposits of Rajasthan suggest the existence of an even earlier advanced civilization (>12.82 ka). This Vedic civilization, which flourished along the banks of the Sarasvati River, is older than the Harappan and Mohenjo-Daro civilizations (5 ka) of the Indus Valley [*Weiss et al.*, 1993; *Rashid et al.*, 2011]. However, the lack of high-resolution radiocarbon dates prevents correlation of these terrestrial records to Indian summer monsoon records [*Thompson et al.*, 1997; *Rashid et al.*, 2011], which is essential to evaluate the impact of climate changes on the fate of these ancient civilizations.

From Sierra Nevada's Coburn Lake, *Wathen* [this volume] reconstructed an 8500 year fire history using charcoal microparticles as a proxy. In addition to published data from the Sierra Nevada and surroundings, the author correlates his records with the long-distance fire records from Lake Francis of eastern Canada and soot records of Greenland ice cores. Wathen suggests that severe fires at Coburn Lake occurred at the beginning of severe droughts, consistent with the Greenland soot records. Furthermore, the 8.2 ka and 5.2 ka climatic events identified in the Lake Coburn microcharcoal record are inferred to be related to large-scale shifts in the major precipitation belts: the Intertropical Convergence Zone (ITCZ), the Subtropical Desert Zone, and the Polar Front. Wathen hypothesizes that most instances of severe fires and erosion at Coburn Lake occurred in response to stresses on vegetation that were adapted to colder and wetter conditions in response to abrupt climate events. A similar mechanism is suggested for the Coburn Lake fire history of the last 1800 years, although some major climatic settings, such as the position of the ITCZ, were different in the late Holocene.

Accumulation of past climate records over the last decade shows that during the past 5000 years when climatic conditions were similar to recent (prior to the rise in anthropogenic greenhouse gases), Earth experienced at least two abrupt global-scale climate events (5.2 ka and 4.2 ka). *Thompson* [this volume] provides a detailed synthesis of these events using ice core records from the world's highest mountains combined with other published paleoclimate histories. These events, which persisted for a few centuries, were related to profound changes in the hydrological cycle in the tropics and midlatitudes [*Magny and Haas*, 2004] and appear to correlate with the decline of ancient civilizations [*Weiss et al.*, 1993; *MacDonald*, 2011; *Rashid et al.*, 2011; *Thompson*, this volume]. However, the forcing mechanisms for these abrupt changes remain elusive thus far due in large measure to insufficient spatial coverage.

The 5.2 ka event is preserved in diverse paleoclimate archives that include ice core oxygen isotope records from Kilimanjaro in Tanzania and Huascarán in northern Peru, methane records from Antarctica and Greenland, marine sediment records from the Bay of Bengal, and deposition of Saharan dust in the South American Andes. Other ice core proxies including dust and soluble chemical species such as Cl^- and SO_4^{2-} also show drastic changes associated with the 5.2 ka event. Terrestrial records such as trees preserved in a standing position hundreds of feet below the surface of Lake Tahoe [*Lindström*, 1990], plants emerging from the retreating margins of an Andean glacier [*Thompson et al.*, 2006], and the "Tyrolean ice man" or "Özti" in the Eastern Alps are other notable observations concomitant with the 5.2 ka event. There are not many deep-sea records that contain the 5.2 ka event at present. However, all of these paleoclimatic records point to a near-global low- to mid-low-latitude abrupt climate event that appears to have impacted civilizations on three continents.

The other abrupt climate event during the late Holocene is dated between 4.0 and 4.5 ka and seems to be very widespread as it is found in many tropical and extratropical paleoclimate archives. It is associated with prominent dust peaks in the Huascarán and Kilimanjaro ice cores and is thus interpreted as a major drought interval during the late Holocene [*Davis and Thompson*, 2006]. Sediment cores from the Gulf of Oman and Bay of Bengal, as well as lake records from the Gharwal Himalayas and northern Africa also indicate dry climatic conditions around 4.2 ka. These events are also expressed in archeological data from the Euphrates and Tigris drainage basins. *Thompson* [this volume] further synthesizes many North African, Middle Eastern, South Pacific, and South Asian paleoclimate records centered around the 4.2 ka event. Like the 5.2 ka event, it is also believed to have lasted at least a few centuries. If such a rapid and sustained drought as during the 4.2 ka event were to occur today, with Earth's population rapidly approaching 7 billion, the impact would be very troubling.

4. RECOMMENDATIONS

Participants recommended several major areas in which improved approaches to understanding abrupt climate change are needed. These include the following: (1) collecting more high-accumulation-rate and high-resolution data and improving coordination of paleoclimate proxy and modeling approaches; (2) concentrating on a few key time intervals (such as 5.2 ka and 4.2 ka events) but using multiple proxies; (3) developing novel paleoclimatic proxies such as clumped isotopes, a promising tool for an independent paleotemperature record, and biomarker proxies for sea ice and temperature; (4) reconstructing a history of sea ice distribution that provides a fast and powerful feedback in polar and subpolar regions such as the North Atlantic and sub-Antarctic; and (5) facilitating deep drilling in the Indian Ocean, an under investigated ocean region that is critical to understanding the history of the Indian monsoon as well as the erosion and uplift of the Himalayas, with the recent recovery of a nearly a million yearlong climate record from the EPICA Dome C ice core raising the scientific necessity of attaining an equally long record from the southern Indian Ocean.

Many meeting participants emphasized that more paleoclimatic records with improved temporal resolution and "near-zero uncertainty" dating are required for state-of-the-art model to data comparison studies. Plans are urgently needed for the paleoclimatic community to expand the number and spatial distribution of longer high-quality proxy data sets to complement those that already exist for the last glacial cycle. Data sets that include multiple evolutions of glacial and interglacial conditions are essential for current efforts to project climate change under a warmer Earth scenario as is currently underway by the Intergovernmental Panel on Climate Change.

To contribute to climate change prediction efforts, it is especially important to understand the mechanisms driving late Holocene abrupt climate events such as those at circa 5.2 and 4.2 ka. These events are found mainly in low-latitude climate archives related to the hydrological cycle history and appear to be contemporaneous with collapses of civilizations in the Middle East, Indian subcontinent, and Mesoamerica. As more than half of the humanity lives in the tropical belt, the significance of any change in regional hydrological cycles cannot be overemphasized.

REFERENCES

Adkins, J. F., K. McIntyre, and D. P. Schrag (2002), The salinity, temperature, and $\delta^{18}O$ of the glacial deep ocean, *Science, 298,* 1769–1773, doi:10.1126/science.1076252.

Anderson, R. F., et al. (2009), Wind-driven upwelling in the Southern Ocean and the deglacial rise in atmospheric CO_2, *Science, 323,* 1443–1448.

Andrews, J. T., A. E. Jennings, M. Kerwin, M. Kirby, W. Manley, G. H. Miller, G. Bond, and B. MacLean (1995), A Heinrich-like event, H-0 (DC-0): Source(s) for detrital carbonate in the North Atlantic during the Younger Dryas chronozone, *Paleoceanography, 10,* 943–952.

Applegate, P. J., and R. B. Alley (2011), Challenges in the use of cosmogenic exposure dating of moraine boulders to trace the geographic extents of abrupt climate changes: The Younger Dryas example, in *Abrupt Climate Change: Mechanisms, Patterns, and Impacts, Geophys. Monogr. Ser.*, doi:10.1029/2010GM001029, this volume.

Basak, C., E. E. Martin, K. Horikawa, and T. M. Marchitto (2010), Southern Ocean source of ^{14}C-depleted carbon in the North Pacific Ocean during the last deglaciation, *Nat. Geosci., 3,* 770–773.

Berger, A., and M. F. Loutre (1991), Insolation values for the climate of the last 10 million years, *Quat. Sci. Rev., 10,* 297–317.

Blunier, T., and E. J. Brook (2001), Timing of millennial-scale climate change in Antarctica and Greenland during the last glacial period, *Science, 291,* 109–112.

Bond, G., et al. (1997), A pervasive millennial-scale cycle in North Atlantic Holocene and glacial climates, *Science, 278,* 1257–1266.

Bond, G., et al. (2001), Persistent solar influence on North Atlantic climate during the Holocene, *Science, 294,* 2130–2136.

Broecker, W. S. (2009), The mysterious ^{14}C decline, *Radiocarbon, 51,* 109–119.

Broecker, W. S., and S. Barker (2007), A 190% drop in atmosphere's $\Delta^{14}C$ during the "Mystery Interval" (17.5 to 14.5 kyr, *Earth Planet Sci. Lett., 256,* 90–99.

Broecker, W. S., and E. Clark (2010), Search for a glacial-age ^{14}C-depleted ocean reservoir, *Geophys. Res. Lett., 37,* L13606, doi:10.1029/2010GL043969.

Broecker, W. S., J. P. Kennett, B. P. Flower, J. T. Teller, S. Trumbore, G. Bonani, and W. Wolfi (1989), Routing of meltwater from the Laurentide Ice Sheet during the Younger Dryas cold episode, *Nature, 341,* 318–321.

Bryan, S. P., T. M. Marchitto, and S. J. Lehman (2010), The release of ^{14}C-depleted carbon from the deep ocean during the last deglaciation: Evidence from the Arabian Sea, *Earth Planet Sci. Lett., 298,* 244–254.

Chapman, D. C. (2000), Boundary layer control of buoyant coastal currents and the establishment of a shelf break front, *J. Phys. Oceanogr., 30,* 2941–2955.

Charles, C. D., et al. (2010), Millennial scale evolution of the Southern Ocean chemical divide, *Quat. Sci. Rev., 29,* 399–409.

Cheng, J., Z. Liu, F. He, B. L. Otto-Bliesner, E. C. Brady, and M. Wehrenberg (2011), Simulated two-stage recovery of Atlantic meridional overturning circulation during the last deglaciation, in *Abrupt Climate Change: Mechanisms, Patterns, and Impacts, Geophys. Monogr. Ser.*, doi:10.1029/2010GM001014, this volume.

Clement, A. C., and L. C. Peterson (2008), Mechanisms of abrupt climate change of the last glacial period, *Rev. Geophys., 46,* RG4002, doi:10.1029/2006RG000204.

Clement, A. C., A. Hall, and A. J. Broccoli (2004), The importance of precessional signals in the tropical climate, *Clim. Dyn., 22,* 327–341.

Davis, M. E., and L. G. Thompson (2006), An Andean ice-core record of a middle Holocene mega-drought in North Africa and Asia, *Ann. Glaciol., 43,* 34–41.

Denton, G. H., W. S. Broecker, and R. B. Alley (2006), The Mystery Interval 1.5 to 14.5 kyrs ago, *PAGES Newsl., 14*(2), 14–16.

de Vernal, A., C. Hillaire-Marcel, J. L. Turon, and J. Matthiessen (2000), Reconstruction of sea-surface temperature, salinity, and sea-ice cover in the northern North Atlantic during the last glacial maximum based on dinocyst assemblages, *Can. J. Earth Sci., 37,* 725–750.

EPICA Community Members et al. (2006), One-to-one coupling of glacial climate variability in Greenland and Antarctica, *Nature, 444,* 195–198.

Flower, B. P., D. W. Hastings, H. W. Hill, and T. M. Quinn (2004), Phasing of deglacial warming and Laurentide Ice Sheet meltwater in the Gulf of Mexico, *Geology, 32,* 597–600.

Flower, B. P., C. Williams, H. W. Hill, and D. W. Hastings (2011), Laurentide Ice Sheet meltwater and the Atlantic meridional overturning circulation during the last glacial cycle: A view from the Gulf of Mexico, in *Abrupt Climate Change: Mechanisms, Patterns, and Impacts, Geophys. Monogr. Ser.*, doi: 10.1029/2010GM001016, this volume.

Gajewski, K., and A. E. Viau (2011), Abrupt climate changes during the Holocene across North America from pollen and paleolimnological records, in *Abrupt Climate Change: Mechanisms, Patterns, and Impacts, Geophys. Monogr. Ser.*, doi: 10.1029/2010GM001015, this volume.

Ganssen, G. M., and D. Kroon (2000), The isotopic signature of planktonic foraminifera from NE Atlantic surface sediments: Implications for the reconstruction of past oceanic conditions, *J. Geol. Soc. London, 157,* 693–699.

Goosse, H., H. Renssen, F. M. Selten, R. J. Haarsma, and J. D. Opsteegh (2002), Potential causes of abrupt climate events: A numerical study with a three-dimensional climate model, *Geophys. Res. Lett., 29*(18), 1860, doi:10.1029/2002GL014993.

Harada, N., M. Sato, and T. Sakamoto (2008), Freshwater impacts recorded in tetraunsaturated alkenones and alkenone sea surface temperatures from the Okhotsk Sea across millennial-scale cycles, *Paleoceanography*, *23*, PA3201, doi:10.1029/2006PA001410.

Hill, H. W., B. P. Flower, T. M. Quinn, D. J. Hollander, and T. P. Guilderson (2006), Laurentide Ice Sheet meltwater and abrupt climate change during the last glaciation, *Paleoceanography*, *21*, PA1006, doi:10.1029/2005PA001186.

Jaeschke, A., C. Rühlemann, H. Arz, G. Heil, and G. Lohmann (2007), Coupling of millennial-scale changes in sea surface temperature and precipitation off northeastern Brazil with high-latitude climate shifts during the last glacial period, *Paleoceanography*, *22*, PA4206, doi:10.1029/2006PA001391.

Johnson, R. G., and B. T. McClure (1976), A model for Northern Hemisphere continental ice sheet variation, *Quat. Int.*, *6*, 325–353.

Kawamura, K., et al. (2007), Northern Hemisphere forcing of climatic cycles in Antarctica over the past 360,000 years, *Nature*, *448*, 912–916.

Keigwin, L. D., J. P. Sachs, Y. Rosenthal, and E. A. Boyle (2005), The 8200 year B.P. event in the slope water system, western subpolar North Atlantic, *Paleoceanography*, *20*, PA2003, doi:10.1029/2004PA001074.

Kennett, J. P., K. G. Cannariato, I. L. Hendy, and R. J. Behl (2003), *Methane Hydrates in Quaternary Climate Change: The Clathrate Gun Hypothesis*, AGU, Washington, D. C.

Laepple, T., M. Werner, and G. Lohmann (2011), Synchronicity of Antarctic temperatures and local solar insolation on orbital time-scales, *Nature*, *471*, 91–94.

Levac, E., C. F. M. Lewis, and A. A. L. Miller (2011), The impact of the final Lake Agassiz flood recorded in northeast Newfoundland and northern Scotian shelves based on century-scale palynological data, in *Abrupt Climate Change: Mechanisms, Patterns, and Impacts*, *Geophys. Monogr. Ser.*, doi:10.1029/2010GM001051, this volume.

Li, C., D. S. Battisti, and C. M. Bitz (2010), Can North Atlantic sea ice anomalies account for Dansgaard-Oeschger climate signals?, *J. Clim.*, *23*, 5457–5475.

Lindström, S. (1990), Submerged tree stumps as indicators of mid-Holocene aridity in the Lake Tahoe Basin, *J. Calif. Great Basin Anthropol.*, *12*, 146–157.

Lisiecki, L. E., and M. E. Raymo (2005), A Pliocene-Pleistocene stack of 57 globally distributed benthic $\delta^{18}O$ records, *Paleoceanography*, *20*, PA1003, doi:10.1029/2004PA001071.

Liu, Z., et al. (2009), Transient simulation of last deglaciation with a new mechanism for Bølling-Allerød warming, *Science*, *325*, 310–314.

Lowell, T. V., et al. (2005), Testing the Lake Agassiz meltwater trigger for the Younger Dryas, *Eos Trans. AGU*, *86*(40), 365.

Lynch-Stieglitz, J., et al. (2007), Atlantic meridional overturning circulation during the last glacial maximum, *Science*, *316*, 66–69.

MacDonald, G. (2011), Potential influence of the Pacific Ocean on the Indian summer monsoon and Harappan decline, *Quat. Int.*, *229*, 140–148.

Magny, M., and J. N. Haas (2004), A major widespread climatic change around 5300 cal. yr BP at the time of the Alpine Iceman, *J. Quat. Sci.*, *19*, 423–430.

Marchitto, T. M., S. J. Lehman, J. D. Ortiz, J. Fluckiger, and A. van Geen (2007), Marine radiocarbon evidence for the mechanism of deglacial atmospheric CO_2 rise, *Science*, *316*, 1456–1459.

Martarat, B., et al. (2007), Four climate cycles of recurring deep and surface water destabilizations on the Iberian margin, *Science*, *317*, 502–507.

McManus, J. F., R. Francois, J.-M. Gherardi, L. D. Keigwin, and S. Brown-Leger (2004), Collapse and rapid resumption of Atlantic meridional circulation linked to deglacial climate changes, *Nature*, *428*, 834–837.

Mercer, J. H. (1969), The Allerød Oscillation: A European climatic anomaly?, *Arct. Alp. Res.*, *1*, 227–234.

Monnin, E., A. Indermühle, A. Dällenbach, J. Flückiger, B. Stauffer, T. F. Stocker, D. Raynaud, and J.-M. Barnola (2001), Atmospheric CO_2 concentrations over the last glacial termination, *Science*, *291*, 112–114, doi:10.1126/science.291.5501.112.

North Greenland Ice Core Project Members (2004), High-resolution record of Northern Hemisphere climate extending into the last interglacial period, *Nature*, *431*, 147–151.

Nowaczyk, N. R., M. Melles, and P. S. Minyuk (2007), A revised age model for core PG1351 from Lake El'gygytgyn, Chukotka, based on magnetic susceptibility variations tuned to Northern Hemisphere insolation variations, *J. Paleolimnol.*, *37*, 65–76.

Pahnke, K., and R. Zahn (2005), Southern Hemisphere water mass conversion linked with North Atlantic climate variability, *Science*, *307*, 1741–1746.

Paliwal, B. S. (2011), Abrupt Holocene climatic change in northwestern India: Disappearance of the Sarasvati River and the end of Vedic civilization, in *Abrupt Climate Change: Mechanisms, Patterns, and Impacts*, *Geophys. Monogr. Ser.*, doi:10.1029/2010GM001028, this volume.

Pflaumann, U., J. Duprat, C. Pujol, and L. D. Labeyrie (1996), SIMMAX: A modern analog technique to deduce Atlantic sea surface temperatures from planktonic foraminifera in deep-sea sediments, *Paleoceanography*, *11*, 15–35.

Rahmstorf, S. (2002), Ocean circulation and climate during the past 120,000 years, *Nature*, *419*, 207–214.

Rashid, H., S. Lodestro, J. Gruzner, A. Voelker, and B. P. Flower (2009), New assessment for the last four glacial cycles' Northern Hemispheric ice-sheet variability from the IODP Site U1313 of the North Atlantic, paper presented at Chapman Conference on Abrupt Climate Change, AGU, Columbus, Ohio.

Rashid, H., E. England, L. G. Thompson, and L. Polyak (2011), Late glacial to Holocene Indian summer monsoon variability from the Bay of Bengal sediment records, *Terr. Atmos. Oceanic Sci.*, *22*(2), 215–228.

Rashid, H., D. J. W. Piper, and B. P. Flower (2011), The role of Hudson Strait outlet in Younger Dryas sedimentation in the Labrador Sea, in *Abrupt Climate Change: Mechanisms, Patterns, and Impacts*, *Geophys. Monogr. Ser.*, doi:10.1029/2010GM001011, this volume.

Reimer, P. J., et al. (2009), IntCal09 and Marine09 radiocarbon age calibration curves, 0-50,000 years cal BP, *Radiocarbon*, *51*, 1111–1150.

Renold, M., C. C. Raible, M. Yoshimori, and T. F. Stocker (2010), Simulated resumption of the North Atlantic meridional overturning circulation – Slow basin-wide advection and abrupt local convection, *Quat. Sci. Rev.*, *29*, 101–112.

Rial, J. A., and R. Saha (2011), Modeling abrupt climate change as the interaction between sea ice extent and mean ocean temperature under orbital insolation forcing, in *Abrupt Climate Change: Mechanisms, Patterns, and Impacts*, *Geophys. Monogr. Ser.*, doi:10.1029/2010GM001027, this volume.

Ruddiman, W. F., and A. McIntyre (1981), The North-Atlantic Ocean during the last deglaciation, *Palaeogeogr. Palaeoclimatol. Palaeoecol.*, *35*, 145–214.

Russell, J. M., et al (2009), Abrupt climate change in the southeast African tropics between 0 and 60,000 yr BP: Testing the influence of ITCZ migration on temperature and precipitation in the southern tropics, paper presented at Chapman Conference on Abrupt Climate Change, AGU, Columbus, Ohio.

Ruzmaikin, A., and J. Feynman (2011), The 1500 year quasiperiodicity during the Holocene, in *Abrupt Climate Change: Mechanisms, Patterns, and Impacts*, *Geophys. Monogr. Ser.*, doi: 10.1029/2010GM001024, this volume.

Salgueiro, E., et al. (2010), Temperature and productivity changes off the western Iberian margin during the last 150 ky, *Quat. Sci. Rev.*, *29*, 680–695.

Saltzman, B., A. Sutera, and A. Evanson (1981), Structural stochastic stability of a simple auto-oscillatory climatic feedback system, *J. Atmos. Sci.*, *38*, 494–503.

Severinghaus, J. P., R. Beaudette, M. A. Headly, K. Taylor, and E. J. Brook (2009), Oxygen-18 of O_2 records the impact of abrupt climate change on the terrestrial biosphere, *Science*, *324*, 1431–1434, doi:10.1126/science.1169473.

Shackleton, N. J., M. A. Hall, and E. Vincent (2000), Phase relationships between millennial-scale events 64,000–24,000 years ago, *Paleoceanography*, *15*, 565–569.

Stenni, B., et al. (2010), The deuterium excess records of EPICA Dome C and Dronning Maud Land ice cores (East Antarctica), *Quat. Sci. Rev.*, *29*, 146–159.

Stott, L., and A. Timmermann (2011), Hypothesized link between glacial/interglacial atmospheric CO_2 cycles and storage/release of CO_2-rich fluids from deep-sea sediments, in *Abrupt Climate Change: Mechanisms, Patterns, and Impacts*, *Geophys. Monogr. Ser.*, doi:10.1029/2010GM001052, this volume.

Stott, L., J. Southon, A. Timmermann, and A. Koutavas (2009), Radiocarbon age anomaly at intermediate water depth in the Pacific Ocean during the last deglaciation, *Paleoceanography*, *24*, PA2223, doi:10.1029/2008PA001690.

Stouffer, R. J., et al. (2006), Investigating the causes of the response of the thermohaline circulation to past and future climate changes, *J. Clim.*, *19*, 1365–1387.

Stuiver, M., and P. M. Grootes (2000), GISP2 oxygen isotope ratios, *Quat. Res.*, *53*, 277–284.

Svensson, A., et al. (2008), A 60,000 year Greenland stratigraphic ice core chronology, *Clim. Past*, *4*, 47–57.

Teller, J. T., and L. H. Thorleifdson (1983), The Lake Agassiz-Lake Superior connection, in *Glacial Lake Agassiz*, edited by J. T. Teller and L. Clayton, *Geol. Assoc. Can. Spec. Pap.*, *26*, 261–290.

Teller, J. T., M. Boyd, Z. Yang, P. S. G. Kor, and A. Mokhtari-Fard (2005), Alternative routing of Lake Agassiz overflow during the Younger Dryas: New dates, paleotopography, and a re-evaluation, *Quat. Sci. Rev.*, *24*, 1890–1905.

Thompson, L. G. (2011), Abrupt climate change: A paleoclimate perspective from the world's highest mountains, in *Abrupt Climate Change: Mechanisms, Patterns, and Impacts*, *Geophys. Monogr. Ser.*, doi:10.1029/2010GM001015, this volume.

Thompson, L. G., T. Yao, M. E. Davis, K. A. Henderson, E. Mosley-Thompson, P.-N. Lin, J. Beer, H.-A. Synal, J. Cole-Dai, and J. F. Bolzan (1997), Tropical climate instability: The last glacial cycle from a Qinghai-Tibetan ice core, *Science*, *276*, 1821–1825, doi:10.1126/science.276.5320.1821.

Thompson, L. G., E. Mosley-Thompson, H. Brecher, M. E. Davis, B. León, D. Les, P.-N. Lin, T. Mashiotta, and K. Mountain (2006), Abrupt tropical climate change: Past and present, *Proc. Natl. Acad. Sci. U. S. A.*, *103*, 10,536–10,543, doi:10.1073/pnas.0603900103.

Tierney, J. E., et al. (2008), Northern Hemisphere controls on tropical southeast African climate during the last 60,000 years, *Science*, *322*, 252–255.

Toggweiler, J. R. (2009), Shifting westerlies, *Science*, *323*, 1434–1435.

Vacco, D. A., R. B. Alley, and D. Pollard (2009), Modeling dependence of moraine deposition on climate history: The effect of seasonality, *Quat. Sci. Rev.*, *28*, 639–646.

Viau, A. E., K. Gajewski, M. C. Sawada, and P. Fines (2006), Millennial-scale temperature variations in North America during the Holocene, *J. Geophys. Res.*, *111*, D09102, doi:10.1029/2005JD006031.

Voelker, A. H. L., and L. de Abreu (2011), A review of abrupt climate change events in the northeastern Atlantic Ocean (Iberian margin): Latitudinal, longitudinal, and vertical gradients, in *Abrupt Climate Change: Mechanisms, Patterns, and Impacts*, *Geophys. Monogr. Ser.*, doi:10.1029/2010GM001021, this volume.

Voelker, A. H. L., et al. (2002), Global distribution of centennial-scale records for marine isotope stage (MIS) 3: A database, *Quat. Sci. Rev.*, *21*, 1185–1212.

Voelker, A. H. L., L. de Abreu, J. Schönfeld, H. Erlenkeuser, and F. Abrantes (2009), Hydrographic conditions along the western Iberian margin during marine isotope stage 2, *Geochem. Geophys. Geosyst.*, *10*, Q12U08, doi:10.1029/2009GC002605.

Wang, X.-F., A. S. Auler, R. L. Edwards, H. Cheng, E. Ito, and M. Solheid (2006), Interhemispheric antiphasing of rainfall during the last glacial period, *Quat. Sci. Rev.*, *25*, 3391–3403.

Wang, Y., H. Cheng, R. L. Edwards, X.-G. Kong, X. Shao, S. Chen, J.-Y. Wu, X.-Y. Jiang, X.-F. Wang, and Z.-S. An (2008), Millennial- and orbital-scale changes in the East Asian monsoon over the past 224,000 years, *Nature*, *451*, 1090–1093, doi:10.1038/nature06692.

Wathen, S. F. (2011), Evidence for climate teleconnections between Greenland and the Sierra Nevada of California during the Holocene, including the 8200 and 5200 climate events, in *Abrupt Climate Change: Mechanisms, Patterns, and Impacts*, *Geophys. Monogr. Ser.*, doi:10.1029/2010GM001022, this volume.

Weiss, H., M.-A. Courty, W. Wetterstrom, F. Guichard, L. Senior, R. Meadow, and A. Curnow (1993), The genesis and collapse of third millennium north Mesopotamian civilization, *Science*, *261*, 995–1004, doi:10.1126/science.261.5124.995.

Williams, C., B. P. Flower, D. W. Hastings, T. P. Guilderson, K. A. Quinn, and E. A. Goddard (2010), Deglacial abrupt climate change in the Atlantic Warm Pool: A Gulf of Mexico perspective, *Paleoceanography*, *25*, PA4221, doi:10.1029/2010PA001928.

E. Mosley-Thompson, L. Polyak, and H. Rashid, Byrd Polar Research Center, Ohio State University, Columbus, OH 43210, USA. (rashid.29@osu.edu)

A Review of Abrupt Climate Change Events in the Northeastern Atlantic Ocean (Iberian Margin): Latitudinal, Longitudinal, and Vertical Gradients

Antje H. L. Voelker

Unidade de Geologia Marinha, Laboratorio Nacional de Energia e Geologia, Amadora, Portugal
CIMAR Associate Laboratory, Porto, Portugal

Lucia de Abreu[1]

Godwin Laboratory for Palaeoclimate Research, Department of Earth Sciences, University of Cambridge, Cambridge, UK
Unidade de Geologia Marinha, Laboratorio Nacional de Energia e Geologia, Amadora, Portugal
CIMAR Associate Laboratory, Porto, Portugal

The Iberian margin is a key location to study abrupt glacial climate change, and regional variability is studied combining published and new records. Looking at the trend from marine isotope stage (MIS) 10 to 2, the planktic foraminifer data, conforming to *Martrat et al.* [2007], show that abrupt events, especially Heinrich events, became more frequent and their impacts stronger during the last glacial cycle. However, there were two older periods with strong impacts on the Atlantic meridional overturning circulation: the Heinrich-type event associated with termination IV and the one occurring during MIS 8 (269 to 265 ka). During Heinrich stadials, the Polar Front reached the northern Iberian margin (approximately 41°N), while the Arctic Front was located in the vicinity of 39°N. During all glacial periods, there existed a boundary at the latter latitude, either the Arctic Front during extreme cold events or the Subarctic Front during less strong coolings or warmer glacials. Along with the fronts, sea surface temperature (SST) increased southward by about 1°C per 1° latitude leading to steep SST gradients. Glacial hydrographic conditions were similar during MIS 2 and 4 but much different during MIS 6. MIS 6 was a warmer glacial with subtropical waters reaching as far north as 40.6°N. In the vertical structure, Greenland-type oscillations were recorded down to 2465 m during Heinrich stadials, i.e., deeper than in the western basin, due to the admixing of Mediterranean Outflow Water. It is evident that latitudinal, longitudinal, and

[1]Now at Camberley, UK.

Abrupt Climate Change: Mechanisms, Patterns, and Impacts
Geophysical Monograph Series 193

10.1029/2010GM001021

vertical gradients existed along the Iberian margin, i.e., in a relatively restricted area, but sufficient paleodata now exist to validate regional climate models for abrupt climate change events.

1. INTRODUCTION

The western Iberian margin is a focal location for studying the impact and intensity of abrupt climate change variability. Sediment cores retrieved there at a depth of more than 2200 m showed that the $\delta^{18}O$ of planktic foraminifer exhibits changes similar to those found in Greenland ice core records (e.g., $\delta^{18}O_{ice}$), whereas the $\delta^{18}O$ record of benthic foraminifer varies in a manner more reminiscent of the Antarctic temperature signal [*Shackleton et al.*, 2000]. Thus, core sites retrieved at this margin allow studying interhemispheric linkages in the climate system. In addition, the southern edge of the North Atlantic's ice-rafted detritus (IRD) belt [*Hemming*, 2004; *Ruddiman*, 1977] intercepted with the margin, so that melting icebergs reached the margin during Heinrich and Greenland stadials of the last glacial cycle and during ice-rafting events of preceding glacials [*Baas et al.*, 1997; *Bard et al.*, 2000; *de Abreu et al.*, 2003; *Moreno et al.*, 2002; *Naughton et al.*, 2007; *Sánchez-Goñi et al.*, 2008; *Zahn et al.*, 1997]. Following the work of *Sánchez-Goñi and Harrison* [2010] who documented that on the Iberian margin, the duration of the related surface water cooling and the Heinrich ice-rafting event per se can differ, Greenland stadials associated with Heinrich events are referred to as Heinrich stadials. Otherwise, the Greenland stadial and Greenland interstadial nomenclature in this chapter follows the Integration of Ice-core, Marine and Terrestrial Records (INTIMATE) group [*Lowe et al.*, 2001, 2008] and the *North Greenland Ice Core Project members* [2004]. Only during Heinrich stadials did the Polar Front reach the Iberian margin [*Eynaud et al.*, 2009] associated with abrupt and intense cooling in the sea surface temperature (SST) [*Bard et al.*, 2000; *Cayre et al.*, 1999; *de Abreu et al.*, 2003; *Martrat et al.*, 2007; *Naughton et al.*, 2009; *Vautravers and Shackleton*, 2006; *Voelker et al.*, 2006]. Using the records of three core sites, *Salgueiro et al.* [2010] were the first to show that while cooling was recorded at all sites during Heinrich events, there existed a clear boundary between 40°N and 38°N that affected not only the SST but also productivity. They attributed this boundary to a stronger influence of subtropical surface and subsurface waters in the southern region, which is in accordance with evidence from nannofossils [*Colmenero-Hidalgo et al.*, 2004; *Incarbona et al.*, 2010] and planktic foraminifer stable isotope data [*Rogerson et al.*, 2004; *Voelker et al.*, 2009]. *Voelker et al.* [2009], furthermore, showed that upper water column stratification was diminished during the Heinrich events of marine isotope stage (MIS) 2, especially along the western margin.

The Heinrich and Greenland stadials left their imprints also farther down in the water column related to changes in the Atlantic meridional overturning circulation (AMOC) strength. One well-documented change was the increased influence of lesser ventilated southern sourced waters, in particular, the Antarctic Bottom Water (AABW), due to the shoaling of the interface between Glacial North Atlantic Intermediate Water (GNAIW) and AABW [*Margari et al.*, 2010; *Shackleton et al.*, 2000; *Skinner and Elderfield*, 2007; *Skinner et al.*, 2003] when AMOC was reduced or shut off. Along with this change, ventilation of the deeper water column was reduced [*Baas et al.*, 1998; *Schönfeld et al.*, 2003; *Skinner and Shackleton*, 2004], and nutrient levels were raised [*Willamowski and Zahn*, 2000]. In the middepth range, another water mass is also important on the Iberian margin: the Mediterranean Outflow Water (MOW). Evidence for MOW changes mainly come from core sites in the Gulf of Cadiz, i.e., the southern margin. *Voelker et al.* [2006] showed that the lower MOW core reacted to abrupt climatic changes and was stronger during most parts of the Heinrich stadials and during Greenland stadials in accordance with evidence for deep convection in the Mediterranean Sea [*Kuhnt et al.*, 2008; *Schmiedl et al.*, 2010; *Sierro et al.*, 2005]. Similar evidence also emerged for the upper MOW core [*Llave et al.*, 2006; *Toucanne et al.*, 2007], and for MIS 2, it has been shown that the MOW was not only strengthened, but settled significantly deeper, as deep as 2000 m, in the water column [*Rogerson et al.*, 2005; *Schönfeld and Zahn*, 2000]. Thus, abrupt climatic changes affected all levels of the water column on the western Iberian margin.

During the last decades, many cores have been retrieved from this region and studied in high-resolution, but the records were seldom combined for a comprehensive regional reconstruction. In this review, records from several cores are being compiled to look at regional variability in the response to abrupt climate change events and to trace latitudinal, longitudinal, and vertical gradients during the last glacial cycle. All of this is important information needed for model/data comparisons to validate how well climate models reproduce past conditions [e.g., *Kjellström et al.*, 2010] and which local phenomena might have to be included in regional models to correctly represent the past conditions. Thus, this study aims to describe how hydrographic conditions changed along with the abrupt climate events and to relate them to the potential driving mechanisms. After having identified gradients during the last cycle, their existence at the same position and with the same intensity during previous glacial cycles will be tested.

Hereby, one focus will be on the glacial upper water column structure as this will allow identifying boundaries between subpolar- and subtropical-dominated waters with implications for the position of hydrographic fronts.

2. MODERN HYDROGRAPHIC SETTING

The western Iberian margin represents the northern part of the Canary/northwest African eastern boundary upwelling system, and its upper water column hydrography is marked by seasonally variable currents and countercurrents (Figure 1a). Upwelling and its associated features (Figure 1b) dominate the hydrography generally from late May/early June to late September/early October [*Haynes et al.*, 1993] and is driven by the northward displacement of the Azores high-pressure cell and the resulting northerly winds. Intense upwelling on the western margin is linked to topographic features like Cape Finisterre, Cape Roca, and Cape São Vicente (Figure 1b) or submarine canyons [*Sousa and Bricaud*, 1992]. The Lisbon plume, linked to Cape Roca, can either extend westward as in Figure 1b or southward toward Cape Sines. During intense upwelling events, the filament off Cape São Vicente extends southward and is fed by the Portugal Coastal Current (PCC) [*Fiúza*, 1984]). The more persistent feature, however, is an eastward extension of the filament along the southern Portuguese shelf break and slope [*Relvas and Barton*, 2002] where, when westerly winds prevail, the waters merge with locally upwelled waters (Figure 1b).

The Portugal Current (PC), which branches of the North Atlantic Drift off Ireland, consists of the PC per se in the open ocean and the PCC along the slope during the upwelling season. The PC advects surface and subsurface waters slowly equatorward [*Perez et al.*, 2001; *van Aken*, 2001] and is centered west of 10°W in winter (Figure 1a) [*Peliz et al.*, 2005]. The PC's subsurface component is the Eastern North Atlantic Central Water (ENACW) of subpolar (sp) origin, which is formed by winter cooling in the eastern North Atlantic Ocean [*Brambilla et al.*, 2008; *McCartney and Talley*, 1982]. The PCC, on the other hand, is a jet-like upper slope current transporting the upwelled waters southward [*Alvarez-Salgado et al.*, 2003; *Fiúza*, 1984]. At Cape São Vicente, a part of this jet turns eastward and enters the Gulf of Cadiz [*Sanchez and Relvas*, 2003]. In the Gulf of Cadiz, it flows along the upper slope toward the Strait of Gibraltar [*Garcia-Lafuente et al.*, 2006], then called the Gulf of Cadiz Slope Current [*Peliz et al.*, 2007]. This current either forms an anticyclonic meander in the eastern Gulf of Cadiz or

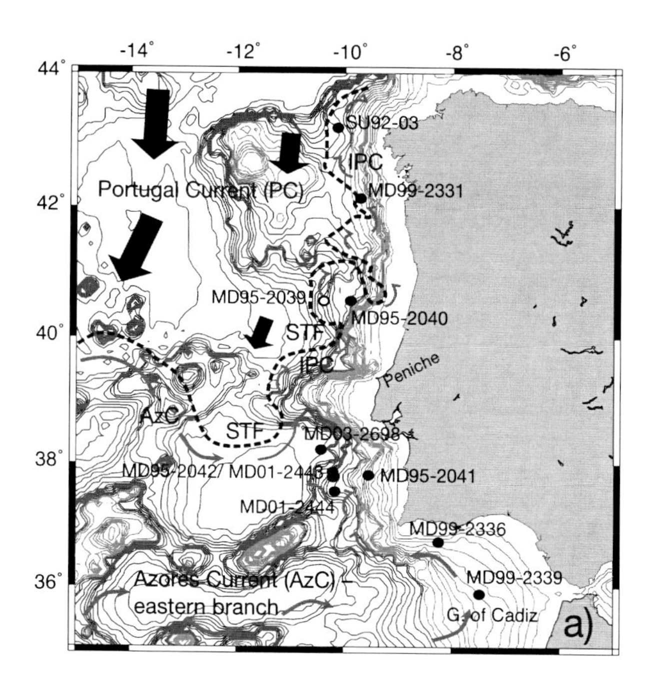

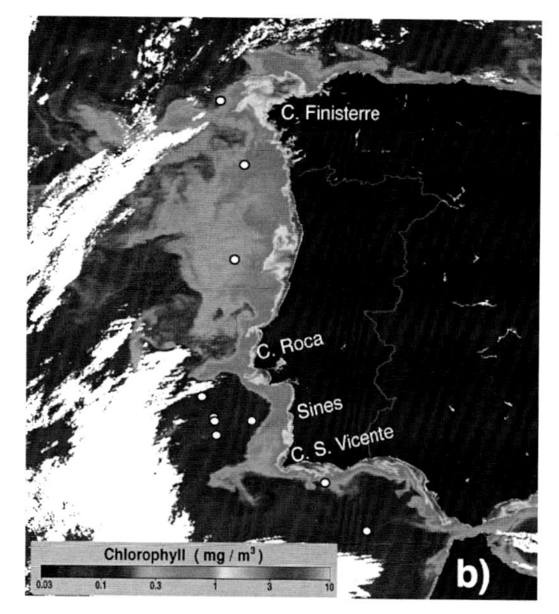

Figure 1. (a) Map of the western Iberian margin with core sites and surface water circulation in winter as summarized by *Peliz et al.* [2005]. The location of core MD95-2039 (white circle), which is mentioned in the text but for which no data are shown, is also indicated. (b) NASA Aqua MODIS satellite-derived chlorophyll *a* picture (http://oceancolor.gsfc.nasa.gov/FEATURE/gallery.html) for 13 September 2005 showing the regions most affected by upwelling along the Iberian margin (lighter shades) and the extensive filaments off capes Finisterre, Roca, and São Vicente. Dots mark the same core locations as in Figure 1a (except for MD95-2039).

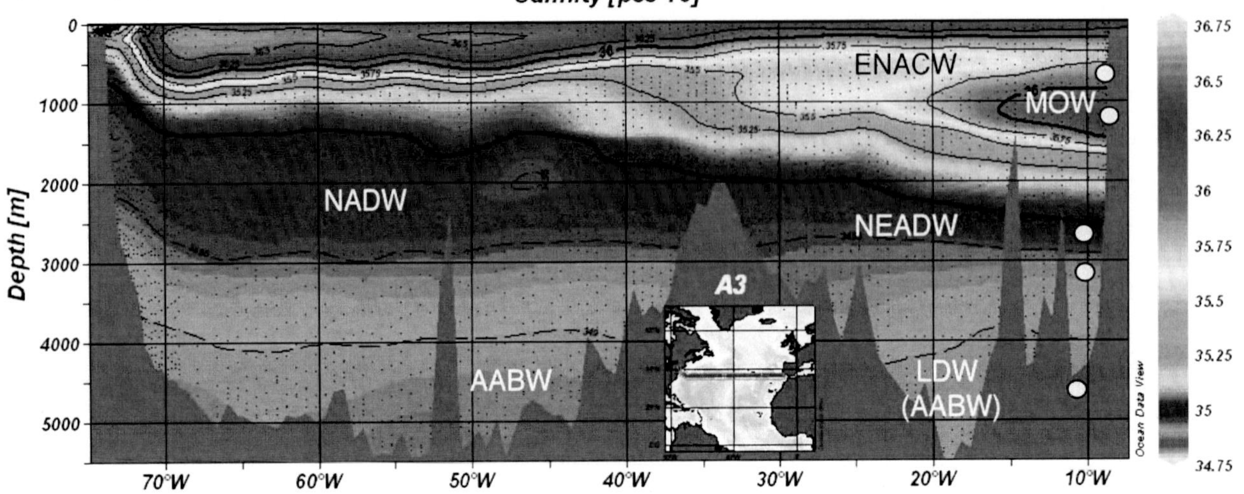

Figure 2. Salinity profile of World Ocean Circulation Experiment (WOCE) transect A3 [*Schlitzer*, 2000] (http://www.ewoce.org/) with dots marking from top to bottom depths of core sites MD99-2336, MD99-2339 (both not on correct longitude), MD01-2444, MD95-2042/MD99-2334K, and MD03-2698. Water mass abbreviations are ENACW, Eastern North Atlantic Central Water; MOW, Mediterranean Outflow Water; NADW, North Atlantic Deep Water; NEADW, Northeastern Atlantic Deep Water; AABW, Antarctic Bottom Water; LDW, Lower Deep Water.

enters the Mediterranean Sea as Atlantic inflow [*Garcia-Lafuente et al.*, 2006; *Sanchez and Relvas*, 2003].

The Azores Current (AzC), another current branching of the Gulf Stream/North Atlantic Drift, and the associated Subtropical Front reveal large meanders between 35°N and 37°N in the eastern North Atlantic. While most of the AzC recirculates southward, its eastern branch flows into the Gulf of Cadiz [*Johnson and Stevens*, 2000; *Peliz et al.*, 2005; *Vargas et al.*, 2003], where it feeds the offshore flow (Figure 1a). Ocean models indicate that the AzC flow into the Gulf is quite significant [*Penduff et al.*, 2001], and they link the existence of the current itself to the entrainment of surface to subsurface waters into the MOW [*Jia*, 2000; *Oezgoekmen et al.*, 2001]. Between October and March, when the Iberian Poleward Current (IPC) (Figure 1a), also a branch of the AzC, becomes a prominent feature off western Iberia, the thermal Subtropical Front (at ~17°C) is shifted northward and reaches the SW Iberian margin [*Pingree et al.*, 1999]. Along with this shift, AzC waters tend to recirculate from the Gulf of Cadiz into the region off Sines (Figure 1a). *Peliz et al.* [2005] observe a recurrent frontal system, the Western Iberia Winter Front, which follows the thermal Subtropical Front in the south but then meanders northward and separates the IPC from the PC (Figure 1a). The IPC, extending down to 400 m, transports warm and salt-rich waters of subtropical origin [*Frouin et al.*, 1990; *Haynes and Barton*, 1990] and can be traced into the Bay of Biscay. The IPC's subsurface or undercurrent part conveys ENACW of subtropical (st) origin poleward year round. $ENACW_{st}$, which is formed by strong evaporation and winter cooling along the Azores Front [*Fiúza*, 1984; *Rios et al.*, 1992], is poorly ventilated, warmer, and saltier than its subpolar counterpart. ENACW is the source for the water upwelled from May to September, and in general, $ENACW_{st}$ is upwelled south of 40°N and $ENACW_{sp}$ north of 45°N. In between, either water mass can be upwelled depending on the strength of the wind forcing.

Between 500 and 1500 m, the water column along the western Iberian margin is dominated by the warm, salty MOW (Figure 2) that is formed in the Gulf of Cadiz by mixing of Mediterranean Sea with Atlantic water, the above mentioned entrainment. Owing to the mixing, the MOW splits into two cores centered at about 800 and 1200 m [*Ambar and Howe*, 1979], which flow as undercurrents northward along the western Iberian margin. Facilitated by the margin's topography (e.g., canyons, capes, and seamounts), the MOW cores shed many eddies [*Richardson et al.*, 2000; *Serra and Ambar*, 2002], called meddies, which greatly contribute to the MOW's admixing into the wider North Atlantic Basin. Below the MOW at a depth around 1600 m, Labrador Sea Water (LSW), the uppermost component of the North Atlantic Deep Water (NADW), can be found on the margin north of 40.5°N [*Alvarez et al.*, 2004; *Fiuza et al.*, 1998]. Deeper down in the water column, Northeastern Atlantic Deep Water (NEADW) and Lower Deep Water (LDW) are found. LDW (>4000 m) is warmed AABW that enters the eastern Atlantic Basin through the

Vema fracture zone at 11°N and the Iberian and Tagus abyssal plains partly as intensified current through the Discovery Gap near 37°N [*Saunders*, 1987]. The NEADW is a mixture between Iceland-Scotland Overflow Water, LSW, LDW, and MOW with the contributions of LDW and MOW increasing to the south [*van Aken*, 2000]. The admixing of MOW into the NEADW explains why salinities (Figure 2) and temperatures are higher in the eastern than in the western basin for equivalent depths down to 2500 m.

3. MATERIAL AND METHODS

Most of the records shown here are from Calypso piston cores retrieved with R/V *Marion Dufresne* II (IPEV) during the first IMAGES cruise in 1995 (MD95-) [*Bassinot and Labeyrie*, 1996], the fifth IMAGES cruise in 1999 (MD99-) [*Labeyrie et al.*, 2003], the Geosciences cruise in 2001 (MD01-), and the PICABIA cruise in 2003 (MD03-). Details on core locations and respective water depths are given in Table 1.

Planktic foraminifer census counts were done in the fraction >150 μm. In general, SST data were calculated with the SIMMAX transfer function [*Pflaumann et al.*, 1996] using an extended (1066 samples) version of the *Salgueiro et al.* [2010] database that is well suited for SST reconstructions in the eastern North Atlantic. The additional samples are located mostly off NW Africa, and for those cores for which we recalculated SST (MD95-2040, MD01-2443, MD01-2444, MD99-2339), interstadial and interglacial temperatures are often slightly (≈0.2°C) warmer than those previously published. We present only summer (July/August/September) temperatures (SST_{su}), but temperatures for the other seasons as well as the standard deviations derived from the minimum and maximum values of the selected nearest neighbors are available from the World Data Centre-Mare through the parent link http://doi.pangaea.de/10.1594/PANGAEA.737449. The methodical error for the SIMMAX-based SST reconstructions is ±0.8°C [*Salgueiro et al.*, 2010]. For core MD99-2331 and the MIS 3 section of core MD95-2042, SST values depicted in Figure 3 are from the work of *Sánchez-Goñi et al.* [2008] and represent August SST. For core MD95-2042, the *Sánchez-Goñi et al.* [2008] August SST do not differ significantly from those obtained by *Salgueiro et al.* [2010] for the lower-resolution counts done by *Cayre et al.* [1999], so that the records of MD95-2042 and MD99-2331 are comparable to the other ones shown in Figure 3.

Foraminifer-based stable isotope data were measured either at Marum, University Bremen (Germany), in the Godwin Laboratory, Cambridge University (United Kingdom), in the Leibniz Laboratory for Radiometric Dating and Stable Isotope Research or at IfM-Geomar, the latter two in Kiel

Table 1. List of Core Sites and References for Data and Age Models

Core Number	Longitude	Latitude	Water Depth (m)	Data Sources	Age Model
SU92-03	43.20°N	10.11°W	3005	*Salgueiro et al.* [2010]	*Salgueiro et al.* [2010]: GISP2
MD99-2331	42.15°N	9.68°W	2110	*Sánchez-Goñi et al.* [2008]	*Sánchez-Goñi et al.* [2008]: NGRIP tuned
MD95-2040	40.58°N	9.86°W	2465	*de Abreu et al.* [2003]; *Schönfeld et al.* [2003]; *Pailler and Bard* [2002]; this study	*Salgueiro et al.* [2010] for MIS 1-3, MIS 4-5 tuned to MD95-2042; MIS 6: *Margari et al.* [2010]; ≥MIS 7: tuned to LR04
MD03-2698	38.24°N	10.39°W	4602	*Lebreiro et al.* [2009]	*Lebreiro et al.* [2009]
MD95-2041	37.83°N	9.52°W	1123	*Voelker et al.* [2009]; this study	*Voelker et al.* [2009] and tuning to MD95-2042 for >30 ka
MD95-2042	37.80°N	10.17°W	3146	*Cayre et al.* [1999]; *Shackleton et al.* [2000]; *Sánchez-Goñi et al.* [2008]; this study (SIMMAX SST)	*Shackleton et al.* [2000]: GISP2
MD99-2334K	37.80°N	10.17°W	3146	*Skinner et al.* [2003]	*Skinner et al.* [2003]: GISP2
MD01-2443	37.88°N	10.18°W	2941	*de Abreu et al.* [2005]; *Tzedakis et al.* [2004]; *Martrat et al.* [2007]; this study	*Tzedakis et al.* [2009]: tuned to EDC3
MD01-2444	37.57°N	10.13°W	2656	*Vautravers and Shackleton* [2006]; *Martrat et al.* [2007]; [*Skinner and Elderfield* [2007]; this study (SIMMAX SST)	*Vautravers and Shackleton* [2006] modified to GISP2 ages for MIS 3 and *Martrat et al.* [2007]
MD99-2336	36.72°N	8.26°W	690	*Voelker et al.* [2009]; this study	*Voelker et al.* [2009] and tuning to MD95-2042 for MIS 4
MD99-2339	35.89°N	7.53°W	1170	*Voelker et al.* [2006, 2009]; this study	*Voelker et al.* [2006]

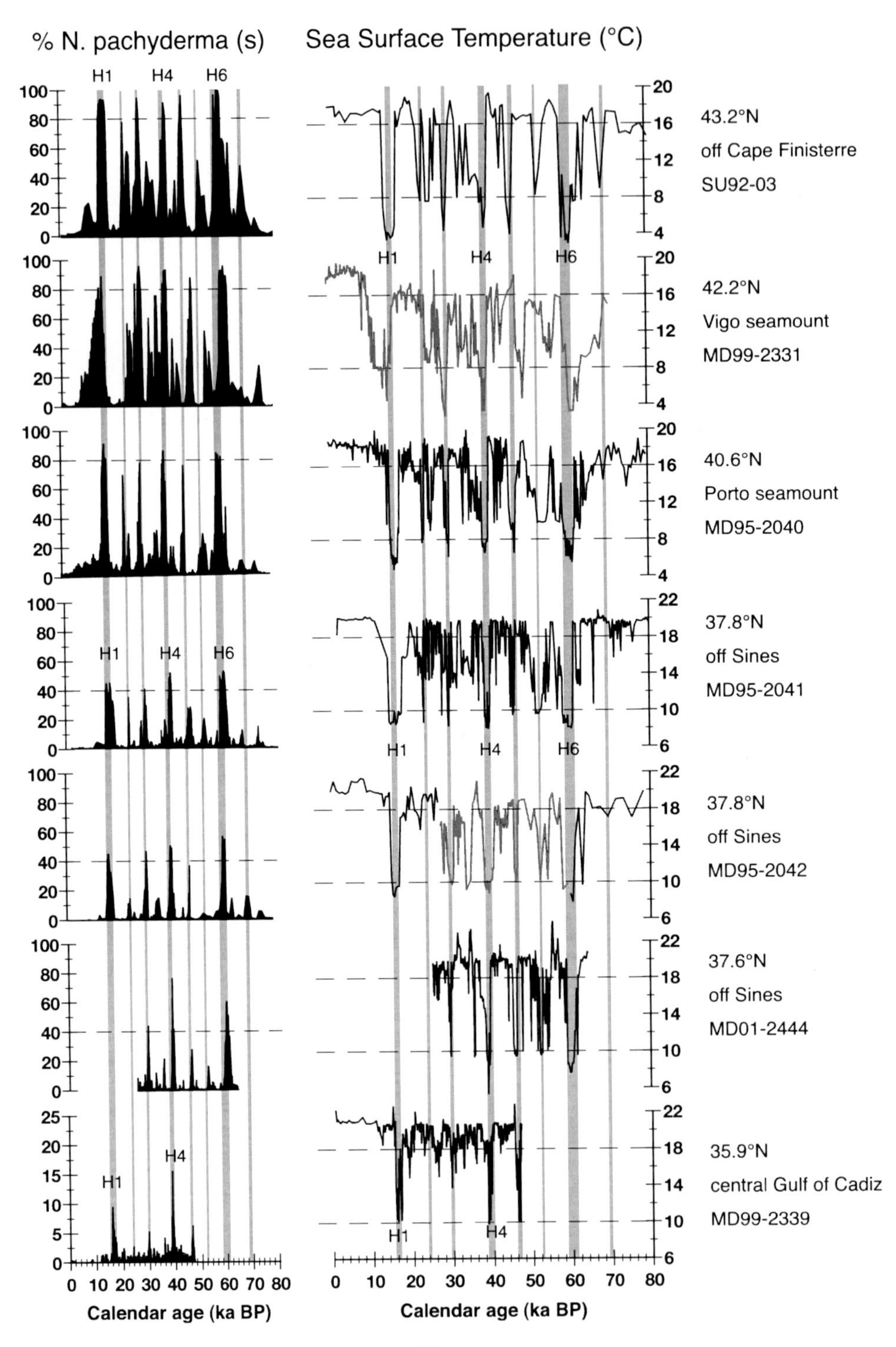
% N. pachyderma (s)
Sea Surface Temperature (°C)
H1
H4
H6
43.2°N
off Cape Finisterre
SU92-03
42.2°N
Vigo seamount
MD99-2331
40.6°N
Porto seamount
MD95-2040
37.8°N
off Sines
MD95-2041
37.8°N
off Sines
MD95-2042
37.6°N
off Sines
MD01-2444
35.9°N
central Gulf of Cadiz
MD99-2339
Calendar age (ka BP)
Calendar age (ka BP)

Figure 3

(Germany) (for details see original references listed in Table 1 and *Voelker et al.* [2009]). The benthic $\delta^{18}O$ record of core MD95-2040 combines values corrected to the *Uvigerina* level of the following foraminifer species: *Cibicidoides wuellerstorfi*, *Cibicidoides kullenbergi*, *Cibicidoides* sp., *Uvigerina peregrina*, *Uvigerina pygmea*, *Melonis* sp., and *Globobulimina affinis* (only in MIS 6). Correction factors are those listed by *de Abreu et al.* [2005]. The benthic $\delta^{13}C$ record, on the other hand, only includes *Cibicidoides*-derived values. For details on the MD01-2443 benthic records, the reader is referred to the work of *Martrat et al.* [2007].

Following the work of *Voelker et al.* [2009], planktic foraminifer species for which stable isotope data were obtained for the MIS 3, 4, and 6 intervals are *Globigerina bulloides*, *Globigerinoides ruber* white, *Neogloboquadrina pachyderma* (r) or (s), *Globorotalia inflata*, *Globorotalia scitula*, and *Globorotalia truncatulinoides* (r) or (s). *G. truncatulinoides* (s) values are generally only shown for MIS 4 when this species dominates over the right-coiling variety in the assemblage. For the MIS 3 section of core MD99-2339, on the other hand, samples of either coiling direction were analyzed, and a combined record, sometimes based on mean values from double measurements, is shown. The combined $\delta^{18}O$ and $\delta^{13}C$ records for these species, which cover calcification depths from 50 to 400 m [see *Voelker et al.*, 2009, Table 4], are used to reconstruct conditions in the upper water column during the respective glacial intervals. Following the work of *Ganssen and Kroon* [2000], the $\delta^{18}O$ difference between *G. bulloides* and *G. inflata* is used to evaluate seasonality. As discussed by *Voelker et al.* [2009], none of the planktic foraminifer isotope values are corrected because regional correction factors do not exist, yet.

Data and age models of cores MD95-2040, -2041, MD99-2336, -2339, MD01-2443, and -2444 used in this study are available from the WDC-Mare through the parent link http://doi.pangaea.de/10.1594/PANGAEA.737449.

4. CHRONOSTRATIGRAPHIES

For many of the cores, for which data from the last glacial cycle are shown, the (initial) age model was linked to the Greenland Ice Sheet Project 2 (GISP2) ice core chronology either by direct tuning or by calibrating accelerator mass spectrometry ^{14}C ages with the *Hughen et al.* [2004] data. Because the focus of this study is on amplitudes and timing between the different records rather than absolute ages, GISP2-linked chronologies were kept instead of revising to North Greenland Ice Core Project (NGRIP) or Hulu Cave-based calibration data. Thus, data of core MD95-2042 are shown on the *Shackleton et al.* [2000] chronology using the GISP2 correlation points, while cores MD99-2334K, MD99-2339, MD03-2698, and SU92-03 are shown on their original published timescales (Table 1). For MIS 3 data of core MD01-2444, correlation points to the GRIP chronology given by *Vautravers and Shackleton* [2006] were converted to GISP2-based ages. However, the alkenone-derived SST record of this core is shown on the age scale in the work of *Martrat et al.* [2007], which is related to the NGRIP ice core on the GICC05 (back to 60 ka) and ss09sea chronologies for the last 120 kyr. The age model of core MD95-2041, except for the MIS 2 section, where the age model of *Voelker et al.* [2009] is applied, was established by correlating its *G. bulloides* $\delta^{18}O$ record to the one of core MD95-2042 taking the positions of percent *N. pachyderma* (s) maxima that are marking Heinrich stadials into account. The percent *N. pachyderma* (s) maxima were especially relevant to identify Heinrich stadials 3 to 5 and 8. Also, the age model for the MIS 4 section of core MD99-2336 is based on correlating its *G. bulloides* $\delta^{18}O$ record [*Llave et al.*, 2006] to core MD95-2042, while the MIS 2 section follows the work of *Voelker et al.* [2009].

Data of core MD01-2443 are shown on the age model established by *Tzedakis et al.* [2009], who, following the work of *Shackleton et al.* [2000], tuned the benthic $\delta^{18}O$ record of this core to the δD record of the EPICA Dome C (EDC) ice core on its EDC3 chronology. *Salgueiro et al.* [2010] recently published a chronology of core MD95-2040 back to the top of MIS 6. However, when compiling the figures for this chapter, we noted that with the *Salgueiro et al.* [2010] age model, the percent *N. pachyderma* (s) maximum/SST minimum of Heinrich stadial 8 is significantly older than the one in core MD95-2042. Thus, a revised

Figure 3. (opposite) Latitudinal gradients in the abundance of percent *Neogloboquadrina pachyderma* (s) (percent) (left column) and sea surface temperature (SST) derived from planktic foraminifer assemblages (°C) (right column) over the last 80 kyr. SST values refer to summer (black) or August (gray). The transect from north to south consists of cores SU92-03 [*Salgueiro et al.*, 2010], MD99-2331 [*Sánchez-Goñi et al.*, 2008], MD95-2040 [*de Abreu et al.*, 2003] with SST recalculated for this study, MD95-2041 [*Voelker et al.*, 2009, this study], MD95-2042 with SST recalculated based on counts of *Cayre et al.* [1999] (black SST line and percent *N. pachyderma* (s)) and SST data from *Sánchez-Goñi et al.* [2008] (gray line), MD01-2444 [*Vautravers and Shackleton*, 2006] with SST recalculated for this study, and MD99-2339 [*Voelker et al.*, 2006, 2009, this study]. Note the change in the percent *N. pachyderma* (s) scale for core MD99-2339 to adjust for the gradient of more than 90% in the north to the maxima of just 16% in the south. A shift by 2°C in the SST scale is also observed for core MD95-2041 and the records below. Gray bars mark Heinrich stadials.

stratigraphy for MIS 4 and late 5 was established by correlating the *G. bulloides* $\delta^{18}O$ records of cores MD95-2040 and MD95-2042. Stratigraphic control within MIS 6 follows the work of *Margari et al.* [2010], whereas for the section older than MIS 6, the new benthic $\delta^{18}O$ record was tuned to the LR04 stack [*Lisiecki and Raymo*, 2005]. The core now has a bottom age of 360 ka (MIS 10/11 boundary) that is significantly younger than the age obtained by *Thouveny et al.* [2004] through extrapolation.

5. SURFACE WATER GRADIENTS AND IMPLICATIONS FOR THE POLAR FRONT POSITION

5.1. Conditions During the Last 80 kyr

The compilation of the existing high-resolution planktic foraminifer derived-SST and percent *N. pachyderma* (s) records (Figure 3) visibly reveals the strong impact the Heinrich stadials had in this region and the temperature gradients that existed within a latitudinal band of only 7°. The Heinrich stadials are clearly distinguished by maxima in percent *N. pachyderma* (s) and the coldest SST in all records. The coldest SST during Heinrich and Greenland stadials were recorded at the two northernmost sites SU92-03 and MD99-2331 with SST_{su} in the range of 4°C to 6°C during the Heinrich stadials. Cooling at these sites occurred during the whole period of a Heinrich stadial, and percent *N. pachyderma* (s) generally exceeded 90%, values today associated with polar water masses [*Eynaud et al.*, 2009]. While these values are the coldest/highest in our compilation, conditions were more extreme just 2° farther to the north in the Bay of Biscay [*Sánchez-Goñi et al.*, 2008; *Toucanne et al.*, 2009] where percent *N. pachyderma* (s) was close to 100% during the Heinrich stadials and most Greenland stadials. If one compares the records of cores MD99-2331 and MD95-2040 (Figure 3), also separated by about 2° latitude, another gradual change appears. At the latter site at 40.6°N, maximal percent *N. pachyderma* (s) was more in the range of 80% to 90% resulting in 2° warmer surface waters. Despite the warmer conditions at site MD95-2040, the overall shape in the percent *N. pachyderma* (s) and SST curves is similar in the three sites north of 40°N setting them apart from the ones farther to the south and confirming the hydrographic boundary between 40°N and 38°N described by *Salgueiro et al.* [2010] as a robust feature.

South of 38°N, the percent *N. pachyderma* (s) values, with the exception of one data point during Heinrich stadial 4 in core MD01-2444, did not exceed 60% (Figure 3), and coldest SST during Heinrich stadials were in the range of 8°C to 10°C, i.e., 2° or more warmer than at site MD95-2040. The percent *N. pachyderma* (s) values are those associated with Arctic waters in the Nordic Seas [*Eynaud et al.*, 2009], but the reconstructed SST values are more in the range of the modern subpolar gyre. Nevertheless, the data clearly show that the Arctic or even Subarctic Front was located in the range of 39°N during Heinrich and Greenland stadials, while the hydrographic Polar Front seems to have been located somewhere close to 41°N [*Eynaud et al.*, 2009]. Such a close spacing of hydrographic fronts, but on a more longitudinal scale, is today observed off New Foundland and in the Norwegian Sea [*Dickson et al.*, 1988], i.e., in regions where Atlantic surface waters come in close vicinity to (sub)polar waters. On the latitudinal scale, such steep temperature gradients are known from the Last Glacial Maximum (LGM) [*Pflaumann et al.*, 2003]. The front near 39°N would generally mark the southern edge of the Heinrich IRD belt, in accordance with evidence from the western basin [*Hemming*, 2004], but this does not mean that icebergs did not cross this boundary and deposited their IRD farther to the south [*Bard et al.*, 2000; *Toucanne et al.*, 2007; *Voelker et al.*, 2006; *Zahn et al.*, 1997]. From the percent *N. pachyderma* (s) records, however, it becomes quickly obvious that Heinrich stadials 1, 4, and 6 had a stronger impact on the hydrography in the Sines region than in the Gulf of Cadiz (site MD99-2339) (Figure 3) where percent *N. pachyderma* (s) values were significantly lower (<16%).

Even smaller-scale regional differences can be investigated using the three records off Sines (MD95-2041, MD95-2042, and MD01-2444, Figure 3). The two core sites at 10°W, i.e., farther offshore, tend to record slightly warmer SST not only during the cold climate events but also during some Greenland interstadials (for an explanation see section 5.2). Site MD01-2444 especially reveals warmer SST during the Greenland interstadials indicating that this site was more strongly influenced by subtropical AzC waters than the other two sites, either from being located underneath a northward-extending meander of the Azores Front or from being influenced by a paleo-IPC (Figure 1a). Sporadically, the SST was even warmer than those recorded farther to the south at site MD99-2339. On the other hand, sites MD01-2444 and MD99-2339 experienced the colder conditions during the first half of Greenland interstadial 8 more strongly than sites MD95-2042 and MD95-2041. This cooling was more pronounced at the three northern sites (MD95-2040, MD9-2331, and SU92-03, Figure 3) and must therefore have been advected from the north to the south, most likely with the more offshore-located PC. Thus, high regional variability linked to the position and shape of fronts and/or upwelling system dynamics also occurred under glacial climate conditions and needs to be taken into account when impacts of abrupt climate change are discussed and compared to climate model results.

In summary, the core transect along the western Iberian margin reveals that during the Heinrich and Greenland stadials of the last 80 kyr, SST increased from north to south by about 1°C along with latitudinal shifts of about 1°. During Heinrich events, SST_{su} minima were around 4°C between 42°N and 43° N, near 6°C at 40.6°N, between 8°C and 9°C near 38°N, and near 10°C at 36°N. The Polar Front was most likely located near 41°N and the (Sub)Arctic Front with the atmospheric Polar Front at about 39°N. In accordance with the temperature gradients and frontal positions, climate conditions were more severe in the north than in the south with subsequent impacts on the vegetation [*Fletcher et al.*, 2010; *Naughton et al.*, 2009; *Roucoux et al.*, 2005; *Sánchez-Goñi et al.*, 2008].

5.2. Longitudinal Differences off Sines 38°N: The Upwelling Influence

The above-mentioned upwelling system dynamics that might drive regional variability off the Sines coast can best be seen by comparing the records of cores MD95-2042 and MD95-2041 (Figure 4). The *G. bulloides* $\delta^{18}O$ record of core MD95-2041, located closer to the coast (Figure 1a), differs from the one of MD95-2042, especially during MIS 3, with a less clear imprint of Greenland stadial and interstadial cycles. Thus, at site MD95-2041, a different hydrographic signal was recorded. The $\delta^{13}C$ and SST records further support this. Site MD95-2041 experienced a much higher SST_{su} variability with frequent short to longer-lasting coolings in the range of 3°C to 6°C between the Heinrich stadials (Figure 4e), a variability that persists in relation to the higher-resolution SST_{Aug} record of *Sánchez-Goñi et al.* [2008] for core MD95-2042 shown in Figure 3. Along with the colder SST, *G. bulloides* $\delta^{13}C$ values are generally higher at site MD95-2041 than at MD95-2042 (Figure 4c). If one excludes temperature [*Bemis et al.*, 2000] as cause, the difference would indicate that nutrient concentrations were lower in the near-shore waters. Fewer nutrients together with the SST variability indicate that site MD95-2041 experienced periods of intense upwelling in the intervals between the Heinrich stadials with the associated high surface water productivity

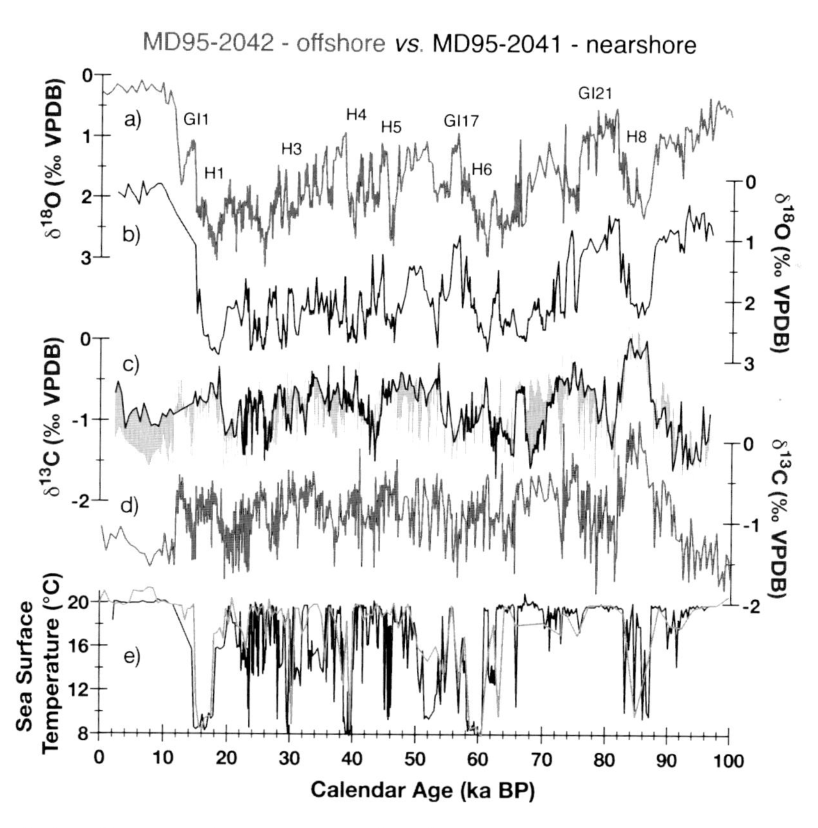

Figure 4. Longitudinal gradients in surface water properties off Sines between offshore site MD95-2042 (10.17°W, gray lines [*Cayre et al.*, 1999; *Shackleton et al.*, 2000] SST [*Salgueiro et al.*, 2010]) and nearshore site MD95-2041 (9.52°W, black lines) [*Voelker et al.*, 2009, this study]. (a and b) Respective *Globigerina bulloides* $\delta^{18}O$ records and (c and d) *G. bulloides* $\delta^{13}C$ records with the gray shading in Figure 4c representing the offset between the two records. (e) Respective foraminifer-based summer SST records. H1 to H8 mark Heinrich stadials 1 to 8, and GI indicates Greenland interstadials.

depleting the nutrients. Today, site MD95-2041 is more strongly influenced by the filament often extending southward from Lisbon than site MD95-2042 and during glacial times, when due to the lower sea level, the coastline was displaced farther offshore; also, the local upwelling along the Sines coast (Figure 1b) would be in the vicinity of site MD95-2041. High glacial productivity at this site outside of the ice-rafting events of MIS 2 was also observed by *Voelker et al.* [2009]. In consequence, these two closely spaced sites reveal that a local phenomenon like upwelling can strongly modify the paleodata and result in locally different signals that are not related to the millennial-scale climate variability.

5.3. Comparison Between the Last and Previous Glacial Cycles

To verify if the temperature gradients described in section 5.1 and the associated frontal positions also existed during previous glacials, we are using the records of core MD95-2040, the site located north of the front, and spliced records from the offshore sites off the Sines coast (Figure 5). The planktic and benthic stable isotope records indicate that both sites reliably recorded the glacial/interglacial cycles and experienced millennial-scale variability in the surface and deep-water hydrography.

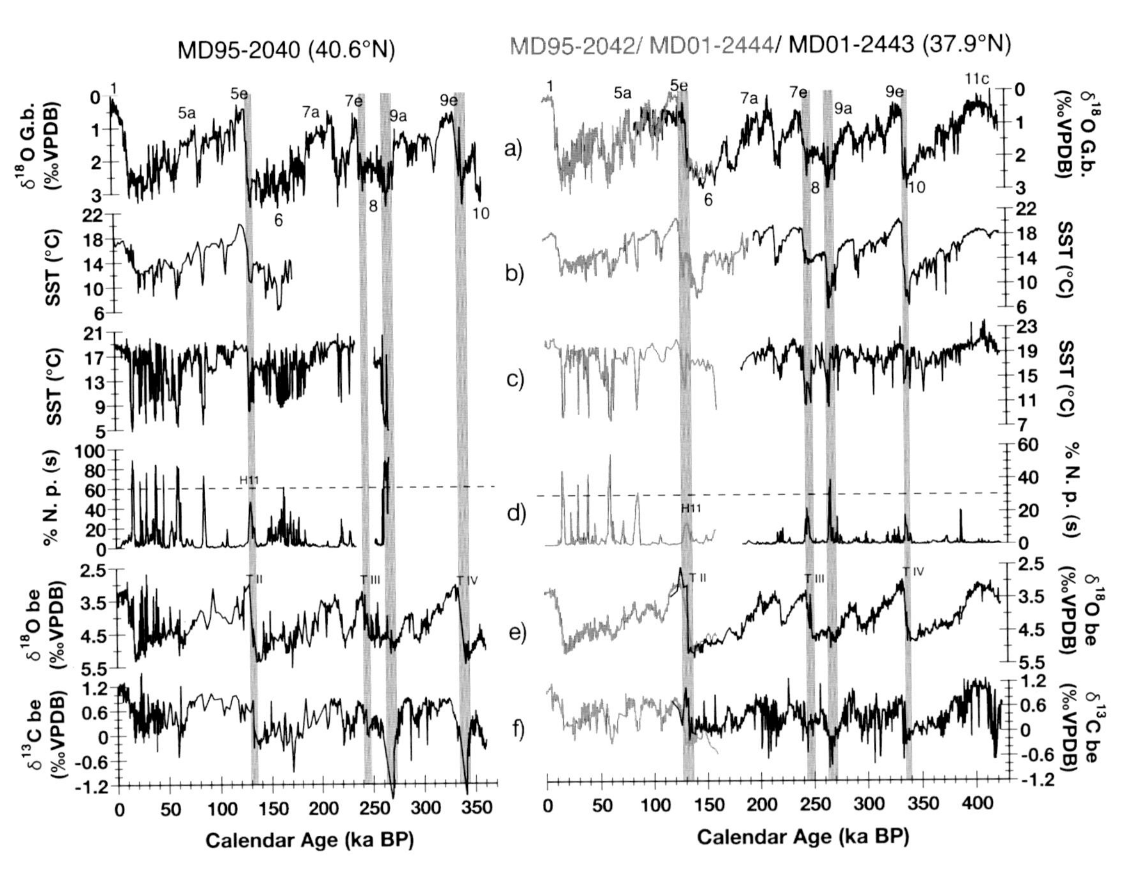

Figure 5. Latitudinal gradients in surface and deep-water properties between the Porto seamount (core MD95-2040) and the Sines coast (MD95-2042 in gray, MD01-2444 in gray, and MD01-2443 in black) over the last 420 kyr. (a) The $\delta^{18}O$ of *G. bulloides* records of cores MD95-2040 [*de Abreu et al.*, 2003; this study], MD95-2042 [*Cayre et al.*, 1999; *Shackleton et al.*, 2000], and MD01-2443 [*de Abreu et al.*, 2005; *Martrat et al.*, 2007]. (b) Alkenone-based mean annual SST records for cores MD95-2040 [*Pailler and Bard*, 2002], MD01-2444, and MD01-2443 [*Martrat et al.*, 2007]. (c) Foraminifer assemblage-based summer SST. (d) Percent *N. pachyderma* (s) records of cores MD95-2040 [*de Abreu et al.*, 2003; this study], MD95-2042 [*Cayre et al.*, 1999; *Salgueiro et al.*, 2010], and MD01-2443 [*de Abreu et al.*, 2005; this study]. Benthic (e) $\delta^{18}O$ and (f) $\delta^{13}C$ records of cores MD95-2040 [*de Abreu et al.*, 2003; *Schönfeld et al.*, 2003; this study], MD95-2042 [*Shackleton et al.*, 2000], and MD01-2443 [*de Abreu et al.*, 2005; *Martrat et al.*, 2007]. Numbers mark marine isotope stages (MIS) and terminations (T) II, III, and IV. H11 indicates Heinrich event 11. Gray bars highlight events discussed in the text.

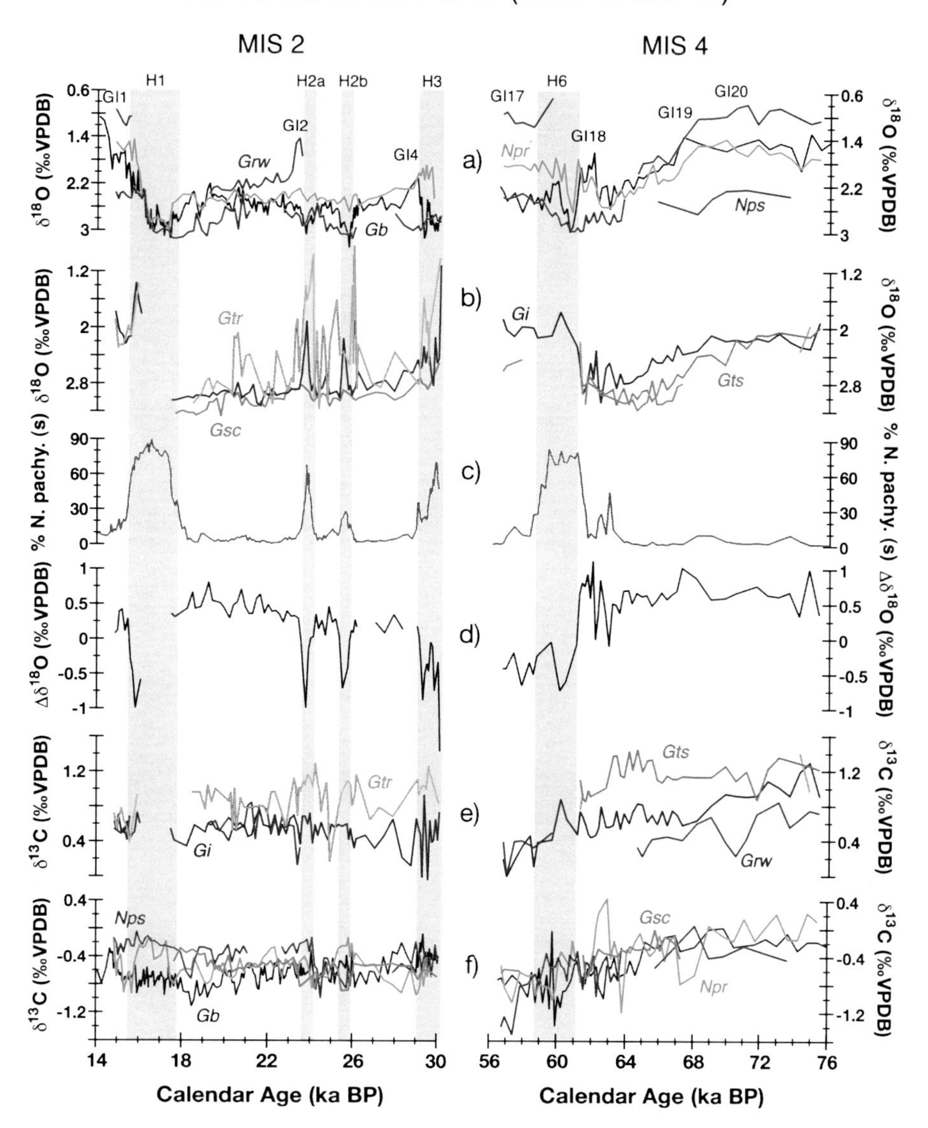

Plate 1. Vertical gradients in the upper water column at site MD95-2040 during MIS 2 [*Voelker et al.*, 2009] and 4 [*de Abreu et al.*, 2003; this study]. (a) The $\delta^{18}O$ records of surface to thermocline dwelling species *Globigerinoides ruber* white (red, Grw), *G. bulloides* (black, Gb), *N. pachyderma* (r) (cyan, Npr), and *N. pachyderma* (s) (magenta, Nps). (b) The $\delta^{18}O$ records of winter mixed layer species *Globorotalia inflata* (dark blue, Gi) and deep dwellers *Globorotalia scitula* (green, Gsc), *Globorotalia truncatulinoides* (r) (light orange, Gtr) and *G. truncatulinoides* (s) (dark orange, Gts; only MIS 4). (e and f) Respective $\delta^{13}C$ values. (c) Percent *N. pachyderma* (s) data. (d) Difference between $\delta^{18}O$ of *G. inflata* and $\delta^{18}O$ of *G. bulloides* reflecting seasonality. Blue bars and H1, H2a, H2b, H3, and H6 mark the respective Heinrich stadials. GI indicates respective Greenland interstadial.

The percent *N. pachyderma* (s) and SST_{su} records of core MD95-2040 clearly indicate that glacial MIS 6 differed not only in absolute values but also in the intensity (percent *N. pachyderma* (s); SST) of the abrupt climate change variability as previously described by *de Abreu et al.* [2003]. MIS 6 percent *N. pachyderma* (s) values in core MD95-2040 are comparable to the levels recorded in the cores off Sines during the last glacial cycle, meaning SST were significantly warmer. Although hampered by a data gap, the same can be said for the Sines area where percent *N. pachyderma* (s) values were about half of those of core MD95-2040 (Figure 5d), especially during Heinrich event 11, and more in the range

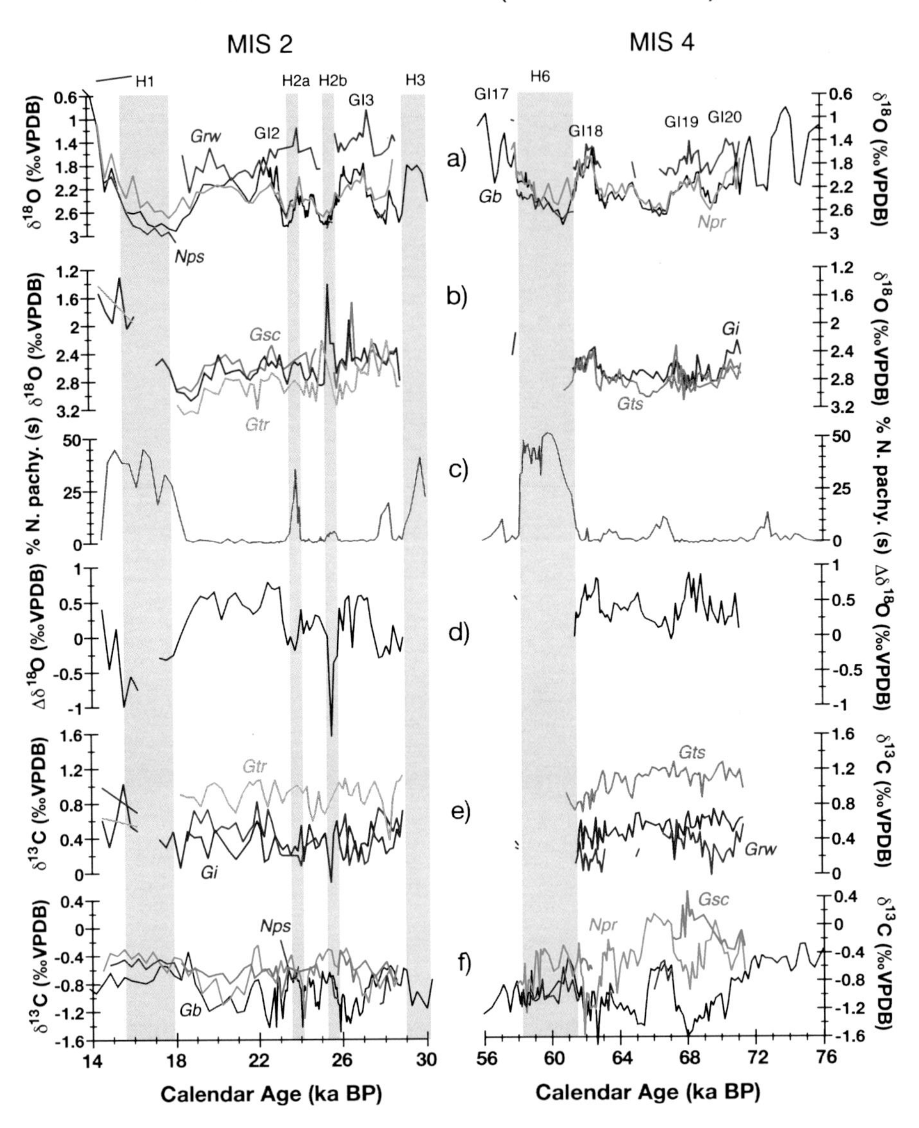

Plate 2. Vertical gradients in the upper water column at site MD95-2041 during MIS 2 [*Voelker et al.*, 2009] and 4 [this study]. Plots and foraminifer species are as in Plate 1.

of the core MD99-2339 during Heinrich stadial 4 (Figure 3). Thus, a boundary again existed between 38°N and 40°N, but this time, it was clearly the Subarctic Front. Percent *N. pachyderma* (s) values during Heinrich event 11 reached 90% at site SU92-03 [*Salgueiro et al.*, 2010] and 100% in the Bay of Biscay [*Toucanne et al.*, 2009]. So the Polar Front still reached the Iberian margin but only in the northernmost (≥43°N) section and was located farther to the north than during the last glacial cycle.

For glacial MIS 8, data for core MD95-2040 exist only between 253 and 266 ka. Within this interval, percent *N. pachyderma* (s) levels exceeded 70% and reached 90% and thus were in the range of those recorded during the Heinrich stadials of the last glacial cycle. Conditions stayed cold for an extended period (261.6 to >266 ka), lasting longer than a typical Heinrich event. Long-lasting cold was also recorded in the Bay of Biscay [*Toucanne et al.*, 2009], mostly with levels close to 100% *N. pachyderma* (s) and reminding of the hydrographic conditions observed for the Heinrich stadials. Within age constraints, this interval coincided with a Heinrich-type ice rafting event recorded at Integrated Ocean Drilling Program Site U1308 [*Hodell et al.*, 2008], so that similar forcing mechanisms and responses in the AMOC can be assumed. The low benthic $\delta^{13}C$ values recorded at sites

MD95-2040 and MD01-2443 (Figure 5f) between 271 and 262 ka clearly indicate that the AMOC was reduced or shut off. The more depleted signal recorded in core MD95-2040 is most likely related to marine snow [*Mackensen et al.*, 1993] during a period of high productivity [*Thomson et al.*, 2000]. In comparison to the northern areas, cooling in the surface waters off Sines was reduced, similar to the previous glacials, but this time, the peak cooling in the south was significantly shorter in the planktic foraminifer records (Figures 5c and 5d). Its duration was, however, comparable in the alkenone SST record (Figure 5b), indicating decoupling in the response of the two plankton groups. On the other hand, the second abrupt cold event within MIS 8 (242.5–246.5 ka) is only evident in the foraminifer records and not in the alkenone SST. This event, associated with termination III, had a lesser impact on the AMOC because the benthic $\delta^{13}C$ values were less depleted and thus indicate a GNAIW/AABW boundary deeper in the water column than during the previous event.

During glacial MIS 10, only the Heinrich-type event associated with termination IV [*Hodell et al.*, 2008; *Stein et al.*, 2009] had a pronounced impact on the hydrography off Iberia. The percent *N. pachyderma* (s) levels at site MD01-2443 were again lower than during the last glacial cycle and during the MIS 8 Heinrich-type event (Figure 5d), but they were comparable to Heinrich event 11. The benthic $\delta^{13}C$ levels at both sites were, however, similar to the MIS 8 event (Figure 5f) and again indicate a much reduced AMOC and a modified signal at site MD95-2040 due to high productivity [*Thomson et al.*, 2000]. Including evidence from other cold stages such as MIS 7d, it is clear that a boundary, sometimes the Arctic, sometimes the Subarctic Front, always separated the two core sites during abrupt cooling events. The longer records show, moreover, that the strong coolings associated with the Heinrich stadials of the last glacial cycle were close to unique and had only two counterparts during the last 420 kyr.

6. IMPACTS ON THE GLACIAL UPPER WATER COLUMN

6.1. The Last Glacial Cycle: MIS 4 and MIS 2

Abrupt climate events not only affected the uppermost waters but also left their imprints in the subsurface waters [*Rashid and Boyle*, 2007; *Voelker et al.*, 2009], information on which is often sparse. The structure of the water column from 0 to about 400 m can be investigated by combining the isotope data of various planktic foraminifer species (for details see *Voelker et al.* [2009, Table 4]). Here we focus on the last three glacial periods but using the MIS 2 data only for comparison because they have been discussed in detail by *Voelker et al.* [2009]. Data from north of the front existing between 38°N and 40°N, i.e., core MD95-2040, are compared to records from south of the front, i.e., core MD95-2041 for the last glacial cycle and core MD01-2443 for MIS 6. One peculiarity associated with the deep dwelling foraminifers used is the dominant coiling direction of *G. truncatulinoides*. During MIS 2 and 6, the right-coiling variety, which is known from just one genotype [*de Vargas et al.*, 2001], dominated, while during MIS 4, the left-coiling variety that can be attributed to all four known genotypes is more abundant. Since the genotype often found in the subtropical waters of the Sargasso and Mediterranean Sea [*de Vargas et al.*, 2001] is the only one with both coiling directions, we assume that our species belong to the same genotype.

The hydrography during the glacial maxima of MIS 2 and 4 at site MD95-2040 was similar (Plate 1), with the IPC, as indicated by the *G. ruber* white values, being absent during the latest part and during the deglaciations, i.e., Heinrich stadials 1 and 6, respectively. The interval when *G. ruber* white was absent during late MIS 4 is also the one when *N. pachyderma* (s), and thus subpolar waters, were continuously present. Along with the rise in percent *N. pachyderma* (s), just prior to Greenland interstadial 18, seasonality ($\Delta\delta^{18}O$, Plate 1d) increased, but the highest seasonal contrast was associated with Heinrich stadial 6. Then, seasonality was in the same range as the values observed during the MIS 2 Heinrich stadials. Greenland interstadial 18 was associated with warming (lower $\delta^{18}O$ values) in the surface to subsurface waters shown in particular by *G. bulloides*, *N. pachyderma* (r) and *G. inflata* (Plates 1a and 1b). The earlier Greenland interstadials 19 and 20 are poorly resolved, but the presence of *G. ruber* white and reduced seasonality indicates relative warm and stable conditions. This constancy also referred to the subsurface waters as indicated by the relative stable $\delta^{18}O$ records of *G. inflata* and *G. truncatulinoides* (Plate 1b). The MIS 4 deep-dweller records are clearly different from MIS 2 when extremely light values were measured [*Voelker et al.*, 2009]. During this early part of MIS 4, the subsurface waters were well ventilated, especially the $ENACW_{st}$ recorded in the *G. truncatulinoides* $\delta^{13}C$ values (Plate 1e). We relate the *G. truncatulinoides* data of core MD95-2040 to the $ENACW_{st}$, and thus a signal transported northward because of the similar isotopic levels observed in both MD95-2041 (Plate 2e) and MD95-2040. The good ventilation in the subsurface waters is also common to both glacial periods. Another difference to MIS 2 or more specifically to the younger Heinrich stadials at site MD95-2040 is, however, that *G. inflata* was present during some intervals of Heinrich event 6 with the light $\delta^{18}O$ values pointing to lower salinities in the subsurface waters.

South of the Arctic Front at site MD95-2041 (Plate 2), the planktic $\delta^{18}O$ and for *G. bulloides* and *N. pachyderma* (r) also, the $\delta^{13}C$ records show distinct millennial-scale oscillations that were related to the Greenland stadial/interstadial cycles 18 to 20. During all the interstadials, warming is observed in the surface to subsurface waters (Plates 2a and 2b). Conditions in the subsurface waters appear to have been very stable, and the $\delta^{18}O$ values of all three deep-dwelling species were close together (Plate 2b). Thus, subsurface water conditions on the southwestern Iberian margin were more stable during MIS 4 than during MIS 2. Seasonality (Plate 2d) seems to have been a bit more variable at site MD95-2041 than at MD95-2040 and increased during Greenland stadial 19. During this stadial, percent *N. pachyderma* (s) rose slightly (Plate 2c), and $\delta^{13}C$ of *N. pachyderma* (r) and *G. bulloides* (Plate 2f) indicated fewer nutrients in the surface waters. Since this site was highly influenced by upwelling with increased productivity during some of the MIS 2 stadials [*Voelker et al.*, 2009], all these signals are interpreted as being upwelling related.

Overall, hydrographic conditions north and south of the front were similar during MIS 4 and 2, respectively. Differences between the two glacial periods were more restricted to the subsurface waters, especially the $ENACW_{st}$, where conditions were more stable during the older glacial period. The presence of $ENACW_{st}$ and in sections also of *G. ruber* white indicate that AzC-derived waters were present during much of the glacial periods, even if potentially restricted to a circulation pattern similar to the modern winter circulation (Figure 1a). This further implies that the Azores Front most likely extended toward the southern Iberian margin during both MIS 2 [*Rogerson et al.*, 2004] and 4 and might be the front observed between 38°N and 40°N.

6.2. The Penultimate Glacial: MIS 6

As already indicated by the percent *N. pachyderma* (s) evidence discussed in section 5.3, MIS 6 differed from the two younger glacial periods. This is further supported by the multispecies stable isotope evidence of cores MD95-2040 (Figure 6) and MD01-2443 (Figure 7). The subtropical species *G. ruber* white was always present at the southern location and nearly continuously also at site MD95-2040, indicating a northward heat transport stronger than during the last glacial cycle. This heat flux most likely occurred with the IPC, similar to the LGM [*Eynaud et al.*, 2009; *Pflaumann et al.*, 2003; *Voelker et al.*, 2009]. Ventilation of those waters was, however, highly variable (Figure 6e), much more so than during any of the younger glacial periods. Seasonality variations (Figure 6d), on the other hand, were much higher than during MIS 2 and 4 (Plate 1d). Seasonal contrasts were driven by the relatively more stable conditions in the winter mixed layer (Figure 6b). Longer-lasting seasonality extremes were associated with Heinrich event 11, similar to the younger Heinrich stadials and occurred during the intervals from 158.9 to 163 ka and from 168.2 to 173.2 ka, while shorter

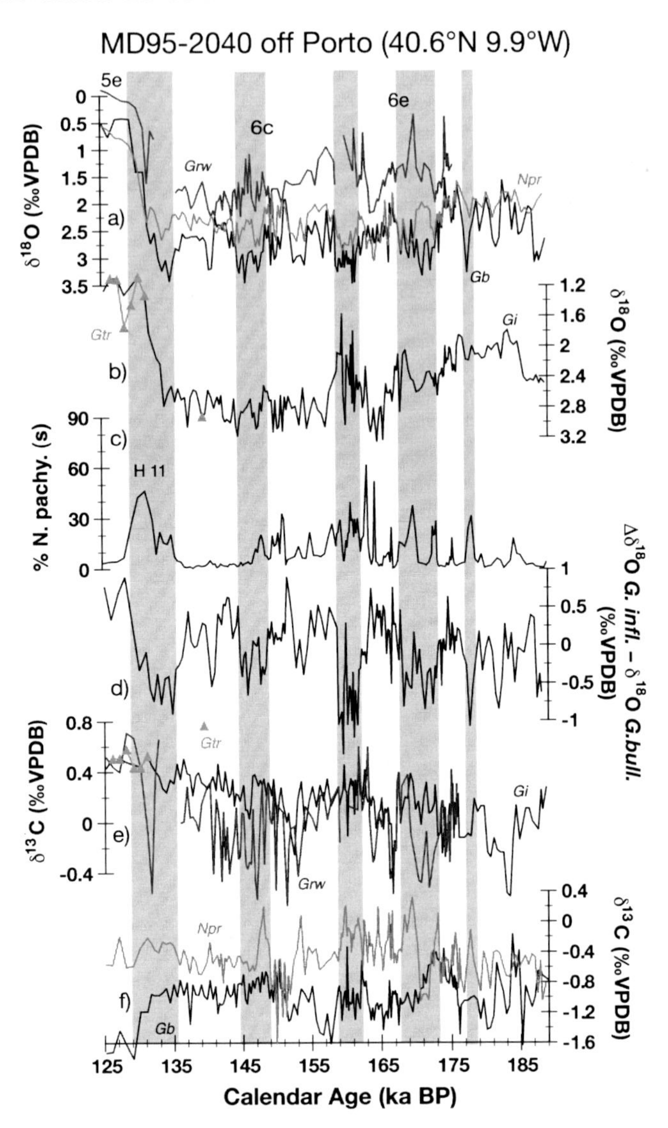

Figure 6. Vertical gradients in the upper water column at site MD95-2040 during MIS 6 [*de Abreu et al.*, 2003; this study]. (a) The $\delta^{18}O$ records of surface to thermocline dwelling species *G. ruber* white (dark gray, Grw), *G. bulloides* (black, Gb), and *N. pachyderma* (r) (light gray, Npr). (b) The $\delta^{18}O$ records of winter mixed layer species *G. inflata* (black, Gi) and deep-dweller *G. truncatulinoides* (r) (light gray and triangles, Gtr). (e and f) Respective $\delta^{13}C$ values. (c) Percent *N. pachyderma* (s) data. (d) Difference between $\delta^{18}O$ of *G. inflata* and $\delta^{18}O$ of *G. bulloides* reflecting seasonality. Gray bars mark periods with increased seasonality. Numbers refer to MIS substages; H11 is Heinrich event 11.

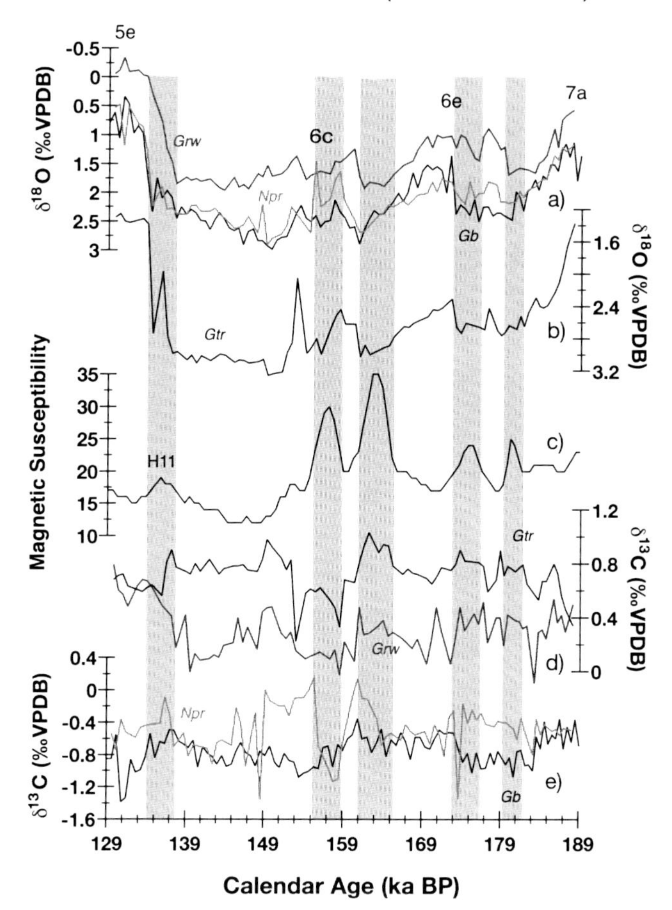

Figure 7. Vertical gradients in the upper water column at site MD01-2443 during MIS 6 (this study). (a) The $\delta^{18}O$ records of *G. ruber* white (dark gray, Grw); *G. bulloides* (black, Gb), and *N. pachyderma* (r) (light gray, Npr). (b) The $\delta^{18}O$ record of deep-dweller *G. truncatulinoides* (r) (Gtr). (d and e) Respective $\delta^{13}C$ values. (c) Magnetic susceptibility record with peaks (also marked by gray bars) indicating ice-rafting events. Numbers refer to MIS substages; H11 is Heinrich event 11.

oscillations marked the beginning of MIS 6 (177–188 ka, Figure 6d). In particular, the interval from 158.9 to 163 ka was associated with higher abundances of *N. pachyderma* (s) (Figure 6c), the presence of IRD [*de Abreu et al.*, 2003], and a reduced tree cover on land [*Margari et al.*, 2010] supporting relative harsher climate conditions.

As described for the previous glacial periods, conditions on the southwestern Iberian margin were more stable (Figure 7). The *G. truncatulinoides* $\delta^{18}O$ record shows hardly any change, and the respective $\delta^{13}C$ values indicate a well-ventilated $ENACW_{st}$ (Figure 7d), in contrast to the subtropical surface waters reflected in the *G. ruber* white $\delta^{13}C$ values. Millennial-scale oscillations were limited and restricted to the earlier part of MIS 6. However, the two older cooling events within MIS 6e had no major impact on the water column structure, and the nearly flat *G. truncatulinoides* $\delta^{18}O$ record indicates that conditions must have been similar to the ones of the penultimate glacial maximum (Figure 7b). In the $ENACW_{st}$, the most pronounced changes occurred between 153 and 161 ka, when ventilation was reduced, and *G. truncatulinoides* $\delta^{18}O$ values were lower. It needs to be seen in the future if these lighter $\delta^{18}O$ values were a temperature and/or salinity signal. Overall, the same picture as for the previous glacial periods emerges with the southern area being strongly affected by subtropical waters and the Azores Front located nearby. This clearly indicates that this pattern is a robust feature independent of the overall climate forcing.

7. IMPRINTS THROUGHOUT THE WHOLE WATER COLUMN

As mentioned in the introduction, the impacts of the abrupt climate change events can be traced down into the intermediate and deep water levels. To emphasize this, records from cores off the Sines coast or in the Gulf of Cadiz are combined in Figures 8 and 9. The hydrographic evidence for the last 65 kyr is based on planktic and benthic foraminifer $\delta^{18}O$ data, the mean grain size as evidence for MOW variability and deep-water temperature (DWT) records. Ventilation status (Figure 9) is assessed from planktic and benthic $\delta^{13}C$ records.

7.1. *Hydrography*

The records clearly show that the Greenland-type millennial-scale variability impacted the entire water column from the sea surface down to 2465 m, i.e., the water depth of site MD01-2444. The planktic foraminifer records all show warming (interstadial) and cooling (stadial) cycles that were contemporary in the water depths from 0 to 400 m with similar amplitudes in the *G. bulloides* (Figure 8c) and *G. truncatulinoides* (Figure 8d) records and a smaller amplitude in the *G. ruber* white data (Figure 8b). The *G. truncatulinoides* data of core MD99-2336 from the southern Portuguese margin (gray lines in Figure 8d) even indicate the presence of subtropical ENACW in the region during Heinrich stadials 1 and 6 in accordance with nannofossil evidence, i.e., maxima of the subtropical, deep-dwelling coccolithophore *Florisphaera profunda* [*Colmenero-Hidalgo et al.*, 2004; *Incarbona et al.*, 2010]. The presence of *G. ruber* white during these periods (Figure 8b) even points to the presence of subtropical surface waters [*Voelker et al.*, 2009]. The stadial/interstadial cyclicity is also recorded in the MOW strength (Figure 8e) with enhanced bottom current speeds (higher mean grain size values) during the cold periods [*Voelker et al.*, 2006]. Temperatures in the upper NADW (Figure 8f) [*Skinner and Elderfield*, 2007] generally also follow the Greenland-type pattern with warmer DWT during the interstadials and colder ones during the stadials as to be expected by changes in NADW or AABW predominantly bathing the site, respectively. The DWT record of site MD01-2444, however, also shows short-term warming events during Heinrich stadials 4 and 5 that *Skinner and Elderfield* [2007] attribute to the potential admixing of MOW, which are similar to the last deglaciations, could have reached as deep down as 2200 m on the Sines margin [*Schönfeld and Zahn*, 2000]. Admixing of deeper flowing MOW into depths of 2465 m could also explain some of the signals seen in the benthic stable isotope records of core MD95-2040 (Figures 5e, 5f) farther to the north. The shift from a Greenland- to an Antarctic-type climate signal occurred somewhere between 2500 and 3100 m water depth where the benthic $\delta^{18}O$ signal

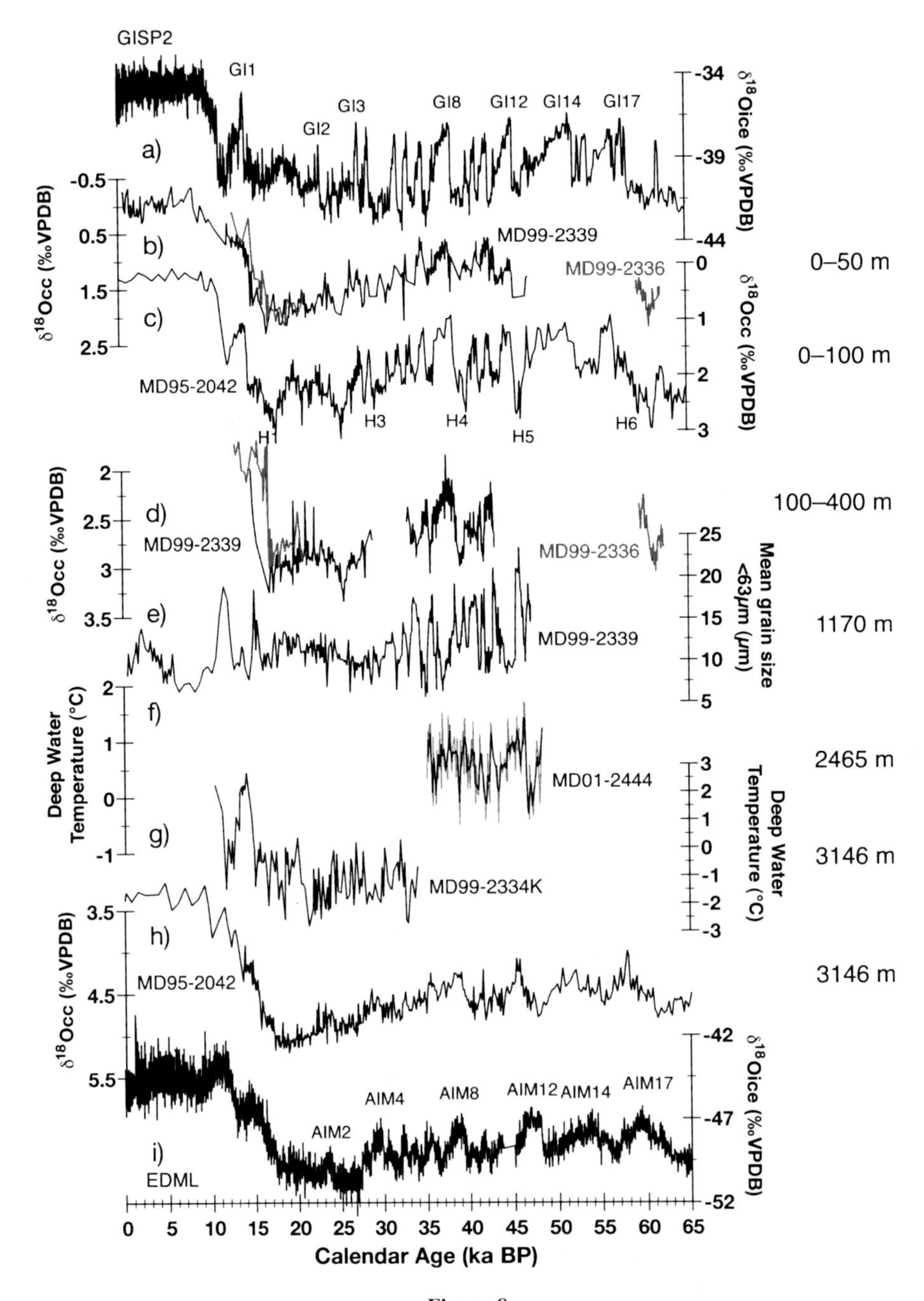

Figure 8

of core MD95-2042 (Figure 8h) [*Shackleton et al.*, 2000] clearly reflects the oscillations depicted in the EPICA Dronning Maud Land ice core record (Figure 8i) [*EPICA Community Members*, 2006]. The GNAIW/AABW boundary was therefore located deeper in the water column off southern Iberia than in the western Atlantic Basin [*Curry and Oppo*, 2005] and the northeastern Atlantic [*Sarnthein et al.*, 2001], most probably due to the presence of the deeper flowing MOW.

7.2. Water Column Ventilation

In the upper 400 m of the water column, nutrient levels and thus ventilation of the respective water mass (Figures 9b–9d) were not driven by the millennial-scale variability seen in the $\delta^{18}O$ records. For *G. ruber* white, glacial values tend to be lower than the Holocene ones reflecting the oligotrophic waters in the central Gulf of Cadiz. During the glacial and deglacial section, the lower values probably mirror local conditions with periods of stronger winter mixing, the time for refurbishing nutrients in the Gulf of Cadiz [*Navarro and Ruiz*, 2006]. For *G. bulloides*, the trend is opposite with higher values during the glacial. Since the *G. bulloides* record is from core MD95-2042 off Sines and thus from a region potentially experiencing upwelling, the *G. bulloides* $\delta^{13}C$ record was most likely modified by the productivity conditions in this region. Glacial productivity, and thus nutrient consumption, was higher in this region than during the Holocene [*Salgueiro et al.*, 2010]. The glacial subthermocline waters (100–400 m) were mostly well ventilated and contained few nutrients hinting to $ENACW_{st}$ as prevailing water mass. Only during Heinrich stadials 1 and 4 were lower $\delta^{13}C$ values recorded that could indicate that either less ventilated $ENACW_{sp}$ penetrated into the Gulf of Cadiz along with the melting icebergs [*Voelker et al.*, 2006] or that Antarctic Intermediate Water (AAIW) was mixed into the subtropical ENACW. Small amounts of AAIW can be found in the Gulf of Cadiz waters today [*Cabeçadas et al.*, 2003], and paleoceanographic studies have shown that AAIW penetrated farther northward during glacial times [*Pahnke et al.*, 2008].

Millennial-scale oscillations in the ventilation of the water column were finally recorded in the intermediate to bottom waters, i.e., those water masses directly reflecting the status of the overturning circulation either in the Mediterranean Sea or in the Atlantic Ocean (Figures 9e–9h). The record of core MD99-2339 bathed by the lower MOW core (Figures 2, 9e) shows clear cyclicity with relatively poorer ventilation during the Greenland interstadials and better during the Greenland stadials [*Voelker et al.*, 2006], similar to the pattern observed for the Western Mediterranean Deep Water [*Cacho et al.*, 2000; *Sierro et al.*, 2005]. Records from the Mediterranean Sea's eastern and western basins indicate that intermediate and deep waters were well oxygenated during Greenland stadials and during the greater parts of the Heinrich stadials [*Bassetti et al.*, 2010; *Cacho et al.*, 2000; *Schmiedl et al.*, 2010; *Sierro et al.*, 2005]. Thus, the poor ventilation of the MOW during the Heinrich stadials must result from the admixing of poorly ventilated Atlantic waters such as the $ENACW_{st}$ reflected in the *G. truncatulinoides* data (Figure 9d) and potentially also AAIW. In the upper NADW/GNAIW level at 2465 m (Figure 9e) and deeper down, the ventilation status was primarily driven by the well-known up and down movement of the NADW/AABW interface with better ventilation (equal to NADW) during the interstadials, when AMOC was strong, and poorer ventilation during the stadials (equal to AABW). When NADW was present, benthic $\delta^{13}C$ values were similar from 2465 to 3146 m water depth and during glacial times also not much different at 4602 m (Figures 9f–9h), indicating a homogeny in the deeper-water column during the interstadials that is also seen today [*Alvarez et al.*, 2004]. During the LGM and most Heinrich stadials, the 2465 m data show, however, excursions to higher $\delta^{13}C$ values that were in the range of those recorded in the lower MOW core at site MD99-2339 (Figures 9e, 9f). Thus, the benthic $\delta^{13}C$ data confirm what the

Figure 8. (opposite) Vertical gradients in the hydrography at the southwestern Iberian margin over the last 65 kyr in comparison to the (a) Greenland (Greenland Ice Sheet Project 2 (GISP2)) [*Grootes and Stuiver*, 1997] and (i) Antarctic (EPICA Dronning Maud Land) [*EPICA Community Members*, 2006] ice core records. (b and c) Uppermost water column conditions as reflected in the *G. ruber* white $\delta^{18}O$ values of cores MD99-2339 (black) [*Voelker et al.*, 2009; this study] and MD99-2336 (gray) [*Voelker et al.*, 2009; this study] and in the *G. bulloides* $\delta^{18}O$ data of core MD95-2042 [*Cayre et al.*, 1999; *Shackleton et al.*, 2000]. (d) ENACW-level subsurface water conditions based on the $\delta^{18}O$ of *G. truncatulinoides* from cores MD99-2339 (black) [*Voelker et al.*, 2009; this study] and MD99-2336 (gray) [*Voelker et al.*, 2009; this study]. (e) Response in the lower MOW's flow strength (core MD99-2339 [*Voelker et al.*, 2006]) to the millennial-scale variability. Deep water temperature changes at (f) 2465 m (MD01-2444 [*Skinner and Elderfield*, 2007]) and (g) 3146 m (MD99-2334K [*Skinner et al.*, 2003]). For core MD01-2444, the original data are shown in gray, and a three-point moving average record is shown in black. (h) Benthic $\delta^{18}O$ record of core MD95-2042 [*Shackleton et al.*, 2000]. GI, H, and AIM refer to Greenland interstadials, Heinrich stadials, and Antarctic isotope maxima, respectively. Depth ranges on the right refer to the living depths of the respective planktic foraminifer (Figures 8a–8c) [*Voelker et al.*, 2009] or to the depth of the respective core site(s).

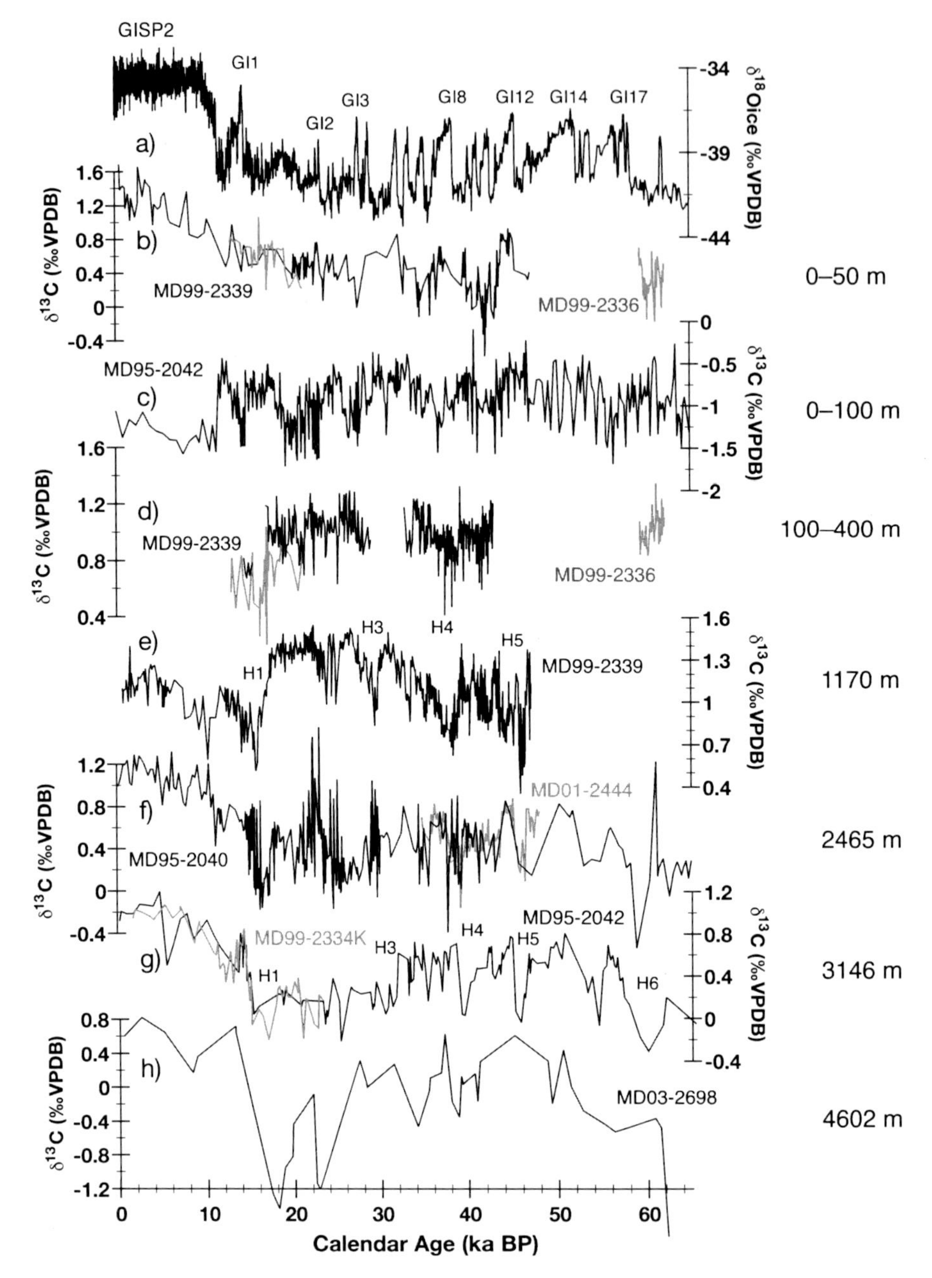

Figure 9. Vertical gradients in water column ventilation at the southwestern Iberian margin over the last 65 kyr in comparison to the (a) Greenland (GISP2) [*Grootes and Stuiver*, 1997]) ice core record. Uppermost water column conditions as reflected in the (b) *G. ruber* white $\delta^{13}C$ values of cores MD99-2339 (black) [*Voelker et al.*, 2009, this study]) and MD99-2336 (gray) [*Voelker et al.*, 2009, this study] (c) and in the *G. bulloides* $\delta^{13}C$ data of core MD95-2042 [*Cayre et al.*, 1999; *Shackleton et al.*, 2000]. (d) ENACW-level subsurface water conditions based on the $\delta^{13}C$ of *G. truncatulinoides* from cores MD99-2339 (black) [*Voelker et al.*, 2009, this study] and MD99-2336 (gray) [*Voelker et al.*, 2009; this study]. (e) Ventilation changes in the lower MOW level (core MD99-2339 [*Voelker et al.*, 2006]). (f) Benthic $\delta^{13}C$ data of cores MD95-2040 (black) [*de Abreu et al.*, 2003; *Schönfeld et al.*, 2003; this study] and MD01-2444 (gray) [*Skinner and Elderfield*, 2007]. (g) Benthic $\delta^{13}C$ records of cores MD95-2042 (black) [*Shackleton et al.*, 2000] and MD99-2334K (gray) [*Skinner and Shackleton*, 2004]. (h) Benthic $\delta^{13}C$ record of core MD03-2698 [*Lebreiro et al.*, 2009]. Nomenclature and depth ranges are as in Figure 8.

DWT already implied: the deeper flowing MOW was admixed into the GNAIW and led sometimes to a better ventilation of the intermediate-depth water column along the western Iberian margin. Extremely low benthic $\delta^{13}C$ values, on the other hand, were recorded at 4602 m during MIS 2 and 4 (Figure 9h). This record is from core MD03-2698 [*Lebreiro et al.*, 2009] located in the Tagus Abyssal Plain and shows that bottom waters in this deep basin were hardly renewed during the glacial maxima. Even today, the deep abyssal plains off Iberia can only be ventilated by flows through a few deep gaps [*Saunders*, 1987]. However, additional modification of the benthic $\delta^{13}C$ signal due to remineralization of organic matter transported down the canyons by the frequent turbidites [*Lebreiro et al.*, 2009] could also have played a role.

8. CONCLUSIONS

The combination of various high-resolution records allowed studying how events of abrupt climate change affected the water column along the western Iberian margin and which latitudinal and vertical boundaries existed. The abundance of records from the Sines region on the southwestern margin, furthermore, permitted assessing signal modification due to upwelling.

The surface water records from the Iberian margin clearly reveal that two fronts intercepted with the margin during the Heinrich stadials of the last glacial cycle and during extreme cold events of previous glacial periods. During the last glacial cycle, the Polar Front was located near 41°N, leading to the harshest climate conditions in the northern regions. The Arctic Front was located about 2° farther to the south, near 39°N, and might have coincided with the Azores Front. The latitudinal positioning of these fronts led to steep temperature gradients along the margin with SST_{su} increasing by 1°C per degree of latitude, i.e., from 4°C at 42°N to 10°C at 36°N. The foraminifer data, furthermore, showed that Heinrich events became more frequent and had stronger hydrographic impacts during the last glacial cycle. Similar events were recorded only with Heinrich event 11 during termination II and along with the Heinrich-type events during early MIS 8 and termination IV. MIS 6 was overall a warm glacial with subtropical waters dominating the hydrography along the southern Iberian margin and penetrating at least as far north as 40.6°N during much of the period. Hydrographic conditions north and south of the 39°N boundary were, in general, similar during MIS 2 and 4, but MIS 4 like MIS 6 seems to have more strongly been affected by subtropical subsurface waters. MIS 6, however, differed from its younger counterparts in regard to the increased seasonality and the extreme variations in the nutrient level of the subtropical surface waters.

Because the western Iberian margin is an upwelling area, upwelling can always affect the climate records. A clear indication for upwelling events, modifying climate records and thus leading to different paleodata for the same events in close vicinity, is given by the differences observed between cores MD95-2041 and MD95-2042 on the Sines coast. Core MD95-2041, located closer to the coast, experienced much more variability in its *G. bulloides* stable isotope records that can only be explained by upwelling. Upwelling also seems to be driving some of the variations observed in the planktic stable isotope records of core MD95-2040 during MIS 6.

A second "local" hydrographic phenomenon affecting the water column in the eastern North Atlantic is the MOW. During glacial times and especially the Heinrich events and Greenland stadials, the MOW settled deeper in the water column allowing it to be admixed into the intermediate-depth water masses. Thus, records from 2465 m water depth indicate imprints of MOW by warming events, so far confirmed for Heinrich events 4 and 5, and by a better ventilation of these water depths relative to the western basin. Owing to the admixing, the boundary between GNAIW and AABW was located between 2465 and 3100 m on the Iberian margin. As a consequence, Greenland-type climate oscillations can be traced down to this level, while the deeper sites follow the Antarctic-type of climate change. The deepest basins on the Iberian margin apparently experienced periods of reduced water mass renewal during MIS 2 and 4.

For the last glacial cycle, we now have a comprehensive picture regarding latitudinal and vertical gradients in the water column along the Iberian margin, and it is hoped that the existing data can serve as grounds for regional climate models of abrupt climate change events.

Acknowledgments. We are indebted to Yvon Balut, IPEV, and the crew of R/V *Marion Dufresne* as well as the IMAGES project for the recovery of excellent core material. The EU Access to Research Infrastructure PALEOSTUDIES program is acknowledged for the financial support that allowed the multispecies stable isotope analyses. Monika Segl and the Geosciences Department (FB 5) of the University of Bremen are thanked for hosting A. V. and L. A. during their respective PALEOSTUDIES stays. Additional thanks for excellent stable isotope results go to Helmut Erlenkeuser (Leibniz Labor, University Kiel) and Mike Hall and James Rolfe (Godwin Laboratory, University of Cambridge). A. Rebotim is thanked for her help in completing the benthic isotope records of core MD95-2040. The Fundação de Ciência e Tecnologia (FCT) supported this research through the MOWFADRI, SEDPORT and INTER-TRACE projects and postdoctoral fellowships to A. V. and L. A. A. V. furthermore acknowledges her Ciência 2007 grant. Finally, L. A. would like to remember all the dedication, incentive, and enthusiasm of the late Nick Shackleton and his guidance and collaboration throughout many of the studies involving these particular Iberian margin cores.

REFERENCES

Alvarez, M., F. F. Perez, H. Bryden, and A. F. Rios (2004), Physical and biogeochemical transports structure in the North Atlantic subpolar gyre, *J. Geophys. Res.*, *109*(C3), C03027, doi:10.1029/2003JC002015.

Alvarez-Salgado, X. A., et al. (2003), The Portugal coastal counter current off NW Spain: New insights on its biogeochemical variability, *Prog. Oceanogr.*, *56*(2), 281–321.

Ambar, I., and M. R. Howe (1979), Observations of the Mediterranean Outflow: 1. Mixing in the Mediterranean Outflow, *Deep Sea Res., Part A*, *26*(5), 535–554.

Baas, J. H., J. Mienert, F. Abrantes, and M. A. Prins (1997), Late Quaternary sedimentation on the Portuguese continental margin: Climate-related processes and products, *Palaeogeogr. Palaeoclimatol. Palaeoecol.*, *130*, 1–23.

Baas, J. H., J. Schönfeld, and R. Zahn (1998), Mid-depth oxygen drawdown during Heinrich events: Evidence from benthic foraminiferal community structure, trace-fossil tiering, and benthic $\delta^{13}C$ at the Portuguese margin, *Mar. Geol.*, *152*, 25–55.

Bard, E., F. Rostek, J.-L. Turon, and S. Gendreau (2000), Hydrological impact of Heinrich events in the subtropical northeast Atlantic, *Science*, *289*, 1321–1324.

Bassetti, M. A., P. Carbonel, F. J. Sierro, M. Perez-Folgado, G. Jouit, and S. Berne (2010), Response of ostracods to abrupt climate changes in the western Mediterranean (Gulf of Lions) during the last 30 kyr, *Mar. Micropaleontol.*, *77*(1–2), 1–14.

Bassinot, F., and L. Labeyrie (1996), IMAGES MD 101 A coring cruise of the R/V *Marion Dufresne* in the North Atlantic and Norwegian Sea, *Rep. 96-1*, 217 pp., Inst. Fr. pour la Rech. et la Technol. Polaires, Plouzane, France.

Bemis, B. E., H. J. Spero, D. W. Lea, and J. Bijma (2000), Temperature influence on the carbon isotopic composition of *Globigerina bulloides* and *Orbulina universa* (planktonic foraminifera), *Mar. Micropaleontol.*, *38*(3–4), 213–228.

Brambilla, E., L. D. Talley, and P. E. Robbins (2008), Subpolar Mode Water in the northeastern Atlantic: 2. Origin and transformation, *J. Geophys. Res.*, *113*, C04026, doi:10.1029/2006JC004063.

Cabeçadas, G., M. J. Brogueira, and C. Gonçalves (2003), Intermediate water masses off south-southwest Portugal: Chemical tracers, *J. Mar. Res.*, *61*(4), 539–552.

Cacho, I., J. O. Grimalt, F. J. Sierro, N. Shackleton, and M. Canals (2000), Evidence for enhanced Mediterranean thermohaline circulation during rapid climatic coolings, *Earth Planet. Sci. Lett.*, *183*, 417–429.

Cayre, O., Y. Lancelot, E. Vincent, and M. A. Hall (1999), Paleoceanographic reconstructions from planktonic foraminifera off the Iberian margin: Temperature, salinity, and Heinrich events, *Paleoceanography*, *14*, 384–396.

Colmenero-Hidalgo, E., J.-A. Flores, F. J. Sierro, M. A. Barcena, L. Loewemark, J. Schönfeld, and J. O. Grimalt (2004), Ocean surface water response to short-term climate changes revealed by coccolithophores from the Gulf of Cadiz (NE Atlantic) and Alboran Sea (W Mediterranean), *Palaeogeogr. Palaeoclimatol. Palaeoecol.*, *205*(3–4), 317–336.

Curry, W. B., and D. W. Oppo (2005), Glacial water mass geometry and the distribution of $\delta^{13}C$ of ΣCO_2 in the western Atlantic Ocean, *Paleoceanography*, *20*, PA1017, doi:10.1029/2004PA001021.

de Abreu, L., N. J. Shackleton, J. Schönfeld, M. Hall, and M. Chapman (2003), Millennial-scale oceanic climate variability off the western Iberian margin during the last two glacial periods, *Mar. Geol.*, *196*(1–2), 1–20.

de Abreu, L., F. F. Abrantes, N. J. Shackleton, P. C. Tzedakis, J. F. McManus, D. W. Oppo, and M. A. Hall (2005), Ocean climate variability in the eastern North Atlantic during interglacial marine isotope stage 11: A partial analogue to the Holocene?, *Paleoceanography*, *20*, PA3009, doi:10.1029/2004PA001091.

de Vargas, C., S. Renaud, H. Hilbrecht, and J. Pawlowski (2001), Pleistocene adaptive radiation in *Globorotalia truncatulinoides*: Genetic, morphologic, and environmental evidence, *Paleobiology*, *27*(1), 104–125.

Dickson, R. R., J. Meincke, S.-A. Malmberg, and A. J. Lee (1988), The "great salinity anomaly" in the northern North Atlantic 1968–1982, *Prog. Oceanogr.*, *20*, 103–151.

EPICA Community Members (2006), One-to-one coupling of glacial climate variability in Greenland and Antarctica, *Nature*, *444*, 195–198.

Eynaud, F., et al. (2009), Position of the Polar Front along the western Iberian margin during key cold episodes of the last 45 ka, *Geochem. Geophys. Geosyst.*, *10*, Q07U05, doi:10.1029/2009GC002398.

Fiúza, A. F. G. (1984), Hidrologia e Dinamica das Aguas Costeiras de Portugal, Ph.D. thesis, 294 pp., Univ. de Lisboa, Lisbon.

Fiuza, A. F. G., M. Hamann, I. Ambar, G. D. del Rio, N. Gonzalez, and J. M. Cabanas (1998), Water masses and their circulation off western Iberia during May 1993, *Deep Sea Res., Part I*, *45*(7), 1127–1160.

Fletcher, W. J., et al. (2010), Millennial-scale variability during the last glacial in vegetation records from Europe, *Quat. Sci. Rev.*, *29*(21–22), 2839–2864.

Frouin, R., A. F. G. Fiuza, I. Ambar, and T. J. Boyd (1990), Observations of a poleward surface current off the coasts of Portugal and Spain during winter, *J. Geophys. Res.*, *95*(C1), 679–691.

Ganssen, G. M., and D. Kroon (2000), The isotopic signature of planktonic foraminifera from NE Atlantic surface sediments: Implications for the reconstruction of past oceanic conditions, *J. Geol. Soc. London*, *157*, 693–699.

Garcia-Lafuente, J., J. Delgado, F. Criado-Aldeanueva, M. Bruno, J. del Rio, and J. Miguel Vargas (2006), Water mass circulation on the continental shelf of the Gulf of Cadiz, *Deep Sea Res., Part II*, *53*(11–13), 1182–1197.

Grootes, P. M., and M. Stuiver (1997), $^{18}O/^{16}O$ variability in Greenland snow and ice with 10^{-3} to 10^{5} year time resolution, *J. Geophys. Res.*, *102*(C12), 26,455–26,470.

Haynes, R., and E. D. Barton (1990), A poleward flow along the Atlantic Coast of the Iberian Peninsula, *J. Geophys. Res.*, *95*(C7), 11,425–11,441.

Haynes, R., E. D. Barton, and I. Pilling (1993), Development, persistence, and variability of upwelling filaments off the Atlantic coast of the Iberian Peninsula, *J. Geophys. Res.*, *98*(C12), 22,681–22,692.

Hemming, S. R. (2004), Heinrich events: Massive late Pleistocene detritus layers of the North Atlantic and their global climate imprint, *Rev. Geophys.*, *42*, RG1005, doi:10.1029/2003RG000128.

Hodell, D. A., J. E. T. Channell, J. H. Curtis, O. E. Romero, and U. Röhl (2008), Onset of 'Hudson Strait' Heinrich events in the eastern North Atlantic at the end of the middle Pleistocene transition (~640 ka)?, *Paleoceanography*, *23*, PA4218, doi:10.1029/2008PA001591.

Hughen, K., S. Lehman, J. Southon, J. Overpeck, O. Marchal, C. Herring, and J. Turnbull (2004), ^{14}C activity and global carbon cycle changes over the past 50,000 years, *Science*, *303*, 202–207.

Incarbona, A., B. Martrat, E. Di Stefano, J. O. Grimalt, N. Pelosi, B. Patti, and G. Tranchida (2010), Primary productivity variability on the Atlantic Iberian margin over the last 70,000 years: Evidence from coccolithophores and fossil organic compounds, *Paleoceanography*, *25*, PA2218, doi:10.1029/2008PA001709.

Jia, Y. (2000), Formation of an Azores Current due to Mediterranean overflow in a modeling study of the North Atlantic, *J. Phys. Oceanogr.*, *30*, 2342–2358.

Johnson, J., and I. Stevens (2000), A fine resolution model of the eastern North Atlantic between the Azores, the Canary Islands and the Gibraltar Strait, *Deep Sea Res., Part I*, *47*(5), 875–899.

Kjellström, E., J. Brandefelt, J. O. Näslund, B. Smith, G. Strandberg, A. H. L. Voelker, and B. Wohlfarth (2010), Simulated climate conditions in Europe during the marine isotope stage 3 stadial, *Boreas*, *39*(2), 436–456.

Kuhnt, T., G. Schmiedl, W. Ehrmann, Y. Hamann, and N. Andersen (2008), Stable isotopic composition of Holocene benthic foraminifers from the eastern Mediterranean Sea: Past changes in productivity and deep water oxygenation, *Palaeogeogr. Palaeoclimatol. Palaeoecol.*, *268*(1–2), 106–115.

Labeyrie, L., E. Jansen, and E. Cortijo (Eds.) (2003), MD 114/IMAGES V, á bord du *Marion Dufresne*, Fort de France, 11 Juin 1999-Marseille, 20 Septembre 1999, *Rep. OCE/2003/02*, pp. 1– 380 and 381–849, Inst. Polaire Fr. Paul-Emile Victor, Brest, France.

Lebreiro, S. M., A. H. L. Voelker, A. Vizcaino, F. G. Abrantes, U. Alt-Epping, S. Jung, N. Thouveny, and E. Gracia (2009), Sediment instability on the Portuguese continental margin under abrupt glacial climate changes (last 60 kyr, *Quat. Sci. Rev.*, *28*(27–28), 3211–3223.

Lisiecki, L. E., and M. Raymo (2005), A Pliocene-Pleistocene stack of 57 globally distributed benthic $\delta^{18}O$ records, *Paleoceanography*, *20*, PA1003, doi:10.1029/2004PA001071.

Llave, E., J. Schönfeld, F. J. Hernandez-Molina, T. Mulder, L. Somoza, V. Diaz del Rio, and I. Sanchez-Almazo (2006), High-resolution stratigraphy of the Mediterranean outflow contourite system in the Gulf of Cadiz during the late Pleistocene: The impact of Heinrich events, *Mar. Geol.*, *227*(3–4), 241–262.

Lowe, J. J., W. Z. Hoek, and INTIMATE group (2001), Inter-regional correlation of palaeoclimatic records for the last glacial-interglacial transition: A protocol for improved precision recommended by the INTIMATE project group, *Quat. Sci. Rev.*, *20*(11), 1175–1187.

Lowe, J. J., S. O. Rasmussen, S. Björck, W. Z. Hoek, J. P. Steffensen, M. J. C. Walker, and Z. C. Yu (2008), Synchronisation of palaeoenvironmental events in the North Atlantic region during the last termination: A revised protocol recommended by the INTIMATE group, *Quat. Sci. Rev.*, *27*(1–2), 6–17.

Mackensen, A., H.-W. Hubberten, T. Bickert, G. Fischer, and D. K. Fütterer (1993), The $\delta^{13}C$ in benthic foraminiferal tests of *Fontbotia wuellerstorfi* (Schwager) relative to the $\delta\ ^{13}C$ of dissolved inorganic carbon in Southern Ocean deep water: Implications for glacial ocean circulation models, *Paleoceanography*, *8*(5), 587–610.

Margari, V., L. C. Skinner, P. C. Tzedakis, A. Ganopolski, M. Vautravers, and N. J. Shackleton (2010), The nature of millennial-scale climate variability during the past two glacial periods, *Nat. Geosci.*, *3*(2), 127–131.

Martrat, B., J. O. Grimalt, N. J. Shackleton, L. de Abreu, M. A. Hutterli, and T. F. Stocker (2007), Four climate cycles of recurring deep and surface water destabilizations on the Iberian margin, *Science*, *317*, 502–507.

McCartney, M. S., and L. D. Talley (1982), The sub-polar mode water of the North-Atlantic Ocean, *J. Phys. Oceanogr.*, *12*(11), 1169–1188.

Moreno, E., N. Thouveny, D. Delanghe, I. N. McCave, and N. J. Shackleton (2002), Climatic and oceanographic changes in the northeast Atlantic reflected by magnetic properties of sediments deposited on the Portuguese margin during the last 340 ka, *Earth Planet. Sci. Lett.*, *202*(2), 465–480.

Naughton, F., M. F. Sanchez Goni, S. Desprat, J. L. Turon, J. Duprat, B. Malaize, C. Joli, E. Cortijo, T. Drago, and M. C. Freitas (2007), Present-day and past (last 25 000 years) marine pollen signal off western Iberia, *Mar. Micropaleontol.*, *62*(2), 91–114.

Naughton, F., et al. (2009), Wet to dry climatic trend in north-western Iberia within Heinrich events, *Earth Planet. Sci. Lett.*, *284*(3–4), 329–342.

Navarro, G., and J. Ruiz (2006), Spatial and temporal variability of phytoplankton in the Gulf of Cadiz through remote sensing images, *Deep Sea Res., Part II*, *53*(11–13), 1241–1260.

North Greenland Ice Core Project members (2004), High-resolution record of Northern Hemisphere climate extending into the last interglacial period, *Nature*, *431*(7005), 147–151.

Oezgoekmen, T. M., E. P. Chassignet, and C. G. H. Rooth (2001), On the connection between the Mediterranean Outflow and the Azores Current, *J. Phys. Oceanogr.*, *31*, 461–480.

Pahnke, K., S. L. Goldstein, and S. R. Hemming (2008), Abrupt changes in Antarctic Intermediate Water circulation over the past 25,000 years, *Nat. Geosci.*, *1*(12), 870–874.

Pailler, D., and E. Bard (2002), High frequency palaeoceanographic changes during the past 140000 yr recorded by the organic matter in sediments of the Iberian margin, *Palaeogeogr. Palaeoclimatol. Palaeoecol.*, *181*(4), 431–452.

Peliz, A., J. Dubert, A. M. P. Santos, P. B. Oliveira, and B. Le Cann (2005), Winter upper ocean circulation in the western Iberian Basin—Fronts, eddies and poleward flows: An overview, *Deep Sea Res., Part I*, *52*(4), 621–646.

Peliz, A., J. Dubert, P. Marchesiello, and A. Teles-Machado (2007), Surface circulation in the Gulf of Cadiz: Model and mean flow structure, *J. Geophys. Res.*, *112*, C11015, doi:10.1029/2007JC004159.

Penduff, T., A. C. de Verdiere, and B. Barnier (2001), General circulation and intergyre dynamics in the eastern North Atlantic from a regional primitive equation model, *J. Geophys. Res.*, *106*(C10), 22,313–22,329.

Perez, F. F., C. G. Castro, X. A. Alvarez-Salgado, and A. F. Rios (2001), Coupling between the Iberian basin-scale circulation and the Portugal boundary current system: A chemical study, *Deep Sea Res., Part I*, *48*(6), 1519–1533.

Pflaumann, U., J. Duprat, C. Pujol, and L. D. Labeyrie (1996), SIMMAX: A modern analog technique to deduce Atlantic sea surface temperatures from planktonic foraminifera in deep-sea sediments, *Paleoceanography*, *11*(1), 15–36.

Pflaumann, U., et al. (2003), Glacial North Atlantic: Sea-surface conditions reconstructed by GLAMAP 2000, *Paleoceanography*, *18*(3), 1065, doi:10.1029/2002PA000774.

Pingree, R. D., C. Garcia-Soto, and B. Sinha (1999), Position and structure of the Subtropical/Azores Front region from combined Lagrangian and remote sensing (IR/altimeter/SeaWiFS) measurements, *J. Mar. Biol. Assoc. U.K.*, *79*(5), 769–792.

Rashid, H., and E. A. Boyle (2007), Mixed-layer deepening during Heinrich Events: A multi-planktonic foraminiferal $\delta^{18}O$ approach, *Science*, *318*(5849), 439–441.

Relvas, P., and E. D. Barton (2002), Mesoscale patterns in the Cape São Vicente (Iberian Peninsula) upwelling region, *J. Geophys. Res.*, *107*(C10), 3164, doi:10.1029/2000JC000456.

Richardson, P. L., A. S. Bower, and W. Zenk (2000), A census of meddies tracked by floats, *Progr. Oceanogr.*, *45*, 209–250.

Rios, A. F., F. F. Perez, and F. Fraga (1992), Water masses in the upper and middle North-Atlantic Ocean east of the Azores, *Deep Sea Res., Part A*, *39*(3–4A), 645–658.

Rogerson, M., E. J. Rohling, P. P. E. Weaver, and J. W. Murray (2004), The Azores Front since the Last Glacial Maximum, *Earth Planet. Sci. Lett.*, *222*(3–4), 779–789.

Rogerson, M., E. J. Rohling, P. P. E. Weaver, and J. W. Murray (2005), Glacial to interglacial changes in the settling depth of the Mediterranean Outflow plume, *Paleoceanography*, *20*, PA3007, doi:10.1029/2004PA001106.

Roucoux, K. H., L. de Abreu, N. J. Shackleton, and P. C. Tzedakis (2005), The response of NW Iberian vegetation to North Atlantic climate oscillations during the last 65 kyr, *Quat. Sci. Rev.*, *24*(14–15), 1637–1653.

Ruddiman, W. F. (1977), Late Quaternary deposition of ice-rafted sand in the subpolar North Atlantic (lat 40° to 65°N), *Geol. Soc. Am. Bull.*, *88*, 1813–1827.

Salgueiro, E., A. H. L. Voelker, L. de Abreu, F. Abrantes, H. Meggers, and G. Wefer (2010), Temperature and productivity changes off the western Iberian margin during the last 150 ky, *Quat. Sci. Rev.*, *29*(5–6), 680–695.

Sanchez, R. F., and P. Relvas (2003), Spring-summer climatological circulation in the upper layer in the region of Cape St. Vincent, southwest Portugal, *ICES J. Mar. Sci.*, *60*(6), 1232–1250.

Sánchez-Goñi, M. F., and S. P. Harrison (2010), Millennial-scale climate variability and vegetation changes during the last glacial: Concepts and terminology, *Quat. Sci. Rev.*, *29*(21–22), 2823–2827.

Sánchez-Goñi, M. F., A. Landais, W. J. Fletcher, F. Naughton, S. Desprat, and J. Duprat (2008), Contrasting impacts of Dansgaard-Oeschger events over a western European latitudinal transect modulated by orbital parameters, *Quat. Sci. Rev.*, *27*(11–12), 1136–1151.

Sarnthein, M., et al. (2001), Fundamental modes and abrupt changes in North Atlantic circulation and climate over the last 60 ky—Numerical modelling and reconstruction, in *The Northern North Atlantic: A Changing Environment*, edited by P. Schäfer et al., pp. 365–410, Springer, Heidelberg, Germany.

Saunders, P. M. (1987), Flow through Discovery Gap, *J. Phys. Oceanogr.*, *17*, 631–643.

Schlitzer, R. (2000), Electronic atlas of WOCE hydrographic and tracer data now available, *Eos Trans. AGU*, *81*(5), 45.

Schmiedl, G., T. Kuhnt, W. Ehrmann, K.-C. Emeis, Y. Hamann, U. Kotthoff, P. Dulski, and J. Pross (2010), Climatic forcing of eastern Mediterranean deep-water formation and benthic ecosystems during the past 22 000 years, *Quat. Sci. Rev.*, *29*(23–24), 3006–3020.

Schönfeld, J., and R. Zahn (2000), Late Glacial to Holocene history of the Mediterranean Outflow. Evidence from benthic foraminiferal assemblages and stable isotopes at the Portuguese margin, *Palaeogeogr. Palaeoclimatol. Palaeoecol.*, *159*, 85–111.

Schönfeld, J., R. Zahn, and L. de Abreu (2003), Surface and deep water response to rapid climate changes at the western Iberian margin, *Global Planet. Change*, *36*(4), 237–264.

Serra, N., and I. Ambar (2002), Eddy generation in the Mediterranean undercurrent, *Deep Sea Res., Part II*, *49*(19), 4225–4243.

Shackleton, N. J., M. A. Hall, and E. Vincent (2000), Phase relationships between millennial-scale events 64,000-24,000 years ago, *Paleoceanography*, *15*(6), 565–569.

Sierro, F. J., et al. (2005), Impact of iceberg melting on Mediterranean thermohaline circulation during Heinrich events, *Paleoceanography*, *20*, PA2019, doi:10.1029/2004PA001051.

Skinner, L. C., and H. Elderfield (2007), Rapid fluctuations in the deep North Atlantic heat budget during the last glacial period, *Paleoceanography*, *22*, PA1205, doi:10.1029/2006PA001338.

Skinner, L. C., and N. J. Shackleton (2004), Rapid transient changes in northeast Atlantic deep water ventilation age across termination I, *Paleoceanography, 19*, PA2005, doi:10.1029/2003PA000983.

Skinner, L. C., N. J. Shackleton, and H. Elderfield (2003), Millennial-scale variability of deep-water temperature and $\delta^{18}O_{dw}$ indicating deep-water source variations in the northeast Atlantic, 0–34 cal. ka BP, *Geochem. Geophys. Geosyst., 4*(12), 1098, doi:10.1029/2003GC000585.

Sousa, F. M., and A. Bricaud (1992), Satellite-derived phytoplankton pigment structures in the Portuguese upwelling area, *J. Geophys. Res., 97*(C7), 11,343–11,356.

Stein, R., J. Hefter, J. Grützner, A. Voelker, and B. D. A. Naafs (2009), Variability of surface-water characteristics and Heinrich-like events in the Pleistocene mid-latitude North Atlantic Ocean: Biomarker and XRD records from IODP Site U1313 (MIS 16–9), *Paleoceanography, 24*, PA2203, doi:10.1029/2008PA001639.

Thomson, J., S. Nixon, C. P. Summerhayes, E. J. Rohling, J. Schönfeld, R. Zahn, P. Grootes, F. Abrantes, L. Gaspar, and S. Vaqueiro (2000), Enhanced productivity on the Iberian margin during glacial/interglacial transitions revealed by barium and diatoms, *J. Geol. Soc. London, 157*(3), 667–677.

Thouveny, N., J. Carcaillet, E. Moreno, G. Leduc, and D. Nerini (2004), Geomagnetic moment variation and paleomagnetic excursions since 400 kyr BP: A stacked record from sedimentary sequences of the Portuguese margin, *Earth Planet. Sci. Lett., 219*, 377–396.

Toucanne, S., T. Mulder, J. Schoenfeld, V. Hanquiez, E. Gonthier, J. Duprat, M. Cremer, and S. Zaragosi (2007), Contourites of the Gulf of Cadiz: A high-resolution record of the paleocirculation of the Mediterranean outflow water during the last 50,000 years, *Palaeogeogr. Palaeoclimatol. Palaeoecol., 246*(2–4), 354–366.

Toucanne, S., et al. (2009), Timing of massive 'Fleuve Manche' discharges over the last 350 kyr: Insights into the European ice-sheet oscillations and the European drainage network from MIS 10 to 2, *Quat. Sci. Rev., 28*(13–14), 1238–1256.

Tzedakis, P. C., K. H. Roucoux, L. de Abreu, and N. J. Shackleton (2004), The duration of forest stages in southern Europe and interglacial climate variability, *Science, 306*, 2231–2235.

Tzedakis, P. C., H. Pälike, K. H. Roucoux, and L. de Abreu (2009), Atmospheric methane, southern European vegetation and low-mid latitude links on orbital and millennial timescales, *Earth Planet. Sci. Lett., 277*(3–4), 307–317.

van Aken, H. M. (2000), The hydrography of the mid-latitude northeast Atlantic Ocean—Part I: The deep water masses, *Deep Sea Res., Part I, 47*, 757–788.

van Aken, H. M. (2001), The hydrography of the mid-latitude northeast Atlantic Ocean—Part III: The subducted thermocline water mass, *Deep Sea Res., Part I, 48*(1), 237–267.

Vargas, J. M., J. Garcia-Lafuente, J. Delgado, and F. Criado (2003), Seasonal and wind-induced variability of Sea surface temperature patterns in the Gulf of Cadiz, *J. Mar. Syst., 38*(3–4), 205–219.

Vautravers, M. J., and N. J. Shackleton (2006), Centennial-scale surface hydrology off Portugal during marine isotope stage 3: Insights from planktonic foraminiferal fauna variability, *Paleoceanography, 21*, PA3004, doi:10.1029/2005PA001144.

Voelker, A. H. L., S. M. Lebreiro, J. Schönfeld, I. Cacho, H. Erlenkeuser, and F. Abrantes (2006), Mediterranean outflow strengthening during northern hemisphere coolings: A salt source for the glacial Atlantic?, *Earth Planet. Sci. Lett., 245*(1–2), 39–55.

Voelker, A. H. L., L. de Abreu, J. Schönfeld, H. Erlenkeuser, and F. Abrantes (2009), Hydrographic conditions along the western Iberian margin during marine isotope stage 2, *Geochem. Geophys. Geosyst., 10*, Q12U08, doi:10.1029/2009GC002605.

Willamowski, C., and R. Zahn (2000), Upper ocean circulation in the glacial North Atlantic from benthic foraminiferal isotope and trace element fingerprinting, *Paleoceanography, 15*, 515–527.

Zahn, R., J. Schönfeld, H. Kudrass, M. Park, H. Erlenkeuser, and P. Grootes (1997), Thermohaline instability in the North Atlantic during meltwater events: Stable isotope and ice-rafted detritus records from core SO75-26KL, Portuguese margin, *Paleoceanography, 12*(5), 696–710.

L. de Abreu, 18 Upland Rd., Camberley GU15 4JW, UK. (luciaabreu@yahoo.com)

A. H. L. Voelker, Unidade de Geologia Marinha, Laboratorio Nacional de Energia e Geologia (LNEG), Estrada da Portela, Zambujal, P-2610-143 Amadora, Portugal. (antje.voelker@lneg.pt)

Laurentide Ice Sheet Meltwater and the Atlantic Meridional Overturning Circulation During the Last Glacial Cycle: A View From the Gulf of Mexico

B. P. Flower, C. Williams, and H. W. Hill[1]

College of Marine Science, University of South Florida, St. Petersburg, Florida, USA

D. W. Hastings

Marine Science Discipline, Eckerd College, St. Petersburg, Florida, USA

Meltwater input from the Laurentide Ice Sheet (LIS) has often been invoked as a cause of proximal sea surface temperature (SST) and salinity change in the North Atlantic and of regional to global climate change via its influence on the Atlantic meridional overturning circulation (AMOC). Here we review the evidence for meltwater inflow to the Gulf of Mexico and its reduction relative to the onset of the Younger Dryas, compare inferred meltwater inflow during marine isotope stage 3 (MIS 3), and thereby assess the role of LIS meltwater routing as a trigger of abrupt climate change. We present published and new Mg/Ca and $\delta^{18}O$ data on the planktic foraminifer *Globigerinoides ruber* from four northern Gulf of Mexico sediment cores that provide detailed records of SST and $\delta^{18}O$ of seawater ($\delta^{18}Osw$) for most of the last glacial cycle (48–8 ka). These results generally support models that suggest meltwater rerouting away from the Gulf of Mexico and directly to the North Atlantic may have caused Younger Dryas cooling via AMOC reduction. Alternatively, southern meltwater input may simply have been reduced during the Younger Dryas. Indeed, Dansgaard-Oeschger cooling events must have had a different cause because southern meltwater input during MIS 3 does not match their number or timing. Furthermore, the relationships between Gulf of Mexico meltwater input, Heinrich events, Antarctic warm events, and AMOC variability suggest bipolar warming and enhanced seasonality during meltwater episodes. We formulate a "meltwater capacitor" hypothesis for understanding enhanced seasonality during abrupt climate change in the North Atlantic region.

1. INTRODUCTION

Understanding past abrupt climate change is an important societal challenge and requires documenting changes ranging from global sea level, regional surface temperature, and ice sheet meltwater input, among many other changes. Proxy records of high-latitude air temperature have been particularly valuable in defining past abrupt temperature change. Greenland ice sheet records have demonstrated a series of abrupt

[1]Now at Department of Geological Sciences, University of Michigan, Ann Arbor, Michigan, USA.

Abrupt Climate Change: Mechanisms, Patterns, and Impacts
Geophysical Monograph Series 193

10.1029/2010GM001016

shifts of 6°C–10°C termed Dansgaard-Oeschger (D-O) events that have a repeat time of 1–3 kyr [*Dansgaard et al.*, 1993; *Severinghaus et al.*, 1998]. These events extended to large areas of the North Atlantic Ocean [*Bond et al.*, 1993; *Bard et al.*, 2000] and also affected the Asian monsoon system [*Wang et al.*, 2001]. Correlation to Antarctic air temperature records by methane synchronization revealed that warming events were not in phase [*Blunier et al.*, 1997]. This finding led to the hypothesis that a bipolar seesaw was operating, in which North Atlantic Deep Water (NADW) reduction was associated with Southern Hemisphere warming [*Broecker*, 1998; *Stocker*, 1998]. Marine proxy records [*Lamy et al.*, 2004, 2007; *Barker et al.*, 2009] and ocean circulation modeling [*Crowley*, 1992; *Manabe and Stouffer*, 1995] support this idea. Recently, a very detailed record of Antarctic air temperature from the European Project for Ice Coring in Antarctica (EPICA) Dronning Maud Land (EDML) [*EPICA Community Members et al.*, 2006] synchronized via methane to North Greenland Ice Core Project [*Rasmussen et al.*, 2006] reveals a complex relationship between Antarctic and Greenland air temperature (Figure 1). Significantly, each Greenland interstadial appears to have a corresponding warming event in the EDML core, which is termed an Antarctic isotope maxima (AIM). AIM events seem to precede onset of Greenland warming but peak at about the same time, and their magnitude correlates with the duration of Greenland interstadials at least during marine isotope stage 3 (MIS 3) [*EPICA Community Members et al.*, 2006].

The D-O events are episodically punctuated by ice-rafting events known from North Atlantic sediments, termed Heinrich events [*Heinrich*, 1988; *Hemming*, 2004]. The youngest event occurs during the Younger Dryas [*Andrews et al.*, 1995], an enigmatic cold snap that occurred during the last deglaciation [*Broecker and Denton*, 1989; *Alley*, 2000]. The leading hypothesis for the cause of the Younger Dryas is a rerouting of Laurentide Ice Sheet (LIS) meltwater away from the Gulf of Mexico to the North Atlantic via a route that is a topic of current debate. *Broecker et al.* [1989] suggested that ice sheet recession in North America allowed eastward routing through the St. Lawrence Seaway, which found some support in reconstructions of LIS meltwater routing [*Teller*, 1990; *Licciardi et al.*, 1999] and ocean modeling [*Manabe and Stouffer*, 1997]. However, marine evidence for meltwater input via this route is equivocal [e.g., *Rodrigues and Vilks*, 1994; *Keigwin and Jones*, 1995; *deVernal et al.*, 1996; *Carlson et al.*, 2007; *Peltier et al.*, 2007]. Furthermore, a field campaign in eastern Canada to search for geomorphic and geologic evidence for meltwater flow during the Younger Dryas determined that meltwater conduits were formed after the Younger Dryas [*Lowell et al.*, 2005; *Teller et al.*, 2005], highlighting the uncertainty of meltwater rerouting

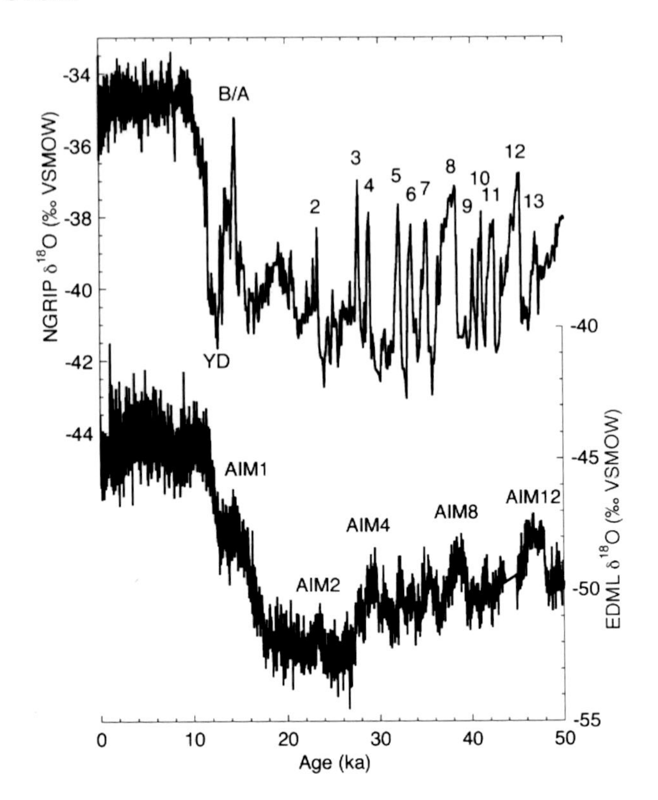

Figure 1. Comparison of polar ice core records for the 0–50 ka interval including North Greenland Ice Core Project (NGRIP) $\delta^{18}O$ [*Rasmussen et al.*, 2006] and EDML $\delta^{18}O$ [*EPICA Community Members et al.*, 2006]. Younger Dryas (YD), Bølling/Allerød (B/A), and numbered Dansgaard-Oeschger (D-O) events are labeled on the NGRIP $\delta^{18}O$ record, and Antarctic isotope maxima (AIM) events are labeled on the EDML $\delta^{18}O$ record.

[*Broecker*, 2006a]. More recently, *Murton et al.* [2010] presented evidence for northward meltwater flow into the Arctic Ocean, which supports earlier modeling studies [*Tarasov and Peltier*, 2006]. *Rashid et al.* [this volume] present evidence from the Labrador Sea for ice rafting and potential meltwater from under the LIS and out through Hudson Strait. In whatever fashion LIS meltwater entered the Atlantic Ocean, geochemical records clearly indicate a reduction of Atlantic meridional overturning circulation (AMOC) during the Younger Dryas, although the magnitude may have been less severe than during Heinrich stadial 1 [*Hughen et al.*, 1998, 2000; *Keigwin*, 2004; *McManus et al.*, 2004; *Keigwin and Boyle*, 2008].

Proxy records from high accumulation rate sediment cores in the Gulf of Mexico can help test the routing hypothesis as the cause of the Younger Dryas and the D-O cooling events. Here we present published and new Mg/Ca and stable isotope measurements on the planktic foraminifer *Globigerinoides ruber* from four northern Gulf of Mexico sediment cores that provide detailed records of sea surface temperature (SST) and $\delta^{18}O$ of seawater ($\delta^{18}O_{sw}$) for most of the last glacial cycle (approximately 48–8 ka).

2. PREVIOUS STUDIES OF DEGLACIAL MELTWATER IN THE GULF OF MEXICO

Deglacial meltwater was first documented in the Gulf of Mexico sediment cores based on $\delta^{18}O$ in planktic foraminifera and tied to LIS decay [*Kennett and Shackleton*, 1975; *Emiliani et al.*, 1975]. Planktic foraminiferal $\delta^{18}O$ reached minimum values of −2.6‰ in core K97 from the north central Gulf of Mexico (Figures 2 and 3). Attempts to constrain the thickness of the inferred meltwater layer by comparing deeper-dwelling planktic foraminiferal $\delta^{18}O$ from *Neogloboquadrina dutertrei*, which today lives throughout 0–200 m [*Bé*, 1982], indicated minimal influence in deeper waters [*Kennett and Shackleton*, 1975; *Leventer et al.*, 1983; *Flower and Kennett*, 1990]. However, because planktic foraminifera may change their depth preferences in order to avoid low-salinity meltwater, the thickness is not well constrained.

The geographic distribution of planktic foraminiferal $\delta^{18}O$ has allowed assessment of the areal extent of inferred meltwater. An earlier compilation suggested decreased influence with distance from the Mississippi Delta junction [*Williams*, 1984]. However, there is conflicting evidence in the northeastern Gulf of Mexico near DeSoto Canyon. Deglacial planktic foraminiferal $\delta^{18}O$ reached minimum values of −3.5‰ in a core from the western flank [*Emiliani et al.*, 1978], −2.2‰ in one core from the eastern flank [*Emiliani et al.*, 1975], yet −1.2‰ in a nearby core [*Nürnberg et al.*, 2008]. The relatively high $\delta^{18}O$ values in the latter core (MD02-2575) led to the conclusion that meltwater did not affect the eastern Gulf of Mexico [*Nürnberg et al.*, 2008]. To re-examine the geographic extent, we compile and plot surface-dwelling planktic foraminiferal $\delta^{18}O$ minima versus linear distance from the Mississippi Delta junction (Figure 4 and Table 1). We restrict our comparison to records based on the white variety of *G. ruber*. The data come from nine locations, some of which have several cores (e.g., Orca Basin has three cores). One location, LOUIS on the Louisiana Slope, has multiple cores [*Aharon*, 2003], and we chose the core with the best data resolution and age control.

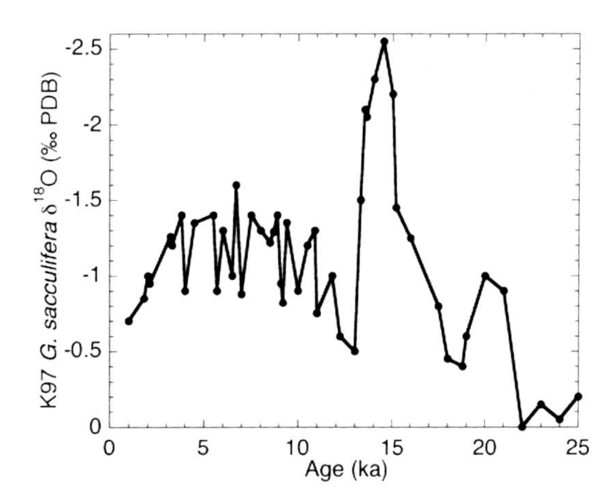

Figure 3. Sigsbee Plain core K97 *Globigerinoides sacculifera* $\delta^{18}O$ record [*Kennett and Shackleton*, 1975] that first documented the major deglacial meltwater spike centered at about 14.5 ka.

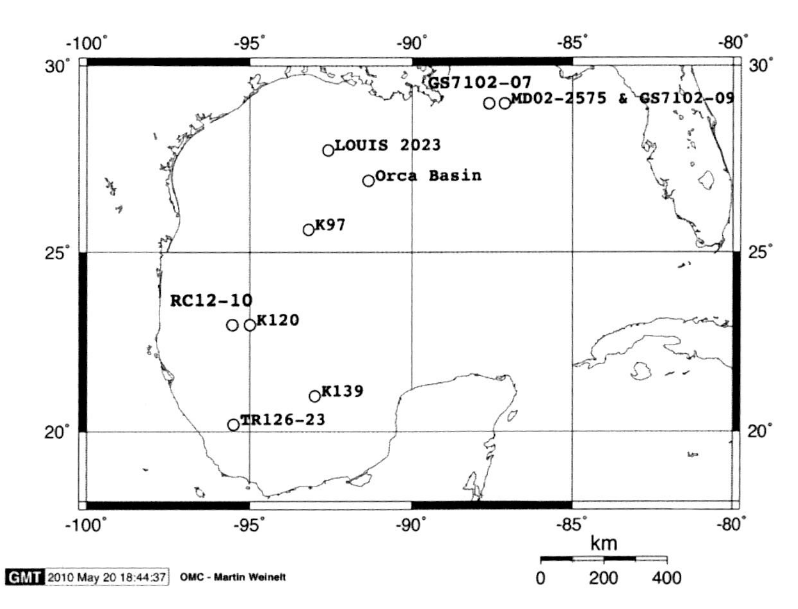

Figure 2. Map showing core locations in the Gulf of Mexico. Core information is given in Table 1.

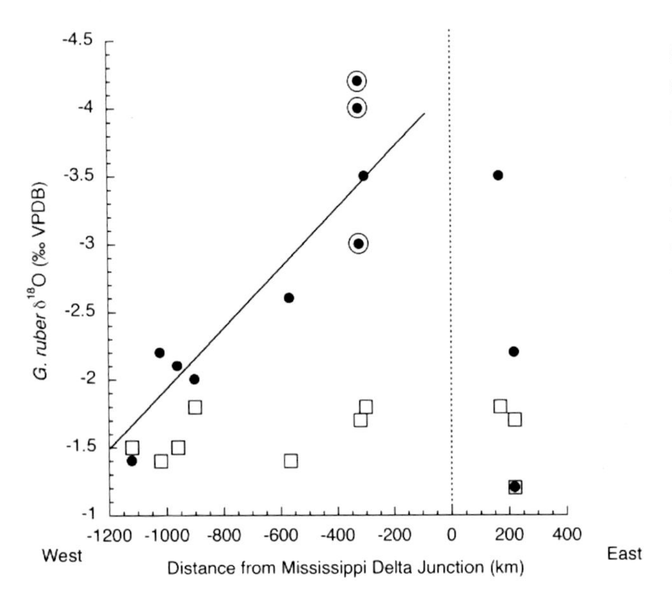

Figure 4. Distance $-\delta^{18}O$ plot comparing white *Globigerinoides ruber* $\delta^{18}O$ values for the Holocene (open squares) and peak deglacial meltwater spike (solid circles) versus distance from the Mississippi Delta junction. Deglacial Orca Basin values are circled. A best fit line is drawn to highlight the increased anomaly in meltwater spike $\delta^{18}O$ values with lesser distance. Core information is given in Table 1.

Mean Holocene $\delta^{18}O$ values (open squares) are in the −1.2‰ to −1.8‰ range for each core, with only slightly lower $\delta^{18}O$ values at lesser distance from the Mississippi Delta junction (Figure 4). In contrast, deglacial $\delta^{18}O$ minima exhibit a clear trend versus distance, with lowest values found in Orca Basin core MD02-2550 (C. Williams et al., Abrupt climate change and Laurentide Ice Sheet meltwater routing to the Gulf of Mexico, submitted to *Geology*, 2011, hereinafter referred to as Williams et al., submitted manuscript, 2011). There is quite a range of deglacial $\delta^{18}O$ minima even within Orca Basin cores (−3‰ to −4.2‰). However, the range is even greater near DeSoto Canyon (−1.2‰ to −3.5‰) as discussed earlier. These ranges may relate to differential temporal resolution in the cores and/or different numbers of specimens analyzed. Nevertheless, there is a clear trend toward higher $\delta^{18}O$ values with distance west of the Mississippi Delta junction, in agreement with a previous compilation [*Williams*, 1984]. East of the junction, the evidence is equivocal. What is clear is that meltwater influence was less than at corresponding distances west of the junction. But in light of conflicting data from three DeSoto Canyon area cores, the contention that LIS meltwater did not influence the northeastern Gulf of Mexico [*Nürnberg et al.*, 2008] remains to be confirmed by additional data.

3. ORCA BASIN CLIMATE RECORDS

Some of the most detailed records of the last glacial cycle come from Orca Basin in the north central Gulf of Mexico. Located ~300 km from the present-day Mississippi River delta junction, Orca Basin is well-suited for high-resolution paleoclimatology because of unusually high accumulation rates (~30–300 cm kyr^{-1}) and a ~200 m thick brine layer (salinity >250) that preserves sedimentary laminations [*Shokes et al.*, 1977; *Pilcher and Blumstein*, 2007]. Abundant pteropods throughout the glacial to present section of the core indicate minimal dissolution of carbonate material [*Leventer et al.*, 1983]. Orca Basin lies within the Western Hemisphere Warm Pool, which develops in late summer with SST reaching >28.5°C in the Caribbean and Gulf of Mexico.

Stable isotope data from Orca Basin cores are available for most of MIS 3 through the last deglaciation [*Leventer et al.*, 1982; *Flower and Kennett*, 1990, 1995; *Flower et al.*, 2004;

Table 1. Selected Gulf of Mexico Core Sites

Core	Latitude	Longitude	Water Depth (m)	Reference
GS7102-07	29°	87°36′	1569	*Emiliani et al.* [1978]
GS7102-09	29°	87°	695	*Emiliani et al.* [1975]
EN32-PC4[a]	26°56.1′	91°21.7′	2260	*Flower and Kennett* [1990]
EN32-PC6[a]	26°56.8′	91°21.0′	2280	*Leventer et al.* [1982]
K97	25°37′	93°12′	3408	*Kennett and Shackleton* [1975]
K120	23°17.8′	96°10.8	2537	*Kennett and Shackleton* [1975]
K139	20°30′	92°37′	2462	*Kennett and Shackleton* [1975]
TR126-23	20.8°	95.5°	2200	*Williams* [1984]
MD02-2550[a]	26°56.77′	91°20.74′	2248	*Williams et al.* [2010]
MD02-2551[a]	26°56.78′	91°20.75′	2255	*Hill et al.* [2006]
MD02-2575	29°0.1′	87°7.13′	847	*Nürnberg et al.* [2008]
LOUIS 2023	27.76°	92.59°	401	*Aharon* [2003]
RC12-10	23°	95.53°	3054	*Poore et al.* [2003]

[a]Orca Basin cores.

Hill et al., 2006; Williams et al., submitted manuscript, 2011; this study]. Age control for these cores is based on accelerator mass spectrometry ^{14}C dates. All raw ^{14}C dates were recalibrated to calendar years before present using the MARINE09 calibration and the Calib 6.0 program (http://intcal.qub.ac.uk/calib/) that includes an assumed constant 405 year reservoir correction [*Stuiver et al.*, 1998; *Reimer et al.*, 2009]. The lack of age reversals in the ^{14}C data set lends some support to the assumed constant reservoir age, but we later relax this assumption in examining the timing of changes near the onset of the Younger Dryas.

4. CESSATION OF MELTWATER INPUT AND THE YOUNGER DRYAS EVENT

Shortly after minimum $\delta^{18}O$ values were reached during the last deglaciation, there was a sharp increase that is widely interpreted as a reduction in southerly LIS meltwater flow (Figure 3). Because sea level was still rising [*Fairbanks*, 1989; *Bard et al.*, 2010], a redirection of LIS meltwater has been inferred. *Kennett and Shackleton* [1975] suggested that the sharp $\delta^{18}O$ increase "represents a major change in the direction of meltwater flow from southward to eastward." *Broecker et al.* [1989] linked this "cessation event" to the onset of the Younger Dryas and postulated that eastward flow could have reduced AMOC and caused regional cooling in the North Atlantic.

Critical to testing the routing hypothesis is the age of the cessation event in the Gulf of Mexico. The most detailed records of this event come from Orca Basin cores EN32-PC4, EN32-PC6, and MD02-2550. Each core exhibits a $\delta^{18}O$ increase of 2‰–3‰ over a thickness of 5–10 cm (Figure 5). Our radiocarbon ages place this event from 11,200 to 10,900 ^{14}C years ago, assuming a constant 400 year reservoir age, which is synchronous with the onset of the Younger Dryas determined in European lake sequences [*Ammann and Lotter*, 1989]. Conversion of a central date of 11,050 ^{14}C years ago to calendar years yields 13.0 ka based on the new MARINE09 calibration (Calib 6.0) [*Reimer et al.*, 2009].

The duration of the cessation event may also provide insight into the mechanisms of reduction of southern meltwater and its potential routing elsewhere. Because the $\delta^{18}O$ increase involves a combination of SST cooling and salinity increase, a more precise definition of the cessation event is based on $\delta^{18}O_{sw}$ data (Williams et al., submitted manuscript, 2011). In $\delta^{18}O$, it consists of a temporary increase after local minimum $\delta^{18}O$ values, a recovery, and then a large, rapid $\delta^{18}O$ increase of 1.4‰–3‰ (Figure 5). The total thickness is 20, 20, and 11 cm in cores EN32-PC4, EN32-PC6, and MD02-2550, respectively. The rapid $\delta^{18}O$ increase is 10, 10, and 5.5 cm thick, respectively. These relative thicknesses

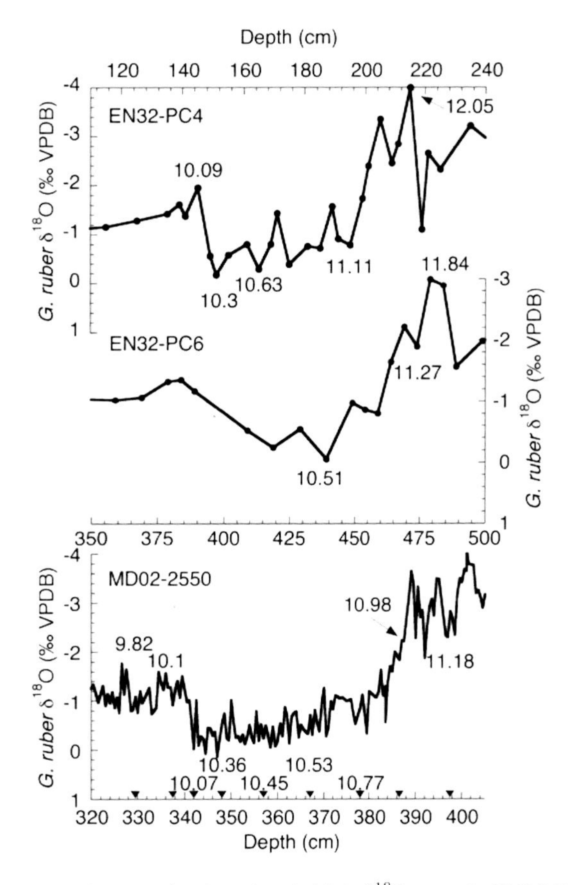

Figure 5. Orca Basin *G. ruber* (white) $\delta^{18}O$ records EN32-PC4 [*Flower and Kennett*, 1990], EN32-PC6 [*Leventer et al.*, 1982; *Flower et al.*, 2004], and MD02-2550 (Williams et al., submitted manuscript, 2011) showing cessation event and subsequent YD interval. Following minimum $\delta^{18}O$ values (−3‰ to −4‰), note cessation event marked by a 2‰–3‰ increase. Triangles indicate reservoir-corrected accelerator mass spectrometry ^{14}C dates.

are consistent with the fact that the first two cores were piston cores, while the last was a giant gravity core, and hence, the sediments were probably differentially expanded/compressed during core retrieval.

Radiocarbon age control can be used to estimate the duration of the cessation event. In cores EN32-PC4 and EN32-PC6, ^{14}C dates converted to calendar years constrain the accumulation rate at about 40 cm kyr^{-1} yielding a total duration of about 500 years and the rapid portion at about 250 years. In MD02-2550, the dating is sufficiently detailed to bracket the temporary increase and the rapid increase separately with three ^{14}C dates. This is because of a ^{14}C date at the inflection point. Accepting the calibrated ^{14}C dates at face value, the calculated total duration is 362 years, and the rapid $\delta^{18}O$ increase of 3.1‰ took about 184 years. Interestingly, this interval lies within a section of continuously laminated sediment, indicating no obvious changes in oxygenation associated with the cessation event.

Overall, the rapidity and magnitude of the cessation event appears consistent with a cessation of meltwater to the GOM [*Broecker et al.*, 1989] but not a sudden, complete diversion. It is difficult to differentiate whether the cessation indicates a "diversion" or a "reduction" of LIS meltwater. Because sea level continued to rise during the Younger Dryas [*Fairbanks*, 1989; *Bard et al.*, 2010], it seems unlikely that melting stopped along the entirety of the southern LIS margin unless significant melting continued elsewhere including Antarctica. There was at least a local readvance in the Lake Superior region [*Lowell et al.*, 1999]. Although Lake Agassiz may have ceased its southward flow during the Marquette readvance that corresponds closely with the Younger Dryas, summer melting along the southern margin of the LIS may have continued. However, the rapidity of the cessation event (possibly <200 years) is difficult to explain by shutting off meltwater sources. In combination with recent evidence for northward flow during the Younger Dryas [*Murton et al.*, 2010], a diversion of LIS meltwater remains the leading possibility. Indeed, recent records from the North Atlantic are interpreted to indicate eastern flow of LIS meltwater [*Thornalley et al.*, 2010]. Regardless of whether the cessation event indicates a diversion or a reduction of LIS meltwater, the timing (11,000 ^{14}C years ago, 12.9 ka) seems to coincide with the onset of the Younger Dryas, whether based on comparison to ^{14}C ages in European lake sequences [*Ammann and Lotter*, 1989] or to layer-counted ages in a German lake [*Brauer et al.*, 2008] and in Greenland ice [*Rasmussen et al.*, 2006].

On the other hand, independent evidence from the Cariaco Basin for reservoir age changes may complicate age determination of the cessation event. Comparison of the German tree ring ^{14}C chronology and the Cariaco Basin ^{14}C chronology suggest the oceanic reservoir age was reduced by 100–400 years near the start of the Younger Dryas event [*Kromer et al.*, 2004; *Muscheler et al.*, 2008; *Reimer et al.*, 2009; *Hua et al.*, 2009]. A reduction in reservoir age would increase the age of the cessation event by a similar amount and increase its lead over the Younger Dryas (Williams et al., submitted manuscript, 2011). To the extent that the reservoir age in the Gulf of Mexico was close to that in Cariaco Basin as it is today [*Wagner et al.*, 2009], the ages of the cessation event and AMOC reduction (based on $\Delta^{14}C$ in Cariaco Basin) would remain similar. However, the reduced reservoir age may be unique to Cariaco Basin [*Reimer et al.*, 2009; *Hua et al.*, 2009]. By whatever route the meltwater entered the North Atlantic Ocean (discussed above), it seems that the cessation event coincided with AMOC reduction within a few hundred years [*Hughen et al.*, 1998, 2000; *Keigwin*, 2004; *McManus et al.*, 2004; *Keigwin and Boyle*, 2008].

5. ISOLATING δ^{18}Osw USING PAIRED δ^{18}O AND Mg/Ca DATA

One complication in this analysis is the fact that foraminiferal $\delta^{18}O$ is a function of temperature and ice volume, as well as δ^{18}Osw. Recent studies have attempted to separate the temperature and δ^{18}Osw components of Orca Basin planktic foraminiferal $\delta^{18}O$ by paired $\delta^{18}O$ and Mg/Ca measurements [*Flower et al.*, 2004; *Hill et al.*, 2006; Williams et al., submitted manuscript, 2011). At least 30 specimens of *G. ruber* from the 250–355 μm size fraction were gently crushed and aliquots taken for stable isotopic (~80 μg) and elemental (~300 μg) analyses. Stable isotope data were generated on a ThermoFinnigan Delta Plus XL light stable isotope ratio mass spectrometer (SIRMS) with a Kiel III automated carbonate preparation device at the College of Marine Science, University of South Florida. Long-term external precision for this instrument is 0.09 and 0.06‰ for $\delta^{18}O$ and $\delta^{13}C$, respectively, based on >4000 analyses of NBS-19 run since July 2000.

The elemental splits were cleaned for Mg/Ca analysis using the Cambridge method that does not include a reductive cleaning step [*Barker et al.*, 2003]. Samples were dissolved in weak HNO_3 to yield calcium concentrations of ~20 ppm to minimize calcium concentration effects. Elemental ratio data were generated on a Perkin Elmer 4300 dual-view inductively coupled plasma (ICP)-optical emission spectrometer. A standard instrument-drift correction technique [*Schrag*, 1999] was routinely used. Precision based on replicates was 0.16 mmol mol^{-1} (±0.4°C) for EN32PC6 [*Flower et al.*, 2004] and 0.09 mmol mol^{-1} (±0.3°C) for MD02-2551 [*Hill et al.*, 2006]. Our new data on MD02-2550 were generated on an Agilent 7500cx ICP–mass spectrometer (MS), and precision was 0.08 mmol mol^{-1} (±0.3°C) [*Williams et al.*, 2010]. Al/Ca and Mn/Ca data showed no correlation to Mg/Ca, indicating minimal influence of insufficient clay removal or Mn-Fe overgrowths on our Mg/Ca values [*Williams et al.*, 2010]. The $\delta^{18}O$ and Mg/Ca data will be available at the National Climate Data Center, NOAA at http://www.ncdc.noaa.gov/paleo/paleo.html.

Salinity has long been a potential problem in the calibration of foraminiferal Mg/Ca data to temperature. Early culture experiments on *Globigerinoides sacculifera* revealed a substantial increase in Mg/Ca of 110% when salinity was increased by 10 psu [*Nürnberg et al.*, 1996]. Smaller salinity changes (<3 psu) produced no change in foraminiferal Mg/Ca. Other culturing experiments with *Globigerinoides bulloides* and *Orbulina universa* found a 4% increase in Mg/Ca per psu [*Lea et al.*, 1999]. Results from culturing *G. ruber* are not yet available. *Ferguson et al.* [2008] provided evidence from sediment traps that high salinity in the eastern Mediterranean Sea was associated with elevated foraminiferal Mg/Ca values, although high Mg/Ca calcite overgrowths were later suggested as the cause [*Hoogakker et al.*, 2009]. Sediment core-top data suggest that Mg/Ca may depend on several factors in addition to temperature, including salinity, but that the effects are strongest only at salinities >36.5 psu [*Arbuszewski et al.*, 2010]. Because modern annual mean salinity in the vicinity of Orca Basin is 35.5 psu [*Levitus and Boyer*, 1994], and increased freshwater would further lower the salinity, the salinity effect on Mg/Ca-SST is likely minimal (as discussed by *Williams et al.* [2010]).

On the other hand, one concern is that Mississippi River water, which is known to have a low dissolved Mg/Ca value of about 0.49 mol mol^{-1} compared to the Gulf of Mexico value of about 5.1 mol mol^{-1}, might lower foraminiferal Mg/Ca values. Nevertheless, a simple box model calculation shows that the concentrations of both Mg and Ca are too low in Mississippi River water (425 and 870 μM, respectively) [*Briggs and Ficke*, 1978] to affect seawater (53 and 10.3 mM, respectively) or foraminiferal Mg/Ca values. Calculations show that a 25% dilution of surface seawater would only decrease seawater Mg/Ca by <3%, which is within measurement error [*Flower et al.*, 2004]. Therefore low-Mg/Ca meltwater is unlikely to have affected our SST determinations based on *G. ruber*.

Two commonly used equations for conversion of Mg/Ca values to temperature [*Dekens et al.*, 2002; *Anand et al.*, 2003] yield different absolute temperatures but identical relative changes. Only the latter study supplied specific equations for white and pink *G. ruber* (Mg/Ca = 0.449 × exp(0.09 × SST) and Mg/Ca = 0.381 × exp(0.09 × SST, respectively). We have chosen to use the white *G. ruber* equation for internal consistency and because core-top Mg/Ca values yield SST values of 25.4°C [*LoDico et al.*, 2006; *Richey et al.*, 2007, 2009], equivalent to mean annual SST in the northern Gulf of Mexico [*Earth System Research Laboratory*, 1994]. Paired Mg/Ca SST and $\delta^{18}O$ values were used to calculate $\delta^{18}Osw$ values based on a paleotemperature equation (SST(°C) = 14.9 − 4.8(δc − δw)) [*Bemis et al.*, 1998] that is appropriate for *G. ruber* [*Thunell et al.*, 1999]. Adding 0.27‰ yields values on the Vienna standard mean ocean water scale [*Hut*, 1987].

Paired Mg/Ca SST and $\delta^{18}O$ data from core EN32-PC6 [*Flower et al.*, 2004], updated to the MARINE09 timescale [*Reimer et al.*, 2009] are compared to polar ice core records (Figure 6). The original Mg/Ca data [*Flower et al.*, 2004] have here been calibrated to SST using the white *G. ruber* equation from *Anand et al.* [2003] instead of from *Dekens et al.* [2002] for three reasons: (1) for consistency with our other Gulf of Mexico Mg/Ca work (2) because early Holocene Mg/Ca SST from this core (~26°C) is close to modern mean annual SST and (3) because early Holocene $\delta^{18}Osw$ is close to modern values [*Fairbanks et al.*, 1992; *Schmidt et al.*, 1999]. Calculated $\delta^{18}Osw$ values have also been corrected for changing ice volume effects using deglacial sea level records [*Fairbanks*, 1989; *Bard et al.*, 1990] and a

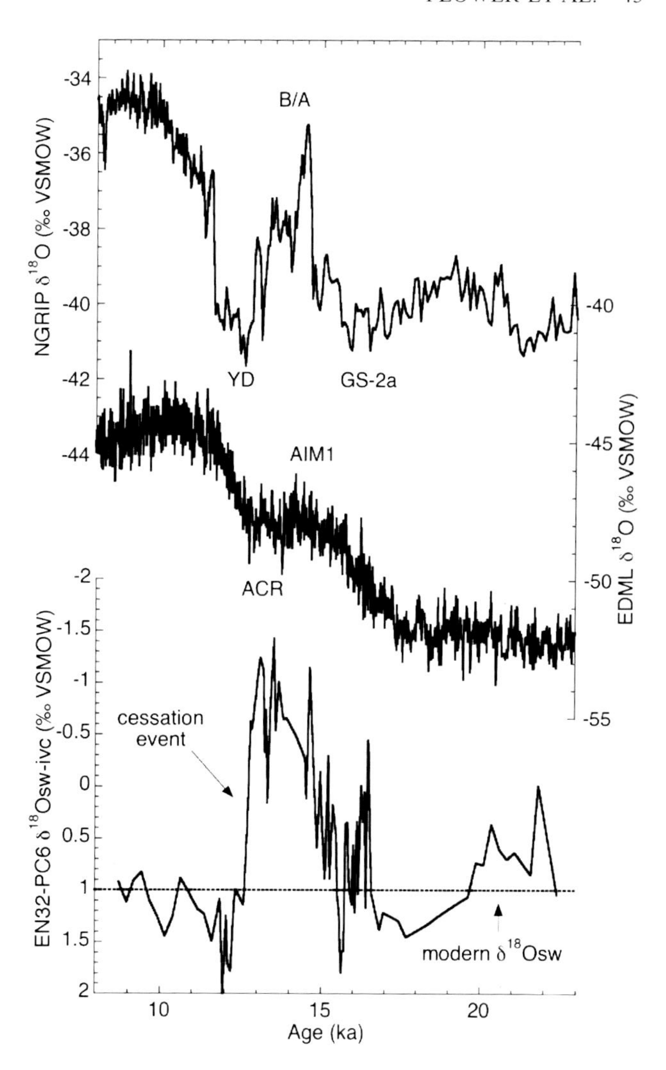

Figure 6. Comparison of polar ice core records for the 8–23 ka interval. NGRIP $\delta^{18}O$ [*Rasmussen et al.*, 2006] and EDML $\delta^{18}O$ [*EPICA Community Members et al.*, 2006] are compared to the ice volume-corrected $\delta^{18}Osw$ record from EN32-PC6 [*Flower et al.*, 2004] updated to the MARINE09 ^{14}C calibration data set [*Reimer et al.*, 2009] using Calib 6.0 (http://intcal.qub.ac.uk/calib/). YD, B/A, Greenland stadial 2a (GS-2a), Antarctic Cold Reversal (ACR), and AIM 1 are labeled on the ice core records. Modern Gulf of Mexico $\delta^{18}Osw$ is indicated by horizontal line.

δ^{18}Osw/ice volume calibration [*Schrag et al.*, 2002] as described earlier [*Flower et al.*, 2004]. Accordingly, deglacial δ^{18}Osw values can be compared directly with modern δ^{18}Osw of about 1‰, in contrast to the δ^{18}Osw values originally published. The new ice volume-corrected δ^{18}Osw data (δ^{18}Osw-ivc) confirm major episodic negative spikes from approximately 16.5 to 12.9 ka followed by a rapid increase of 2.3‰ at approximately 12.9 ka inferred to represent the cessation event (Figure 6). Clearly, both the major meltwater spike and the cessation event are preserved in δ^{18}Osw.

Although diversion of meltwater from the Gulf of Mexico to the North Atlantic may have triggered the Younger Dryas, such a diversion does not appear to have triggered the Oldest Dryas event, known in Greenland as stadial 2a [*Walker et al.*, 1999] and in the North Atlantic as the Heinrich 1 stadial [*Barker et al.*, 2009]. Simple rerouting of meltwater to the North Atlantic might be expected to yield no southern routed meltwater during the Oldest Dryas. Yet we record at least two major negative δ^{18}Osw excursions of >1.5 ‰ between 17 and 14.5 ka, interrupted by a brief return to modern δ^{18}Osw values of about 1‰ at approximately 15.4 ka (Figure 6). The earlier excursion appears to be accompanied by distinctive illite clay mineralogy derived from the eastern LIS ablation zone (Ohio River drainage), while other excursions marked by smectite clay mineralogy were derived from the Missouri River drainage [*Sionneau et al.*, 2010]. Sea level is known to have risen by about 20 m during this interval [*Fairbanks*, 1989], but Antarctica is considered to have contributed <14 m throughout the last deglaciation [*Denton and Hughes*, 2002]. Furthermore, the southern Laurentide ice margin was far south of potential eastern outlets [*Lowell et al.*, 1999, 2009]. Accordingly, we suggest that the Gulf of Mexico served as a major outlet for LIS meltwater and an integrator of the Mississippi River drainage basin, throughout the last deglaciation until the onset of the Younger Dryas.

6. TESTING THE MELTWATER ROUTING HYPOTHESIS DURING MIS 3

Could meltwater from the LIS and European ice sheets have caused D-O coolings during MIS 3? *Clark et al.* [2001] summarized the evidence from the last deglaciation that meltwater routing was controlled by the southern extent of the LIS and suggested that the D-O events have a similar cause. Specifically, when the southern LIS margin was south of about 43°–49°N, meltwater would have flowed south and allowed normal AMOC and hence warm climates in the North Atlantic region. But when the southern LIS margin was north of these latitudes, a switch was triggered whereby meltwater entered the North Atlantic and caused AMOC reduction [*Clark et al.*, 2001]. Regardless of where meltwater was routed during stadials, this hypothesis predicts meltwater flow to the Gulf of Mexico during D-O interstadials.

Hill et al. [2006] presented paired Mg/Ca and δ^{18}O data from Orca Basin core MD02-2551 from approximately 45 to 28 ka that argued against the original version of this hypothesis. Calculated δ^{18}Osw values indicated five minima that did not appear to match the nine D-O interstadials in this interval. Indeed, the δ^{18}Osw record exhibited greater similarity to global sea level change [*Siddall et al.*, 2003] and Antarctic air temperature based on Byrd ice core δ^{18}O on the Greenland Ice Sheet Project 2 timescale [*Johnsen et al.*, 1972; *Blunier and Brook*, 2001]. In particular, the largest and longest negative δ^{18}Osw excursion seemed to correspond to the A1 Antarctic warming centered at 38.5 ka [*Hill et al.*, 2006]. The minimum values (about −0.5 ‰) are considerably higher than during the last deglaciation, which may relate to different ages of the ice melted and their associated δ^{18}O compositions [*Marshall*, 2009]. Here we extend the MD02-2551 data set, update the chronology using the MARINE09 ^{14}C calibration [*Reimer et al.*, 2009], and compare to the new EDML ice core δ^{18}O record [*EPICA Community Members et al.*, 2006] to reassess these findings.

The new δ^{18}Osw record on the MARINE09 timescale confirms five δ^{18}Osw minima during the 48–28 ka interval that do not match D-O interstadials 4–12 (Figure 7). Instead, the five δ^{18}Osw minima better match the five AIM events from the EDML ice core δ^{18}O record [*EPICA Community Members et al.*, 2006]. Furthermore, these AIM events are associated with distinct CO_2 maxima from the Byrd ice core [*Ahn and Brook*, 2008]. The match between Antarctic warm events, CO_2 maxima, and inferred LIS meltwater input to the Gulf of Mexico suggests episodic bipolar warming enhanced by greenhouse feedbacks, at least during Northern Hemisphere summers that controlled LIS ablation. Southern Hemisphere warming is reflected in Antarctic air temperature, but synchronous Northern Hemisphere warming is inferred here from coeval LIS meltwater input to the Gulf of Mexico. These findings are consistent with far-field manifestations of the "Antarctic climate signature" in the equatorial Pacific Ocean [*Stott et al.*, 2002; *Lea et al.*, 2006] and in the Greenland dust record [*Barker and Knorr*, 2007]. Overall, our observation of LIS ablation during AIM events raises the possibility that summer LIS melting may be influenced by the Antarctic climate signal including greenhouse gas changes. Taken together, the deglacial and MIS 3 records support our inference that the Gulf of Mexico served as a major outlet for LIS meltwater throughout the last glacial cycle until the onset of the Younger Dryas.

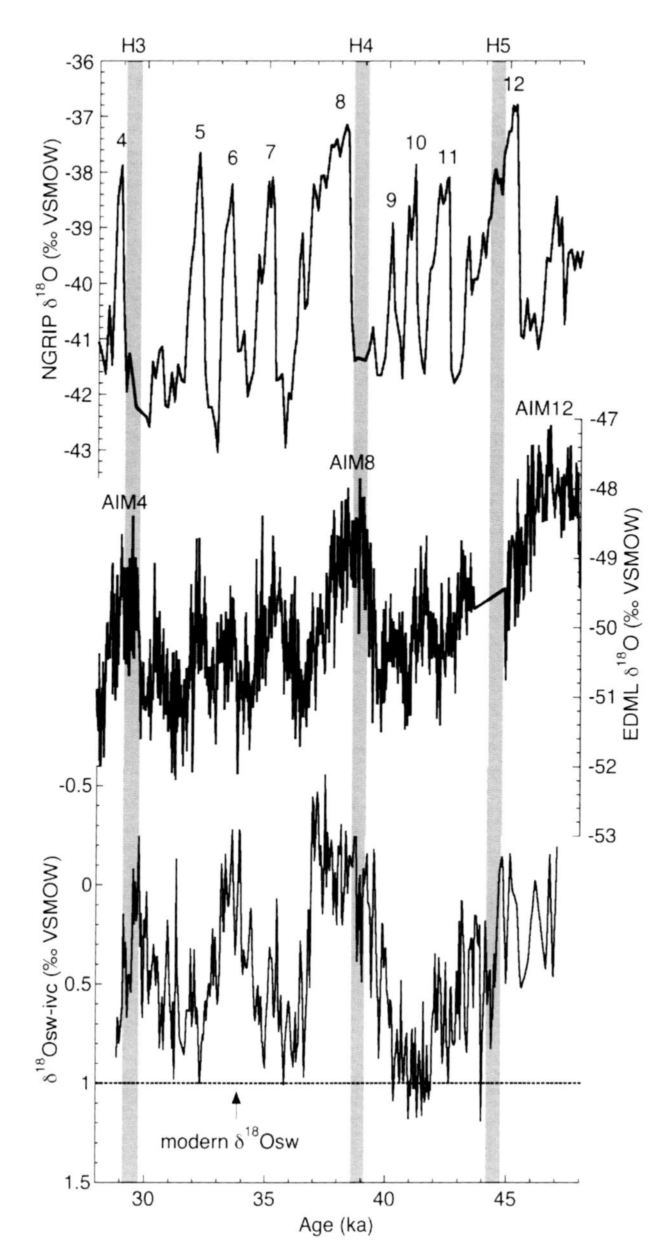

Figure 7. Comparison of ice volume-corrected δ^{18}Osw from MD02-2551 [*Hill et al.*, 2006; this study] to NGRIP δ^{18}O and EDML δ^{18}O from 48 to 28 ka. Also shown are Heinrich events 3–5, D-O events 4–12 and AIM 4, 8, and 12. Modern Gulf of Mexico δ^{18}Osw is indicated by horizontal line.

7. ICE SHEET MELTWATER AND THE ATLANTIC OCEAN DEEP CIRCULATION

Modeling studies and North Atlantic sediment records have suggested that changes in deep circulation were linked to the D-O events [*Dokken and Jansen*, 1999; *Bond et al.*, 1999; *Sarnthein*, 2000; *Hagen and Keigwin*, 2002; *Rahmstorf*, 2002; *Clark et al.*, 2002; *Hoogakker et al.*, 2007]. Other studies have inferred that circulation changes were greatest during Heinrich stadials [*Boyle and Keigwin*,1987; *Vidal et al.*, 1997; *Curry et al.*, 1999; *McManus et al.*, 2004]. In particular, AMOC reduction during Heinrich stadial 1 appears to have been more severe than during the Younger Dryas [*McManus et al.*, 2004]. It is important to stress that there are uncertainties in the degree of inferred Atlantic Ocean circulation change based on different proxies, including the Pa/Th proxy data for AMOC. Opal production may significantly affect sedimentary Pa/Th and lead to overestimates of AMOC reduction [*Chase et al.*, 2002; *Keigwin and Boyle*, 2008; *Lippold et al.*, 2009]. Nevertheless, it is clear from ^{14}C records in Cariaco Basin [*Hughen et al.*, 2000] and ^{14}C [*Keigwin*, 2004; *Robinson et al.*, 2005; *Keigwin and Boyle*, 2008] and Nd records in the deep Atlantic Ocean [*Roberts et al.*, 2010] that significant AMOC reduction occurred during the H1 stadial and the Younger Dryas relative to the Bølling/Allerød and the Holocene. This contention agrees with older nutrient-based proxy data for NADW volume reduction and Southern Component Water increase, such as Cd/Ca and δ^{13}C records in benthic foraminifera from the deep western North Atlantic [*Boyle and Keigwin*, 1987; *Keigwin et al.*, 1991; *Curry and Oppo*, 1997]. The latter records do not require, but are consistent with, AMOC reduction. In particular, decreased benthic δ^{13}C values are recorded on Ceara Rise during the Heinrich 3, 4, and 5 stadials and interpreted as decreased NADW production [*Curry and Oppo*, 1997].

It is now clear that the Heinrich stadials correlate with some of the major AIM warming events in Antarctica [*EPICA Community Members et al.*, 2006]. Indeed, AIM 1, 4, 8, and 12 are associated with Heinrich events 1, 3, 4, and 5, respectively. Heinrich events are classically interpreted as increased iceberg discharge during particularly severe D-O stadials [*Bond et al.*, 1993; *Hemming*, 2004]. Although Heinrich events appear to have discharged into a cold North Atlantic, subtropical warming has been observed [*Mix et al.*, 1986; *Rühlemann et al.*, 1999; *Arz et al.*, 1999; *Grimm et al.*, 2006; *Williams et al.*, 2010], and "hidden warming" in the Northern Hemisphere has long been suspected as a contributor to cryospheric instability (S. Hemming, personal communication, 2009). Our reconstructions from the Gulf of Mexico suggest that LIS meltwater spanned the AIM events that were associated with Heinrich events 1, 3, 4, and 5 (Figure 8). These records represent perhaps the best evidence for bipolar warming during AIM events.

Comparing the temporal patterns of meltwater input during the last deglaciation and MIS 3 leads to the observation that major LIS melting was associated with AMOC reduction

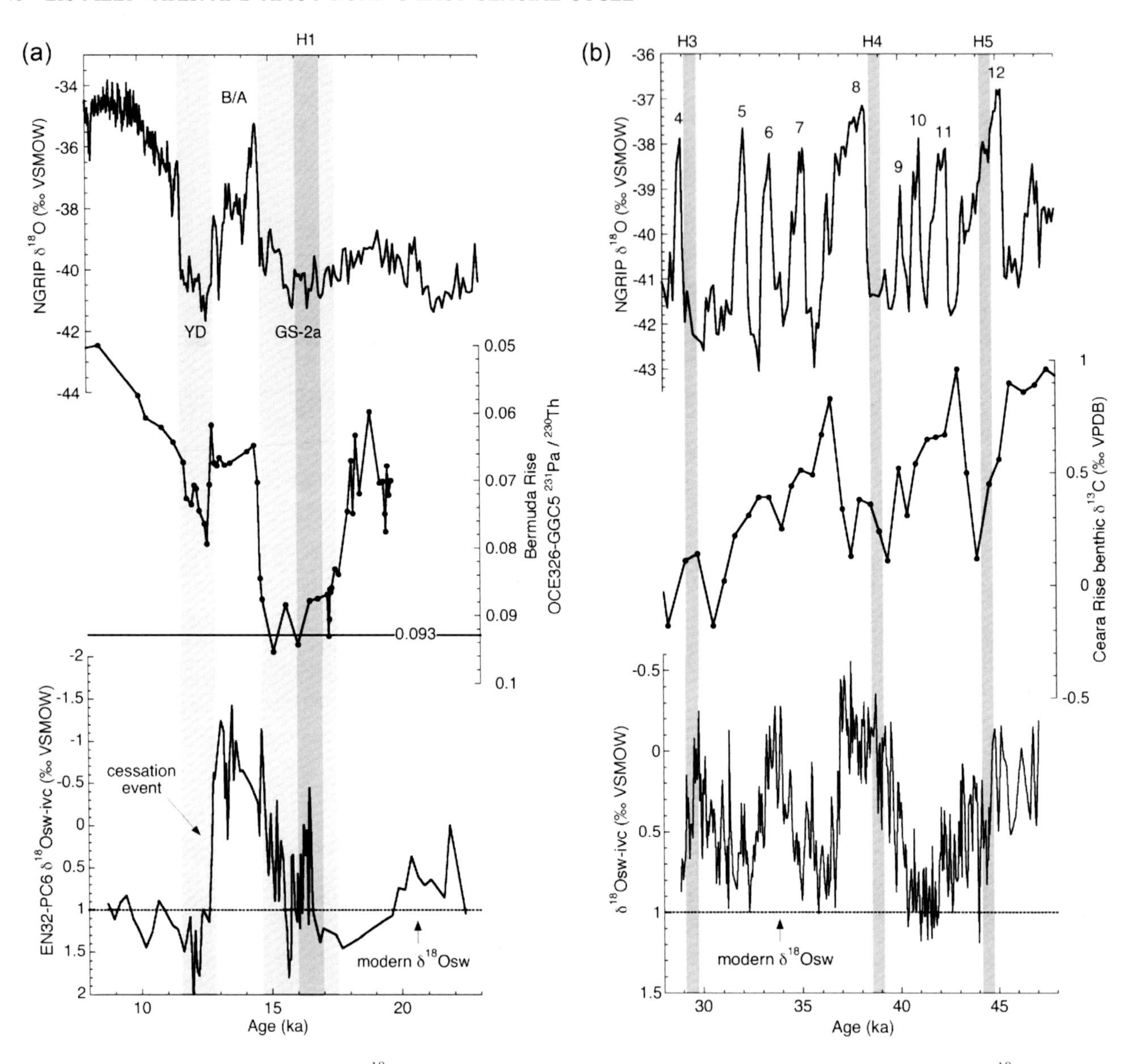

Figure 8. (a) Ice volume-corrected δ^{18}Osw from EN32-PC6 [*Flower et al.*, 2004; this study] compared to NGRIP δ^{18}O and Bermuda Rise Pa/Th [*McManus et al.*, 2004]. (b) Ice volume-corrected δ^{18}Osw from MD02-2551 [*Hill et al.*, 2006; this study] is compared to NGRIP δ^{18}O and Ceara Rise benthic δ^{13}C [*Curry and Oppo*, 1997]. Decreased AMOC or North Atlantic Deep Water (NADW) is oriented down in Figures 8a and 8b. YD, B/A, GS-2a, and Heinrich events are shown. Modern Gulf of Mexico δ^{18}Osw is indicated by horizontal lines. Note significant meltwater during Heinrich stadials of both the last deglaciation and MIS 3 and its association with AMOC or NADW reduction.

(Figure 8). For example, although the initiation of LIS meltwater input is not resolved in core EN32-PC6, significant input occurred during the Heinrich 1 stadial (also known as GS-2a in Greenland ice) that is considered an interval of reduced AMOC [*McManus et al.*, 2004; *Robinson et al.*, 2005; *Keigwin and Boyle*, 2008; *Roberts et al.*, 2010]. However, LIS meltwater input apparently increased through the onset of the Bølling and peaked during the Allerød (Figure 8) [*Flower et al.*, 2004]. Yet the AMOC increased strongly at the onset of the Bølling [*McManus et al.*, 2004; *Knorr and Lohmann*, 2007; *Liu et al.*, 2009]. These findings argue against a simple rerouting of meltwater from northern to southern routes at the onset of the Bølling [*Clark et al.*, 2001; *Liu et al.*, 2009]. One possible resolution to this conundrum is that glacial meltwater was unable to weaken the AMOC during the Bølling because of extensive salt storage in the low-latitude Atlantic Ocean during the preceding H1 stadial [*Vellinga and Wu*, 2004; *Schmidt et al.*, 2004; *Weldeab et al.*, 2006; *Carlson et al.*, 2008].

During MIS 3, there is also a correspondence between inferred LIS meltwater input, Heinrich stadials, and NADW reduction (Figure 8). In particular, the largest and longest inferred meltwater spike coincides with the greatest NADW reduction during the Heinrich 4 stadial. LIS meltwater appears to start during the Heinrich 4 stadial and continue through the subsequent D-O interstadial 8. Similar to the last deglaciation, the persistence of meltwater through interstadial 8 seems contradictory to evidence for strong resumption of NADW [*Curry and Oppo*, 1997] and appears to require salt buildup in the low-latitude Atlantic Ocean to offset the meltwater effect [*Vellinga and Wu*, 2004; *Schmidt et al.*, 2006; *Jaeschke et al.*, 2007]. The cessation of this meltwater spike may coincide with the following stadial, but age control is insufficient to evaluate this possibility. Nevertheless, the number and timing of meltwater events does not match the nine D-O events in this interval, as discussed above, yet they do match the inferred decreases in NADW production.

The following pattern emerges from comparison of proxy data for LIS melting and AMOC reduction during the last glacial cycle (Figure 8): (1) LIS meltwater begins during Heinrich stadials and lasts through the subsequent D-O event. (2) LIS meltwater appears to coincide with major AIM events. (3) LIS meltwater is associated with distinct changes in deep circulation during Heinrich events. The likely trigger for LIS meltwater during the last deglaciation was summer insolation, but insolation changes were much too slow during MIS 3 to initiate LIS melting. The trigger for LIS melting and Heinrich events during MIS 3 remains unknown [*Hemming*, 2004] but may have been enhanced by CO_2 rise via the bipolar seesaw [*Knorr and Lohmann*, 2007; *Barker et al.*, 2009] or via the displacement of the Southern Hemisphere westerlies [*Anderson et al.*, 2009; *Denton et al.*, 2010].

To what extent did LIS meltwater to the Gulf of Mexico directly affect the AMOC? Modeling work has found different sensitivities. A classic paper determined that input to the Gulf of Mexico had four to five times less impact on the thermohaline circulation versus the North Atlantic [*Manabe and Stouffer*, 1997]. A recent paper found that the sensitivity was about half that of northern North Atlantic input; 0.28 Sv to the Gulf of Mexico was sufficient to reduce AMOC by about 40% [*Otto-Bliesner and Brady*, 2010]. Clearly, meltwater input to the northern North Atlantic is more effective at reducing the AMOC.

Accordingly, a key unresolved issue is the contribution of the other circum-North Atlantic ice sheets. Data are limited on the individual Greenland, Iceland, and Fennoscandian ice sheets and other outlets of the LIS. However, ice-rafted debris records and planktic foraminiferal $\delta^{18}O$ records provide some evidence for meltwater during Heinrich events and some interstadials. Data from Deep-Sea Drilling Project 609 in the northeast Atlantic reveal a significant $\delta^{18}O$ minimum recorded in *Neogloboquadrina pachyderma* (sinistral) associated with Heinrich event 1 [*Bond et al.*, 1993]. Other North Atlantic cores also exhibit significant $\delta^{18}O$ excursions and other parameters during Heinrich stadials, particularly those associated with H1 and H4 [*Sarnthein et al.*, 1995; *Maslin et al.*, 1995; *Zahn et al.*, 1997; *Cortijo et al.*, 1997; *Vidal et al.*, 1997; *Labeyrie et al.*, 1999; *Mangerud et al.*, 2004; *Peck et al.*, 2006; *Toucanne et al.*, 2010]. These $\delta^{18}O$ excursions and associated increases in ice-rafted debris are found across the North Atlantic, generally consistent with

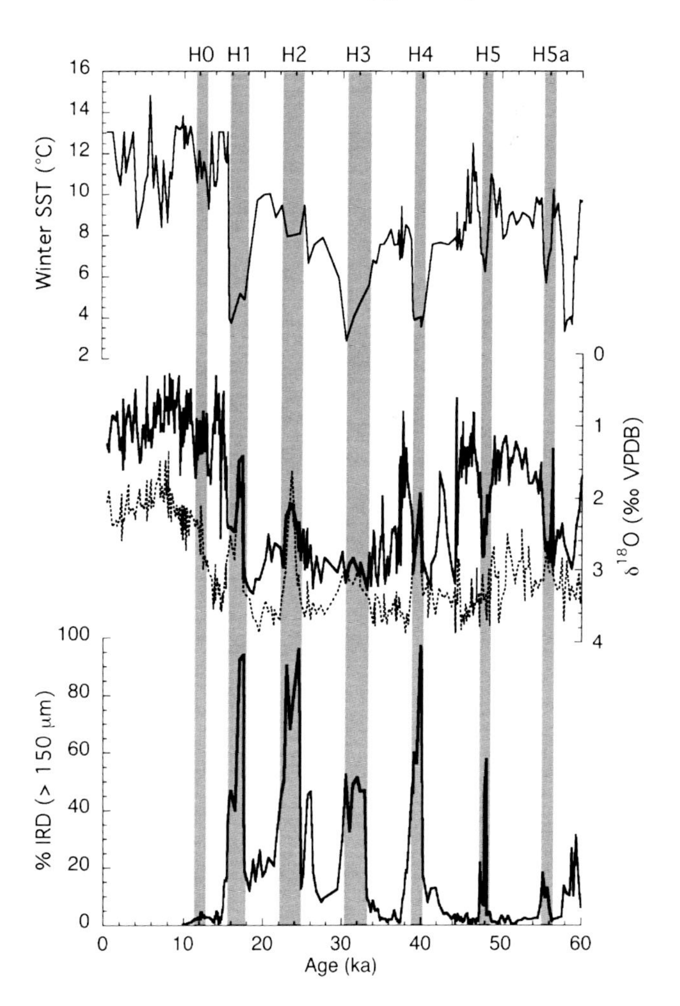

Figure 9. Core CH69-K09 records from offshore Newfoundland [*Labeyrie et al.*, 1999]. Winter SST based on planktic foraminiferal assemblage data is plotted with $\delta^{18}O$ based on *Globigerinoides bulloides* (solid line) and *Neogloboquadrina pachyderma* (s.) (dashed line). Heinrich events based on percent ice-rafted debris are shaded.

meltwater sources from the circum-North Atlantic ice sheets. One record is especially important because it comes from offshore Newfoundland, off the St. Lawrence River system [*Labeyrie et al.*, 1999]. In core CH69-K09, planktic $\delta^{18}O$ data on *G. bulloides* and *N. pachyderma* (s.) exhibit distinct minima associated with Heinrich events 1, 2, and 4 (Figure 9). *Obbink et al.* [2010] also present evidence from this region for LIS meltwater during the H1 stadial and the Bølling-Allerød. Overall, these findings on LIS melting during AMOC reductions highlight the importance of circum-North Atlantic meltwater supply, either by direct or indirect supply to deep convection regions. However, refining the timing of the different sources of meltwater to the North Atlantic Ocean remains a continuing challenge.

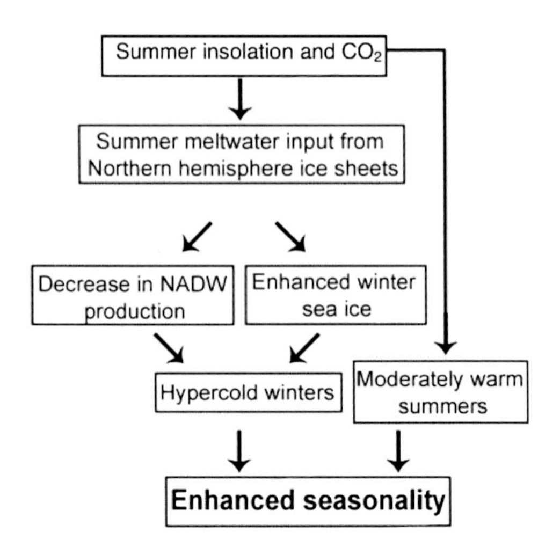

Figure 10. Conceptual model of the "meltwater capacitor" hypothesis for enhanced seasonality during Heinrich stadials.

8. THE "MELTWATER CAPACITOR" HYPOTHESIS AND SEASONALITY IN THE NORTH ATLANTIC

Our observations on LIS meltwater history are consistent with the "deglacial seasonality" hypothesis [*Denton et al.*, 2005; *Broecker*, 2006b] in which Heinrich stadials and the Younger Dryas are marked by greatly enhanced seasonality in the North Atlantic region. Marshalling evidence from Greenland ice core records, snowline records from east Greenland, northern Europe and North America, and floral and faunal assemblage records from northern Europe, *Denton et al.* [2005] documented an important mismatch between inferred summer and mean annual temperatures. For example, Greenland ice cores indicated a Younger Dryas mean annual cooling of ~16°C [*Severinghaus et al.*, 1998], yet snowline records suggested only 4°C–6°C. The likely resolution to this seeming mismatch is that snowlines reflect summer conditions and that winters must have been ~26°C–28°C colder. The only viable way to produce such cold winters is greatly expanded sea ice, which would limit heat release from the upper limb of the AMOC and increase the continentality of the region. Summer meltwater is one way to enhance sea ice formation the following winter and may be the link between extreme seasonality and AMOC reduction.

We propose a scenario to help explain enhanced seasonality during Heinrich stadials: the "meltwater capacitor" hypothesis (Figure 10). The hypothesis has the following four components: (1) Northern Hemisphere ice sheets start decaying early in the deglacial sequence, mainly during summers, in response to increased summer insolation and atmospheric CO_2 rise. (2) Ice sheet meltwater from the circum-North Atlantic ice sheets decreases salinity in the North Atlantic Ocean, whose signal is preserved during subsequent winters. (3) Low-density meltwaters inhibit winter deep convection in the high-latitude North Atlantic, resulting in reduction of AMOC and associated heat transport. (4) Expanded winter sea ice increases the continentality of the region and promotes hypercold winters. Just as an electronic capacitor stores energy, persistent summer meltwater stores a low-salinity signal to inhibit deep convection during subsequent winters. Our hypothesis highlights the role of summer meltwater in enhancing winter sea ice, as well as reducing AMOC and thereby contributing to the enigmatic increased seasonality observed during Heinrich stadials and the Mystery Interval [*Denton et al.*, 2005]. This proposed set of mechanisms was not operating during the Bølling/Allerød or major D-O interstadials despite continued meltwater input because of greatly enhanced AMOC. Glacial meltwater apparently was unable to weaken the AMOC at these times, perhaps because of extensive salt storage in the low-latitude Atlantic Ocean during long intervals of AMOC reduction such as the H1 and H4 stadials [*Vellinga and Wu*, 2004; *Schmidt et al.*, 2004, 2006; *Weldeab et al.*, 2006; *Jaeschke et al.*, 2007; *Knorr and Lohmann*, 2007]. Our hypothesis highlights the need for continued data/model comparisons on the role of Northern Hemisphere ice sheet meltwater in regional seasonality changes and global ocean circulation changes.

9. CONCLUSIONS

LIS meltwater has long been invoked as the trigger of AMOC reduction during the Younger Dryas, by re-routing from the Gulf of Mexico to the North Atlantic. A review of published and new data from three Orca Basin cores confirms that the cessation of southward meltwater flow coincided with the onset of the Younger Dryas at about 11,000 ^{14}C years before present (12.9 ka). In contrast, the timing of LIS inflow during MIS 3 is inconsistent with a simple version of the routing hypothesis applied to explain the D-O events. We find five major episodes of LIS meltwater input that do not match the nine D-O events known from this interval. We consider the Gulf of Mexico record to reflect LIS melting along the entire southern LIS margin on North America and not subject to rerouting changes until the onset of the Younger Dryas.

Episodic LIS meltwater input during the last glacial cycle instead appears to match the five major warm events identified in Antarctic air temperature records as AIM events. Some of these LIS meltwater events have also been observed in the North Atlantic. This correspondence may indicate bipolar warming, at least during Northern Hemisphere summers that control LIS melting. Furthermore, LIS melting is associated with deep Atlantic Ocean circulation changes during Heinrich stadials and likely contributed to its weakening, along with iceberg discharge to regions of deep convection. The observation that meltwater continued during subsequent interstadials, including the Bølling-Allerød, which featured a strong AMOC, may indicate that advection of salt from the low-latitude Atlantic was sufficient to offset a meltwater effect on the AMOC. Finally, we formulate a "meltwater capacitor" hypothesis that may help explain enhanced seasonality during Heinrich stadials, in which summer melting enhances winter sea ice formation and associated hypercold winters, as well as AMOC reduction.

Acknowledgments. We thank the IMAGES program, Viviane Bout-Roumazeilles, Yvon Balut, and Laurent Labeyrie for a productive cruise on the R/V *Marion Dufresne* in 2002. This work was supported in part by the Comer Science and Education Foundation (CP-18 to BPF and DWH) and the National Science Foundation under grants OCE-0318361 and OCE-0903017. We thank Ethan Goddard and Kelly Quinn for invaluable expertise in SIRMS and ICP-MS analysis, George Denton, Julie Richey, and USF Paleo lab members for useful discussions, and Eckerd College summer research interns for help in sample preparation. We also thank two reviewers for helpful comments, and Harunur Rashid, Ellen Mosley-Thompson, and Leonid Polyak for hosting the AGU Chapman Conference on abrupt climate change and for editorial work.

REFERENCES

Aharon, P. (2003), Meltwater flooding events in the Gulf of Mexico revisited: Implications for rapid climate changes during the last deglaciation, *Paleoceanography*, *18*(4), 1079, doi:10.1029/2002PA000840.

Ahn, J., and E. Brook (2008), Atmospheric CO_2 and climate on millennial time scales during the last glacial period, *Science*, *322*, 83–85, doi:10.1126/science.1160832.

Alley, R. B. (2000), The Younger Dryas cold interval as viewed from central Greenland, *Quat. Sci. Rev.*, *19*, 213–226.

Ammann, B., and A. F. Lotter (1989), Late-glacial radiocarbon- and palynostratigraphy on the Swiss Plateau, *Boreas*, *18*, 109–126.

Anand, P., H. Elderfield, and M. H. Conte (2003), Calibration of Mg/Ca thermometry in planktonic foraminifera from a sediment trap time series, *Paleoceanography*, *18*(2), 1050, doi:10.1029/2002PA000846.

Anderson, R. F., S. Ali, L. I. Bradtmiller, S. H. H. Nielsen, M. Q. Fleisher, R. E. Anderson, and L. H. Burckle (2009), Wind-driven upwelling in the Southern Ocean and the deglacial rise in atmospheric CO_2, *Science*, *323*, 1443–1448.

Andrews, J. T., A. E. Jennings, M. W. Kerwin, M. Kirby, W. Manley, G. H. Miller, G. Bond, and B. MacLean (1995), A Heinrich-like event, H-0 (DC-0): Source(s) for detrital carbonate in the North Atlantic during the Younger Dryas chronozone, *Paleoceanography*, *10*(5), 943–952.

Arbuszewski, J., P. deMenocal, A. Kaplan, and E. C. Farmer (2010), On the fidelity of shell-derived $\delta^{18}O_{seawater}$ estimates, *Earth Planet. Sci. Lett.*, *300*, 185–196.

Arz, H. W., J. Pätzold, and G. Wefer (1999), The deglacial history of the western tropical Atlantic as inferred from high resolution stable isotope records off northeastern Brazil, *Earth Planet. Sci. Lett.*, *167*, 105–117.

Bard, E., B. Hamelin, R. G. Fairbanks, and A. Zindler (1990), Calibration of the ^{14}C timescale over the past 30,000 years using mass spectrometric U-Th ages from Barbados corals, *Nature*, *345*, 405–410.

Bard, E., F. Rostek, J.-L. Turon, and S. Gendeau (2000), Hydrological impact of Heinrich events into the subtropical northeast Atlantic, *Science*, *289*, 1321–1323.

Bard, E., B. Hamelin, and D. Delanghe-Sabatier (2010), Deglacial meltwater pulse 1B and Younger Dryas sea levels revisited with boreholes at Tahiti, *Science*, *327*, 1235–1237, doi:10.1126/science.1180557.

Barker, S., and G. Knorr (2007), Antarctic climate signature in the Greenland ice core record, *Proc. Natl. Acad. Sci. U. S. A.*, *104*, 17,278–17,282.

Barker, S., M. Greaves, and H. Elderfield (2003), A study of cleaning procedures used for foraminiferal Mg/Ca paleothermometry, *Geochem. Geophys. Geosyst.*, *4*(9), 8407, doi:10.1029/2003GC000559.

Barker, S., P. Diz, M. J. Vautravers, J. Pike, G. Knorr, I. R. Hall, and W. S. Broecker (2009), Interhemispheric Atlantic seesaw response during the last deglaciation, *Nature*, *457*, 1097–1102, doi:10.1038/nature07770.

Bé, A. W. H. (1982), Biology of planktonic foraminifera, in *Foraminifera: Notes for a Short Course*, edited by T. W. Broadhead, pp. 51–92, Dep. of Geol. Sci., Knoxville, Tenn.

Bemis, B. E., H. J. Spero, J. Bijma, and D. W. Lea (1998), Reevaluation of the oxygen isotopic composition of planktonic foraminifera: Experimental results and revised paleotemperature equations, *Paleoceanography*, *13*(2), 150–160.

Blunier, T., and E. J. Brook (2001), Timing of millennial-scale climate change in Antarctica and Greenland during the last glacial period, *Science*, *291*, 109–112.

Blunier, T., J. Schwander, B. Stauffer, T. Stocker, A. Dällenbach, A. Indermühle, J. Tschumi, J. Chappellaz, D. Raynaud, and J.-M. Barnola (1997), Timing of the Antarctic cold reversal and the atmospheric CO_2 increase with respect to the Younger Dryas event, *Geophys. Res. Lett.*, *24*, 2683–2686.

Bond, G. C., W. S. Broecker, S. Johnsen, J. F. McManus, L. Labeyrie, J. Jouzel, and G. Bonani (1993), Correlations between climate records from North Atlantic sediments and Greenland ice, *Nature*, *365*, 143–147.

Bond, G. C., W. Showers, M. Elliot, M. Evans, R. Lotti, I. Hajdas, G. Bonani, and S. Johnson (1999), The North Atlantic's 1-2 kyr climate rhythm: Relation to Heinrich events, Dansgaard/Oeschger cycles, and the Little Ice Age, in *Mechanisms of Global Climate Change at Millennial Time Scales*, *Geophys. Monogr. Ser.*, vol. 112, edited by P. U. Clark, R. S. Webb, and L. D. Keigwin, pp. 35–58, AGU, Washington D. C.

Boyle, E. A., and L. D. Keigwin (1987), North Atlantic thermohaline circulation during the past 20,000 years linked to high-latitude surface temperature, *Nature*, *330*, 35–40.

Brauer, A., G. H. Haug, P. Dulski, D. M. Sigman, and J. F. W. Negendank (2008), An abrupt wind shift in western Europe at the onset of the Younger Dryas cold period, *Nat. Geosci.*, *1*, 520–523.

Briggs, J. C., and J. F. Ficke (1978), Quality of rivers of the United States, 1975 water year, *U.S. Geol. Surv. Open File Rep.*, *78–200*, 436 pp.

Broecker, W. S. (1998), Paleocean circulation during the last deglaciation: A bipolar seesaw?, *Paleoceanography*, *13*(2), 119–121.

Broecker, W. S. (2006a), Was the Younger Dryas triggered by a flood?, *Science*, *312*, 1146–1148.

Broecker, W. S. (2006b), Abrupt climate change revisited, *Global Planet. Change*, *54*, 211–215.

Broecker, W. S., and G. H. Denton (1989), The role of ocean-atmosphere reorganizations in glacial cycles, *Geochim. Cosmochim. Acta*, *53*, 2465–2501.

Broecker, W. S., J. P. Kennett, B. P. Flower, J. T. Teller, S. Trumbore, G. Bonani, and W. Wolfli (1989), The routing of meltwater from the Laurentide Ice Sheet during the Younger Dryas cold episode, *Nature*, *341*, 318–321.

Carlson, A. E., P. U. Clark, B. A. Haley, G. P. Klinkhammer, K. Simmons, E. J. Brook, and K. J. Meissner (2007), Geochemical proxies of North American freshwater routing during the Younger Dryas cold event, *Proc. Natl. Acad. Sci. U. S. A.*, *104*, 6556–6561.

Carlson, A. E., D. W. Oppo, R. E. Came, A. N. LeGrande, L. D. Keigwin, and W. B. Curry (2008), Subtropical Atlantic salinity variability and Atlantic meridional circulation during the last deglaciation, *Geology*, *36*, 991–994.

Chase, Z., R. F. Anderson, M. Q. Fleisher, and P. W. Kubik (2002), The influence of particle composition and particle flux on scavenging of Th, Pa and Be in the ocean, *Earth Planet. Sci. Lett.*, *204*, 215–229.

Clark, P. U., S. J. Marshall, G. K. C. Clarke, S. W. Hostetler, J. M. Licciardi, and J. T. Teller (2001), Freshwater forcing of abrupt climate change during the last glaciation, *Science*, *293*, 283–287.

Clark, P. U., N. G. Pisias, T. F. Stocker, and A. J. Weaver (2002), The role of the thermohaline circulation in abrupt climate change, *Nature*, *415*, 863–869.

Cortijo, E. L., L. Labeyrie, M. Vidal, M. Vautravers, M. Chapman, J. C. Duplessy, M. Elliot, M. Arnold, J. L. Turon, and G. Auffret (1997), Changes in sea surface hydrology associated with Heinrich event 4 in the North Atlantic Ocean between 40° and 60°N, *Earth Planet. Sci. Lett.*, *146*, 129–145.

Crowley, T. J. (1992), North Atlantic Deep water cools the Southern Hemisphere, *Paleoceanography*, *7*(4), 489–497.

Curry, W. B., and D. W. Oppo (1997), Synchronous, high-frequency oscillations in tropical sea surface temperatures and North Atlantic Deep Water production during the last glacial cycle, *Paleoceanography*, *12*(1), 1–14.

Curry, W. B., T. M. Marchitto, J. F. McManus, D. W. Oppo, and K. L. Laarkamp (1999), Millennial-scale changes in ventilation of the thermocline, intermediate, and deep waters of the glacial North Atlantic, in *Mechanisms of Global Climate Change at Millennial Time Scales*, *Geophys. Monogr. Ser.*, vol. 112, edited by P. U. Clark, R. S. Webb, and L. D. Keigwin, pp. 59–76, AGU, Washington, D. C.

Dansgaard, W., et al. (1993), Evidence for general instability of past climate from a 250-kyr ice-core record, *Nature*, *364*, 218–220.

Dekens, P. S., D. W. Lea, D. K. Pak, and H. J. Spero (2002), Core top calibration of Mg/Ca in tropical foraminifera: Refining paleotemperature estimation, *Geochem. Geophys. Geosyst.*, *3*(4), 1022, doi:10.1029/2001GC000200.

Denton, G. H., and T. J. Hughes (2002), Reconstructing the Antarctic Ice Sheet at the Last Glacial Maximum, *Quat. Sci. Rev.*, *21*, 193–202.

Denton, G. H., R. B. Alley, G. C. Comer, and W. S. Broecker (2005), The role of seasonality in abrupt climate change, *Quat. Sci. Rev.*, *24*, 1159–1182.

Denton, G. H., R. F. Anderson, J. R. Toggweiler, R. L. Edwards, J. M. Schaefer, and A. E. Putnam (2010), The last glacial termination, *Science*, *328*, 1652–1656.

deVernal, A., C. Hillaire-Marcel, and G. Bilodeau (1996), Reduced meltwater flow from the Laurentide Ice Sheet during the Younger Dryas, *Nature*, *381*, 774–777.

Dokken, T. M., and E. Jansen (1999), Rapid changes in the mechanism of ocean convection during the glacial period, *Nature*, *401*, 458–461.

Earth System Research Laboratory (1994), NODC (Levitus) World Ocean Atlas Data 1994, January 2006, http://www.cdc.noaa.gov/cdc/data.nodc.woa94.html, Earth Syst. Res. Lab., NOAA, Boulder, Colo.

Emiliani, C., S. Gartner, B. Lidz, K. Eldridge, D. K. Elvey, T. C. Huang, J. J. Stipp, and M. F. Swanson (1975), Paleoclimatological analysis of Late Quaternary cores from the northeastern Gulf of Mexico, *Science*, *189*, 1083–1089.

Emiliani, C., C. Rooth, and J. J. Stipp (1978), The late Wisconsin flood into the Gulf of Mexico, *Earth Planet. Sci. Lett.*, *41*, 159–162.

EPICA Community Members, et al. (2006), One-to-one coupling of glacial climate variability in Greenland and Antarctica, *Nature*, *444*, 195–198.

Fairbanks, R. G. (1989), A 17,000-year glacio-eustatic sea level record: Influence of glacial melting rates on the Younger Dryas event and deep-ocean circulation, *Nature*, *342*, 637–642.

Fairbanks, R. G., C. D. Charles, and J. D. Wright (1992), Origin of global meltwater pulses, in *Radiocarbon After Four Decades*, edited by R. E. Taylor et al., pp. 473–500, Springer, New York.

Ferguson, J. E., G. M. Henderson, M. Kucera, and R. E. M. Rickaby (2008), Systematic change of foraminiferal Mg/Ca ratios across a strong salinity gradient, *Earth Planet. Sci. Lett.*, *265*(1–2), 153–166.

Flower, B. P., and J. P. Kennett (1990), The Younger Dryas cool episode in the Gulf of Mexico, *Paleoceanography*, *5*(6), 949–961.

Flower, B. P., J. P. Kennett (1995), Biotic responses to temperature and salinity changes during the last deglaciation, Gulf of Mexico, in *Effects of Past Global Change on Life*, pp. 209–220, Natl. Acad. Press, Washington, D. C

Flower, B. P., D. W. Hastings, H. W. Hill, and T. M. Quinn (2004), Phasing of deglacial warming and Laurentide Ice Sheet meltwater in the Gulf of Mexico, *Geology*, *32*(7), 597–600.

Grimm, E., W. A. Watts, G. L. Jacobson Jr., B. C. S. Hansen, H. R. Almquist, and A. C. Dieffenbacher-Krall (2006), Evidence for warm wet Heinrich events in Florida, *Quat. Sci. Rev.*, *25*(17–18), 2197–2211.

Hagen, S., and L. D. Keigwin (2002), Sea-surface temperature variability and deep water reorganisation in the subtropical North Atlantic during isotope stage 2–4, *Mar. Geol.*, *189*, 145–162.

Heinrich, H. (1988), Origin and consequences of cyclic ice rafting in the northeast Atlantic Ocean during the past 130,000 years, *Quat. Res.*, *29*, 142–152.

Hemming, S. R. (2004), Heinrich events: Massive late Pleistocene detritus layers of the North Atlantic and their global climate imprint, *Rev. Geophys.*, *42*, RG1005, doi:10.1029/2003RG000128.

Hill, H. W., B. P. Flower, T. M. Quinn, D. J. Hollander, and T. P. Guilderson (2006), Laurentide Ice Sheet meltwater and abrupt climate change during the last glaciation, *Paleoceanography*, *21*, PA1006, doi:10.1029/2005PA001186.

Hoogakker, B. A. A., I. N. McCave, and M. J. Vautravers (2007), Antarctic link to deep flow speed variation during marine isotope stage 3 in the western North Atlantic, *Earth Planet. Sci. Lett.*, *257*, 463–473, doi:10.1016/j.epsl.2007.03.003.

Hoogakker, B. A. A., G. P. Klinkhammer, H. Elderfield, E. J. Rohling, and C. Hayward (2009), Mg/Ca paleothermometry in high salinity environments, *Earth Planet. Sci. Lett.*, *284*(3–4), 583–589.

Hua, Q., M. Barbetti, D. Fink, K. F. Kaiser, M. Friedrich, B. Kromer, V. A. Levchenko, U. Zoppi, A. M. Smith, and F. Bertuch (2009), Atmospheric ^{14}C variations derived from tree rings during the early Younger Dryas, *Quat. Sci. Rev.*, *28*, 2982–2990.

Hughen, K. A., J. T. Overpeck, S. J. Lehman, M. Kashgarian, J. Southon, L. C. Peterson, R. Alley, and D. M. Sigman (1998), Deglacial changes in ocean circulation from an extended radiocarbon calibration, *Nature*, *391*, 65–68.

Hughen, K. A., J. R. Southon, S. J. Lehman, and J. T. Overpeck (2000), Synchronous radiocarbon and climate shifts during the last deglaciation, *Science*, *290*, 1951–1954.

Hut, G. (1987), Consultants' group meeting on stable isotope reference samples for geochemical and hydrological investigations, report , 42 pp., Int. At. Energy Agency, Vienna.

Jaeschke, A., C. Rühlemann, H. Arz, G. Heil, and G. Lohmann (2007), Coupling of millennial-scale changes in sea surface temperature and precipitation off northeastern Brazil with high-latitude shifts during the last glacial period, *Paleoceanography*, *22*, PA4206, doi:10.1029/2006PA001391.

Johnsen, S. J., H. B. Clausen, W. Dansgaard, and C. C. Langway (1972), Oxygen isotope profiles through Antarctic and Greenland ice sheets, *Nature*, *235*, 429–434.

Keigwin, L. D. (2004), Radiocarbon and stable isotope constraints on Last Glacial Maximum and Younger Dryas ventilation in the western North Atlantic, *Paleoceanography*, *19*, PA4012, doi:10.1029/2004PA001029.

Keigwin, L. D., and E. A. Boyle (2008), Did North Atlantic overturning halt 17,000 years ago?, *Paleoceanography*, *23*, PA1101, doi:10.1029/2007PA001500.

Keigwin, L. D., and G. A. Jones (1995), The marine record of deglaciation from the continental margin off Nova Scotia, *Paleoceanography*, *10*(6), 973–985.

Keigwin, L. D., G. A. Jones, S. J. Lehman, and E. A. Boyle (1991), Deglacial meltwater discharge, North Atlantic deep circulation, and abrupt climate change, *J. Geophys. Res.*, *96*, 16,811–16,826.

Kennett, J. P., and N. J. Shackleton (1975), Laurentide Ice Sheet meltwater recorded in Gulf of Mexico deep-sea cores, *Science*, *188*(4184), 147–150.

Knorr, G., and G. Lohmann (2007), Rapid transitions in the Atlantic thermohaline circulation triggered by global warming and meltwater during the last deglaciation, *Geochem. Geophys. Geosyst.*, *8*, Q12006, doi:10.1029/2007GC001604.

Kromer, B., M. Friedrich, K. A. Hughen, F. Kaiser, S. Remmele, M. Schaub, and S. Talamo (2004), Late Glacial ^{14}C ages from a floating, 1382-ring pine chronology, *Radiocarbon*, *46*, 1203–1209.

Labeyrie, L., H. Leclaire, C. Waelbroeck, E. Cortijo, J.-C. Duplessy, L. Vidal, M. Elliot, B. Le Coat, and G. Auffret (1999), Temporal variability of the surface and deep waters of the north west Atlantic Ocean at orbital and millennial scales, in *Mechanisms of Global Climate Change at Milennial Time Scales*, *Geophys. Monogr. Ser.*, vol. 112, edited by P. U. Clark, R. S. Webb, and L. D. Keigwin, pp. 77–98, AGU, Washington, D. C.

Lamy, F., J. Kaiser, U. S. Ninnemann, D. Hebbeln, H. W. Arz, and J. Stoner (2004), Antarctic timing of surface water changes off Chile and Patagonian ice sheet response, *Science*, *304*, 1959–1962.

Lamy, F., J. Kaiser, H. W. Arz, D. Hebbeln, U. S. Ninnemann, O. Timm, A. Timmermann, and J. R. Toggweiler (2007), Modulation of the bipolar seesaw in the southeast Pacific during termination 1, *Earth Planet. Sci. Lett.*, *259*(3–4), 400–413.

Lea, D. W., T. A. Mashiotta, and H. J. Spero (1999), Controls on magnesium and strontium uptake in planktonic foraminifera determined by live culturing, *Geochim. Cosmochim. Acta*, *63*, 2369–2379.

Lea, D. W., D. K. Pak, C. L. Belanger, H. J. Spero, M. A. Hall, and N. J. Shackleton (2006), Paleoclimate history of Galapagos surface waters over the last 135,000 yr, *Quat. Sci. Rev.*, *25*, 1152–1167.

Leventer, A., D. F. Williams, and J. P. Kennett (1982), Dynamics of the Laurentide Ice Sheet during the last deglaciation: Evidence from the Gulf of Mexico, *Earth Planet. Sci. Lett.*, *59*(1), 11–17.

Leventer, A., D. F. Williams, and J. P. Kennett (1983), Relationships between anoxia, glacial meltwater and microfossil preservation in the Orca Basin, Gulf of Mexico, *Mar. Geol.*, *53*, 23–40.

Levitus, S., and T. P. Boyer (1994), *World Ocean Atlas 1994*, vol. 4, *Temperature, NOAA Atlas NESDIS*, vol. 4, 129 pp., NOAA, Silver Spring, Md.

Licciardi, J. M., J. T. Teller, and P. U. Clark (1999), Freshwater routing by the Laurentide Ice Sheet during the last deglaciation, in *Mechanisms of Global Climate Change at Millennial Time Scales*, *Geophys. Monogr. Ser.*, vol. 112, edited by P. U. Clark, R. S. Webb, and L. D. Keigwin, pp. 177–201, AGU, Washington, D. C.

Lippold, J., J. Grützner, D. Winter, Y. Lahaye, A. Mangini, and M. Christl (2009), Does sedimentary $^{231}Pa/^{230}Th$ from the Bermuda Rise monitor past Atlantic meridional overturning circulation?, *Geophys. Res. Lett.*, *36*, L12601, doi:10.1029/2009GL038068.

Liu, Z., et al. (2009), Transient simulation of last deglaciation with a new mechanism for Bølling-Allerød warming, *Science*, *325*, 310–314, doi:10.1126/science.1171041.

LoDico, J. M., B. P. Flower, and T. M. Quinn (2006), Subcentennial-scale climatic and hydrologic variability in the Gulf of Mexico during the early Holocene, *Paleoceanography*, *21*, PA3015, doi:10.1029/2005PA001243.

Lowell, T. V., R. K. Howard, and G. H. Denton (1999), Role of climate oscillations in determining ice-margin position: Hypothesis, examples, and implications, in *Glacial Processes Past and Present*, edited by D. M. Mickelson and J. W. Attig, *Spec. Pap. Geol. Soc. Am.*, *337*, 193–203.

Lowell, T. V., et al. (2005), Testing the Lake Agassiz meltwater trigger for the Younger Dryas, *Eos Trans. AGU*, *86*(40), 365.

Lowell, T. V., T. G. Fisher, I. Hajdas, K. Glover, H. Loope, and T. Henry (2009), Radiocarbon deglaciation chronology of the Thunder Bay, Ontario area and implications for ice sheet retreat patterns, *Quat. Sci. Rev.*, *28*, 1597–1607, doi:10.1016/j.quascirev.2009.02.025.

Manabe, S., and R. J. Stouffer (1995), Simulation of abrupt change induced by freshwater input to the North Atlantic Ocean, *Nature*, *378*, 165–167.

Manabe, S., and R. J. Stouffer (1997), Coupled ocean-atmosphere model response to freshwater input: Comparison to Younger Dryas event, *Paleoceanography*, *12*(2), 321–336.

Mangerud, J., et al. (2004), Ice-dammed lakes and rerouting of the drainage of northern Eurasia during the last glaciation, *Quat. Sci. Rev.*, *23*, 1313–1332.

Marshall, S. (2009), Modeling isotopic discharge from the Laurentide Ice Sheet during D-O cycles and the last deglaciation, paper presented at Chapman Conference on Abrupt Climate Change, AGU, Columbus, Ohio.

Maslin, M. A., N. J. Shackleton, and U. Pflaumann (1995), Surface water temperature, salinity, and density changes in the northeast Atlantic during the last 45,000 years: Heinrich events, deep-water formation, and climatic rebounds, *Paleoceanography*, *10*(3), 527–544.

McManus, J. F., R. Francois, J. Gheradi, L. D. Keigwin, and S. Brown-Leger (2004), Collapse and rapid resumption of Atlantic meridional circulation linked to deglacial climate, *Nature*, *428*, 834–837.

Mix, A. C., W. F. Ruddiman, and A. McIntyre (1986), Late Quaternary paleoceanography of the tropical Atlantic. 1: Spatial variability of annual mean sea-surface temperatures, 0–20,000 years B.P., *Paleoceanography*, *1*(1), 43–66.

Murton, J. B., M. D. Bateman, S. R. Dallimore, J. T. Teller, and Z. Yang (2010), Identification of Younger Dryas outburst flood path from Lake Agassiz to the Arctic Ocean, *Nature*, *464*, 740–743.

Muscheler, R., B. Kromer, S. Björck, A. Svensson, M. Friedrich, K. F. Kaiser, and J. Southon (2008), Tree rings and ice cores reveal ^{14}C calibration uncertainties during the Younger Dryas, *Nat. Geosci.*, *1*, 263–267.

Nürnberg, D., J. Bijma, and C. Hemleben (1996), Assessing the reliability of magnesium in foraminiferal calcite as a proxy for water mass temperatures, *Geochim. Cosmochim. Acta*, *60*(5), 803–814.

Nürnberg, D., M. Ziegler, C. Karas, R. Tiedemann, and M. Schmidt (2008), Interacting Loop Current variability and Mississippi discharge over the past 400 kyrs, *Earth Planet. Sci. Lett.*, *272*, 278–289, doi:10.1016/j.epsl.2008.04.051.

Obbink, E. A., A. E. Carlson, and G. P. Klinkhammer (2010), Eastern North American freshwater discharge during the Bølling-Allerød warm periods, *Geology*, *38*, 171–174, doi:10.1130/G30389.1.

Otto-Bliesner, B., and E. C. Brady (2010), The sensitivity of the climate response to the magnitude and location of freshwater forcing: Last glacial maximum experiments, *Quat. Sci. Rev.*, *29*, 56–73.

Peck, V. L., I. R. Hall, R. Zahn, H. Elderfield, F. Grosset, S. R. Hemming, and J. D. Scourse (2006), High resolution evidence for linkages between NW European ice sheet instability and Atlantic meridional overturning circulation, *Earth Planet. Sci. Lett.*, *243*, 476–488.

Peltier, W. R., A. Vernal, and C. Hillaire-Marcel (2007), Rapid climate change and Arctic Ocean freshening: Comment and reply, *Geology*, *36*, e178, doi:10.1130/G24971Y.1.

Pilcher, R. S., and R. D. Blumstein (2007), Brine volume and salt dissolution rates in Orca Basin, northeast Gulf of Mexico, *AAPG Bull.*, *91*(6), 823–833.

Poore, R. Z., H. J. Dowsett, S. Verardo, and T. M. Quinn (2003), Millennial- to century-scale variability in Gulf of Mexico Holocene climate records, *Paleoceanography*, *18*(2), 1048, doi:10.1029/2002PA000868.

Rahmstorf, S. (2002), Ocean circulation and climate during the past 120,000 years, *Nature*, *419*, 207–214.

Rashid, H., D. J. W. Piper, and B. P. Flower (2011), The role of Hudson Strait outlet in Younger Dryas sedimentation in the Labrador Sea, in *Abrupt Climate Change: Mechanisms, Patterns, and Impacts*, *Geophys. Monogr. Ser.*, doi: 10.1029/2010GM001011, this volume.

Rasmussen, S. O., et al. (2006), A new Greenland ice core chronology for the last glacial termination, *J. Geophys. Res.*, *111*, D06102, doi:10.1029/2005JD006079.

Reimer, P. J., et al. (2009), Incal09 and Marine09 radiocarbon age calibration curves, 0-50,000 years cal B.P, *Radiocarbon*, *51*, 1111–1150.

Richey, J. N., R. Z. Poore, B. P. Flower, and T. M. Quinn (2007), 1400 yr multiproxy record of climate variability from the northern Gulf of Mexico, *Geology*, *35*(5), 423–426.

Richey, J. N., R. Z. Poore, B. P. Flower, T. M. Quinn, and D. J. Hollander (2009), Regionally coherent Little Ice Age cooling in the Atlantic Warm Pool, *Geophys. Res. Lett.*, *36*, L21703, doi:10.1029/2009GL040445.

Roberts, N. R., A. M. Piotrowski, J. F. McManus, and L. D. Keigwin (2010), Synchronous deglacial overturning and water mass source changes, *Science*, *327*, 75–78, doi:10.1126/science.1178068.

Robinson, L., J. F. Adkins, L. D. Keigwin, J. Southon, D. P. Fernandez, S. L. Wang, and D. Scheirer (2005), Radiocarbon variability in the western North Atlantic during the last deglaciation, *Science*, *310*, 1469–1473, doi:10.1126/science.1114832.

Rodrigues, C. G., and G. Vilks (1994), The impact of glacial lake runoff on the Goldthwait and Champlain Seas: The relationship between glacial Lake Agassiz runoff and the Younger Dryas, *Quat. Sci. Rev.*, *13*, 923–944.

Rühlemann, C., S. Mulitza, P. J. Müller, G. Wefer, and R. Zahn (1999), Warming of the tropical Atlantic Ocean and slowdown of thermohaline circulation during the last deglaciation, *Nature*, *402*, 511–514.

Sarnthein, M., et al. (1995), Variations in Atlantic surface ocean paleoceanography, 50°-80°N: A time-slice record of the last 30,000 years, *Paleoceanography*, *10*(6), 1063–1094.

Sarnthein, M., et al. (2000), Fundamental modes and abrupt changes in North Atlantic circulation and climate over the last 60 ky – Concepts, reconstruction, and numerical modelling, in *The Northern North Atlantic: A Changing Environment*, edited by P. Schäfer et al., pp. 365–410, Springer, New York.

Schmidt, G. A., G. R. Bigg, and E. J. Rohling (1999), Global Seawater Oxygen-18 Database, http://data.giss.nasa.gov/o18data/, Goddard Inst. for Space Stud., New York.

Schmidt, M. W., H. J. Spero, and D. W. Lea (2004), Links between salinity variation in the Caribbean and North Atlantic thermohaline circulation, *Nature*, *428*, 160–163.

Schmidt, M. W., M. J. Vautravers, and H. J. Spero (2006), Rapid subtropical North Atlantic salinity oscillations across Dansgaard-Oeschger cycles, *Nature*, *443*, 561–564.

Schrag, D. P. (1999), Rapid analysis of high-precision Sr/Ca ratios in corals and other marine carbonates, *Paleoceanography*, *14*(2), 97–102.

Schrag, D. P., J. F. Adkins, K. McIntyre, J. L. Alexander, D. A. Hodell, C. D. Charles, and J. F. McManus (2002), The oxygen isotopic composition of seawater during the Last Glacial Maximum, *Quat. Sci. Rev.*, *21*, 331–342.

Severinghaus, J., T. Sowers, E. J. Brook, R. B. Alley, and M. L. Bender (1998), Timing of abrupt climate change at the end of the Younger Dryas interval from thermally fractionated gases in polar ice, *Nature*, *391*, 141–146.

Shokes, R. F., P. K. Trabant, B. J. Presley, and D. F. Reid (1977), Anoxic, hypersaline basin in the northern Gulf of Mexico, *Science*, *196*, 1443–1446.

Siddall, M., E. J. Rohling, A. Almogi-Labin, C. Hemleben, D. Meischner, I. Schmelzer, and D. A. Smeed (2003), Sea-level fluctuations during the last glacial cycle, *Nature*, *423*, 853–858.

Sionneau, T., V. Bout-Roumazeilles, B. P. Flower, A. Bory, N. Tribovillard, C. Kissel, B. Van Vliet-Lanoë, and J. C. Montero Serrano (2010), Provenance of freshwater pulses in the Gulf of Mexico during the last deglaciation, *Quat. Res.*, *75*, 235–245, doi:10.1016/j.yqres.2010.07.002.

Stocker, T. F. (1998), The seesaw effect, *Science*, *282*, 61–62.

Stott, L. D. (2002), Super ENSO and global climate oscillations at millennial time scales, *Science*, *297*, 222–226.

Stuiver, M., P. J. Reimer, E. Bard, J. W. Beck, G. S. Burr, K. A. Hughen, B. Kromer, F. G. McCormac, J. Van Der Plicht, and M. Spurk (1998), INTCAL98 radiocarbon age calibration 24,000-0 cal BP, *Radiocarbon*, *40*(3), 1041–1083.

Tarasov, L., and W. R. Peltier (2006), A calibrated deglacial drainage chronology for the North American continent: Evidence of an Arctic trigger for the Younger Dryas, *Quat. Sci. Rev.*, *25*, 659–688.

Teller, J. T. (1990), Meltwater and precipitation runoff to the North Atlantic, Arctic, and Gulf of Mexico from the Laurentide Ice Sheet and adjacent regions during the Younger Dryas, *Paleoceanography*, *5*(6), 897–905.

Teller, J. T., M. Boyd, Z. Yang, P. S. G. Kor, and A. Mokhtari Fard (2005), Alternative routing of Lake Agassiz overflow during the Younger Dryas: New dates, paleotopography, and a reevaluation, *Quat. Sci. Rev.*, *24*, 1890–1905.

Thornalley, D. J. R., I. N. McCave, and H. Elderfield (2010), Freshwater input and abrupt deglacial climate change in the North Atlantic, *Paleoceanography*, *25*, PA1201, doi:10.1029/2009PA001772.

Thunell, R., E. Tappa, C. Pride, and E. Kincaid (1999), Sea-surface temperature anomalies associated with the 1997–1998 El Niño recorded in the oxygen isotope composition of planktonic foraminifera, *Geology*, *27*, 843–846.

Toucanne, S., S. Zaragosi, J.-F. Bourillet, V. Marieu, M. Cremer, M. Kageyama, B. Van Vliet-Lanoc, F. Eynaud, J.-L. Turon, and P. Gibbard (2010), The first estimation of Fleuve Manche palaeoriver discharge during the last deglaciation: Evidence for Fennoscandian ice sheet meltwater flow in the English Channel ca 20–18 ka ago, *Earth Planet. Sci. Lett.*, *290*, 459–473.

Vellinga, M., and P. L. Wu (2004), Low-latitude freshwater influence on centennial variability of the Atlantic thermohaline circulation, *J. Clim.*, *17*, 4498–4511.

Vidal, L., L. Labeyrie, E. Cortijo, M. Arnold, J. C. Duplessy, E. Michel, S. Becqué, and T. C. E. van Weering (1997), Evidence for changes in the North Atlantic Deep Water linked to meltwater surges during the Heinrich events, *Earth Planet. Sci. Lett.*, *146*, 129–145.

Wagner, A., T. P. Guilderson, N. C. Slowey, and J. E. Cole (2009), Pre-bomb surface water radiocarbon of the Gulf of Mexico and Caribbean as recorded in hermatypic corals, *Radiocarbon*, *51*, 947–954.

Walker, M. J. C., S. Björck, J. J. Lowe, L. C. Cwynar, S. Johnsen, K.-L. Knudsen, B. Wohlfarth, and INTIMATE group (1999), Isotopic events in the GRIP ice core: A stereotype for the late Pleistocene, *Quat. Sci. Rev.*, *18*, 1143–1150.

Wang, Y. J., H. Cheng, R. L. Edwards, Z. S. An, J. Y. Wu, C.-C. Shen, and J. A. Dorale (2001), A high-resolution absolute-dated Late Pleistocene monsoon record from Hulu Cave, China, *Science*, *294*, 2345–2348.

Weldeab, S., R. R. Schneider, and M. Kolling (2006), Deglacial sea surface temperature and salinity increase in the western tropical Atlantic in synchrony with high latitude climate instabilities, *Earth Planet. Sci. Lett.*, *241*, 699–706.

Williams, D. F. (1984), Correlation of Pleistocene marine sediments of the Gulf of Mexico and other basins using oxygen isotope stratigraphy, in *Principles of Pleistocene Stratigraphy Applied to the Gulf of Mexico*, edited by N. Healy-Williams, pp. 65–118, Int. Human Resour. Dev. Corp., Boston, Mass.

Williams, C., B. P. Flower, D. W. Hastings, T. P. Guilderson, K. A. Quinn, and E. A. Goddard (2010), Deglacial abrupt climate change in the Atlantic Warm Pool: A Gulf of Mexico perspective, *Paleoceanography*, *25*, PA4221, doi:10.1029/2010PA001928.

Zahn, R., J. Schönfeld, H.-R. Kudrass, M.-H. Park, H. Erlenkeuser, and P. Grootes (1997), Thermohaline instability in the North Atlantic during meltwater events: Stable isotopes and ice-rafted detritus from core SO75-26KL, Portuguese margin, *Paleoceanography*, *12*(5), 696–710.

B. P. Flower and C. Williams, College of Marine Science, University of South Florida, 140 7th Ave. S., St. Petersburg, FL 33701, USA. (bflower@marine.usf.edu)

D. W. Hastings, Marine Science Discipline, Eckerd College, 4200 54th Ave. S., St. Petersburg, FL 33713, USA.

H. W. Hill, Department of Geological Sciences, University of Michigan, 2534 C. C. Little Bldg., 1100 N. University Ave., Ann Arbor, MI 48109-1005, USA.

Modeling Abrupt Climate Change as the Interaction Between Sea Ice Extent and Mean Ocean Temperature Under Orbital Insolation Forcing

J. A. Rial

*Wave Propagation Laboratory, Department of Geological Sciences, University of North Carolina at Chapel Hill
Chapel Hill, North Carolina, USA*

R. Saha

Department of Physics and Astronomy, University of North Carolina at Chapel Hill, Chapel Hill, North Carolina, USA

The Dansgaard-Oeschger (D-O) temperature fluctuations of the last ice age are a prime example of abrupt climate change in the paleoclimate record, and they provide evidence that rapid climate change (warming as well as cooling) can occur within human time scales, making their study highly relevant to the assessment of present and future natural climate variability. Using conceptual climate models, we show that orbital insolation is likely to play an important role in the timing, amplitude, and duration of the D-O fluctuations. We are able to replicate key features of the D-O time series, including the abrupt transition from the last glacial to the present interglacial, using the orbital insolation as the sole external variable forcing of a simple nonlinear Langevin differential equation. A slightly more complex model based on two nonlinear differential equations for sea ice extent and mean ocean temperature reasonably reproduces the past 100 kyr of climate fluctuations recorded in the Greenland ice cores. The model assumes the existence of a free oscillation with a period of ~1500 years, which is then forced with white Gaussian noise and the astronomical insolation. The model also reproduces the time histories and the well-known phase relationship between shallow and deep ocean proxy data, whereby the SST history follows the Greenland climate record, while the deeper ocean follows the Antarctic's. Comparisons with other low-dimensional models and climate models of intermediate complexity (ECBILT-CLIO and U-Victoria ESCM) show consistent results.

1. INTRODUCTION

This chapter describes attempts to understand the causes of abrupt climate change through data analyses and computer modeling of paleoclimate time series, especially from Greenland's ice cores Greenland Ice Core Project (GRIP), Greenland Ice Sheet Project Two (GISP2), and North Greenland Ice Core Project (NGRIP) [*North Greenland Ice Core Project Members*, 2004; *Johnsen et al.*, 1995;

Abrupt Climate Change: Mechanisms, Patterns, and Impacts
Geophysical Monograph Series 193

10.1029/2010GM001027

Dansgaard et al., 1993; *Blunier and Brook*, 2001]. These extraordinary records show abrupt regional temperature increases of up to 15°K occurring in just a few decades, both during the last glacial period and in the transition to the present interglacial [*Severinghaus and Brook*, 1999; *Clark et al.*, 2002; *Alley et al.*, 2003]. During the last ice age, a repeating pattern of abrupt warming, gradual cooling, and abrupt cooling resulted in trains of pulse-like warming events lasting 1–10 kyr (1 kyr = 1000 years) with a distinctive sawtooth appearance, commonly known as the Dansgaard-Oeschger (D-O) oscillations (Figure 1). Abrupt climate change is believed to be the result of the inherently nonlinear nature of the climate system (e.g., instabilities, feedback, and threshold crossings), but despite decades of vigorous research, the causes of abrupt climate change remain elusive, since neither the physical mechanisms involved nor the nature of the nonlinearities are well understood [*Clark et al.*, 2002; *Alley et al.*, 2003; *Rial et al.*, 2004; *Wang and Mysak*, 2006; *Wunsch*, 2006; *Ditlevsen et al.*, 2007].

In this chapter, we use conceptual models based upon simple stochastic differential equations to show that sea ice extent and mean oceanic temperature are key variables in the development of the D-O and that both can be effectively driven by changes in the orbital insolation to create temperature variations fully consistent with the data of the last ice age. One of these conceptual models [*Saltzman et al.*, 1981] consists of a self-sustained relaxation noisy oscillator that reasonably reproduces the past 100 kyr of D-O climate fluctuations. The model, heretofore referred to as sea ice oscillator (SIO), depends on two internal parameters, the free frequency of the oscillator, and the intensity of the positive feedback which controls the abruptness of temperature change, itself a consequence of the rapid response of sea ice. SIO best reproduces the D-O if the free oscillation period is set around ~1.5 ka, a likely free frequency of the climate, as suggested by the work of *Bond et al.* [1997]. At least two types of abrupt events can be identified in the model results: internal, probably caused by convection/diffusion processes in the ocean, and external, as insolation forces the climate

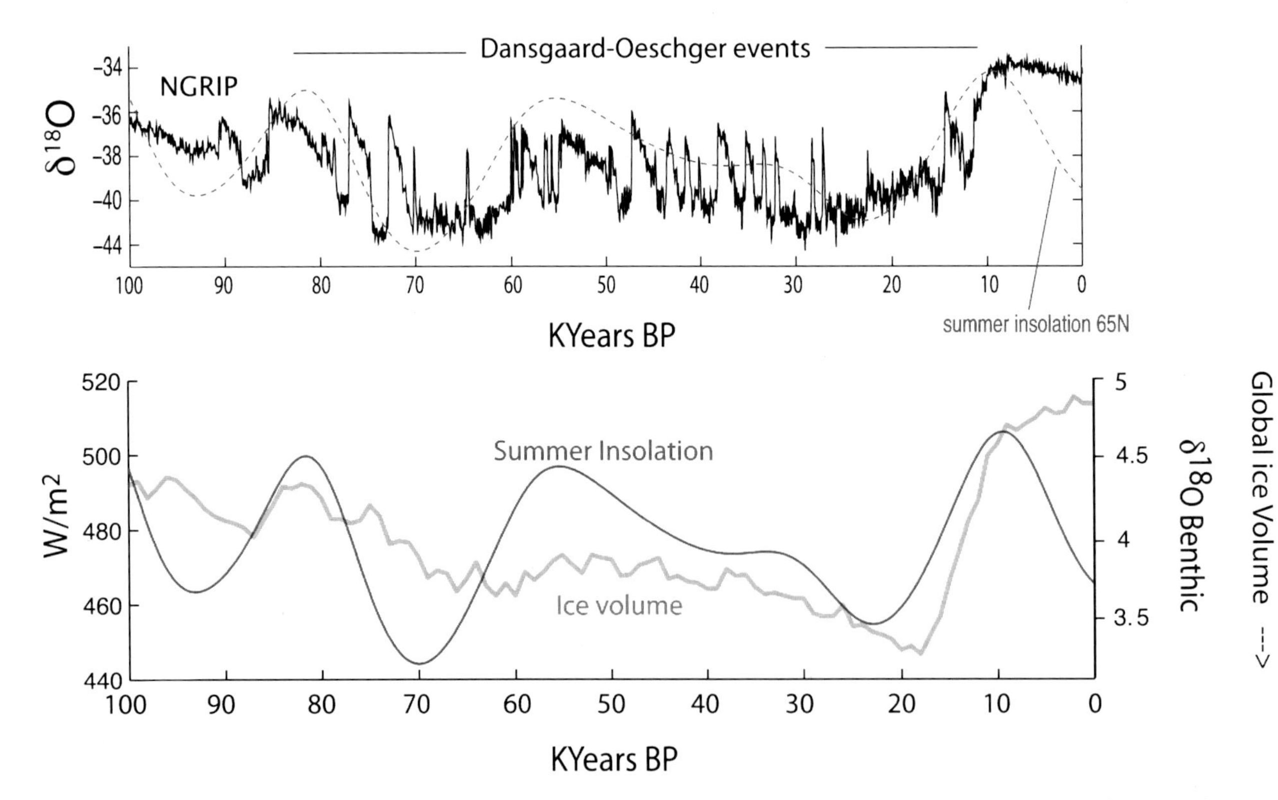

Figure 1. The Dansgaard-Oeschger (D-O) temperature fluctuations (also called D-O oscillations) as recorded in the North Greenland Ice Core Project (NGRIP) ice core [*North Greenland Ice Core Project Members*, 2004]. The orbital induced summer insolation at 65°N [*Laskar et al.*, 2004] appears to influence the long period of the record through amplitude modulation, but it is not clear whether it controls the high frequencies, too. We shall suggest that it does. The ice volume is from the LR04 stack [*Lisiecki and Raymo*, 2005].

system through thresholds. Model validation includes replicating observed features of the D-O time series over the last 100 kyr and corroborating that key results are consistent across models. To serve as a guide into climate behavior beyond the scope of the conceptual models, we use the global atmosphere-ocean-ice coupled climate model of intermediate complexity ECBILT-CLIO [*Opsteegh et al.*, 1998; *Goosse et al.*, 2002]. See Appendix A for a discussion of ECBILT-CLIO'S results.

The chapter is organized as follows: Section 2 describes conditions under which climate models simulate bistable climates that produce D-O-like warming events. Section 3 discusses sea ice extent and mean ocean temperature (the main variables this study focuses on); section 4 describes the Langevin and van der Pol models; section 5 describes the combined use of simplified models, and a discussion of the implications of the modeling results is given in section 6. All models used are described in detail in Appendix A.

2. BISTABLE CLIMATES

Our modeling strategy involves the combined use of climate models to capture some essential physics of the climate response. The most basic of the simplified models we use that describes bistability is a nonlinear stochastic Langevin equation that serves as foundation for the more elaborate, "noisy" SIO. We also use the Earth model of intermediate complexity ECBILT-CLIO, which can sustain D-O-like climate fluctuations [*Schulz et al.*, 2007; *Goosse et al.*, 2002; *Haarsma et al.*, 2001] that can be interpreted as a bistable climate system that transitions at random between two stable states [*Stommel*, 1961; *Lorenz*, 1976]. In Appendix A (Figure A1) we show results of modeling experiments performed for time intervals of 10–20 kyr, allowing ECBILT-CLIO code to produce its built-in responses for a set of boundary conditions that simulates mean climates from preindustrial (PI) to Last Glacial Maximum (LGM).

3. SEA ICE EXTENT AND OCEAN TEMPERATURE DURING ABRUPT CLIMATE CHANGE

Sea ice extent is known to be an important variable in the study of abrupt climate change because the rapid response of floating ice to temperature change and insolation could explain the abruptness of the D-O, and its role as ocean's heat insulator could explain the warming of the covered ocean that eventually drives its retreat [e.g., *Saltzman*, 2002]. In fact, modeling results to be discussed indicate that deep ocean temperature (below ~500 m) interacts with sea ice extent whereby the ocean begins to warm up soon after sea ice extent reaches its maximum and cools off as sea ice retreats. Figure 2 shows an example of this, which is common to any other time interval and boundary condition as long as insolation forcing is kept constant, and D-O-like fluctuations occur. Note that the water column during the time preceding all D-O-like abrupt warming events is warm and salty at depth, while colder and fresher (and lighter) at shallow depths. As shown, ECBILT-CLIO and a much simpler box model [*Colin de Verdiere et al.*, 2006] reproduce the same conditions. One possible interpretation is that under such preconditioning, the water column becomes convectively unstable, with double-diffusion mixing leading to abrupt salt and heat rise, which transforms the stored potential energy of the top-heavy water column into kinetic energy [*Stern*, 1975], while vigorous convection ensues. Note that in Figure 2, the retreat of sea ice begins soon after the deep ocean reaches what will be its maximum temperature. In the actual ocean, this could be the time at which the water column overturns and North Atlantic Deep Water is formed. This apparent causal relationship (more about this in section 4) between warming of deep waters and sea ice retreat appears important and should be testable, for instance, by investigating deep ocean paleotemperature proxies in the Arctic or the change in mean ocean temperature since the beginning of Arctic sea ice retreat in the 1980s. These are also the times, just before an abrupt warming episode, when Heinrich events [*Heinrich*, 1988; *Broecker et al.*, 1992] are likely to occur. The sequence of peak warming and sudden cooling of the deeper ocean near these times may shed some light on the mechanism that accompanies these anomalous detritus deposits [*Bond et al.*, 1993; *Rashid and Boyle*, 2007].

Millennial-scale free oscillations in ocean temperature and salinity due to the interplay of convective instability and turbulent mixing has long been known to theoretically exist [*Bryan*, 1986; *Winton*, 1993; *Haarsma et al.*, 2001; *Sakai and Peltier*, 1997], so the fact that ECBILT-CLIO produces internal, unforced thermohaline oscillations or hesitations (Figure 2) is not necessarily surprising, nor is it surprising that under certain conditions (for instance, if the full ice sheet topography is included), D-O-like oscillations get suppressed [e.g., *Friedrich et al.*, 2010], as discussed in Appendix A. Of course, not all Earth models of intermediate complexity (EMICs) produce identical results. We performed numerous experiments with the University of Victoria's ESCM model that consistently exhibited D-O-like oscillations but with longer periods (~5000 years) and weaker ocean temperature jumps (1K as opposed to 3K in ECBILT-CLIO) under the same boundary conditions. It is not a simple matter to decide which model better captures a semblance of the actual climate. Though we consider this an unresolved issue, the simple box-type models that produce millennial

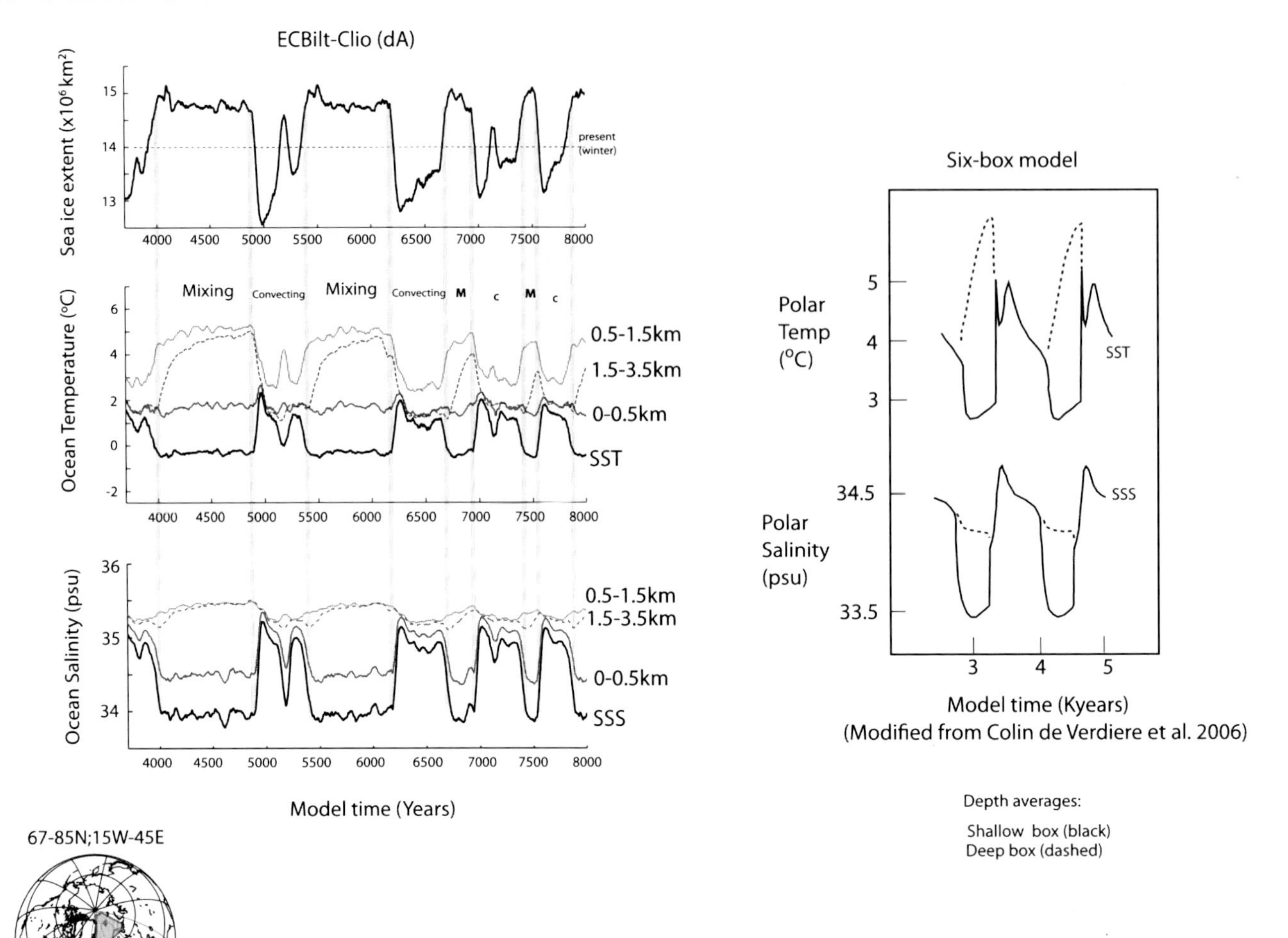

Figure 2. Free oscillations of sea ice extent, ocean temperature, and salinity as computed with ECBILT-CLIO (see Appendix A and Figure A1 for relevant boundary conditions) and the six-box model of *Colin de Verdiere et al.* [2006]. In ECBILT-CLIO, the boundary condition (dA) is intermediate between preindustrial (PI) and last glacial maximum (LGM). Temperature and salinity are averaged over the Arctic area shown. Abrupt changes in all the variables are apparent. Ocean temperature and salinity time histories in the shallow and deep ocean are fully consistent across the two models. SST is sea surface temperature, SSS sea surface salinity. Figure 2 (left) is modified from the work of *Colin de Verdiere et al.* [2006]. Copyright 2006 American Meteorological Society.

oscillations in ocean temperature, salinity, and density requiring no prescribed freshwater forcing or other time-varying forcing [*Colin de Verdiere et al.*, 2006; *Loving and Vallis*, 2005] show millennial oscillations fully consistent with ECBILT-CLIO. As shall be discussed shortly, SIO also produces D-O-like oscillations in sea ice extent and mean ocean temperature qualitatively identical to those produced by EBILT-CLIO. Also, internal oscillations of a modeled Atlantic meridional overturning circulation (AMOC) nearly identical to those computed with ECBILT-CLIO are found in simulations using ocean-alone and coupled models with varied boundary conditions [e.g., *Winton*, 1993; *Winton and Sarachik*, 1993].

4. THE LANGEVIN AND VAN DER POL MODELS

Figure 3 shows two realizations of the nonlinear stochastic Langevin model. The top two plots of Figure 3a show two independent realizations of the noisy model for the same parameter values, followed by the two time series that result when external forcing (summer insolation at 65°N) is included on the right-hand side of the stochastic equation (see

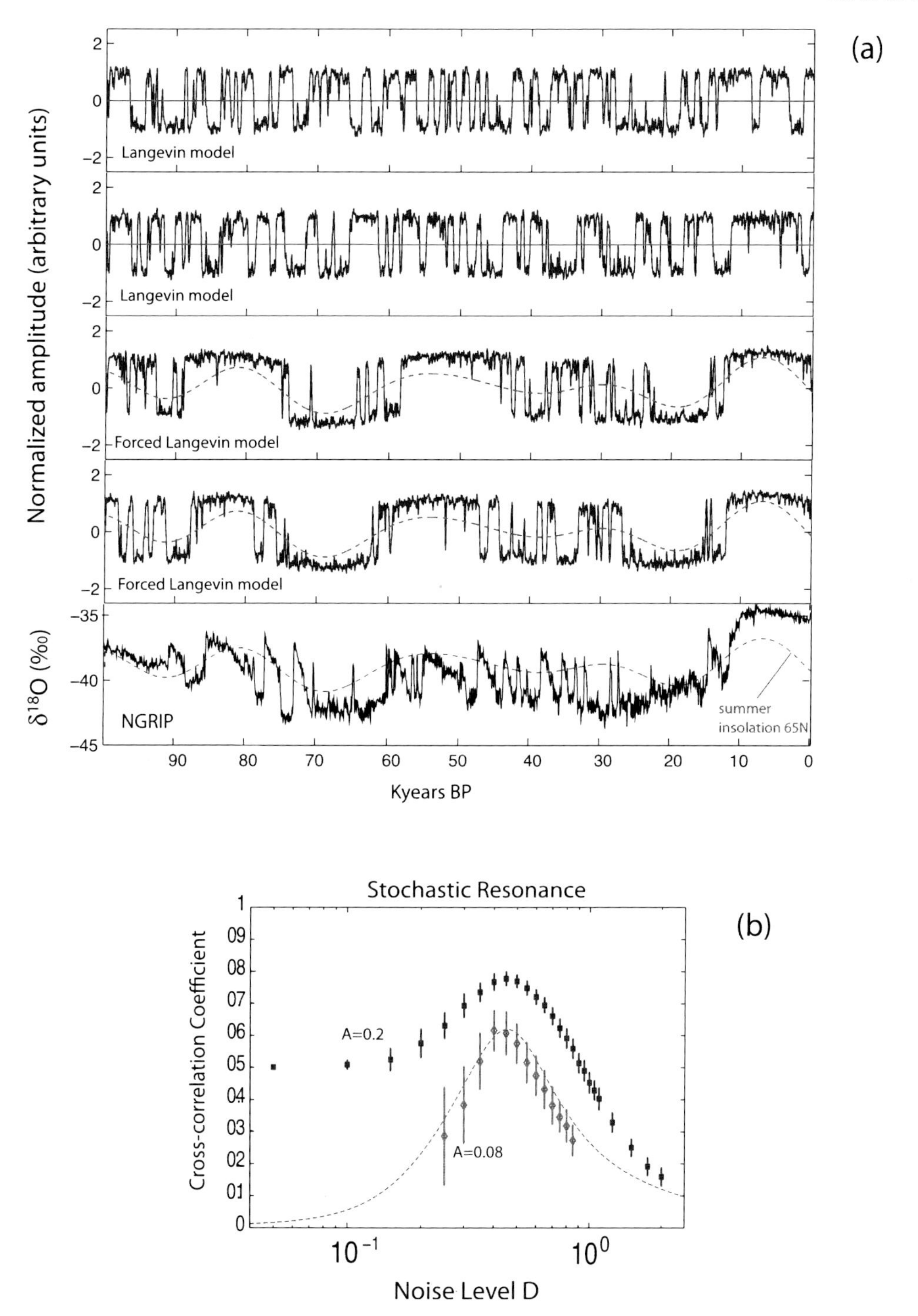

Figure 3. (a) Bistable climate simulations using a Langevin differential equation (see Appendix A) forced by the insolation reproduce some features of the D-O oscillation exhibited by the NGRIP ice core. (b) Correlation coefficients [*Collins et al.*, 1996] reveal the optimal noise level that produces the best shape matching between the insolation forcing and ensembles of realizations of the forced Langevin equation. The diagram compares 100 independent realizations of the forced Langevin simulation to the insolation function for different levels of noise (D) and several values of the orbital forcing level A. Each different symbol shows the mean and vertical bars the standard deviation for constant values of the insolation level A. The amplification of the weak signal ($A = 0.08$) is clearly strongest within a narrow range of noise level. The dashed line is the theoretical shape of the correlation coefficient in the presence of aperiodic stochastic resonance [*Collins et al.*, 1996].

Appendix A). Comparison with the NGRIP time series exemplifies the organizing role of the insolation, as it forces bistable random fluctuations into a time series that bears some similarity with NGRIP, which then loses its apparent random look. Besides frequency modulation, one other important fact this experiment reveals is a form of "aperiodic"

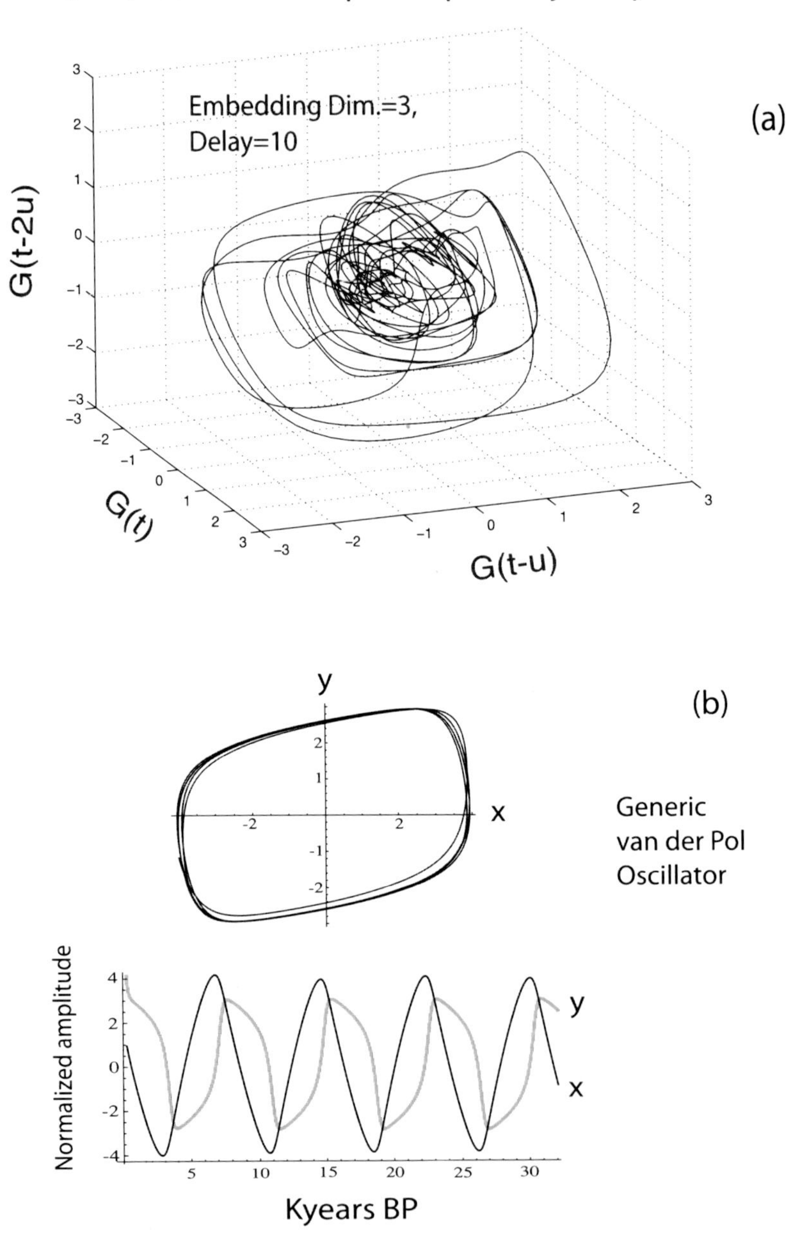

Figure 4. (a) Using Taken's delay method [*Williams*, 1997], the phase-space trajectory of the NGRIP record (band-pass-filtered corner frequencies between 16 and 1 kyr^{-1}) is reconstructed. The resulting embedding dimension is 3 (estimated by the method of near false neighbors), and the delay *u* is 10 sample units or 500 years, determined by mutual information autocorrelation [*Sprott*, 2003]. (b) Synthetic phase-space trajectory computed with a generic, weakly nonlinear van der Pol oscillator. The two-dimensional (2-D) phase-space trajectory of the van der Pol oscillator is very similar to a 2-D projection of Figure 4a. The van der Pol trajectory is a limit cycle that maintains its shape under perturbation. It will, however, be changed from a single loop to multiple loops as a consequence of external forcing (e.g., frequency modulation).

stochastic resonance (SR) [*Collins et al.*, 1996] that could play a role in amplifying the climate system's response to weak external forcing. This is a different mechanism of SR than what has been usually invoked to explain the D-O [*Ganopolski and Rahmstorf*, 2002; *Alley et al.*, 2001]. The latter is the amplification of a weak underlying internal periodic signal by a specific noise level, while in the former, the presence of a particular noise level in the system enhances the effect of the external forcing (insolation), which contributes to transform bistable noise into a time series that mimics the data. As illustrated in Figure 3b, a specific level of noise causes the similarity between the resulting signal and the insolation to increase, even for very weak astronomical forcing.

The implication of Figure 3, however, is not that the ice core records are just the result of astronomically modulated climate noise. Though the similarity between model and data in Figure 3a would appear to suggest thus, there are

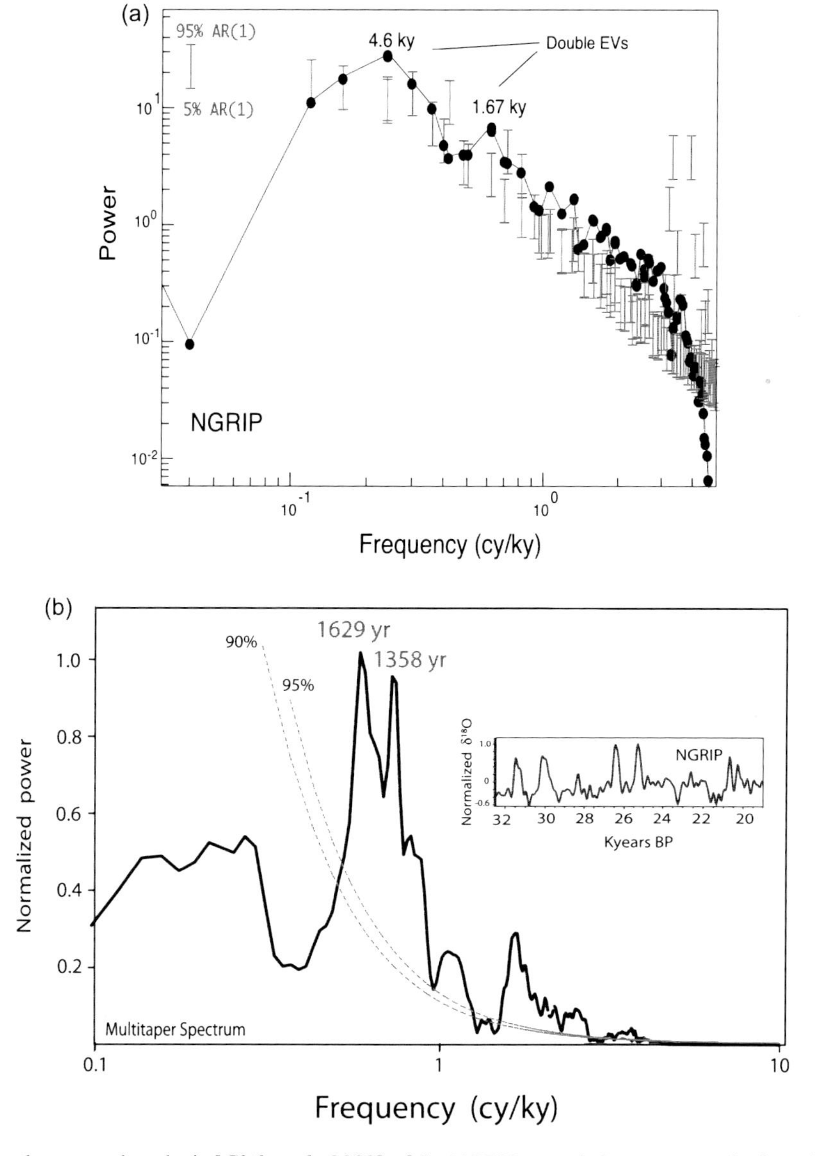

Figure 5. (a) Singular spectral analysis [*Ghil et al.*, 2002] of the NGRIP record. Summary results for other ice cores are given in Table 1. The original time series was high-pass filtered (corner at 0.000067 c yr^{-1}) to eliminate periods at and longer than the insolation band. Two periods in the millennial D-O range (4.6 and1.67 ka) display high power against the red noise null hypothesis at the 95% confidence level. (b) Power spectra (multitaper) of an interval of NGRIP where the influence of insolation variability should be minimal. The 90% and 95% significance levels are marked.

important differences that require further exploration, one of them being the fact that the ice core records (e.g., NGRIP, GRIP, and GISP2) show self-similarity [*Rial and Yang*, 2007], which suggests the presence of internal periodic oscillations. For instance, Figure 4 shows a reconstruction of the three-dimensional phase-space trajectory of NGRIP using Taken's delay method [*Williams*, 1997]. Comparing its phase-space trajectory with the limit cycle (Figure 4b) of a periodic van der Pol oscillator (which is basically SIO, as explained in Appendix A) brings about similarities between the climate processes and the model that we will fully exploit in what follows. As to which periodicities exist in the data, there is some evidence in the singular-spectrum analysis (SSA) of the high-pass-filtered series of several ice core records (Figure 5 and Table 1). The presence of statistically significant oscillatory frequencies corroborates the idea that periodicities are important part of the D-O climate fluctuations. SSA is a nonparametric method [*Ghil et al.*, 2002] that uses data-adaptive basis functions. Data-adaptive eigenmodes can capture the basic periodicity of an irregular waveform without the many overtones needed in a regular Fourier analysis. Accordingly, in the millennial band, two pairs of oscillatory eigenvalues (each pair has the same frequency) with periods ~4.6 and ~1.67 kyr lay outside the 95% percentile, so that the null hypothesis that these eigenvalues are red noise can be rejected with this confidence. However, as we shall see shortly, these particular periods may result from the distortion, for example, through frequency modulation, of the original period of free oscillation if it exists. It has been speculated that a possible source of oscillations in the D-O is the enigmatic and pervasive 1500 ± 500 year climate cycle discovered by *Bond et al.* [1997], though its origin is not known [*Ganopolsky and Rahmstorf*, 2002; *Alley et al.*, 2003]. There is a short segment of the ice core record (32–19 ka) (ka = kiloyear ago), where insolation is nearest its mean value and thus may have its least effect on the climate's intrinsic temperature variability, which means that the underlying ~1.5 kyr oscillation may be free from distortion. Figure 5b shows that the multitaper power spectrum of the NGRIP record in that time interval exhibits peaks at periods of ~1629 and ~1358 years, just within Bond's range.

Table 1. Singular Spectrum Analyses of the Ice Core Data Time Series[a]

Record Name	SSA Window (0.05 kyr sample^{-1})	95% Significant EV Pairs (cycle kyr^{-1})	Period (kyr)
NGRIP	150	0.24/0.60	4.6/1.67
	200	0.23/0.61	4.34/1.64
	250	0.225/0.625	4.44/1.60
GRIPss09	150	0.22–0.24/0.66	4.54–4.16/1.51
	200	0.24/0.68	4.16/1.47
	250	0.24/0.66–0.68	4.16/1.51–1.47
GISP2	150	0.2–0.22/0.68	5–4.54/1.47
	200	0.2/0.68	5.0/1.47
	250	0.2/0.68	5.0/1.47

[a]Monte Carlo singular spectral analysis (SSA) [*Ghil et al.*, 2002] of ice core time series reveals that around ~4.5 and ~1.5 ka, there are significant eigenvalue pairs that indicate nonlinear anharmonic signals for which the null hypothesis (red noise) can be rejected at the 95% level of confidence. The results are robust to changes in window length (should be less than $N/5$, where N is the total number of points). Sampling rate is 50 years and $N = 2000$. Timing errors and different age models account for the main differences in the detected periods among ice cores. Figure 5a shows an example of SSA for NGRIP.

The van der Pol equation-based SIO becomes helpful here, since it is one step higher in complexity from the Langevin model of Figure 3. SIO is a set of two first-order differential equations describing self-sustaining nonlinear oscillations of sea ice extent and mean ocean temperature. From actual climatic parameters, *Saltzman and Moritz* [1980] estimated that the oscillator's natural period would be in the range 1000–3000. Thus, it appears reasonable to prescribe natural periods for the mean ocean temperature in the range 1500 ± 500 years in SIO. When this is done, synthetic time series are obtained that resemble the NGRIP record, as shown in Figures 6 and 7. Here we used noise level $D = 0.45$; insolation level $A = 0.5$; $a = b = 1$; and $c = 0$ (see Appendix A). The best fits with the data are obtained for periods $T = 2\pi/\Omega = 1400$–1600 years, and insolation amplitude and noise levels commensurate with that used in the Langevin model.

That SIO closely mimics the data leads to the idea that few rules may exist governing the long-term evolution of climate and that those could be understood from first principles. If this is true, the D-O just reflects the (nonlinear) interactions among sea ice, ocean temperature, and insolation during the last ice age. It is therefore not unreasonable to state that when it comes to understanding the essential physics, large and complex codes like ECBILT-CLIO may be less useful than much simpler ones.

5. RELATION BETWEEN SURFACE AND DEEP OCEAN TEMPERATURE HISTORIES

SIO produces time series for mean ocean temperature and sea ice extent consistent with well-known properties of proxy data, for instance, from sediment core MD95-2042 collected off Portugal [*Cayre et al.*, 1999]. Figure 8 shows a planktonic $\delta^{18}O$ proxy for SST with the characteristic "square wave" or saw-toothed shape of the Greenland proxies and the benthic $\delta^{18}O$ record with the also characteristic

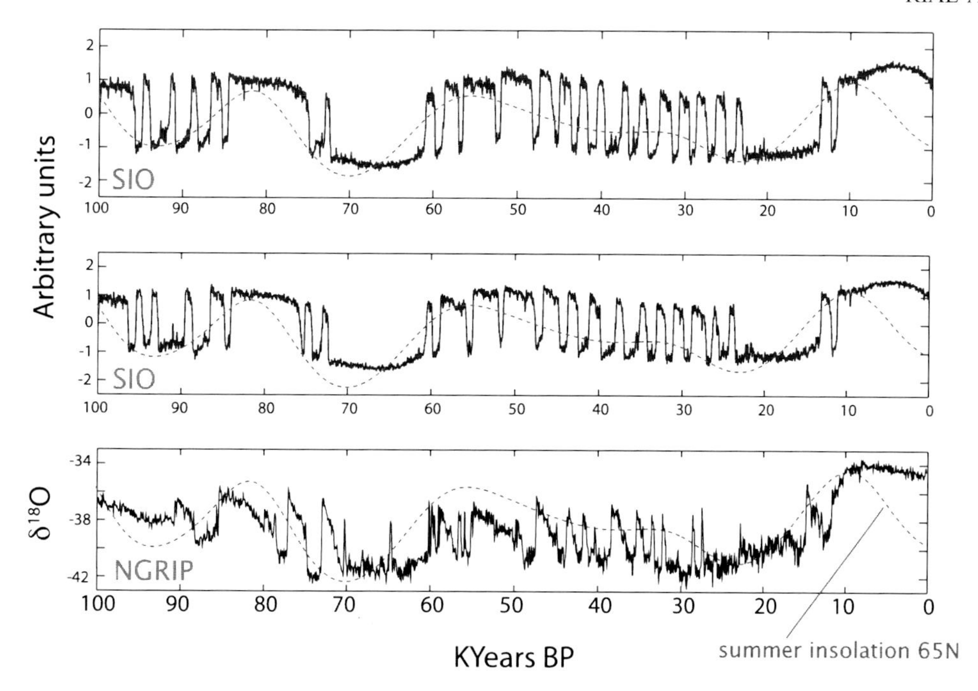

Figure 6. Two of many possible independent realizations that can be obtained with the van der Pol oscillator sea ice oscillator (SIO) (equations (A2a) and (A2b) in Appendix A) reproduce the general structure of the NGRIP data when using a fundamental period of 1500 years in equation (A2a). Frequency modulation distorts this period and creates sidebands with power ~10 to ~1 kyr that control the duration of the warming episodes. Strong-amplitude modulation creates the two cold intervals around 65 and 18 ka, where insolation is low. The time series are robust against 10% changes in the controlling parameters (see also Figure 7). The insolation curve is the summer monthly mean at 65°N [*Laskar et al.*, 2004].

"triangular" waveform that is typical of the Antarctic surface temperature [*Shackleton et al.*, 2000]. Over long time scales, the benthic δ^{18}O proxy is known to be equally sensitive to both global ice volume and deep ocean temperature [*Shackleton*, 2000], but here it is reasonable to infer that since global ice volume does not change rapidly at the millennial scale of the D-O, the benthic record is likely to be a relatively accurate proxy of deep ocean temperature variation (J. Sveringhaus, personal communication, 2010). It is important to emphasize that both the simple SIO model and ECBILT-CLIO reproduce the phase relationship and "waveforms" observed in core MD95-2042 as shown in Figures 8 and 9. It is important because the reason(s) for the peculiar contrasting shape of waveforms in this and in other records of shallow and deep ocean temperatures [*Saikku et al.*, 2009] are not understood and remain a mystery. However, from the data and the model, it is easy to show that the key relationship between the two variables is readily demonstrable: the time integral of the planktonic (SST or air surface temperature proxy) time history is proportional to the benthic (deep oceanic temperature proxy) time history. More physically relevant, deep ocean temperature is $\pi/2$ shifted with respect to sea ice extent variation. This is the same relationship observed between methane-synchronized temperature proxies from Greenland and Antarctica [*EPICA Comm. Members*, 2006; *Blunier and Brook*, 2001]. *Saikku et al.* [2009] confirm that the deep ocean undergoes contemporaneous temperature changes with those in proxies of air surface temperature in Antarctica, while SST and surface air temperature follow the changes seen in Greenland proxies. But these intriguing relations between δ^{18}O planktonic and benthic records are shown in SIO to be the consequence of the coupling between sea ice extent and oceanic temperature [*Saltzman and Moritz*, 1980] that can be described as follows: Regionally, mean ocean temperature begins to increase as sea ice reaches its maximum extent, eventually (several hundred years later) reaching a peak at which sea ice begins to retreat. As sea ice rapidly retreats, the ocean cools until it becomes cold enough that sea ice begins to grow back. This nonlinear dance between rapid sea ice response and slow buildup of temperature of the ocean underneath is a relaxation oscillation described by the van der Pol equation (see Appendix A), and the two variables play a similar role to that of displacement and velocity in a simple linear pendulum oscillation, only that here the oscillation is a self-sustained, nonlinear one.

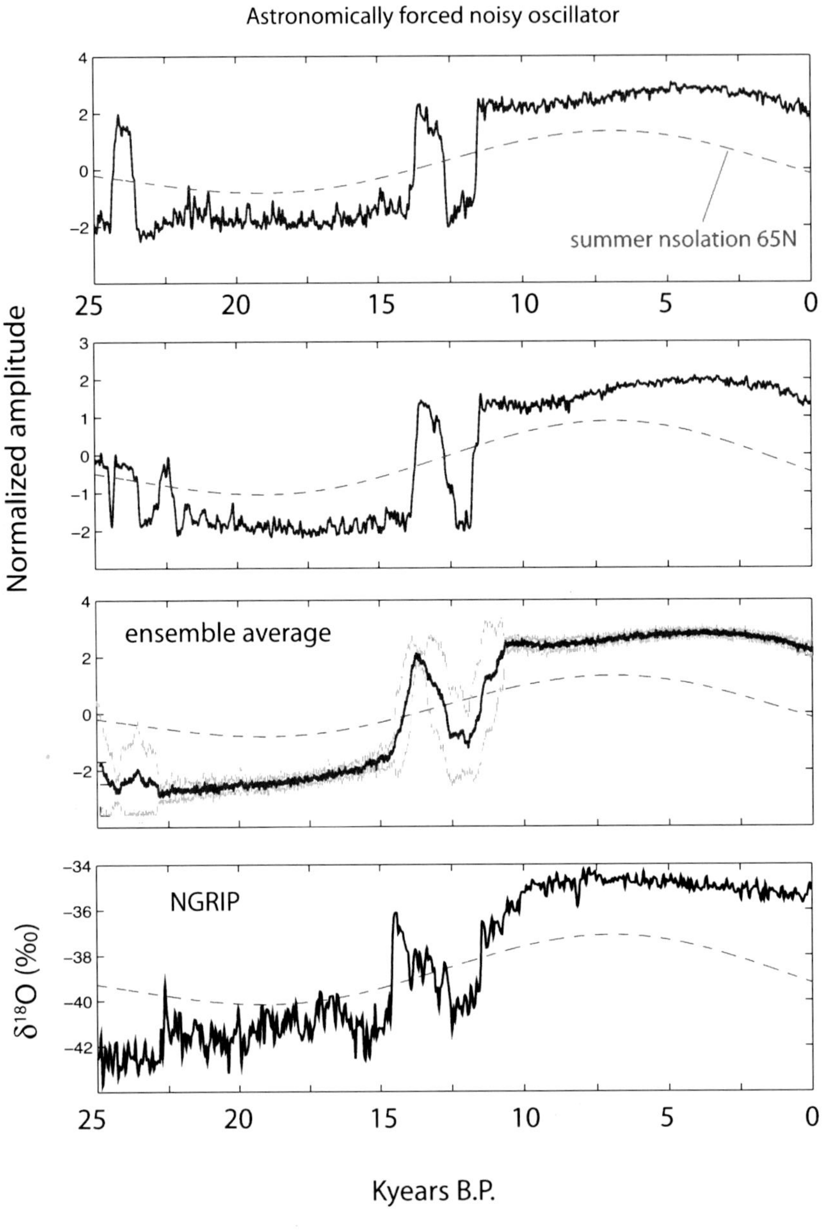

Figure 7. Two independent realizations of the last 25,000 years and an ensemble average (third plot from top) of 10 realizations (one standard deviation intervals are shown in lighter traces) modeled with SIO. The insolation curve is the summer monthly mean at 65°N from the work of *Laskar et al.* [2004]. The effect of melting of the ice sheets is added to the model assuming it responds linearly by warming at the same rate ice volume decreases. After ice volume reaches its Holocene value, this forcing is kept constant for all subsequent times.

However, despite our enthusiasm for the simplified models, all questions concerning the mechanisms by which heat is transferred in the ocean are still difficult to answer. We can speculate that the state of the model ocean just before each abrupt warming is a relatively weak AMOC and a water column with cold fresh layers overlying warm salty (albeit denser) layers (Figure 2). This ocean column preconditioning process (labeled M in Figure 2) begins as sea ice reaches its maximum extent, and the deep ocean rapidly first and then slowly becomes warmer and saltier from the heat and salt advected north from lower latitudes and the tropics. This slowly builds up the water column instability [*Stern*, 1975] until, in time (centuries or millennia), a sudden eruption of raising plumes of heat and salt ensues, the AMOC abruptly

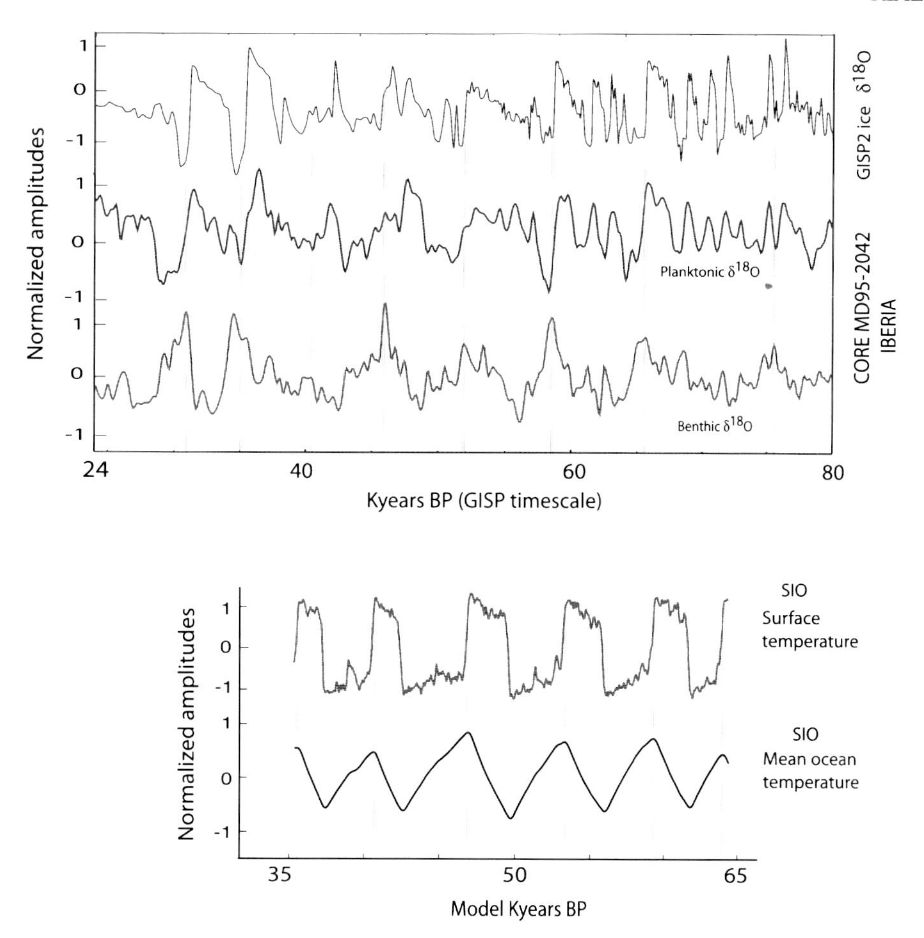

Figure 8. Time series of planktonic and benthic $\delta^{18}O$ measurements from a sediment core off Portugal (both proxies are from the same core and thus relative timing between both series is exact) compared to the simulated surface temperature and mean ocean temperature obtained with SIO. SIO simulates both records with some fidelity and shows that there is indeed a $\pi/2$ phase shift between the two time series. As shown in Figure 9, ECBILT-CLIO also reproduces this relationship. It is easy to demonstrate that the benthic record (a likely proxy for deep ocean temperature or mean ocean temperature) closely correlates with the normalized time integral of the planktonic record (a proxy for SST). Both proxy records are displayed using the Greenland Ice Sheet Project Two age model shown on top. It should be noted that the phase relationship between the planktonic and benthic records is a well-known but unexplained feature of oceanic proxies [*Shackleton*, 2000; *Saikku et al.*, 2009].

strengthens, and large volumes of warm water transfer heat polarward, while the deep ocean cools off and freshens. This is the abrupt warming phase of the relaxation oscillation. ECBILT-CLIO in Figure 2 and SIO and ECBILT-CLIO in Figure 9 show that ocean cooling at depth begins gradually after the sea ice cover retreats accelerated by the positive sea ice-albedo feedback. Vigorous convection follows that keeps the ocean's gradients of temperature salinity and density steep, while the SST slowly cools in the Arctic atmosphere, allowing sea ice to start growing back and eventually reach its maximum extent (this is the abrupt cooling phase of the relaxation oscillation). As the ocean surface is covered up by sea ice, the cycle begins anew.

Of course, models tell but a very incomplete picture of how the ocean system may function. Worse yet, the picture just described may have little to do with reality [e.g., *Wunsch and Ferrari*, 2004]. Nevertheless, it is generally believed that the oceanic thermohaline circulation is driven by high-latitude surface cooling, as the intense heat loss to the atmosphere causes ocean surface water to reach its highest densities at some places in the high latitudes. This cold dense water then sinks into the deep ocean, spreads out along the ocean bottom, and finally returns to the surface by uniform upwelling through the interior ocean. Were this the only driving force, the ocean basins would eventually fill with cold dense water, the sinking of surface water would stop, and the

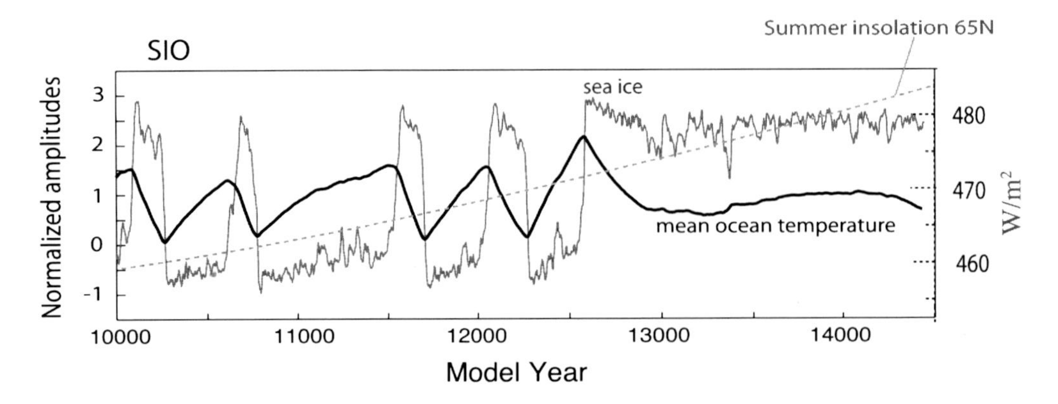

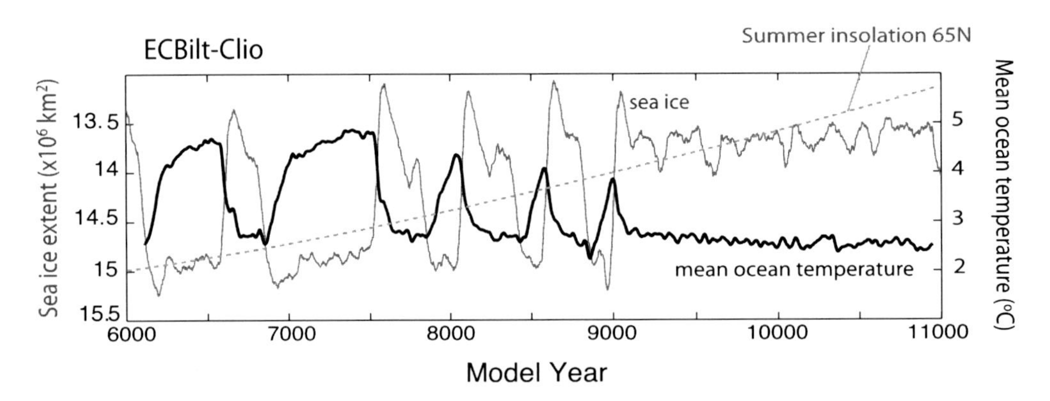

Figure 9. Both SIO and ECBILT-CLIO produce consistent results for astronomically forced sea ice extent and mean ocean temperature. The results are also fully consistent with the Langevin results of Figure 3a in spite of the great difference among models. Note that mean ocean temperature increases as sea ice extent is at its maximum (around 15 million km^2) and forces sea ice to retreat as it reaches a threshold. Both models show ocean cooling following sea ice retreat. Though both EBILT-CLIO and SIO could certainly be wrong, the similarities between the two modeling results and the fact that the data waveforms (Figure 8) are square and triangular, just as in the models, and that the relative phases of sea ice extent and mean ocean temperatures are similar to observations of SST and deep ocean temperature argues for the models to have captured some of the actual physics of the process. If so, it could be surmised that the insolation forcing must be a major influence in creating the timing and duration of the D-O through amplitude and frequency modulation of sea ice feedback and a ~1.5 kyr carrier, as Figures 3, 6, and 7 also suggest. In ECBILT-CLIO, the periodicity is controlled by the boundary conditions (provided no changes in default parameters are made), and in SIO, the periodicity is prescribed to fit that of ECBILT-CLIO.

circulation would come to a halt. Some authors [e.g., *Wunsch*, 2002; *Rahmstorf*, 2006] point out that vertical turbulent mixing plays an essential role in transporting heat to and from the deep ocean, for example, by lowering deep water density and keeping surface water in the high latitudes denser than the water in the deep ocean. If the vertical eddy diffusivity in the ocean is chosen to be large enough (~10^{-4} m^2 s^{-1}, say), models will be very effective in mixing the ocean and transporting heat vertically, but the actual mechanisms through which this happens and realistic values of the diffusivity are incompletely known [*Huybers and Wunsch*, 2010].

6. DISCUSSION

Figure 9 helps to identify at least two types of abrupt events that consistently appear in SIO and ECBILT-CLIO: internal, likely caused by the convection/diffusion processes in the ocean just described, and external, driven by the actual insolation, as it forces the climate system through thresholds. The time series in Figure 9 show both types. The figure illustrates that even the relatively slow increase in summer insolation (~0.0065 W yr^{-1}) can drastically alter the climate system as it forces it to cross a threshold, the timing of which varies with mean climate state (sooner for the warmer climate states). It can be shown that in both models, the system regains bistability when the insolation drops through the same threshold value.

If SIO captures some real aspect of the relationship between sea ice extent and the temperature of the ocean it covers, it would be that the two variables synchronously oscillate as a single dynamic system, consistent with the mechanism proposed by *Saltzman et al.* [1981] and described by equations (A2a) and (A2b) in Appendix A. As shown there, mean ocean temperature (y) is linearly related to the time integral of sea ice extent (x), while x itself is the solution to a nonlinear differential equation. Following *Saltzman et al.* [1981], $x(t)$ is the motion of the sea ice edge relative to an "equilibrium" value and $y(t)$, whose time derivative is proportional to $x(t)$, is the change in mean ocean temperature around that which corresponds to the same equilibrium.

A goal of this investigation is to gain some qualitative understanding of how different parts of the climate system interact with each other to create a burst of abrupt climate change. We have shown that the use of low-order climate models helps greatly, especially to interpret the results of the much more elaborate ECBILT-CLIO. In no experiment or set of experiments are the results inconsistent across models, and in no case was there evidence of propagating errors influencing ECBILT-CLIO'S stability for the crucial pair of variables sea ice extent and mean ocean temperature. In this sense, we usually worked in reverse to the usual modality, using the simple models to check the reliability of the EMICs. The ability of the simple models to highlight the usually counterintuitive nonlinear interactions that would escape even a detailed analysis of any EMIC output is their most important virtue. Physically, sea ice appears as a logical choice for the abruptness of climate change because of its effectiveness as ocean heat insulator and rapid response to temperature changes. These are brought about by the varying orbital insolation and ocean water temperature, whose behavior we have shown is tied (at least in the model) to sea ice by virtue of being the two components of the system's free (nonlinear) oscillation.

APPENDIX A: MODELS

A1. LANGEVIN BISTABLE EQUATION

Our simplest model is a stochastic differential equation of the Langevin type [e.g., *Gitterman*, 2005; *Imkeller*, 2001; *Timmermann and Lohmann*, 2000] given by

$$\frac{\mathrm{d}x}{\mathrm{d}t} = f(x,t) + \xi(t), \qquad \text{(A1)}$$

where x represents the climate variable of interest, $f(x,t) = V'(x) + M(t)$, and $V'(x)$ is the derivative of a two-well potential $V(x) = -(cx + ax^2 - bx^4)$. $M(t)$ is the (zero mean) external (orbital) forcing. The white noise $x(t)$ is Gaussian and such that $\langle\xi(t)\xi(s)\rangle = D\delta(t - s)$, D being the noise level. The cubic nonlinearity in $V'(x)$ is a damping (negative feedback) term that originates the pitchfork bifurcation that sustains the bistable system [*Gitterman*, 2005; *Saltzman*, 2002]. Here we assume these states to be the stadial/interstadial states. Parameters a, b, and c are positive constants that allow different mean climate states to be simulated by altering the geometry of the two-well potential (unless otherwise stated, we use $c = 0$ and $a = b{\sim}1$). We choose $M(t)$ to be the summer insolation function at 65°N [*Laskar et al.*, 2004] and integrate equation (A1) numerically using Euler's method [*Moss and McClintock*, 1989]. Computationally, it is convenient to write $M(t) = Am(t)$ for the forcing function, where A is a parameter that denotes the level of insolation fluctuation and $m(t)$ is the normalized $M(t)$. Equation (A1) is the Newtonian equivalent of a small ball jumping randomly between two contiguous symmetric wells separated by a barrier [*Gitterman*, 2005]. The height of the barrier is $a^2/4b$ or 0.25. This implies that in the absence of noise, the amplitude of the external forcing A less than the threshold value 0.25 will not induce jumps in the system. The noise (which we define as internal forcing) simulates stochastic forcing by weather and plays the important role of exciting jumps across the barrier in either direction, which hence occur at random times only if $A = 0$. If $A \neq 0$ but small, and below threshold, jumps can occur that are modulated by $M(t)$. With one well signifying the cold stadial state and the other the interstadial, the model is a heuristic yet useful representation of a bistable climate system [*Imkeller*, 2001; *Timmermann and Lohmann*, 2000].

A2. NOISY VAN DER POL OSCILLATOR

Simple energy considerations show that in glacial times, the interaction among ocean temperature, sea ice distribution, and sea-atmosphere greenhouse gas fluxes can support nonlinear, self-sustained, relaxation-type thermal oscillations in which the sea ice edge advances and retreats with a millennial-scale period [*Saltzman et al.*, 1981]. *Saltzman and Moritz* [1980] estimated the fundamental period of such thermal oscillator as lying between 300 and 3000 years. From the previous Langevin model, it is easy to build a "noisy" version of Saltzman's oscillator: We add a new coupled variable $y(t)$ to model (A1) to represent zero-mean mean ocean temperature variations, while the variable $x(t)$ now represents zero-mean variation in sea ice extent, and write

$$\frac{\mathrm{d}x}{\mathrm{d}t} = -y + c + ax - bx^3 + \xi(t) \qquad \text{(A2a)}$$

$$\frac{dy}{dt} = \Omega^2 x - Am(t). \qquad \text{(A2b)}$$

The new equation (A2b) adds a new variable to the forced Langevin equation (A1). Without the noise term and with $A = 0$, the system (A2a) and (A2b) transforms into a canonic van der Pol equation on the unknown $\theta(t) = (3b/a)^{1/2}x(t)$, giving

$$\frac{d^2\theta}{dt^2} - a(1-\theta^2)\frac{d\theta}{dt} + \Omega^2\theta = 0. \qquad \text{(A3)}$$

Equations (A2a) and (A2b) include positive feedback due to sea ice albedo and greenhouse gases (lumped into parameter a) and negative feedback that limit sea ice extent (cubic nonlinearity proportional to parameter b). Ω is the natural free circular frequency of the undamped, linearized system. It is important to state that the stochastic component in equation (A2a) makes the oscillator "almost intransitive" [*Lorenz*, 1976], which thus exhibits the bistable behavior of equation (A1) plus the free oscillation. The $Am(t)$ term in the right-hand side of equation (A2b) produces frequency modulation of the periodic relaxation, creating much of the characteristic pattern observed in Greenland's ice cores. Equations (A2) or (A3) represent self-sustained, relaxation oscillations such as found in many nonlinear systems in physics, engineering, biology [*Saltzman*, 2002; *Gitterman*, 2005], and oceanic circulation [e.g., *Marchal et al.*, 2007]. Other nonlinear oscillators with noncircular orbits in phase space can certainly produce similar behavior; however, the van der Pol oscillator is probably the simplest one to do so and thus is favored by many scientists interested in modeling relaxation oscillations.

In practice, two free nondimensional parameters A and a are adjusted to best fit the data, and b is used to control the amplitude range of θ. The period $T = 2\pi/\Omega$ trades off with A, so that as T increases, A must decrease in order to maintain a satisfactory fit to the data. Noise level controls the relative effect of the insolation forcing through SR.

A3. ECBILT-CLIO

We use the global, atmosphere-ocean-ice coupled climate model of intermediate complexity, EBILT-CLIO (version 3). The atmospheric component, ECBILT, is a spectral T21 global quasi-geostrophic model with simple parameterization for the diabatic heating due to radiative fluxes, the release of latent heat, and the exchange of sensible heat with the surface [*Opsteegh et al.*, 1998]. The model contains a full hydrological cycle, which is closed over land by a bucket model for soil moisture. Synoptic variability associated with weather patterns is explicitly computed. The ocean component CLIO is a free-surface ocean general circulation model coupled to a thermodynamic-dynamic sea ice model. The horizontal resolution is 3° by 3°, and there are 20 unevenly spaced vertical levels in the ocean. The coupled model includes realistic topography and bathymetry [*Campin and Goosse*, 1999]. CLIO is a primitive equation and includes a relatively sophisticated parameterization of vertical mixing [*Goosse and Fichefet*, 1999]. A three-layer sea ice model, which takes into account sensible and latent heat storage in the snow-ice system, simulates the changes of snow and ice thickness in response to surface and bottom heat flux. There are three main boundary conditions in ECBILT-CLIO that can be easily kept fixed (continental ice extent (described through albedo extent); orbital parameters (eccentricity, tilt, and precession); and greenhouse gas concentrations), and since each can be chosen to represent either PI or LGM conditions, 2^3 independent combinations can be run with these simple choices of mean climate state (Figure A1). The results for fixed insolation forcing show bistable behavior for "intermediate" mean climates. Though the intermediate cases are only crudely representative of the conditions existing during the last ice age, in terms of global temperature, the results in Figure A1 show temperature variability consistent with the range of Arctic climates estimated from the ice core records. The ECBILT-CLIO results also show that both LGM and PI boundary conditions produce stable climates without D-O-like oscillations, in qualitative agreement with ice core data (Figure 1).

An alert reviewer pointed out that *Friedrich et al.* [2010] made a series of modeling experiments with ECBILT and concluded that the D-O-like oscillations are likely to be an artifact of the code. They show that the oscillations disappear (actually they become damped but are still visible) if the "Hudson Strait outflow" is shut off. This is not very surprising, however, as in our experience, D-O-like oscillations in ECBILT-CLIO fade if the full topography of the ice sheets is used as a boundary condition, or the water column's temperature gradient is very small or zero (D-O-like instability in the code seems to require fresh and cold shallow water overlying salty and warm deeper ocean), or whenever sea ice retreats, allowing the deep ocean to cool (e.g., Figure 9). D-O-like oscillations also peter out and disappear south of Newfoundland and North of the Antarctic Ocean because the sea ice cover is an important control of deep-ocean warming (e.g., Figure 2), and thus, the D-O-like oscillations in the code are mostly a polar region phenomenon. The point being that the example of *Friedrich et al.* [2010] lacks generality because it cannot rule out the possibility that some other "fix" made to ECBILT-CLIO somewhere in its thousands of lines of code, for instance, by changing diffusivity coefficients or its response to fresh water hosing [e.g., *Knutti et al.*, 2004] will fully restore the D-O-like oscillations. We certainly agree that computer simulations should not be taken as proof of natural mechanisms, yet in some cases, models can

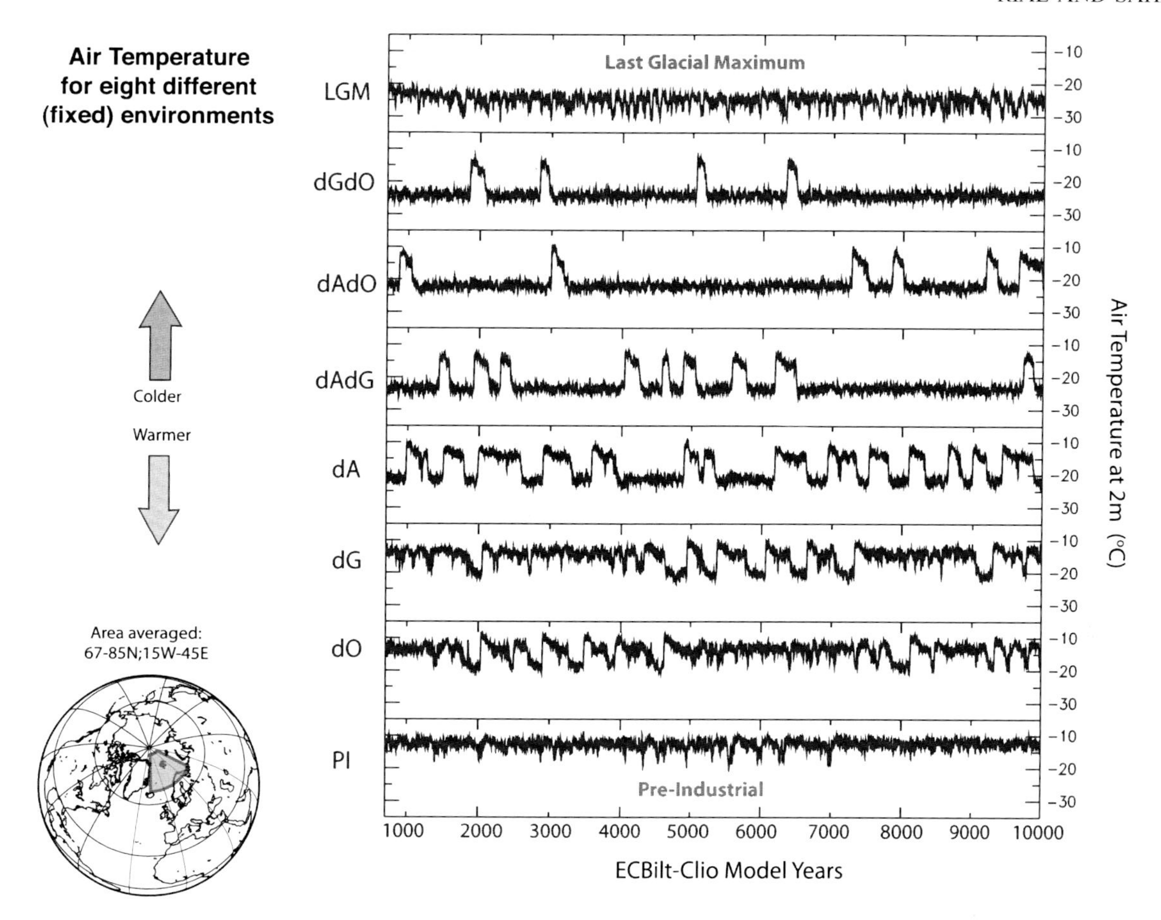

Figure A1. ECBILT-CLIO-computed sawtooth pulses of surface air temperature at 2 m (T2M) for different boundary conditions and averaged over the region shown. No external time-varying forcing is used. Three parameters (albedo dA, orbital dO, and greenhouse gas dG) can be set to either preindustrial (PI) or last glacial maximum (LGM) conditions. The eight combinations are displayed in order of increasing mean temperature from coldest LGM to warmest PI. The time series are labeled according to what PI parameter is switched to the LGM condition. For instance, starting with all conditions being PI, switching only albedo to LGM is labeled dA; switching both albedo and greenhouse gas to LGM is labeled dAdG, and so on. Notice that all but PI and LGM show climate bistability, consistent with the D-O-oscillations. The results are also consistent with the often stated idea that millennial-scale oscillations are most pronounced during times of intermediate ice volume. Sampling rate is 1.0 year^{-1}. A mean (average) smoothing window five-point half width was used in all time series. See details of the water column behavior for the dA condition in Figure 2. Modified from the work of *Rial and Saha* [2008].

guide toward one or can be suggestive of one, as in fact *Friedrich et al.* [2010] do by suggesting that obliquity forcing crucially modulates the strength of feedback between sea ice and convection that affects the Greenland, Icelandic, and Norwegian seas overturning circulation and the AMOC. However, they make these assertions based solely on modeling results, without data to support them.

All climate models are wrong by virtue of being incomplete (some certainly more than others). Therefore, one uses their best features and diligently checks their consistency with other models that one knows in full detail (e.g., SIO) and with the data. For instance, both ECBILT-CLIO and SIO reproduce the subtle phase relationship between shallow and deep ocean proxies (Figures 8 and 9). This is highly suggestive because the relationship is precisely what Saltzman's model [*Saltzman et al.*, 1981] predicts and the phenomenon on which is based. Obviously, there is no Hudson Strait in SIO; neither are there ice sheets.

In our experience, ECBILT-CLIO "tuned" to our needs is self-consistent, has never produced contradictory results, and is remarkably consistent with the data and even with well-known [e.g., *Shackleton et al.*, 2000] but poorly understood observations (Figures 8, 9, and A2). For instance, we have known for years that if the full topography of the ice sheets is

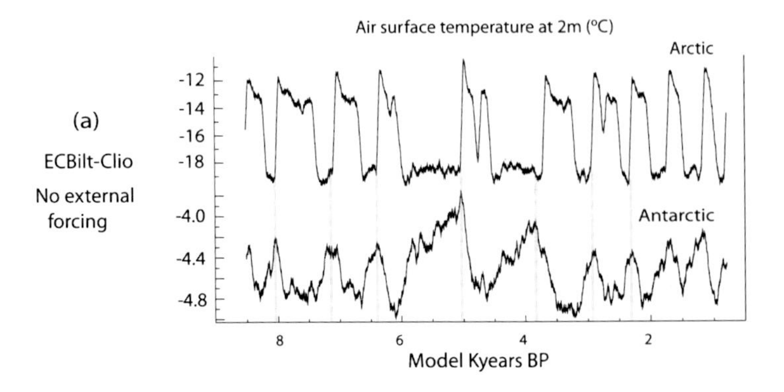

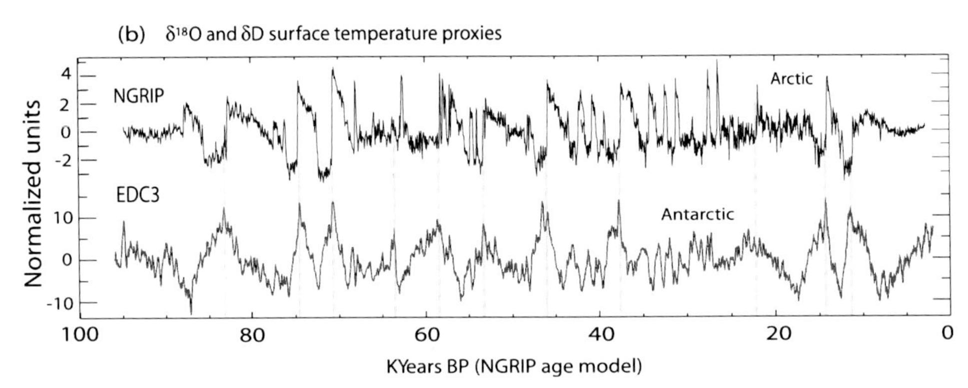

Figure A2. (a) Time series of Arctic and Antarctic surface temperatures obtained with ECBILT-CLIO compared to (b) the actual ice core records (NGRIP and EDC3). Notice that though the simulation is in a different time scale (self-similarity in time is a property of the ice core time series as described by *Rial and Yang* [2007]) than the whole data window (simulation of a 100 kyr long time series takes up to 14 months of continuous computing), the phase relationships and waveforms are fully consistent with those exhibited by the data. In both Figures A2a and A2b, the time integral of the northern record reproduces the southern one, and conversely, the Hilbert transform of the southern record reproduces the northern one. These mathematical operations are just convenient tools; the actual relationship is a $\pi/2$ shift with which Antarctica's temperature leads Greenland [*Steig*, 2006]. Nevertheless, the implications are important because they reveal that there is a fundamental connection between the climates of the polar regions, akin to synchronization (J. A. Rial, Synchronization of climatic change between the polar regions as a feasible origin of the bipolar seesaw, submitted to *Climate Change*, 2011). Note that the ECBILT-CLIO simulation is not forced with the insolation, which would increase the ~1.5 kyr fluctuations to 5 kyr or more, as can be seen in Figures 3, 5, 6, and 7.

included in ECBILT-CLIO, the D-O-like oscillations fade. But since ECBILT-CLIO does not grow or melt continental ice and its ice sheets are fixed boundary conditions, we opt to describe them as flat, high-albedo surface areas because our objective is to study the high-frequency, D-O-like fluctuations that the code produces in this case and how other variables in the code react to these fluctuations. Similarly, it is well known that the Langevin model of Figure 3 needs to be tuned in order to produce bistable fluctuations. Moreover, these are the result of integrating Newton's equations representing random jumps of a small ball between two "potential" wells, yet that hardly prevents its use or that of similar equations [e.g., *Sakai and Peltier*, 1997] as useful simulators of climate variability. *Lorenz* [1976] has clearly illustrated this. Thus, if the D-O-like oscillation in ECBILT-CLIO is an artifact, as far as our experiments are concerned, it must be a benign one.

An extensive literature provides a good source for the scope and limitations of ECBILT-CLIO. A necessarily incomplete list includes *Widmann et al.* [2009], *Timmermann et al.* [2009], *Weber et al.* [2007], *Yanase and Abe-Ouchi* [2007], *Schulz et al.* [2007], *Fluckiger et al.* [2006], and *Knutti et al.* [2004].

Acknowledgments. Steven Meyers, Austin Milt, Bruce Bills, and two anonymous reviewers made the original manuscript intelligible. This research was sponsored by a grant from the National Science Foundation Paleoclimate Program. Supplementary funding was provided by a grant from the National Geographic committee for research and exploration.

REFERENCES

Alley, R. B., S. Anandakrishnan, and P. Jung (2001), Stochastic resonance in the North Atlantic, *Paleoceanography, 16*, 190–198.

Alley, R. B., et al. (2003), Abrupt climate change, *Science, 299*, 2005–2010, doi:10.1126/science.1081056.

Blunier, T., and E. J. Brook (2001), Synchronization of the Byrd and Greenland (GISP2/GRIP) Records, IGBP PAGES/World Data Center for Paleoclimatology, *Data Contrib. Ser. 2001-003*, http://www.ncdc.noaa.gov/paleo, NOAA/NCDC Paleoclimatol. Program, Boulder Colo.

Bond, G., W. Broecker, S. Johnsen, J. McManus, L. Labeyrie, J. Jouzel, and G. Bonani (1993), Correlations between climate records from North Atlantic sediments and Greenland ice, *Nature, 365*, 143–147.

Bond, G., W. Showers, M. Chesby, L. Lotti, P. Almasi, P. deMenocal, P. Priore, H. Cullen, I. Hajdas, and G. Bonani (1997), A pervasive millennial-scale cycle in North Atlantic Holocene and glacial climates, *Science, 278*, 1257–1266.

Broecker, W., G. Bond, M. Klas, E. Clark, and J. McManus (1992), Origin of the northern Atlantic's Heinrich events, *Clim. Dyn., 6*, 265–273.

Bryan, F. (1986), High latitude salinity effects and interhemispheric thermohaline circulation, *Nature, 323*, 301–304.

Campin, J. M., and H. Goosse (1999), A parameterization of density-driven downsloping flow for a coarse-resolution model in z-coordinate, *Tellus, Ser. A, 51*, 412–430.

Cayre, O., Y. Lancelot, E. Vincent, and M. A. Hall (1999), Paleoceanographic reconstructions from planktonic foraminifera off the Iberian margin: Temperature, salinity and Heinrich events, *Paleoceanongraphy, 14*, 384–396.

Clark, P. U., N. G. Pisias, T. F. Stocker, and A. J. Weaver (2002), The role of the thermohaline circulation in abrupt climate change, *Nature, 415*, 863–869.

Colin de Verdiere, A., M. Ben Jelloul, and F. Sévellec (2006), Bifurcation structure of thermohaline millennial oscillations, *J. Clim., 19*, 5777–5795.

Collins, J. J., C. Chow, and T. Imhoff (1996), Aperiodic stochastic resonance in excitable systems, *Phys. Rev. E, 52*, R3321–R3324.

Dansgaard, W., et al. (1993), Evidence for general instability of past climate from a 250-kyr ice-core record, *Nature, 264*, 218–220.

Ditlevsen, P. D., K. K. Andersen, and A. Svensson (2007), The DO-climate events are probably noise induced: Statistical investigation of the claimed 1470 years cycle, *Clim. Past, 3*, 129–134.

EPICA Community Members (2006), One-to-one coupling of glacial climate variability in Greenland and Antarctica, *Nature, 444*, 195–198.

Flückiger, J., R. Knutti, and J. W. C. White (2006), Oceanic processes as potential trigger and amplifying mechanisms for Heinrich events, *Paleoceanography, 21*, PA2014, doi:10.1029/2005PA001204.

Friedrich, T., A. Timmermann, L. Menviel, O. Elison Timm, A. Mouchet, and D. M. Roche (2010), The mechanism behind internally generated centennial-to-millennial scale climate variability in an earth system model of intermediate complexity, *Geosci. Model Dev., 3*, 377–389, doi:10.5194/gmd-3-377-2010.

Ganopolski, A., and S. Rahmstorf (2002), Abrupt glacial climate changes due to stochastic resonance, *Phys. Rev. Lett., 88*, 038501, 4 pp.

Ghil, M., et al. (2002), Advanced spectral methods for climatic time series, *Rev. Geophys., 40*(1), 1003, doi:10.1029/2000RG000092.

Gitterman, M. (2005), *The Noisy Oscillator*, 160 pp., World Sci., Hakensack, N. J.

Goosse, H., and T. Fichefet (1999), Importance of ice-ocean interactions for the global ocean circulation: A model study, *J. Geophys. Res., 104*(C10), 23,337–23,355.

Goosse, H., H. Renssen, F. M. Selten, R. J. Haarsma, and J. D. Opsteegh (2002), Potential causes of abrupt climate events: A numerical study with a three-dimensional climate model, *Geophys. Res. Lett., 29*(18), 1860, doi:10.1029/2002GL014993.

Haarsma, R. J., J. D. Opsteegh, F. M. Selten, and X. Wang (2001), Rapid transitions and ultra-low frequency behavior in a 40 kyr integration with a coupled climate model of intermediate complexity, *Clim. Dyn., 17*, 559–570.

Heinrich, H. (1988), Origin and consequences of cyclic ice rafting in the northeast Atlantic Ocean during the past 130,000 years, *Quat. Res., 29*, 143–152.

Huybers, P., and C. Wunsch (2010), Paleophysical oceanography with an emphasis on transport rates, *Annu. Rev. Mar. Sci., 2*, 1–34.

Imkeller, P. (2001), Energy balance models—Viewed from stochastic dynamics, in *Stochastic Climate Models*, edited by P. Imkeller and J. von Storch, pp. 213–240, Birkhauser, Basel, Switzerland.

Johnsen, S., D. Dahl-Jensen, W. Dansgaard, and N. Gundestrup (1995), Greenland palaeotemperatures derived from GRIP bore hole temperature and ice core isotope profiles, *Tellus, Ser. B, 47*, 624–629.

Knutti, R., J. Fluckiger, T. F. Stoker, and A. Timmermann (2004), Strong hemispheric coupling of glacial climate through freshwater discharge, *Nature, 430*, 851–856.

Laskar, J., P. Robutel, F. Joutel, M. Gastineau, A. C. M. Correia, and B. Levrard (2004), A long-term numerical solution for the insolation quantities of the Earth, *Astron. Astrophys., 428*, 261–285, doi:10.1051/0004-6361:20041335.

Lisiecki, L. E., and M. E. Raymo (2005), A Pliocene-Pleistocene stack of 57 globally distributed benthic $\delta^{18}O$ records, *Paleoceanography, 20*, PA1003, doi:10.1029/2004PA001071.

Lorenz, E. N. (1976), Nondeterministic theories of climatic change, *Quat. Res., 6*, 495–506.

Loving, J. L., and G. K. Vallis (2005), Mechanisms for climate variability during glacial and interglacial periods, *Paleoceanography, 20*, PA4024, doi:10.1029/2004PA001113.

Marchal, O., C. Jackson, J. Nilsson, A. Paul, and T. Stocker (2007), Buoyancy-driven flow and nature of vertical mixing in a zonally averaged model, in *Ocean Circulation: Mechanisms and Impacts, Geophys. Monogr. Ser.*, vol. 173, edited by A. Schmittner, J. C. H. Chiang, and S. R. Hemming, pp. 33–52, AGU, Washington, D. C.

Moss, F., and P. McClintock (1989), *Noise in Nonlinear Dynamical Systems: Theory of Continuous Fokker-Planck Systems, vol. 1*, 365 pp., Cambridge Univ. Press, Cambridge, U. K.

North Greenland Ice Core Project Members (2004), High-resolution record of Northern Hemisphere climate extending into the last interglacial period, *Nature, 431*, 147–151.

Opsteegh, J. D., R. J. Haarsma, F. M. Selten, and A. Kattenberg (1998), ECBILT: A dynamic alternative to mixed boundary conditions in ocean models, *Tellus. Ser. A, 50*, 348–367.

Rahmstorf, S. (2006), Thermohaline ocean circulation, in *Encyclopedia of Quaternary Sciences*, edited by S. A. Elias, pp. 739–750, Elsevier, Amsterdam.

Rashid, H., and E. A. Boyle (2007), Mixed-layer deepening during Heinrich events: A multi-planktonic foraminiferal $\delta^{18}O$ approach, *Science, 318*, 439–441.

Rial, J. A., and R. Saha (2008), Stochastic resonance, frequency modulation and the mechanisms of abrupt climate change in the Arctic, paper presented at First International Symposium on Arctic Research, Sci. Counc. of Jpn., Tokyo, 4–6 Nov.

Rial, J. A., and M. Yang (2007), Is the frequency of abrupt climate change modulated by the orbital insolation?, in *Ocean Circulation: Mechanisms and Impacts, Geophys. Monogr. Ser.*, vol. 173, edited by A. Schmittner, J. C. H. Chiang, and S. R. Hemming, pp. 167–174, AGU, Washington, D. C.

Rial, J. A., et al. (2004), Nonlinearities, feedbacks and critical thresholds within the Earth's climate system, *Clim. Change, 65*, 11–38.

Saikku, R., L. Stott, and R. Thunell (2009), A bi-polar signal recorded in the western tropical Pacific: Northern and Southern Hemisphere climate records from the Pacific warm pool during the last Ice Age, *Quat. Sci. Rev., 28*, 2374–2385, doi:10.1016/j.quascirev.2009.05.007.

Sakai, K., and W. R. Peltier (1997), Dansgaard-Oeschger oscillations in a coupled atmosphere-ocean climate model, *J. Clim., 10*, 949–967.

Saltzman, B. (2002), *Dynamical Paleoclimatology*, 361 pp., Academic, San Diego, Calif.

Saltzman, B., and R. Moritz (1980), A time-dependent climatic feedback system involving sea-ice extent, ocean temperature, and CO_2, *Tellus, 32*, 93–118.

Saltzman, B., A. Sutera, and A. Evanson (1981), Structural stochastic stability of a simple auto-oscillatory climatic feedback system, *J. Atmos. Sci., 38*, 494–503.

Schulz, M., M. Prange, and A. Klocker (2007), Low-frequency oscillations of the Atlantic Ocean meridional overturning circulation in a coupled climate model, *Clim. Past Disc., 2*, 801–830.

Severinhaus, J. P., and E. J. Brook (1999), Abrupt climate change at the end of the last glacial inferred from trapped air in polar ice, *Science, 286*, 930–934.

Shackleton, N. J. (2000), The 100,000-year Ice-Age cycle identified and found to lag temperature, carbon dioxide, and orbital eccentricity, *Science, 289*, 1897–1902.

Shackleton, N. J., M. A. Hall, and E. Vincent (2000), Phase relationships between millennial-scale events 64,000-24,000 years ago, *Paleoceanography, 15*(6), 565–569, doi:10.1029/2000PA000513.

Sprott, J. C. (2003), *Chaos and Time Series Analysis*, 507 pp., Oxford Univ. Press, Oxford, U. K.

Steig, E. J. (2006), The south-north connection, *Nature, 444*, 152–153.

Stern, M. E. (1975), *Ocean Circulation Physics*, 246 pp., Academic, San Diego, Calif.

Stommel, H. (1961), Thermohaline convection with two stable regimes of flow, *Tellus, 13*, 224–228.

Timmermann, A., and G. Lohmann (2000), Noise-induced transitions in a simplified model of the thermohaline circulation, *J. Phys. Oceanogr., 30*, 1891.

Timmermann, A., O. Timm, L. Scott, and L. Menviel (2009), The roles of CO_2 and orbital forcing in driving southern hemispheric temperature variations during the last 21000 yr, *J. Clim., 22*, 1626–1640.

Wang, Z., and L. A. Mysak (2006), Glacial abrupt climate changes and Dansgaard-Oeschger oscillations in a coupled climate model, *Paleoceanography, 21*, PA2001, doi:10.1029/2005PA001238.

Weber, S. L., S. S. Drijfhout, A. Abe-Ouchi, M. Crucifix, M. Eby, A. Ganopolski, S. Murakami, B. Otto-Bliesner, and W. R. Peltier (2007), The modern and glacial overturning circulation in the Atlantic ocean in PMIP coupled model simulations, *Clim. Past, 3*, 51–64.

Widmann, M., H. Goosse, G. van der Schrier, R. Schnur, and J. Barkmeijer (2009), Using data assimilation to study extra-tropical Northern Hemisphere climate over the last millennium, *Clim. Past, 6*, 627–644, doi:10.5194/cp-6-627-2010.

Williams, G. P. (1997), *Chaos Theory Tamed*, 499 pp., J. Henry Press, Washington, D. C.

Winton, M. (1993), Energetics of deep-decoupling oscillations, *J. Phys. Oceanogr., 25*, 420–427.

Winton, M., and E. S. Sarachik (1993), Thermohaline oscillations induced by strong steady salinity forcing of ocean general circulation models, *J. Phys. Oceanogr., 23*, 1389–1410.

Wunsch, C. (2002), What is the thermohaline circulation?, *Science, 298*, 1180–1181.

Wunsch, C. (2006), Abrupt climate change: An alternative view, *Quat. Res., 65*, 191–203.

Wunsch, C., and R. Ferrari (2004), Vertical mixing, energy, and the general circulation of the oceans, *Annu. Rev. Fluid Mech., 36*, 281–314, doi:10.1146/annurev.fluid.36.050802.122121.

Yanase, W., and A. Abe-Ouchi (2007), The LGM surface climate and atmospheric circulation over east Asia and the north Pacific in the PMIP2 coupled model simulations, *Clim. Past, 3*, 439–451.

J. A. Rial, Wave Propagation Laboratory, Department of Geological Sciences, University of North Carolina at Chapel Hill, Chapel Hill, NC 27599-3315, USA. (jose_rial@unc.edu)

R. Saha, Department of Physics and Astronomy, University of North Carolina at Chapel Hill, Chapel Hill, NC 27599-3255, USA.

Simulated Two-Stage Recovery of Atlantic Meridional Overturning Circulation During the Last Deglaciation

Jun Cheng,[1] Zhengyu Liu,[2,3] Feng He,[2] Bette L. Otto-Bliesner,[4] Esther C. Brady,[4] and Mark Wehrenberg[2]

A two-stage recovery of Atlantic meridional overturning circulation (AMOC) during Bølling Allerød (BA) is revealed in the first transient simulation of last deglaciation in a fully coupled general circulation model (Transient Simulation of Climate Evolution over the Last 21,000 Years (TraCE-21000)). After being suppressed during the Heinrich 1 event, North Atlantic Deep Water (NADW) formation first reinitiates in the Labrador Sea in stage 1 and then reinitiates in the Greenland-Iceland-Norwegian (GIN) seas in stage 2. This feature is derived from the investigation of NADW formation in its two origins with a newly developed method and is confirmed by the comprehensive analysis of relative variables. A new mechanism is proposed to interpret the northward asynchronous reinitiation of NADW during BA. In addition, our work also points out that the generation of the AMOC overshoot is associated with the reinitiation of NADW in GIN seas.

1. INTRODUCTION

Paleoclimate records and climate model simulations suggest that climate change on centurial-millennial time scales is largely connected with the variation of Atlantic meridional overturning circulation (AMOC) [*Broecker*, 1990; *Manabe and Stouffer*, 1995; *Vellinga and Wood*, 2002; *Levermann et al.*, 2005]. In addition to the slowdown/shutdown of AMOC, sometimes the pronounced abrupt climate change is also associated with AMOC recovery after fresh meltwater discharge in the North Atlantic [*Ganopolski and Rahmstorf*, 2001; *Rahmstorf*, 2002; *Clarke et al.*, 2002; *McManus et al.*, 2004; *Lippold et al.*, 2009]. In the first transient simulation of the last deglaciation with a fully coupled model (Transient Simulation of Climate Evolution over the Last 21,000 Years (TraCE-21000)), the abrupt onset of Bølling Allerød (BA) warming was captured, and it was found that the onset of BA warming is dominated by the recovery of the AMOC. A 10°C air temperature increase over Greenland during BA onset accompanies a 16.5 Sv increase of AMOC intensity within 300 years [*Liu et al.*, 2009].

The details of the recovery process of AMOC have not been fully explained in the literature so far. Some studies simply define one latitude band in the North Atlantic as the important North Atlantic Deep Water (NADW) formation region to explain AMOC recovery and do not consider the details of NADW reinitiation in multiple origins [*Krebs and Timmermann*, 2007a, 2007b; *Hu et al.*, 2007; *Mignot et al.*, 2007]. This simplified method relies upon two points: (1) the low resolution of the ocean model that employed a longer integration for AMOC study and (2) the broad convection area in the model simulation that was not only limited to the

[1]Key Laboratory of Meteorological Disaster of Ministry of Education and College of Atmospheric Sciences, Nanjing University of Information Science and Technology, Nanjing, China.

[2]Center for Climatic Research and Department of Atmospheric and Oceanic Sciences, University of Wisconsin-Madison, Madison, Wisconsin, USA.

[3]Also at Laboratory for Climate and Ocean-Atmosphere Studies and Department of Atmospheric and Oceanic Sciences, Peking University, Beijing, China.

[4]Climate and Global Dynamics Division, National Center for Atmospheric Research, Boulder, Colorado, USA.

Abrupt Climate Change: Mechanisms, Patterns, and Impacts
Geophysical Monograph Series 193

10.1029/2010GM001014

Labrador Sea or the Greenland-Iceland-Norwegian (GIN) seas. Despite the dependence of model formulation and internal variability of the AMOC fluctuation under control runs, the AMOC changes in perturbation experiments (water-hosing experiments) are mainly associated with the convection changes in the Labrador Sea and GIN seas with asynchronous and different contributions [*Krebs and Timmermann*, 2007a, 2007b; *Mignot et al.*, 2007; *Vellinga and Wood*, 2002]. This mechanism of AMOC variability is found in the widely used Community Climate System Model version 3 (CCSM3) under control and water-hosing experiments in glacial, preindustrial, and present-day climate conditions [*Hu et al.*, 2007; *Renold et al.*, 2009]. So the details of NADW reinitiation in multiple areas of origin are important for the understanding of the AMOC recovery process. Other works have touched upon the asynchronicity of the reinitiation of deepwater formation at different locations, but these studies provide divergent conclusions. Starting with an artificially "turned off" state of AMOC in model HadCM3 under modern climate conditions, *Vellinga and Wood* [2002] found that NADW formation reinitiates first in the GIN seas and then moves southward to the Labrador Sea. This work also proposed that the northward salt transport determines the order of the NADW reinitiation. In contrast, *Renold et al.* [2009] found that the reinitiation of the NADW first occurred in the Labrador Sea and then in the GIN seas, based on a series of water-hosing experiments with CCSM3. Renold et al. proposed that the northward salt and heat transport determines the northward reinitiation of NADW beneath the retreat of sea ice. The contradictory results of *Vellinga and Wood* [2002] and *Renold et al.* [2009] rely upon the attendance of sea ice change during the AMOC recovery period and maybe the dependence of the models. However, the results of *Renold et al.* [2009] seem to simulate the real world more closely because of the more realistic experiment scheme.

Renold et al. [2009] also found two stages in the recovery of AMOC, as suggested by the evolutionary characteristics of the AMOC intensity. However, in many studies using different models with different complexities, a two-stage mechanism of AMOC recovery is not always evident in a single time series of AMOC intensity [*Manabe and Stouffer*, 1997; *Knutti et al.*, 2004; *Mignot et al.*, 2007; *Arzel et al.*, 2008]. The two-stage feature of AMOC recovery seems to rely on the asynchronicity of NADW reinitiation in multiple origins according to *Renold et al.* [2009]. The asynchronous feature of the NADW reinitiation in multiple origins could induce the intensity series to present the two-stage feature but may not always do so. This suggests that a single measure of AMOC intensity maybe not be a good choice for judging whether the reinitiation of NADW in multiple origins is asynchronous or not. A detailed investigation of the AMOC recovery process should take a broader view of the geographical heterogeneity of the NADW formation in the North Atlantic and the regional contributions to the total AMOC intensity.

Another deficiency in studies of the AMOC recovery is that idealized water-hosing experiments are not typically run for long enough to study the entire recovery process. Previous modeling studies usually emphasized the prerecovery period of the simulation, ending when the AMOC intensity resumes its unperturbed level for the first time and neglecting whether the AMOC system has fully recovered or not [*Vellinga and Wood*, 2002; *Bitz et al.*, 2007; *Hu et al.*, 2008; *Renold et al.*, 2009]. Actually, numerous experiments with different models have shown that the AMOC will keep increasing after its initial recovery and continue to increase in intensity before returning to its unperturbed level, exhibiting an "overshoot" phenomenon [*Manabe and Stouffer*, 1997; *Knutti et al.*, 2004; *Stouffer et al.*, 2006; *Mignot et al.*, 2007; *Krebs and Timmermann*, 2007a, 2007b; *Schmittner and Galbraith*, 2008; *Arzel et al.*, 2008; *Liu et al.*, 2009]. The historic occurrence of an AMOC overshoot during the BA event has been validated with combined observational and model evidence. That this is the effect of AMOC overshoot in deep currents of the North Atlantic regionally is shown in model simulations, and this feature is consistent with the distribution of reconstructed proxies such as cores GGC5 and TTR-451 (J. Cheng et al., Model-proxy comparison for overshoot phenomenon of Atlantic thermohaline circulation at Bølling-Allerød, unpublished manuscript, 2011). So the postrecovery stage, or overshoot stage, of the AMOC system should not be ignored.

The mechanisms for AMOC recovery are still not well understood. With an intermediate complexity model (CLIMBER-3α), *Wright and Stocker* [1991] and *Mignot et al.* [2007] thought the recovery of AMOC derived from the destabilized stratification with warmed abyssal water in the North Atlantic. Using the intermediate complexity model ECBILT-CLIO, *Goosse et al.* [2002] proposed that the recovery of AMOC comes from a stochastic process of air-sea interaction. With a coupled general circulation model (GCM) such as LSG-ECHAM3/T42, *Knorr and Lohmann* [2003] proposed that a weakened Antarctic Bottom Water cell contributes to the recovery of AMOC. With different complexity models, *Vellinga and Wood* [2002] (HadCM3 and coupled GCM), *Yin et al.* [2006] (UIUC and coupled GCM), and *Krebs and Timmermann* [2007a, 2007b] (intermediate complexity model and ECBILT-CLIO) hypothesized that the northward advection of the salinity anomaly is the key factor for AMOC recovery, and this hypothesis is supported by paleorecords [*Carlson et al.*, 2008].

Because of our lack of understanding of the dominant processes underlying past shifts in the AMOC and the increasing research interest in global warming in the past [*Carlson et al.*, 2008; *Milne and Mitrovica*, 2008; *Barker et al.*, 2009; *Liu et al.*, 2009], a detailed investigation of the processes controlling AMOC recovery during the last deglaciation is still needed. TraCE-21000 simulation provides a basis to address this issue [*Liu et al.*, 2009]. This simulation focus on a historic global warming period (last deglaciation) and its results should be helpful to understand the more general processes controlling AMOC variation and to project the future behavior of possible AMOC changes under current and projected global warming.

In this study, we investigate the recovery process of AMOC during BA onset in TraCE-21000 simulation, emphasizing the processes underlying the reinitiation of NADW in its two origins. A detailed analysis of surface and subsurface variables provides a framework for discussing the different stages involved in the AMOC recovery process. A description of the model and simulation setup is given in section 2. The two-stage feature of AMOC recovery is presented in section 3, while the processes underlying the recovery stages are discussed in section 4. Discussion and conclusions follow in section 5.

2. MODEL AND EXPERIMENTAL SETUP

The CGCM used in this study is the National Center for Atmospheric Research (NCAR) CCSM3 with a dynamic global vegetation module. CCSM3 is a global, coupled ocean–atmosphere–sea ice–land surface climate model without flux adjustment [*Collins et al.*, 2006]. The atmospheric model is the Community Atmospheric Model 3 (CAM3) with horizontal resolution of about 3.75°×3.75° and 26 vertical hybrid coordinate levels. The land model is the Community Land Model version 3 (CLM3) with the same resolution as the atmosphere. The ocean model is the NCAR implementation of the Parallel Ocean Program (POP) with vertical z coordinate with 25 levels. The longitudinal resolution is 3.6°, and the latitudinal resolution is variable, with finer resolution in the tropics. The sea ice model is the NCAR Community Sea Ice Model (CSIM). The resolution of CSIM is identical to that of POP. CCSM3 has previously been widely used in equilibrium and transient simulations of the glacial/interglacial climate state [*Shin et al.*, 2003a; *Otto-Bliesner et al.*, 2006; *Hu et al.*, 2007, 2008; *Liu et al.*, 2009; *Renold et al.*, 2009].

TraCE-21000 is the first try at representing the transient evolution of Earth's climate system over the last 21,000 years with the low-resolution version (T31_gx3v5) of CCSM3 [*Liu et al.*, 2009]. TraCE-21000 simulation includes two parallel runs DGL-A and DGL-B with different experimental schemes. Compared with DGL-B run, the DGL-A run more successfully represented global climate evolution according to reconstructed paleoclimate proxies (sea surface level, AMOC intensity, surface air temperature (SAT) in Antarctica and Greenland, and sea surface temperature (SST) in the tropical Atlantic Ocean) through a sudden termination of meltwater discharge into the North Atlantic at 14.67 ka. During the integration of the DGL-A run, external forcing such as meltwater discharge, insolation intensity, greenhouse gas (GHG) concentration, and land–ice sheet topography all varied in time based on reconstructed data [*Liu et al.*, 2009]. The key factor for the successful transient simulation is setting the scenario of meltwater discharge in high North Atlantic latitudes (50°–70°N) and the Gulf of Mexico from 19 to 14.67 ka. The total meltwater discharge is equivalent to a 50.1 m rising of the global sea level (the North Atlantic and Gulf of Mexico contribute 45.35 m and 4.75 m, respectively). The mean intensity of meltwater discharge is about 0.133 Sv. In the DGL-A run, AMOC is suppressed to a "turned off" state during Heinrich 1(H1) event (17 ka) with the increasing meltwater discharge into North Atlantic, and the "turned off"

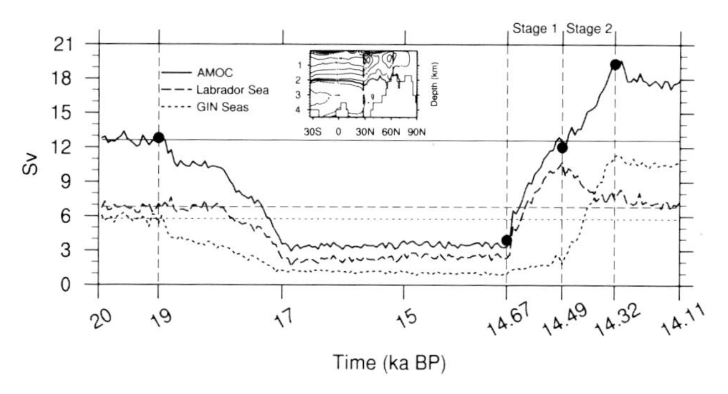

Figure 1. Two-stage feature of AMOC recovery (solid line) based on regional NADW formation volumes in the Labrador Sea (dashed line) and the GIN seas (dotted line) of model run DGL-A. AMOC intensity is defined as the maximum stream function of the Atlantic section below depths of 500 m. Regional NADW formation volume in the GIN seas is defined as the maximum stream function at its southern edge at 62°N. Regional NADW formation volume in the Labrador Sea is defined as the difference of AMOC intensity and regional volume in the GIN seas. Circles show the time of four climate states: the glacial state (GLA, 19 ka B.P.), pre-BA event (pre-BA, 14.67 ka B.P.), recovery (REC, 14.49 ka B.P.) and BA (14.32 ka B.P.). Horizontal dashed lines stand for the Last Glacial Maximum value of AMOC intensity and NADW formation volume in the Labrador Sea and GIN seas. Time periods of the two recovery stages are shown with "stage 1" (pre-BA–REC) and "stage 2" (REC-BA). Inset plot shows the latitude edges of each NADW region used to calculate regional NADW formation.

state of AMOC is maintained until the termination of meltwater discharge. In order to study the recovery process of AMOC from an initialized glacial state, we adapt the DGL-A run of TraCE-21000 simulation.

3. TWO-STAGE FEATURE OF AMOC RECOVERY

Inspired by the "alternative measuring method" of evaluating the AMOC, developed by *Gent* [2001], here we induce a new method to extract regional NADW formation volume in the Labrador Sea and GIN seas from the zonal mean stream function of the Atlantic. Regional NADW formation volume is calculated from the difference of vertical maximum stream function at the latitude edges of each NADW origin. Our method can present the individual volume of NADW formation in two origins quantificationally. The sum of NADW formation volume in two origins is equal to the value of AMOC intensity if we propose that AMOC intensity stand for the total NADW formation volume in the North Atlantic. Compared to the traditional method using maximum mixed

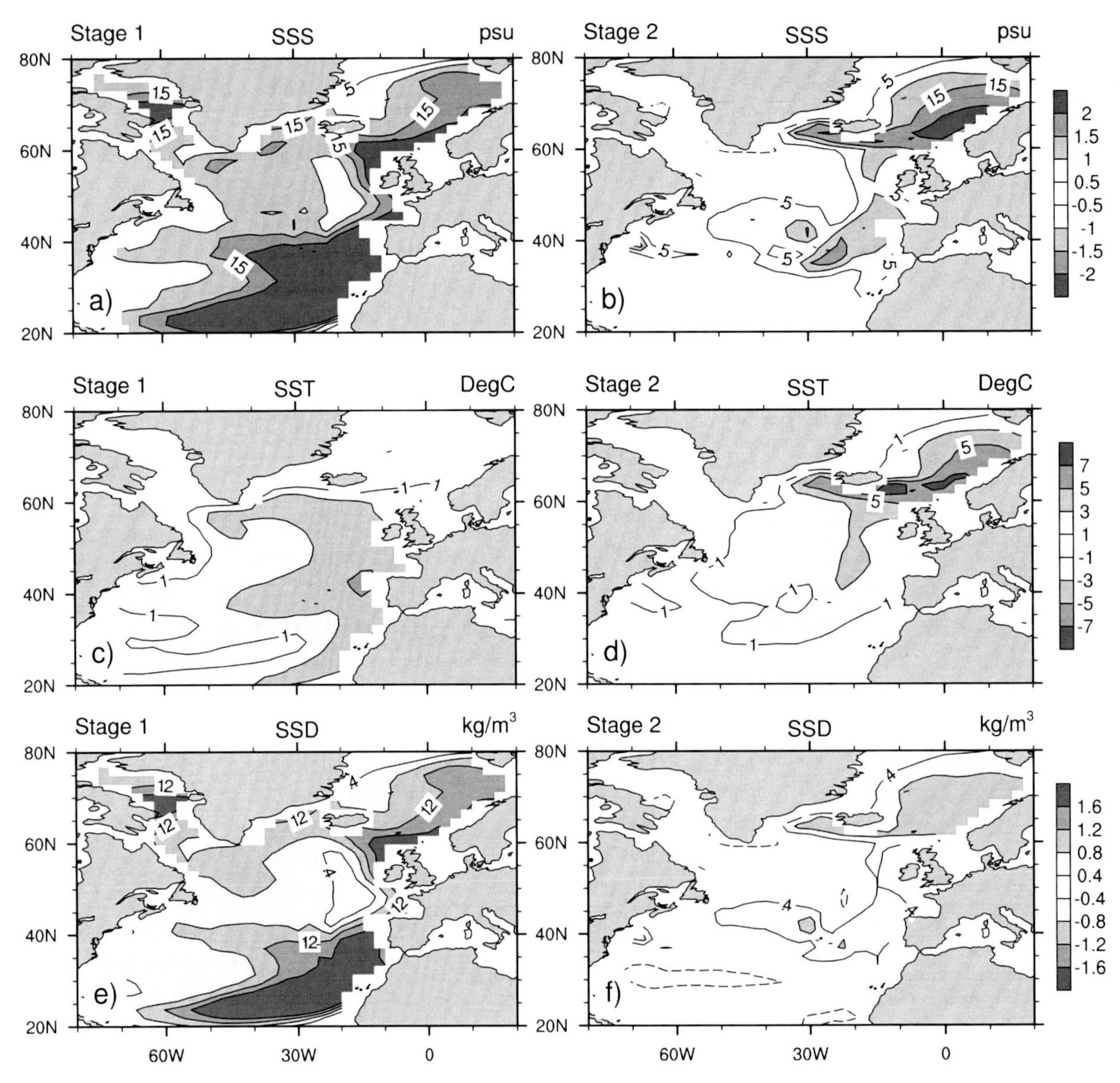

Figure 2. Change in (a and b) annual mean sea surface salinity (SSS), (c and d) sea surface temperature (SST), and (e and f) sea surface density (SSD) during (left) stage 1 and (right) stage 2. Solid and dashed contour lines indicate postive and negative values, respectively. Values of pre-BA/REC/BA are all presented with ±10 year means here and in Figures 3–5.

layer depths [*Renold et al.*, 2009] or area mean vertical velocity [*Vellinga and Wood*, 2002], this method is one alternative way to precisely quantify the NADW formation volume and to detect variation of the regional volume of NADW formation in its multiple origins.

A two-stage feature of AMOC recovery is revealed in this regional analysis of NADW formation during the AMOC recovery period (BA onset), based on the asynchronous reinitiation of NADW in the Labrador Sea and GIN seas (Figure 1). After meltwater discharge in the North Atlantic, NADW formation in the Labrador Sea and GIN seas is suppressed to nearly a "turned off" state after the H1 event, the state in which nearly no NADW was formed. After that, during the recovery period of the AMOC (starting from 14.67 ka), NADW formation first reinitiates in the Labrador Sea, defined here as stage 1 of AMOC recovery (from pre-BA to REC), and subsequently reinitiates in the GIN seas, which constitutes stage 2 (from REC to BA).

During stage 1, once the meltwater forcing in the North Atlantic is suddenly halted, NADW formation in the Labrador

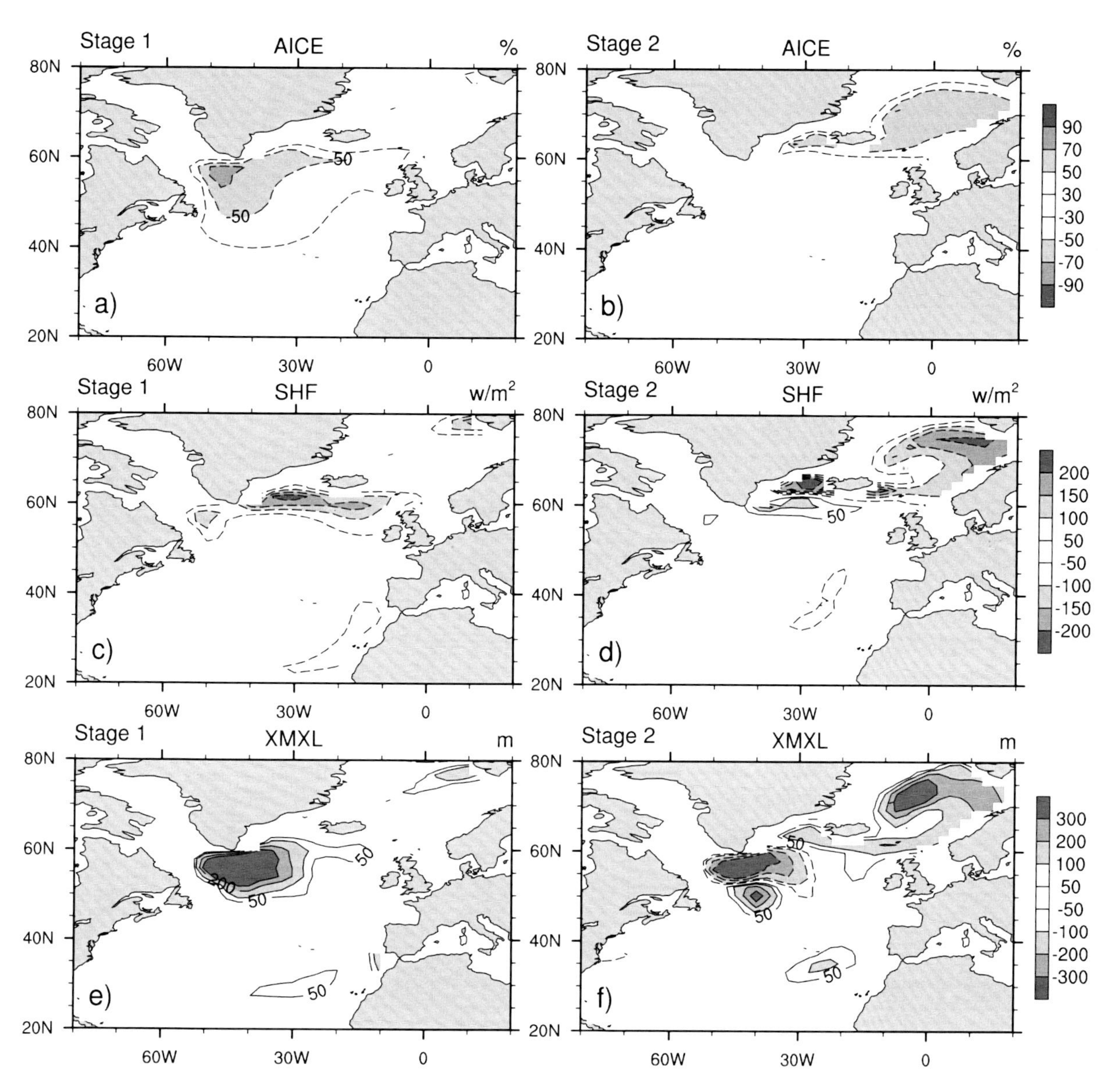

Figure 3. Change in (a and b) annual mean sea ice concentration (AICE), (c and d) surface heat flux (SHF), and (e and f) maximum mixed layer depth (XMXL) during (left) stage 1 and (right) stage 2. SHF is downward; negative value means oceanic heat loss.

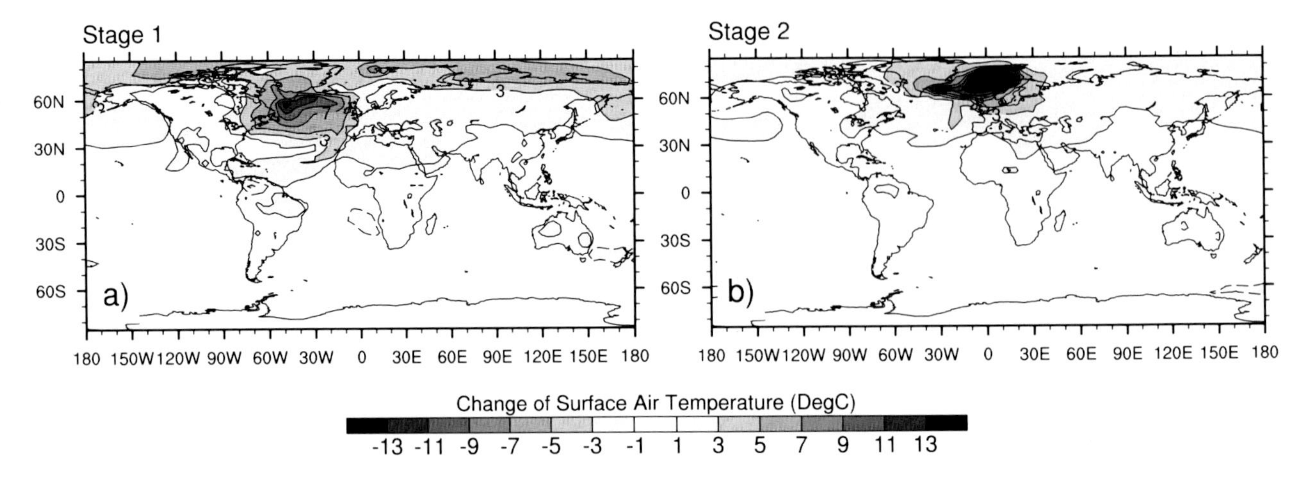

Figure 4. Change in annual mean surface air temperature (SAT) during (a) stage 1 and (b) stage 2. Units are in degrees Celsius (°C).

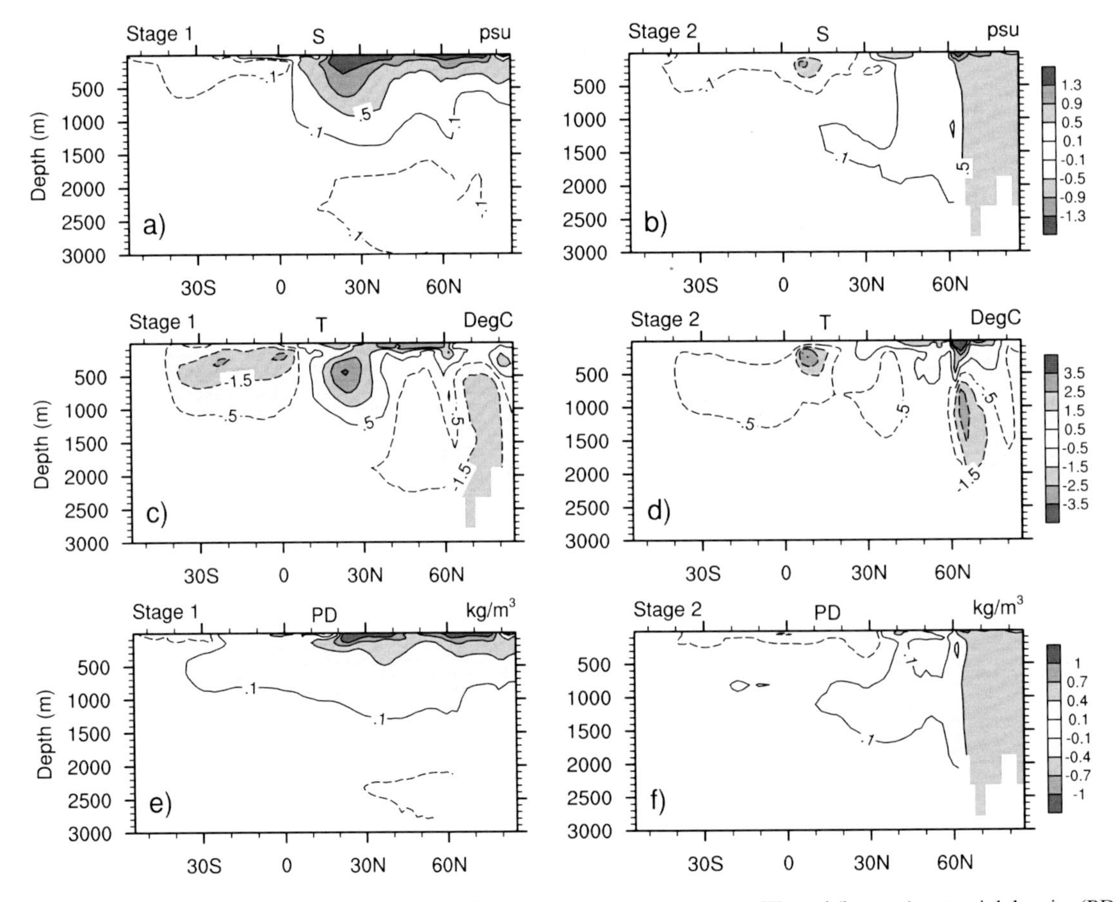

Figure 5. Change in (top) annual mean salinity (S), (middle) potential temperature (T), and (bottom) potential density (PD) during (left) stage 1 and (right) stage 2 in the zonal averaged Atlantic section.

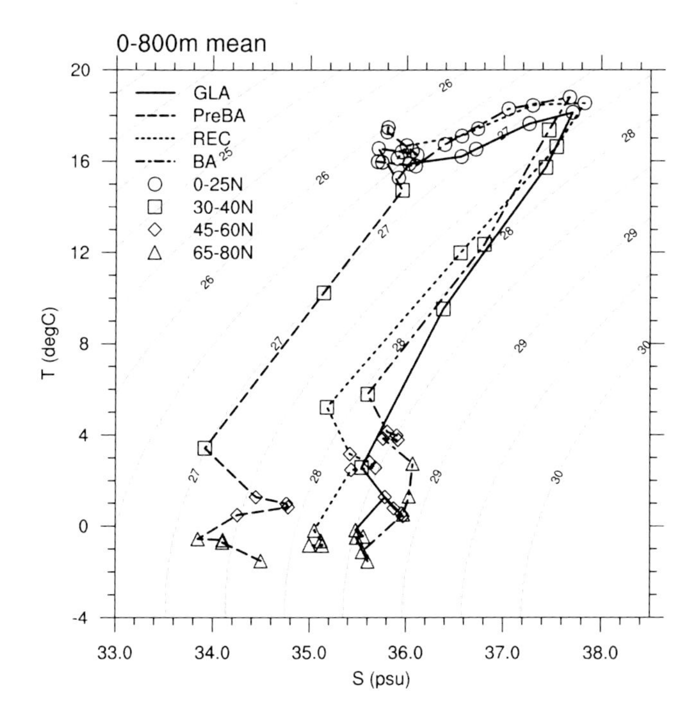

Figure 6. T-S diagram of the North Atlantic upper layer water (0–800 m) for GLA (solid line, mean within 20–19 ka), pre-BA (dashed line, 14.67 ka), REC (dotted line, 14.5 ka) and BA (dash-dotted line, 14.32 ka). Latitude bands are shown with different markers (circle for the tropical North Atlantic, square for the subtropical North Atlantic, diamond for the subpolar North Atlantic which includes the Labrador Sea, and triangle for the GIN seas). Latitude interval is 5°.

Sea is activated from the "turned off" state to the enhanced state (above initialized volumes) within 150 years. This has the result that while NADW formation in the GIN seas has not fully recovered by the end of stage 1, the total volume of NADW formation in the North Atlantic (AMOC intensity) reaches its original glacial level at the end of stage 1 (REC). Thereafter, the continuing recovery of NADW formation in the GIN seas in stage 2 pushes the AMOC intensity to an even higher level (about 20 Sv) within 170 years. Continuous increasing of AMOC intensity results in a robust AMOC overshoot phenomenon on a time scale of hundreds of years and a mean state transition of AMOC from 12.5 Sv at the Last Glacial Maximum to about 17.5 Sv at interglacial state (stable AMOC intensity during BA, similar to its modern simulated value of about 17 Sv in CCSM3 according to *Renold et al.* [2009]). On the basis of the precondition from stage 1, it is evident that the rapid recovery and subsequent overshoot of the AMOC can largely be attributed to the speed and magnitude of enhanced NADW formation in the GIN seas during stage 2.

The two-stage feature of AMOC recovery can also be confirmed by changes in associated sea surface variables in the Atlantic Basin (Figures 2 and 3), surface air temperature (SAT) (Figure 4), and zonal mean salinity, temperature, and potential density (Figure 5).

As shown in Figure 2, sea surface salinity (SSS) and sea surface density (SSD) of the whole North Atlantic Basin and SST of south of Greenland significantly increase during stage 1. However, during stage 2 the increase in SSS, SST, and SSD primarily happens in the GIN seas. SSS and SSD do increase somewhat in the GIN seas during stage 1 but not enough to induce robust NADW formation, as shown in Figure 1. The reason why the increased SSS and SSD in the GIN seas during stage 1 did not induce the robust reinitiation of NADW will be addressed in section 4.

The changes in sea ice cover (annual mean sea ice concentration), surface heat flux (SHF) (downward) and maximum mixed layer depths provide a clearer picture of the two-stage feature in the Labrador Sea and GIN seas (Figure 3). Because of the coarse resolution of the model used here and other similar models, the simulated deepwater formation regions are broader than the Labrador Sea and GIN seas. For simple presentation, here we refer to them as the Labrador Sea and GIN seas to present the surrounding changes in general. The Irminger Sea also has similar significant changes to the GIN seas during stage 2. Because of its small area and complex structure (shown in Figure 3), here and in following sections

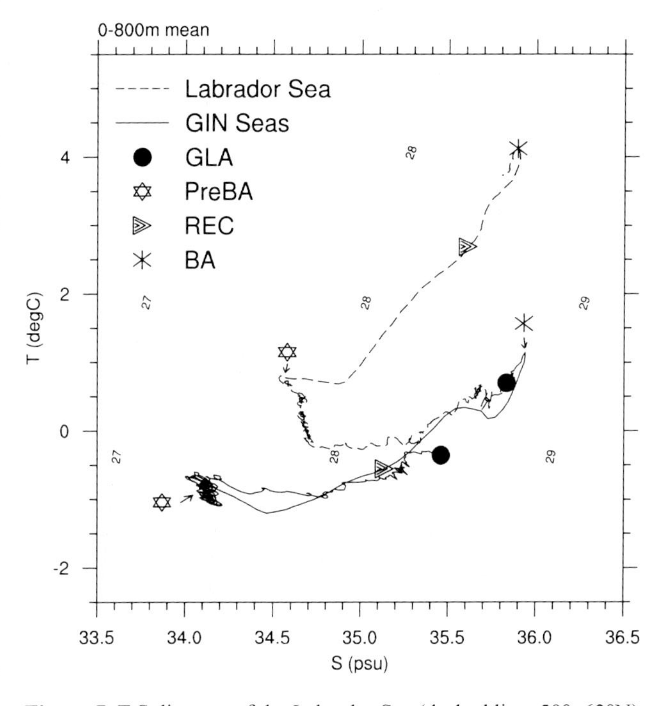

Figure 7. T-S diagram of the Labrador Sea (dashed line, 50°–62°N) and GIN seas (solid line, 62°–80°N) within upper layers (0–800 m). GLA/pre-BA/REC/BA are shown with different symbols.

the Irminger Sea is included with the GIN seas to avoid a too complex discussion. Changes in these three variables are seen primarily around the Labrador Sea during stage 1 and around the GIN seas during stage 2. It is clear that sea ice cover is tightly connected with the increasing SST (Figure 2) in both NADW origins and similarly with oceanic heat loss and convection.

In Figure 4, SAT increases within nearly the whole Northern Hemisphere during stage 1, with a maximum about 15°C around the Labrador Sea during stage 1, and then the warming center moves to the GIN seas during stage 2.

In Figure 5, we see that the zonal mean salinity/temperature/potential density of the whole Atlantic Basin exhibits a simultaneous shift, with salinization/warming/densification of upper layers and the opposite change in deeper layers. This primarily happens in the Labrador Sea during stage 1 and then moves to the GIN seas during stage 2. An exception to these simultaneous shifts in ocean conditions is the subtropical North Atlantic, where Ekman pumping becomes a significant factor. Ekman pumping drives the recovered northward warm and salty water of upper layers to the depth of about 500 m, that is why the maximum change of salinity and temperature at the depth of 500 m in Figure 5, also for the SSS and SSD change in Figure 2, occurs during stage 1 in the subtropical North Atlantic. A more detailed discussion of salinity/temperature/potential density evolution within the two main NADW origins will be the topic of the following section.

The two-stage feature of AMOC recovery becomes apparent through an analysis of NADW formation by region, and it is confirmed by the associated ocean and atmosphere variables. Even though the overall evolution of AMOC intensity lacks an obvious two-stage feature, our regional analysis shows that a two-stage feature exists in the simulated recovery process of AMOC during the last deglaciation.

4. CAUSES OF TWO-STAGE AMOC RECOVERY

The preceding analysis suggests the two-stage evolution of AMOC recovery depends on the asynchronous reinitiation of NADW in the Labrador and GIN seas, but we have yet to determine the reasons for this asynchronicity. In this section, we present the time evolution of salinity/temperature/potential density, give a detailed account of deepwater formation

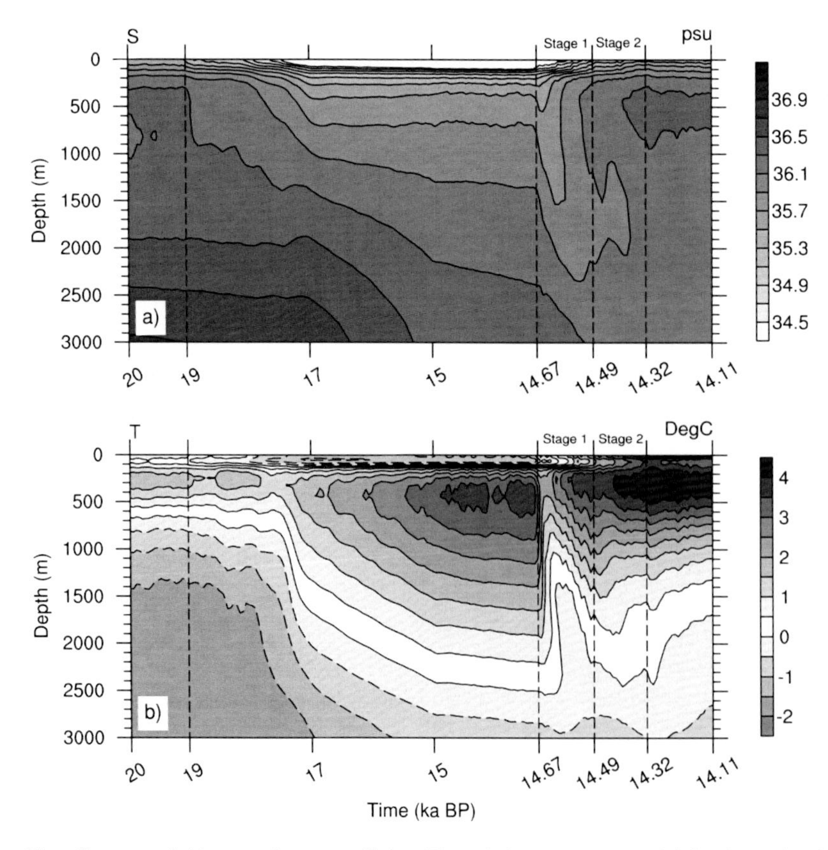

Figure 8. Hovmöller diagram of (a) annual mean salinity (S) and (b) temperature (T) in the Labrador Sea, averaged between 50°N and 62°N.

in each region, reconsider the mechanisms controlling NADW reinitiation, and finally, discuss the role of each NADW formation region in the simulated AMOC recovery and overshoot.

4.1. Evolution of Salinity/Temperature/Potential Density

Figure 6 shows the temperature-salinity (T-S) diagram of North Atlantic upper layers (0–800 m) at GLA, pre-BA, REC, and BA. Generally, the northward increase of potential density within upper ocean layers is controlled more by the temperature gradient than salinity north of 30°N. However, comparing the four consecutive ocean states, the change of potential density during meltwater discharge (GLA to pre-BA) and recovery periods (pre-BA to REC in the Labrador Sea and REC to BA in the GIN seas) is mainly controlled by salinity. During stage 1 (black line to light shaded line), salinity and potential density in the whole North Atlantic (excluding the tropical region) significantly increase, especially for the subtropical North Atlantic. During stage 2 (light shaded line to medium shaded line), salinity and potential density changes mainly happen in the GIN seas. Usually, an increase in salinity is accompanied by warming, but here it is clear that densification of North Atlantic upper layers is controlled more by salinity than temperature. Figure 7 provides another more detailed T-S evolution of upper layers in the Labrador Sea and GIN seas to support the salinity control over potential density change. During the periods of meltwater discharge and recovery, the changes of salinity, both for the Labrador Sea and GIN seas, not only contribute to the change of potential density but also compensate the opposite effect of temperature.

A Hovmöller diagram of area mean salinity and temperature in the Labrador Sea (Figure 8) shows that during the meltwater discharge period (from 19 to 14.67 ka), the Labrador Sea was freshened in all depths, with significant subsurface warming. The warming and freshening signal is mostly inhibited in upper layers (about 0–800 m) through suppressed convection, while turbulent mixing slowly spread the freshening and warming subsurface water to deeper layers. When meltwater discharge is terminated at 14.67 ka, convection in the Labrador Sea is quickly reinitiated, characterized by the sharp downward penetration of freshened water and the release of stored subsurface heat. The depth affected by the

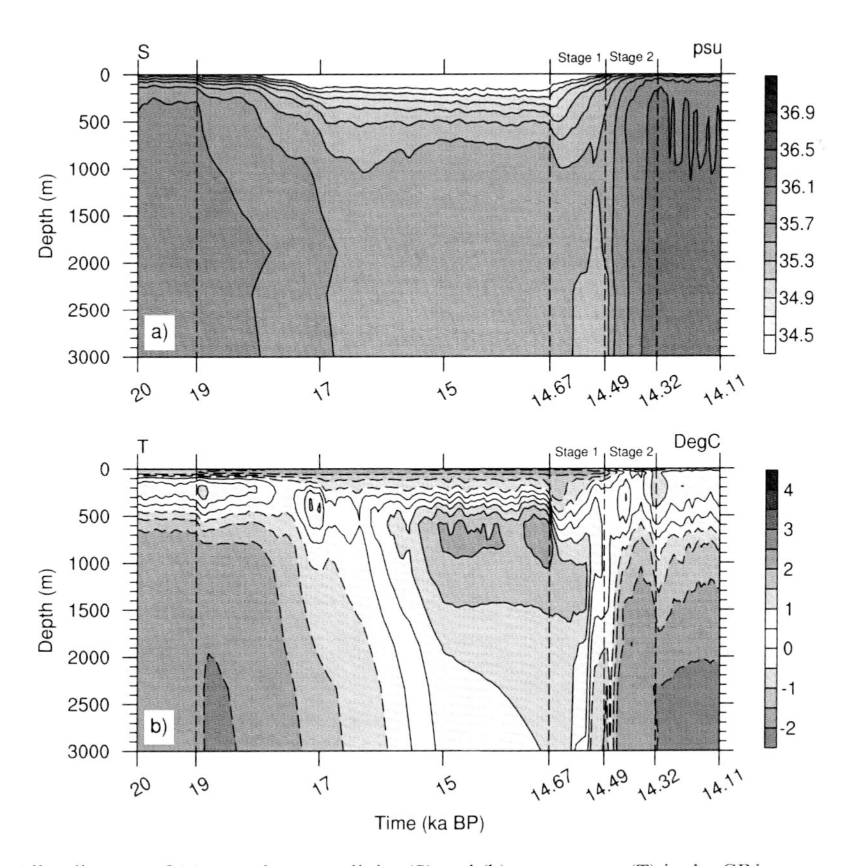

Figure 9. Hovmöller diagram of (a) annual mean salinity (S) and (b) temperature (T) in the GIN seas, averaged between 62°N and 80°N.

reinitiation of convection is limited to the approximately upper 2000 m. After the first sign of convection reinitiation, salinity and temperature both increase noticeably in the upper 2000 m, as a result of the resumed upper layer salt/heat transport from lower latitudes. The deep warm/fresh anomalies in the Labrador Sea soon dissipate, so that the change of salinity and temperature in Labrador Sea mainly happens during stage 1.

Similar to the Labrador Sea, a Hovmöller diagram of area mean salinity and temperature in the GIN seas (Figure 9) shows that the whole depth recovery of salinity and release of stored subsurface heat mainly occurs during stage 2. During stage 1, freshened water in the GIN seas penetrates downward several hundred meters, accompanied by the release of stored subsurface heat within several hundred meters too. However, the dynamic environment during stage 1 is still not sufficient to reinitiate the convection to deeper depths in this region and delays the reinitiation of NADW formation in the GIN seas to stage 2. During stage 2, salinity quickly increases at all depths, ultimately reaching higher levels than the initial glacial state (Figure 9a) and is accompanied by subsurface heat release with warming in the upper layers and cooling in the deeper layers (Figure 9b).

The evolution of potential density at depth in the Labrador Sea and GIN seas confirms the main conclusions from the Hovmöller diagram of salinity and temperature in these two regions (Figure 10). The increasing of potential density in Labrador Sea mainly happens during stage 1 within the upper 1000 m and in the GIN seas during stage 2 within the entire water column. A key development occurs in the GIN seas during stage 1, where a very weak stratification is formed followed by the densification of the entire water column at the onset of stage 2. It is derived by the increasing of the upper layers potential density and the small change in the deep layers. This extremely weak stratification provides a background for subsequent strengthened reinitiation of convection and NADW formation in the GIN seas during stage 2. The physical process of extremely weak stratification and reinitiation of NADW in the GIN seas is similar to the "density threshold" described by *Krebs and Timmermann* [2007a, 2007b] and *Renold et al.* [2009]. Now we will present the physical process in more detail.

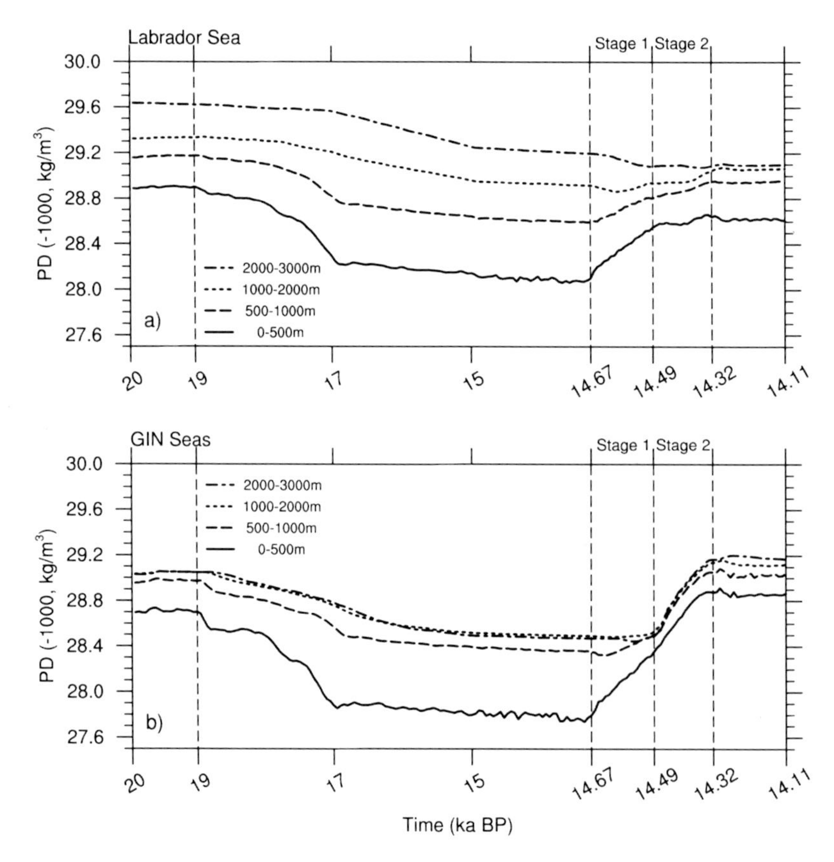

Figure 10. Annual mean potential density (PD) of the (a) Labrador Sea (50°–62°N, 70°–45°W) and (b) GIN seas (62°–80°N, 45°W–20°E), separated into vertical ocean layers.

4.2. *Reinitiation of NADW Formation*

The general picture of NADW formation in multiple origins from an unperturbed state could be described as follows: through air-sea heat and freshwater exchange, salty and warm water is transported to the North Atlantic within the upper ocean layers, which are driven by a meridional steric gradient; fresh and cold water sinks downward through deep convection in both NADW origin regions, i.e., Labrador Sea and GIN seas, and then moves southward driven by a deep ocean meridional pressure gradient.

As shown above, the recovery of AMOC is accompanied by increasing salinity (shown in Figures 6–9) and the retreat of sea ice cover (Figure 3) in the North Atlantic. The change in salinity is due to net salinity advection from low latitudes within the upper layers, and the retreat of sea ice cover significantly affects the local heat–freshwater exchange in the NADW origins. It seems natural that the evolution of these two processes should play a significant role during the AMOC recovery period. On the basis of this preliminary understanding of NADW formation processes, we investigate the simulated NADW reinitiation process during the AMOC recovery period of the last deglaciation.

The net salt transport to the Labrador Sea decreases during meltwater discharge, and it resumes primarily during stage 1 (Figure 11a), consistent with the increase in salinity in this region (Figure 8a). Before stage 1, the water of the Labrador Sea in its entire depth is significantly freshened and lightened through thousands of years of freshwater discharge. During stage 1, the transport of salty and dense water in upper layers into this area (Figures 8a and 10a) could directly induce a reversed vertical density structure (Figure 8a). The reversed density structure is followed by a resumed convection. As convection reinitiates, the net heat transport into the upper layers also increases (Figure 11b). This heat transport comes from both stored subsurface heat release (Figure 8b) and resumed warm water advection from low latitudes. This induces the abrupt retreat of sea ice in the Labrador Sea (Figure 12a) and a corresponding increase in surface heat loss (seen as a decrease in SHF) (Figure 12b).

The local oceanic surface heat loss and nonlocal salt water advection to the Labrador Sea in the upper ocean layers combine to cause upper layer densification, (Figure 10a) and thus the abrupt resumption of convection (Figure 12c). Subsurface warming also contributes to the resumption of convection, favoring a reversed vertical density structure. However, the subsequent convection quickly induces the subsurface heat anomaly to dissipate as it is ventilated (Figure 8b). These three factors simultaneously force NADW formation in the Labrador Sea to an enhanced level beyond its initial intensity (Figure 1, black line).

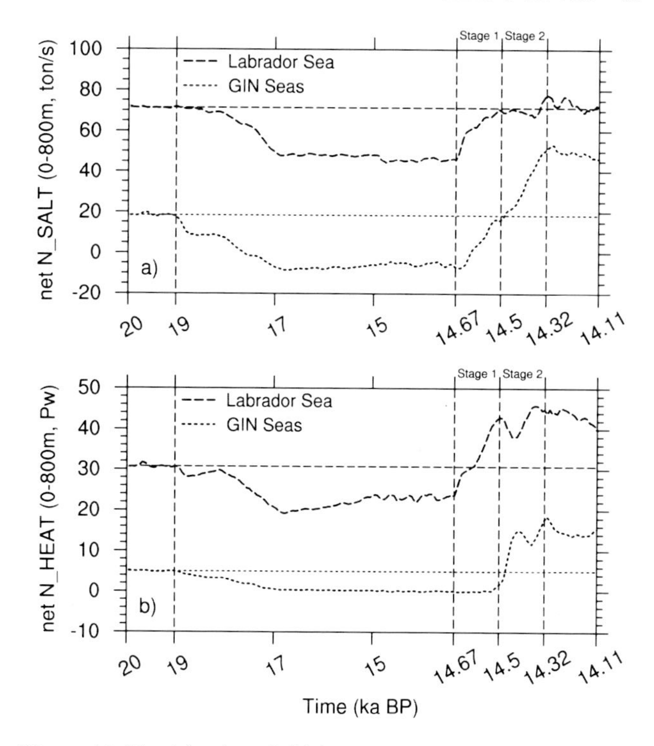

Figure 11. Net (a) salt and (b) heat transport in upper layers (0–800 m) for the Labrador Sea at 50°–62°N, dashed line, and GIN seas, 62°–80°N, dotted line.

The resumption of convection in the Labrador Sea spreads the dense water to nearly its whole depth and induces a meridional steric difference in the surface (Figure 13a, black line) and pressure difference in the deep ocean between the Labrador Sea and its southern ocean (Figure 13b, black line). This further induces resumption of northward salt/heat advection within the upper layers from low latitudes to the Labrador Sea (Figure 11) and southward deepwater flow from the Labrador Sea to low latitudes (Figure 1).

In the GIN seas, the reinitiation process of NADW formation during stage 2 is similar to the Labrador Sea, except the GIN seas first undergo a preconditioning process during stage 1. Net salt transport to the GIN seas within the upper layers resumes during stage 1 (Figure 11a), but net heat transport in the upper layers remains suppressed during this stage (Figure 11b). The net salt transport to the GIN seas induces the densification of its upper layers (Figures 2e and 10b), which causes shallow reinitiation of convection during this stage (Figure 12c) and is characterized by shallow penetration of freshened water and stored subsurface heat release (Figure 9). However, at this time, the dynamical environment is still not sufficient to reinitiate strong deep convection (Figures 9 and 12c) and NADW formation (Figure 1). The resumption of net heat transport to the GIN seas in the upper

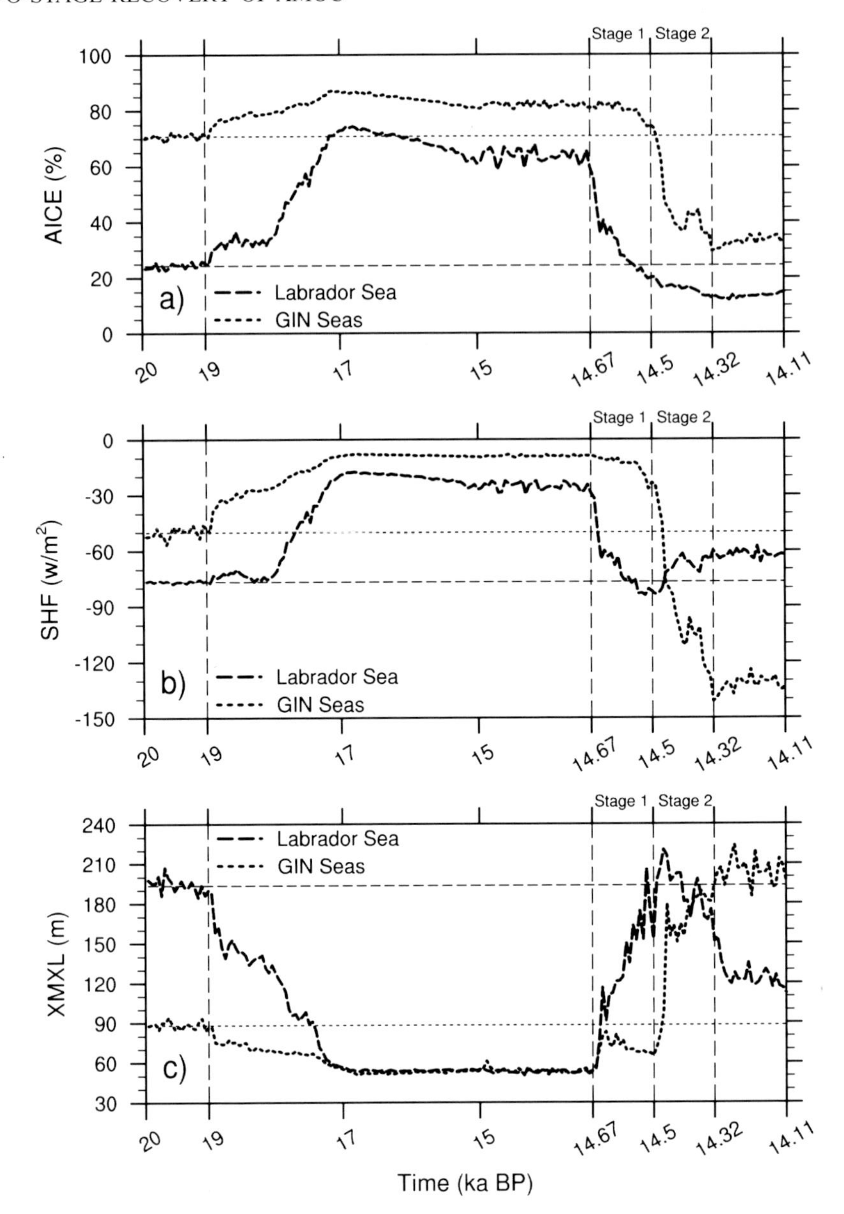

Figure 12. (a) Annual mean sea ice concentration (AICE), (b) surface heat flux (SHF), and (c) maximum mixed layer depth (XMXL). Dashed and dotted lines represent regional mean values for the Labrador Sea and GIN seas, respectively.

layers during stage 2 (Figure 9b) induces sea ice retreat (Figure 12a) and an abrupt shift in SHF (Figure 12b). Only after that, do deep convection (Figure 12c) and NADW formation (Figure 1) resume in the GIN seas. Meanwhile, the GIN seas steric and deep-pressure difference resume a little during stage 1 and then more robustly during stage 2 (Figure 13, shaded line). During stage 2, the relevant variables in the GIN seas can be seen to undergo an extreme shift and are enhanced above initial glacial values by a factor of 2 or more in many cases (Figures 11–13, shaded line). All These changes cause a strong reinitiation of NADW formation (Figure 1, light shaded line) and saltier/denser water in the GIN seas (Figures 9a and 10b). The denser water in the GIN seas at end of stage 2 contributes to the persistence of the AMOC overshoot and the mean state transition of AMOC (Figure 1, black line).

Among three factors, which promote NADW formation in the Labrador Sea and GIN seas, contributions of two factors significantly decrease or disappear at the end of each stage. One is the contribution of salinity advection, which is based on the salinity difference between the advected salty water

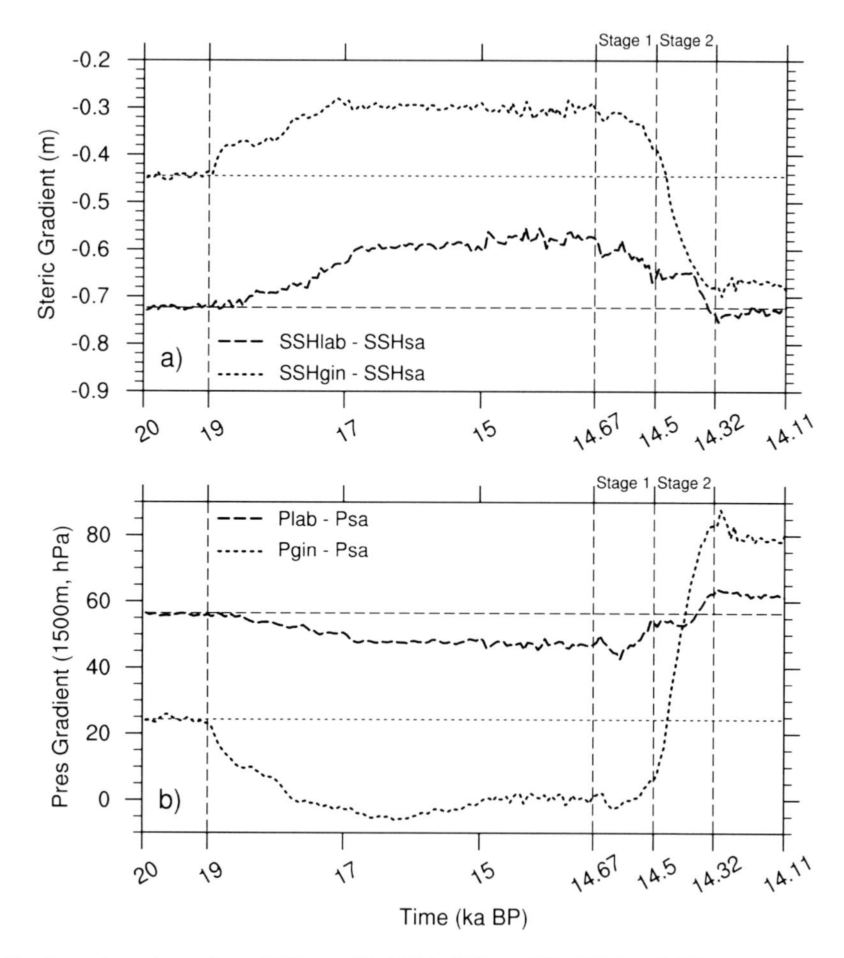

Figure 13. (a) Southward steric gradient (SSH_{lab}, 50°–62°N; SSH_{gin}, 62°–80°N; and SSH_{sa}, 30°S–50°N) and (b) deep ocean pressure gradient (1500 m, hPa) of the Atlantic Basin. Dashed and dotted lines represent values from the Labrador Sea and GIN seas, respectively.

and the local freshened water in the deeper layers of two regions of origin (Figure 14). Another is the contribution of subsurface warming, which disappears when convection is restarted (Figures 8b and 9b). These two processes together induce enhanced NADW formation in each NADW region, and their effects subside at the end of each corresponding stage (Figure 1).

According to the above analysis, contributions of salinity dominate the change of potential density in both NADW regions (Figure 7), and the subsurface heat anomaly provides an additional reversed vertical density structure for convection reinitiation in both stages (Figures 8b and 9b). The main contribution of net heat transport within the upper layers is to melt extended sea ice cover during each stage, which is shown to be of crucial importance in the reinitiation of deep convection in both NADW regions of origin (Figures 11b and 12c).

4.3. A Reconsidered Mechanism for NADW Reinitiation

On the basis of the comprehensive analysis of variables relevant to NADW formation in the DGL-A run, and the previous hypothesized mechanisms [*Vellinga and Wood*, 2002; *Yin et al.* 2006; *Krebs and Timmermann*, 2007a, 2007b], here we reconsider the mechanisms controlling NADW reinitiation with some clarified details (Figure 15).

In our reconsidered mechanism, we identify two dominant processes affecting NADW reinitiation. The first one is the local process as shown in Figure 15 (box 1 to box 2 to box 3 to box 4 to box 5 to box 6 to box 7 to box 1). This process is mainly dominated by the resumption of surface density flux over NADW origins. Resumption of surface density flux, primarily controlled locally by the retreat in sea ice cover, is largely initiated by northward heat transport to NADW formation regions (Figures 11b and 12a), then densified NADW

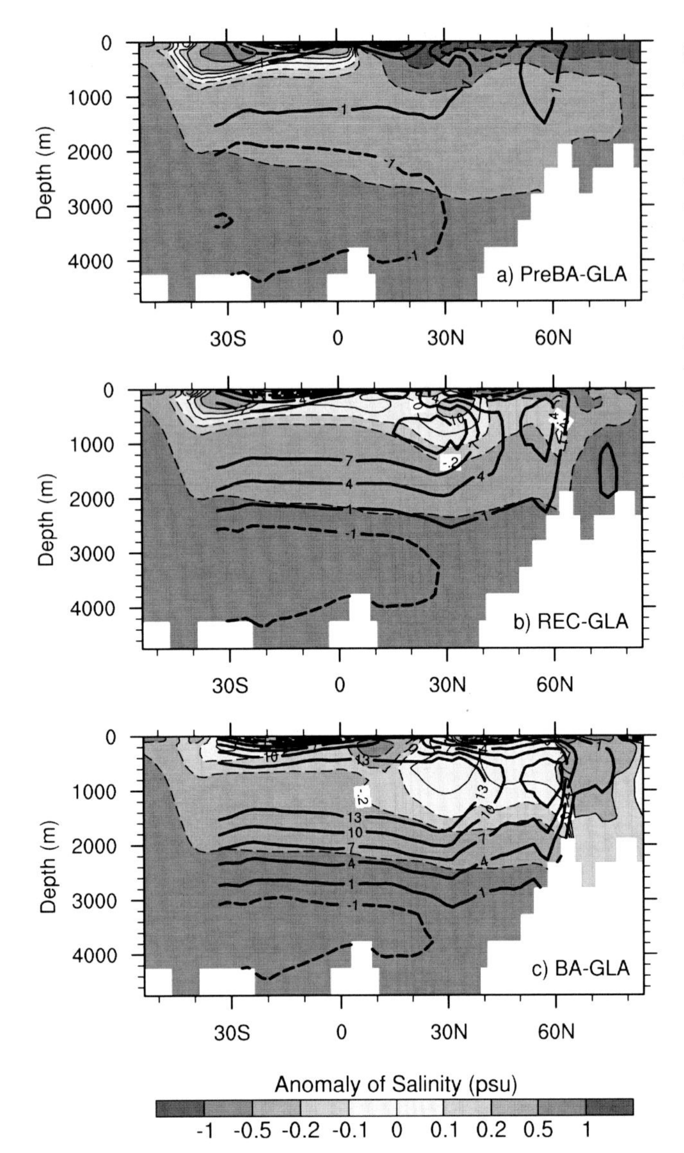

Figure 14. Change in Atlantic zonal averaged salinity anomaly (shading with thin contours) and AMOC stream function (thick contours) during (a) pre-BA–GLA, (b) REC-GLA, and (c) BA-GLA.

origin's surface layer reduces the stratification and starts the reinitiation of NADW. Here we can use SHF as a representation of surface density flux because its contribution to the density flux is dominant [*Shin et al.*, 2003b]. Second is the nonlocal process shown in Figure 15 (box 1 to box 4 to box 5 to box 6 to box 7 to box 1). The salty/dense water transport into the upper layers of each region (Figures 11a and 14) overlies fresher and lighter local water in deeper layers (Figure 14), directly reduces the stratification, and starts the reinitiation of NADW. The local deepwater formation from surface density flux and nonlocal dense water transport simultaneously contributes to the total NADW reinitiation during the whole recovery period.

This reconsidered mechanism provides a clearer picture of how NADW reinitiation could occur after meltwater discharge in the North Atlantic, with some important details that were lacking in previous proposed mechanisms such as the nonlocal process [*Vellinga and Wood*, 2002; *Yin et al.*, 2006; *Krebs and Timmermann*, 2007a, 2007b]. The contribution of subsurface warming to the reinitiation of NADW is not found to be a significant positive feedback, it is seen as a trigger for NADW reinitiation because of its effect on sea ice cover in each region, so we find that the other two processes are the factors responsible for the two-stage feature of AMOC recovery in this reconsidered mechanism.

4.4. Nature of Two-Stage Feature

As shown above, the reinitiation process of NADW formation in two main regions of origin is robustly asynchronous, first in the Labrador Sea and then in the GIN seas. So far, the cause of this asynchronous feature is still a critical issue for the understanding of AMOC recovery.

First, we should notice the difference between the NADW formation under an unperturbed background state and the recovery state after the halting of meltwater discharge. The variation of NADW formation during an unperturbed state is dependent on the local ocean surface density flux. However, during the period of recovery, there is a large-scale retreat of extended sea ice cover that directly affects the surface density flux and allows NADW reinitiation to occur. Another difference with the unperturbed state is that during the AMOC recovery period, the North Atlantic is freshened at greater depths than in other oceans, and the AMOC recovery process is accompanied by the resumption of salinity advection in the North Atlantic in the upper layers, enhancing NADW formation directly. So, owing to the unique local and nonlocal processes affecting NADW formation during AMOC recovery, the dynamical ocean environment is very different from the unperturbed state.

Second, two kinds of processes, which are mentioned in the reconsidered mechanism, each induce the asynchronous feature of the NADW reinitiation in its two origins. As discussed above, the local process is mainly affected by the extended sea ice cover, which retreats primarily as a result of heat transport from low latitudes. Since the GIN seas are located northward of the Labrador Sea, the sea ice retreat happens in the Labrador Sea before the GIN seas.

Third, it was also found that the efficiency of the nonlocal dense water transport is mainly controlled by the intensity of advection within the upper layers and meridional salinity gradient (Figure 14). After the freshening of the North

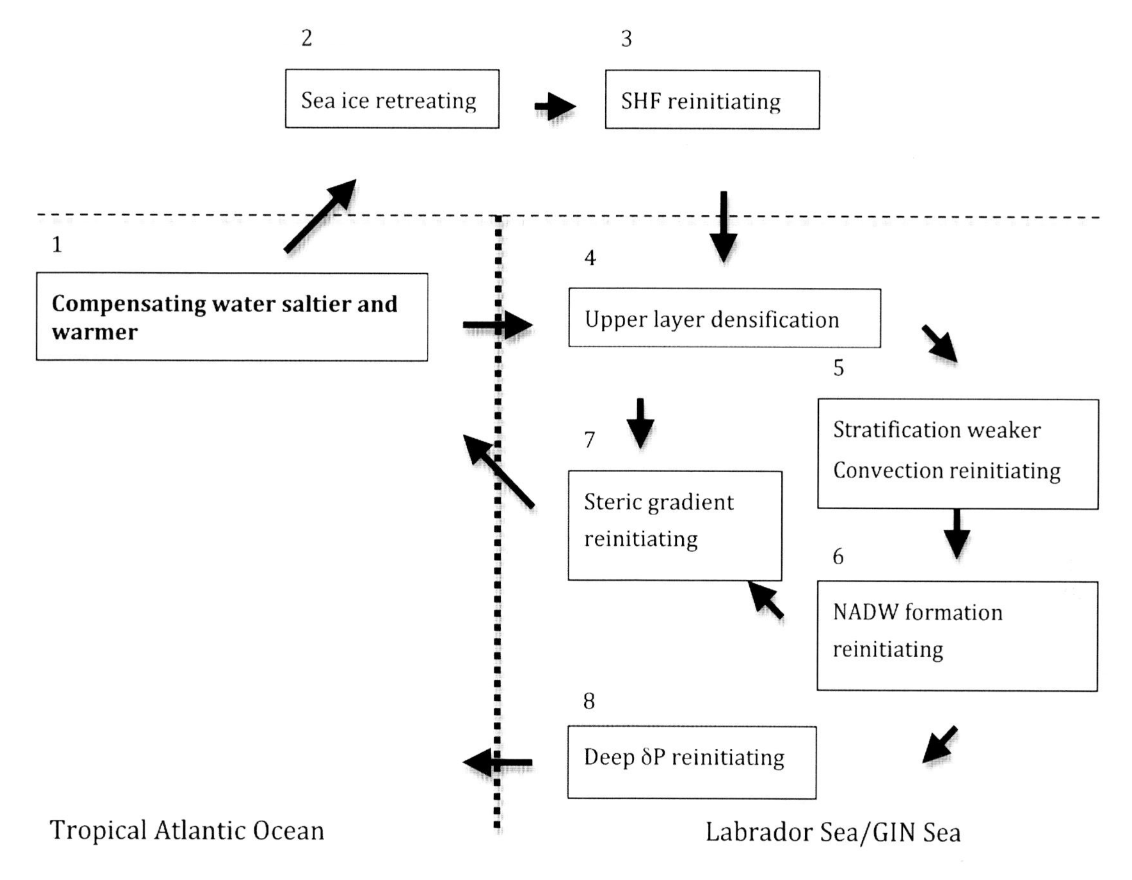

Figure 15. Schematic illustration of the proposed positive feedback process controlling reinitiation of NADW formation in Labrador Sea and GIN seas. Meridional pressure difference is indicated by δP.

Atlantic from meltwater discharge, the whole North Atlantic Basin is nearly homogeneous for salinity, so the salinity gradient at the south boundary of NADW regions is important for salt transport. During the AMOC recovery period, the maximum meridional salinity gradient within the upper layers of the North Atlantic moves northward.

On the basis of our reconsidered mechanism of NADW reinitiation, only after NADW formation is fully reinitiated in the Labrador Sea, does the dynamical environment of the GIN seas result in full convective reinitiation. So the combination of local and nonlocal mechanisms causes NADW formation reinitiation in the Labrador Sea and GIN seas to be asynchronous, and the sequence of reinitiation is northward.

5. DISCUSSIONS AND CONCLUSIONS

In this analysis, we found that the simulated AMOC recovery during the last deglaciation occurred in two stages: first, in the Labrador Sea and then in the GIN seas. We described how the two-stage feature of AMOC recovery depends on the asynchronous reinitiation of NADW formation in the two regions of origin with a reconsidered mechanism. We also suggest that the two-stage feature of AMOC recovery should not just depend on the evolutionary characteristics of a time series of AMOC intensity because a single time series is not sufficient to present the reinitiation process of NADW formation in multiple origins, synchronous or otherwise. Comparing the results of *Renold et al.* [2009], our results provide more physical and geographical analysis with the same model CCSM3 by providing a clearer and more complete mechanism, and these results are doubly interesting because they are based on a particular historical event of abrupt climate change.

Our results, in addition to previous work such as that of *Vellinga and Wood* [2002] and *Renold et al.* [2009], point out that the Labrador Sea and GIN seas both provide a comparable robust contribution to the AMOC recovery. However, some recent work also shows that the role of the multiple origins of NADW formation in AMOC variability and stability, as derived from different models [*Bentsen et al.*, 2004; *Mignot and Frankignoul*, 2005; *Latif et al.*, 2006; *Spence et al.*, 2008], is uncertain. Despite the dependence of model, however, the time scale should not be ignored when addressing the relative contribution of NADW multiple origins to AMOC change. For a time scale of interdecades or shorter in control simulation, the conclusions about the relative contribution do not seem consistent; but for a centuries or longer time scale in water-hosing experiments, the conclusions are consistent because of the big perturbation and large response according to current literature.

In this chapter, we use one new method, which is based on the zonal mean stream function of the Atlantic, to present the two-stage feature of the AMOC recovery in multiple regions of origin of NADW formation primarily. However, the two-stage feature not only depends on the new method but must also be confirmed in the traditional way, through examination of the mixed layer depth in Figure 12c and other different variables (Figures 2–5 and 8–13). The new method is not only capable of representing the change of NADW formation in multiple origins, it can quantify the volume of NADW formation in multiple origins, which cannot be achieved through the traditional way.

The results of the DGL-A run are consistent with an idealized water-hosing experiment under the Last Glacial Maximum state (not shown here). In this idealized water-hosing experiment, AMOC intensity fully recovered by the end of integration (1500 years). In this idealized experiment, stage 2 of AMOC recovery is mainly dominated by the adjustment process of NADW formation in the GIN seas with a time scale of about 500 years. The difference between these two simulations in the recovery of AMOC is that the AMOC intensity of DGL-A jumped to interglacial levels during the recovery process, forced by increased GHG concentration and orbital insolation [*Liu et al.*, 2009]. This difference is not critical to our dynamical analysis of the AMOC recovery process.

In considering the recovery process of the AMOC during last deglaciation, the AMOC overshoot is important in the generation of BA warming in TraCE-21000. So the AMOC overshoot phenomenon may be critical to the understanding of abrupt climate change. As shown above, the development of the AMOC overshoot depends on enhanced NADW formation in the GIN seas during stage 2. Another important aspect of AMOC recovery in DGL-A is the mean state transition of the AMOC from a glacial to an interglacial state. This transition depends significantly on the intensified NADW formation in the GIN seas during BA onset as well. So the GIN seas are not only a key region for the development of an AMOC overshoot but also the key region for mean state transition of AMOC within glacial/interglacial cycles.

With our interpretation of the asynchronous reinitiation of NADW in the Labrador Sea and GIN seas, we speculate that the southward order of NADW reinitiation from the GIN seas to the Labrador Sea by *Vellinga and Wood* [2002] may be caused by an absence of a shift in sea ice cover during periods of "artificially" suppressed AMOC, so the sequence of reinitiation of convection in multiple regions in the Vellinga and Wood work may not be correct. The conclusions of *Renold et al.* [2009] are reasonably correct for their experiments with a "real" AMOC suppressing process. Beyond our results, one point should be mentioned: the contrast reinitiation sequence of NADW is based on two models, northward sequence by *Renold et al.* [2009] and our study from CCSM3 and southward sequence by *Vellinga and Wood* [2002] from HadCM3. Despite the dependence of the experimental scheme we mentioned before, the sequence of NADW reinitiation may also be dependent on models, but this needs further validation.

In our DGL-A model run, the time delay between the onset of stage 1 and stage 2 generally coincides with the recovery of the AMOC to its intensity at the initial glacial state. However, according to *Renold et al.* [2009], under modern conditions the time delay is shorter, showing stage 2 onset before the AMOC has made its initial recovery. This may be an indication that the relationship between the onset time of each stage and the timing of initial AMOC recovery can have an important connection to the climate state.

Acknowledgments. This research was supported by the Paleoclimate Program of NSF, NCAR, DOE, Peking University, Special public sector research of CMA, China (GYHY200906016), and Innovation Plan to Graduate Students in the Universities of Jiangsu Province, China (CX07B_043z) and NUIST (Y602).

REFERENCES

Arzel, O., M. H. England, and W. P. Sijp (2008), Reduced stability of the Atlantic meridional overturning circulation due to wind stress feedback during glacial times, *J. Clim.*, *21*, 6260–6282, doi:10.1175/2008JCLI2291.1.

Barker, S., P. Diz, M. J. Vautravers, J. Pike, G. Knorr, I. R. Hall, and W. S. Broecker (2009), Interhemispheric Atlantic seesaw response during the last deglaciation, *Nature*, *457*, 1097–1102, doi:10.1038/nature07770.

Bentsen, M., H. Drange, T. Furevik, and T. Zhou (2004), Simulated variability of the Atlantic meridional overturning circulation, *Clim. Dyn.*, *22*, 701–720, doi:10.1007/s00382-004-0397-x.

Bitz, C. M., J. C. H. Chiang, W. Cheng, and J. J. Barsugli (2007), Rates of thermohaline recovery from freshwater pulses in modern, Last Glacial Maximum, and greenhouse warming climates, *Geophys. Res. Lett.*, *34*, L07708, doi:10.1029/2006GL029237.

Broecker, W. S. (1990), Salinity history of the northern Atlantic during the last deglaciation, *Paleoceanography*, *5*, 459–467, doi:10.1029/PA005i004p00459.

Carlson, A., D. W. Oppo, R. E. Came, A. N. LeGrande, L. D. Keigwin, and W. B. Curry (2008), Subtropical Atlantic salinity variability and Atlantic meridional circulation during the last deglaciation, *Geology, 36*(12), 991–994, doi:10.1130/G25080A.

Clarke, P. U., N. G. Pisias, T. F. Stocker, and A. J. Weaver (2002), The role of the thermohaline circulation in abrupt climate change, *Nature, 415*, 863–869, doi:10.1038/415863a.

Collins, W. D., et al. (2006), The Community Climate System Model version 3 (CCSM3), *J. Clim., 19*, 2122–2143, doi:10.1175/JCLI3747.1.

Ganopolski, A., and S. Rahmstorf (2001), Rapid changes of glacial climate simulated in a coupled climate model, *Nature, 409*, 153–158, doi:10.1038/35051500.

Gent, P. R. (2001), Will the North Atlantic Ocean thermohaline circulation weaken during the 21st century?, *Geophys. Res. Lett., 28*, 1023–1026, doi:10.1029/2000GL011727.

Goosse, H., H. Renssen, F. M. Selten, R. J. Haarsma, and J. D. Opsteegh (2002), Potential causes of abrupt climate events: A numerical study with a three-dimensional climate model, *Geophys. Res. Lett., 29*(18), 1860, doi:10.1029/2002GL014993.

Hu, A., G. A. Meehl, and W. Han (2007), Role of the Bering Strait in the thermohaline circulation and abrupt climate change, *Geophys. Res. Lett., 34*, L05704, doi:10.1029/2006GL028906.

Hu, A., B. L. Otto-Bliesner, G. A. Meehl, W. Han, C. Morrill, E. Brady, and B. P. Briegleb (2008), Response of thermohaline circulation to freshwater forcing under present day and LGM conditions, *J. Clim., 21*, 2239–2258, doi:10.1175/2007JCLI1985.1.

Knorr, G., and G. Lohmann (2003), Southern Ocean origin for the resumption of Atlantic thermohaline circulation during deglaciation, *Nature, 424*, 532–536, doi:10.1038/nature01855.

Knutti, R., J. Fluckiger, T. F. Stocker, and A. Timmermann (2004), Strong hemispheric coupling of glacial climate through freshwater discharge and ocean circulation, *Nature, 430*, 851–856, doi:10.1038/nature02786.

Krebs, U., and A. Timmermann (2007a), Fast advective recovery of the Atlantic meridional overturning circulation after a Heinrich event, *Paleoceanography, 22*, PA1220, doi:10.1029/2005PA001259.

Krebs, U., and A. Timmermann (2007b), Tropical air–sea interactions accelerate the recovery of the Atlantic meridional overturning circulation after a major shutdown, *J. Clim., 20*, 4940–4956, doi:10.1175/JCLI4296.1.

Latif, M., C. Böning, J. Willebrand, A. Biastoch, J. Dengg, N. Keenlyside, U. Schweckendiek, and G. Madec (2006), Is the thermohaline circulation changing?, *J. Clim., 19*, 4631–4637, doi:10.1175/JCLI3876.1.

Levermann, A., A. Griesel, M. Hofmann, M. Montoya, and S. Rahmstorf (2005), Dynamic sea level changes following changes in the thermohaline circulation, *Clim. Dyn., 24*, 347–354, doi:10.1007/s00382-004-0505-y.

Lippold, J., J. Grützner, D. Winter, Y. Lahaye, A. Mangini, and M. Christl (2009), Does sedimentary ^{231}Pa/^{230}Th from the Bermuda Rise monitor past Atlantic meridional overturning circulation?, *Geophys. Res. Lett., 36*, L12601, doi:10.1029/2009GL038068.

Liu, Z., et al. (2009), Transient simulation of last deglaciation with a new mechanism for Bølling-Allerød warming, *Science, 325*, 310–314, doi:10.1126/science.1171041.

Manabe, S., and R. J. Stouffer (1995), Simulation of abrupt climate change induced by freshwater input to the North Atlantic Ocean, *Nature, 378*, 165–167, doi:10.1038/378165a0.

Manabe, S., and R. J. Stouffer (1997), Coupled ocean-atmosphere model response to freshwater input: Comparison to Younger Dryas event, *Paleoceanography, 12*(2), 321–336, doi:10.1029/96PA03932.

McManus, J. F., R. Francois, J. M. Gherardi, L. D. Keigwin, and S. Brown-Leger (2004), Collapse and rapid resumption of Atlantic meridional circulation linked to deglaciation climate changes, *Nature, 428*, 834–837, doi:10.1038/nature02494.

Mignot, J., and C. Frankignoul (2005), The variability of the Atlantic meridional overturning circulation, the North Atlantic Oscillation, and the El Niño–Southern Oscillation in the Bergen Climate Model, *J. Clim., 18*, 2361–2375, doi:10.1175/JCLI3405.1.

Mignot, J., A. Ganopolski, and A. Levermann (2007), Atlantic subsurface temperatures: Response to a shutdown of the overturning circulation and consequences for its recovery, *J. Clim., 20*, 4884–4898, doi:10.1175/JCLI4280.1.

Milne, G. A., and J. X. Mitrovica (2008), Searching for eustasy in deglacial sea level histories, *Quat. Sci. Rev., 27*, 2292–2302, doi:10.1016/j.quascirev.2008.08.018.

Otto-Bliesner, B. L., E. C. Brady, G. Clauzet, R. Tomas, S. Levis, and Z. Kothavala (2006), Last Glacial Maximum and Holocene climate in CCSM3, *J. Clim., 19*, 2526–2544, doi:10.1175/JCLI3748.1.

Rahmstorf, S. (2002), Ocean circulation and climate during the past 120,000 years, *Nature, 419*, 207–214, doi:10.1038/nature01090.

Renold, M., C. C. Raible, M. Yoshimori, and T. F. Stocker (2009), Simulated resumption of the North Atlantic meridional overturning circulation – Slow basin-wide advection and abrupt local convection, *Quat. Sci. Rev., 29*(1), 101–112, doi:10.1016/j.quascirev.2009.11.005.

Schmittner, A., and E. D. Galbraith (2008), Glacial greenhouse gas fluctuations controlled by ocean circulation changes, *Nature, 456*, 373–376, doi:10.1038/nature07531.

Shin, S.-I., Z. Liu, B. Otto-Bliesner, E. C. Brady, J. E. Kutzbach, and S. Harrison (2003a), A simulation of the Last Glacial Maximum climate using the NCAR-CCSM, *Clim. Dyn., 20*, 127–151, doi:10.1007/s00382-002-0260-x.

Shin, S.-I., Z. Liu, B. L. Otto-Bliesner, J. E. Kutzbach, and S. J. Vavrus (2003b), Southern Ocean sea-ice control of the glacial North Atlantic thermohaline circulation, *Geophys. Res. Lett., 30*(2), 1096, doi:10.1029/2002GL015513.

Spence, J. P., M. Eby, and A. J. Weaver (2008), The sensitivity of the Atlantic meridional overturning circulation to freshwater forcing at eddy-permitting resolutions, *J. Clim., 21*, 2697–2710, doi:10.1175/2007JCLI2103.1.

Stouffer, R. J., et al. (2006), Investigating the causes of the response of the thermohaline circulation to past and future climate changes, *J. Clim., 19*, 1365–1387, doi:10.1175/JCLI3689.1.

Vellinga, M., and R. A. Wood (2002), Global climatic impacts of a collapse of the Atlantic thermohaline circulation, *Clim. Change*, *54*, 251–267.

Wright, D. G., and T. F. Stocker (1991), A zonally averaged ocean model for the thermohaline circulation, Part I: Model development and flow dynamics, *J. Phys. Oceanogr.*, *21*, 1713–1724, doi:10.1175/1520-0485(1991)021<1713:AZAOMF>2.0.CO;2.

Yin, J., M. E. Schlesinger, N. G. Andronova, S. Malyshev, and B. Li (2006), Is a shutdown of the thermohaline circulation irreversible?, *J. Geophys. Res.*, *111*, D12104, doi:10.1029/2005JD006562.

E. C. Brady and B. L. Otto-Bliesner, Climate and Global Dynamics Division, National Center for Atmospheric Research, Boulder, CO 80307-3000, USA.

J. Cheng, Key Laboratory of Meteorological Disaster of Ministry of Education and College of Atmospheric Sciences, Nanjing University of Information Science and Technology, Nanjing, 210044, China. (chengjun@nuist.edu.cn)

F. He, Z. Liu, and M. Wehrenberg, Center for Climatic Research and Department of Atmospheric and Oceanic Sciences, University of Wisconsin-Madison, Madison, WI 53706, USA.

The Role of Hudson Strait Outlet in Younger Dryas Sedimentation in the Labrador Sea

Harunur Rashid

Byrd Polar Research Center, Ohio State University, Columbus, Ohio, USA

David J. W. Piper

Geological Survey of Canada (Atlantic), Bedford Institute of Oceanography, Dartmouth, Nova Scotia, Canada

Benjamin P. Flower

College of Marine Science, University of South Florida, St. Petersburg, Florida, USA

In the Younger Dryas (YD) the circum-North Atlantic region returned to near glacial conditions, widely attributed to the release of fresh water from the Laurentide Ice Sheet (LIS). Hudson Strait was a major outlet for fresh water, icebergs, and sediment from the LIS that deposited Heinrich (H) layers throughout the North Atlantic. Heinrich layer 0 (H0), the YD equivalent ice-rafting event, has been reported to be absent in deep water seaward of the mouth of Hudson Strait. We have identified meter-thick H0 carbonate-rich sediments of YD age seaward of Hudson Strait as nepheloid-flow deposits and as turbidites on the northeast Sohm Abyssal Plain. Otherwise, H0 is generally absent in cores from the Labrador Basin. The thick proximal H0 bed indicates an important supply of fresh water through the Hudson Strait outlet of the LIS during the YD. The sparse distal sediment distribution and values in the planktonic foraminiferal $\delta^{18}O$ in the Labrador Sea, less depleted than during H1, were probably the result of the position of the ice margin, duration of freshwater discharge, and dilution of freshwater signature with the ambient seawater during H0. This change is interpreted to result from differences in the retreat of the ice margin at the mouth of Hudson Strait, resulting in different styles of sediment and freshwater transport during the H0 compared to earlier H events.

1. INTRODUCTION

The last deglaciation was interrupted by a near glacial, millennial-scale Younger Dryas (YD) cooling event in the Northern Hemisphere from 12.9 to 11.6 ka. Evidence for the YD equivalent event is found in ice cores, marine sediment cores, and continental paleoarchives [*Alley and Clark*, 1999]. It is marked by a ~10°C mean annual temperature fluctuation

Abrupt Climate Change: Mechanisms, Patterns, and Impacts
Geophysical Monograph Series 193

10.1029/2010GM001011

in the Greenland ice cores [*Steffensen et al.*, 2008; *Grachev and Severinghaus*, 2005], rapid changes in North Atlantic sea surface temperature, advance of the polar front, and sea ice expansion during the winter in the Northern Hemisphere [*Ruddiman and McIntyre*, 1981; *Sarnthein et al.*, 2003; *Isarin et al.*, 2008]. Deepwater proxy records including the Cd/Ca and Pa/Th ratios [*Boyle and Keigwin*, 1987; *Lippold et al.*, 2009, and references therein] have also shown a weaker ventilation.

Ice advanced during the YD in the Hudson Strait outlet of the Laurentide Ice Sheet (LIS) [*Andrews and MacLean*, 2003]. Earlier ice advances and retreats resulted in Heinrich (H) layers which were dominated by distinctive detrital carbonate being deposited in the Labrador Sea and beyond [*Andrews et al.*, 1993; *Dowdeswell et al.*, 1995; *Stoner et al.*, 1996]. *Andrews et al.* [1995] interpreted a H event (Heinrich layer 0 (H0)) at the time of the YD on the basis of sparse core data on the southeastern Baffin Shelf and the presence of detrital carbonate in cores reported in the literature from more distal sites. These authors inferred that the distribution of the H0 bed was predominantly restricted to the Labrador Shelf.

In this study, we used new northern Labrador Sea cores data to document the presence of a YD H0 bed in deep water seaward of Hudson Strait. We reinterpret published data on the possible extent of the YD signal in the greater Labrador Sea. These data are used to compare the dispersal of the YD freshwater discharge with that of H1 seaward of Hudson Strait and through the Labrador Sea to the North Atlantic and thus its contribution to abrupt climate change through the rapid supply of fresh water [*Tarasov and Peltier*, 2005]. Interpretation of the sedimentology of the H0 and H1 deposits provides evidence for the timing and magnitude of freshwater discharge during H events, of relevance to the coupled ice sheet, ocean, and atmospheric modeling studies.

2. ICE MARGINS, ICE EXTENT DURING EARLIER HEINRICH EVENTS, AND ICE EXTENT AT YD

Glacial ice from the large Hudson Bay Basin drained through Hudson Strait, a narrow graben between Precambrian bedrock of Labrador/Nouveau Québec and Baffin Island (Figure 1a). Hudson Strait is glacially overdeepened locally to 900 m and terminates in a shallow sill ranging from above sea level (Resolution Island) to 450 m depth (Figure 1b). The strait acted as a conduit in dispersing glacigenic sediments and meltwater from Hudson Bay to the Labrador Sea during the last glacial cycle [*Andrews and MacLean*, 2003]. Seaward of the sill, a broad transverse trough up to 600 m deep in Hatton Basin crosses the 250 km wide shelf.

A marine detrital carbonate record corresponding to H0 was recognized by *Andrews et al.* [1995] in three closely spaced cores 20 km off Resolution Island (Figure 1b). *Andrews et al.* [1995] also reported the absence of H0 in two cores (Hu75-55 and Hu75-56) on the lower continental slope seaward of Hudson Strait and pointed out that H0 was not generally recognized in deepwater Labrador Sea cores, despite the presence of H1 and older H layers. These authors also recognized a coeval carbonate-rich layer in several cores off SE Baffin Island, in which a high dolomite/calcite ratio indicated a source from Baffin Island ice [*Andrews et al.*, 1993, 1995].

On the Labrador Shelf, several Holocene high-carbonate beds have been recognized. These beds have been correlated with the Gold Cove and Noble Inlet ice advance/retreat phases in Hudson Strait [*Miller and Kaufman*, 1990; *Kaufman et al.*, 1993; *Andrews et al.*, 1999], the drainage of glacial Lake Agassiz [*Hillaire-Marcel et al.*, 2007], and the final collapse of the Foxe ice dome [*Barber et al.*, 1999; *Utting et al.*, 2007]. The accepted calibrated ages for these events are 10.1–10.4, 8.9–9.2, 8.2–8.4, and 7 ka, respectively [*Kaufman et al.*, 1993; *Thomas et al.*, 2007; *Utting et al.*, 2007; *Rashid et al.*, 2008]. Elsewhere on the shelf seaward of Hudson Strait, any possible H0 layer is either too deep to be penetrated by conventional piston coring or is not preserved in iceberg-scoured terrain or is represented by an unconformity on the inner part of the shelf [*Hall et al.*, 1999].

Earlier studies have shown that H0 is absent from many cores on the Labrador Slope and in the deep Labrador Basin [*Andrews et al.*, 1995; *Rashid and Grosjean*, 2006; *Rashid and Piper*, 2007]. We have compiled the status of H0 in all available cores from the Labrador Sea in Figure 1a. In most of these cores, the absence of H0 is not due to a coring artifact causing core top loss [e.g., *Rashid and Grosjean*, 2006].

3. MATERIALS AND METHODS

In this study, we use cores for which the glacial record has been previously described but in which little attention was paid to the YD record. Sediment cores Hu97048-16 and Hu97048-09 (hereinafter Hu97-16 and Hu97-09) [*Rashid et al.*, 2003a] were retrieved at 1085 and 1575 m water depths, respectively, from the slopes off Hudson Strait and off Saglek Bank (Figure 1b and Table 1). Core Hu98039-05 (hereinafter Hu98-05) was retrieved at 5232 m water depth from the northeastern Sohm Abyssal Plain (SAP), just down flow from the terminus of the Northwest Atlantic Mid-Ocean Channel (NAMOC) [*Piper and Hundert*, 2002] (Figure 1a).

Cores were subsampled at 2.5 to 5 cm intervals with 5 cm^3 plastic vials. Wet sediment samples were dried in an oven at

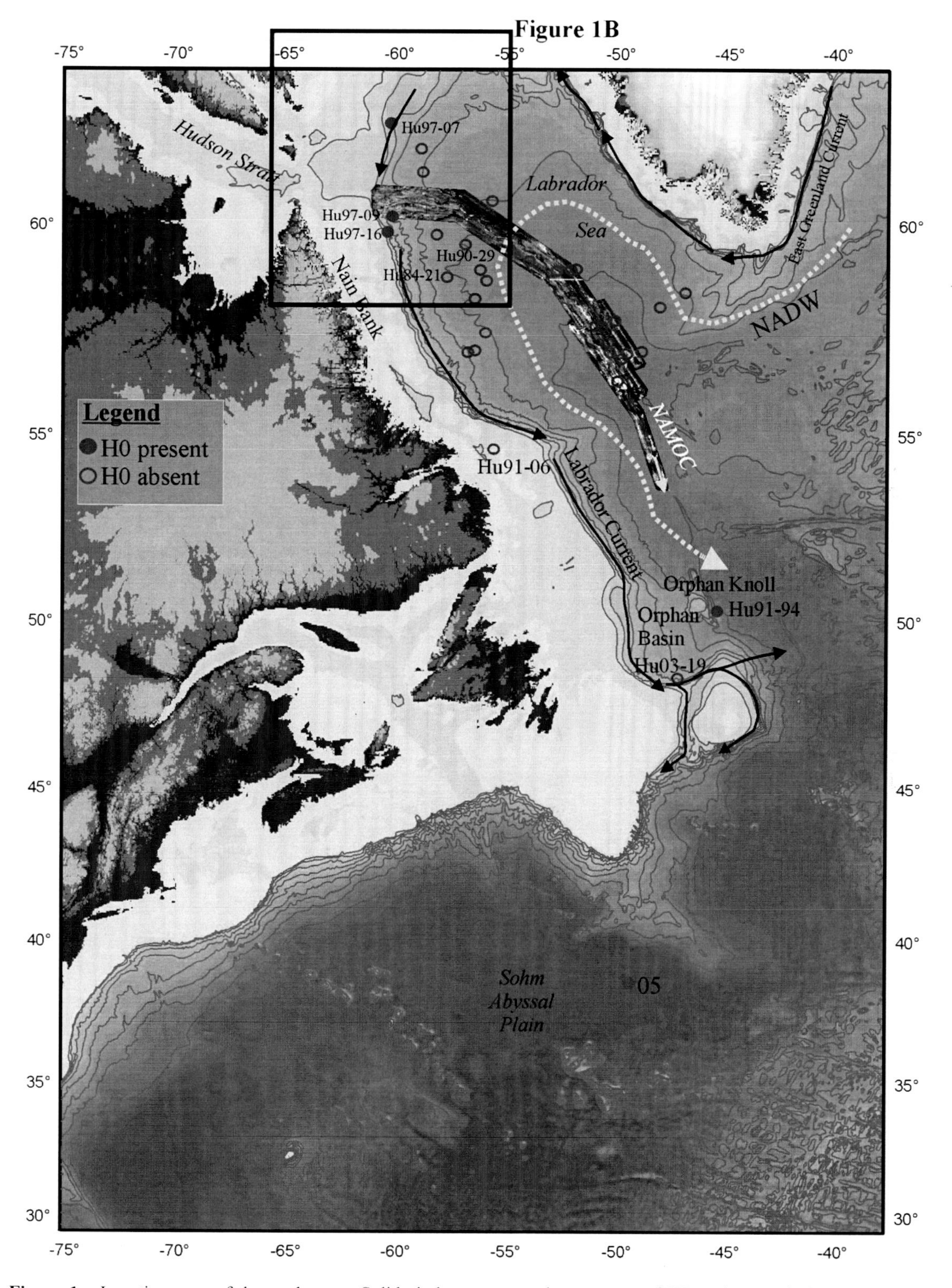

Figure 1a. Location map of the study area. Solid circles represent the presence of H0, and open circles represent the absence of H0. To maintain clarity, we select to plot a few representative cores where the H0 is absent in the Labrador Basin and slope because of the abundances of cores available. The SeaMarc backscatter shows the location of the Northwest Atlantic Mid-Ocean Channel (NAMOC) (the deep-sea channel in the Labrador Sea) [*Hesse et al.*, 1996]. The interpreted Younger Dryas Labrador Current and North Atlantic Deep Water (NADW) paths are also outlined.

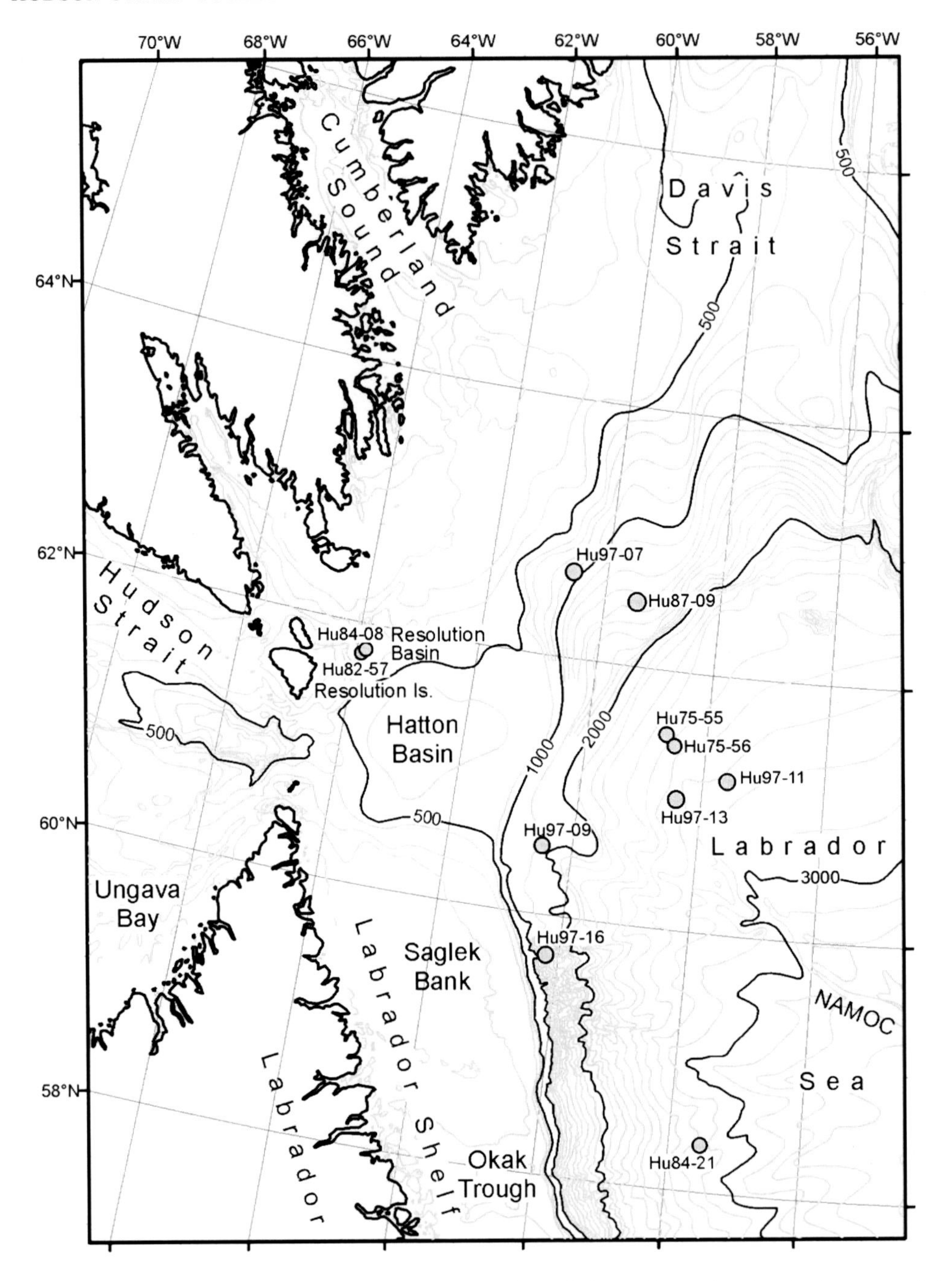

Figure 1b. Close-up view of the Hudson Strait, Resolution Basin, and Hatton Basin showing the locations of the cores discussed in the text.

60°C for 48 h. Bulk carbon and total organic carbon contents were analyzed from 0.5 g of dry sample using a LECO CS-125 IR analyzer. Inorganic carbonate content is calculated as if all the carbonate were $CaCO_3$ (wt %). Abundance of dolomite was estimated by X-ray diffraction. The remaining sample was wet sieved using a >63 μm sieve. The 150–250 μm fraction was used to pick foraminifera for radiocarbon dates and stable isotopes. The >150 μm fraction was used for identification of different components of ice-rafted detritus (IRD) such as limestone and dolostone fragments of different colors and compositions, quartz, granite, basalt, terrigenous sedimentary rock fragments, etc. [*Rashid et al.*, 2003a].

Stable isotopes were measured on 40–50 handpicked planktonic *Neogloboquadrina pachyderma* (s). From all

Table 1. The ^{14}C-AMS Dates in Cores Hu97048-09, Hu97048-16, and Hu98039-05

Subsample	Depth (cm)	Uncorrected ^{14}C Age (years) $\pm 1\sigma$ Error	Calendar Year B.P.[a] (1σ Age Range)	Reference[b]
		Hu97048-09		
1	135–136	10,710 ± 35	11,934–12,065	this study
2	240–242	11,569 ± 75	12,948–13,089	1
3	255–257	11,440 ± 160	12,841–13,069	1
4	380–382	12,770 ± 40	14,082–14,274	this study
5	435–437	13,130 ± 40	14,864–15,095	this study
6	460–462	14,870 ± 160	17,007–17,666	1
7	495–497	13,600 ± 110	15,323–15,755	1
8	865–867	14,620 ± 40	16,705–17,092	this study
9	875–877	15,500 ± 75	18,137–18,361	1
10	1100–1102	20,400 ± 140	23,711–24,095	1
		Hu97048-16		
1	175–176	9,100 ± 40	9,673–9,844	this study
2	300–302	10,150 ± 35	11,106–11,164	this study
3	460–462	11,830 ± 30	13,212–13,277	this study
4	470–472	14,620 ± 170	16,592–17,233	1
5	558–559	14,900 ± 130	17,080–17,672	1
6	559–561	14,990 ± 45	17,338–17,755	this study
7	750–752	20,990 ± 200	24,276–24,924	1
8	840–842	21,200 ± 110	24,833–25,322	1
9	858–859	21,660 ± 245	25,501–25,778	1
10	900–902	22,800 ± 110	26,906–27,172	1
		Hu98039-05		
1	169–172 (strong leach)	9,210 ± 25	9,888–10,037	this study
2	169–172	9,240 ± 25	9,949–10,093	this study
3	249–254	10,070 ± 120	10,868–11,148	this study
4	369–374	12,180 ± 170	13,407–13,746	this study

[a]All ^{14}C-AMS dates were determined on *Neogloboquadrina pachyderma* (sinistral) (Nps), which were converted to calendar age (B.P.) by using Calib 6.0.1 [*Reimer et al.*, 2009] calibration program (see text).

[b]Reference 1 is *Rashid et al.* [2003a].

samples, CO_2 was extracted at 90°C with an ISOCARB™ device online with a VG-PRISM™ mass spectrometer (Center for Research in Isotopic Geochemistry and Geochronology, Université du Québec à Montréal) using a marble standard calibrated against the Carrara marble and other current standard materials of the International Atomic Energy Agency. Results were corrected and converted to the Vienna Peedee belemnite scale. The overall analytical reproducibility, as determined from replicate measurements on the standard material, is routinely better than ±0.04‰ ($\pm 1\sigma$) and ±0.08‰ ($\pm 1\sigma$) for $\delta^{13}C$ and $\delta^{18}O$, respectively.

Accelerator mass spectrometric radiocarbon (^{14}C-AMS) dates were used to constrain the stratigraphy of all cores. These dates were obtained using 800–1200 handpicked tests/sample of *N. pachyderma* (s). All ages reported here were calculated after isotopic normalization (i.e., for a $\delta^{13}C$ value of −25‰) using Libby's ^{14}C half-life as provided by the laboratories (National Ocean Sciences Accelerator Mass Spectrometry Facility at Woods Hole Oceanographic Institution, IsoTrace Laboratory of University of Toronto, National Science Foundation Accelerator Mass Spectrometry Facility at University of Arizona, and Keck Carbon Cycle Laboratory of University of California at Irvine). The ^{14}C-AMS dates were converted to calendar years B.P. (1950) using the CALIB6.0.1 online software program [*Reimer et al.*, 2009] where MARINE09 data sets were used. We have adopted the age with the highest probability, and the calibrated ages are reported as "ka" (see Table 1). We assigned a regional reservoir age correction (ΔR) of 50 years [*Reimer et al.*, 2009] to account for the apparent age of dissolved inorganic carbon in high-latitude sea surface waters of the northwest Labrador Sea not accounted for in the MARINE09 data set [*Reimer et al.*, 2009]. This ΔR may be a minimal estimate for the early Holocene conditions in the northwest Labrador Sea. For example, preindustrial ΔR for infaunal mollusks in Hudson Strait and adjacent areas has been estimated as 334 ±

170 years and for the Labrador Shelf and adjacent areas 140 ± 73 years [*Hillaire-Marcel et al.*, 2007]. *Lewis et al.* [2009] have concurred with the suggestion of *Hillaire-Marcel et al.* [2007] that, along the Labrador margin and Newfoundland margins, the ΔR may range from 200 to 300 years, depending on the duration of the sea ice cover.

4. RESULTS

4.1. H0 Records Seaward of Hudson Strait

On the continental slope immediately south of the mouth of the Hatton Basin transverse trough, a thick H0 bed is present in two cores. New ^{14}C-AMS dates and increases in data resolution since *Rashid et al.* [2003a] have allowed us to confidently identify H0 in cores Hu97-16 and Hu97-09 on the basis of bracketing radiocarbon ages, high carbonate, increase in IRD, and sedimentary structures (Figures 2 and 3). Three ages of 11.99, 13.02, and 12.96 ka bracket H0 in core Hu97-09, whereas a basal date of 13.25 ka at 461 cm and a date of 11.14 ka at 301 cm allowed us to clearly identify H0 in Hu97-16. The sedimentological features of H0 are similar to those described by *Rashid et al.* [2003a] from H1 and H2. At the base of H0 is a 10 to 30 cm thick sediment unit exhibiting a gradual upward increase in randomly oriented dropstones in hemipelagic sediments. The middle unit comprises sediments similar to those interpreted as nepheloid-flow deposits by *Rashid et al.* [2003a], which

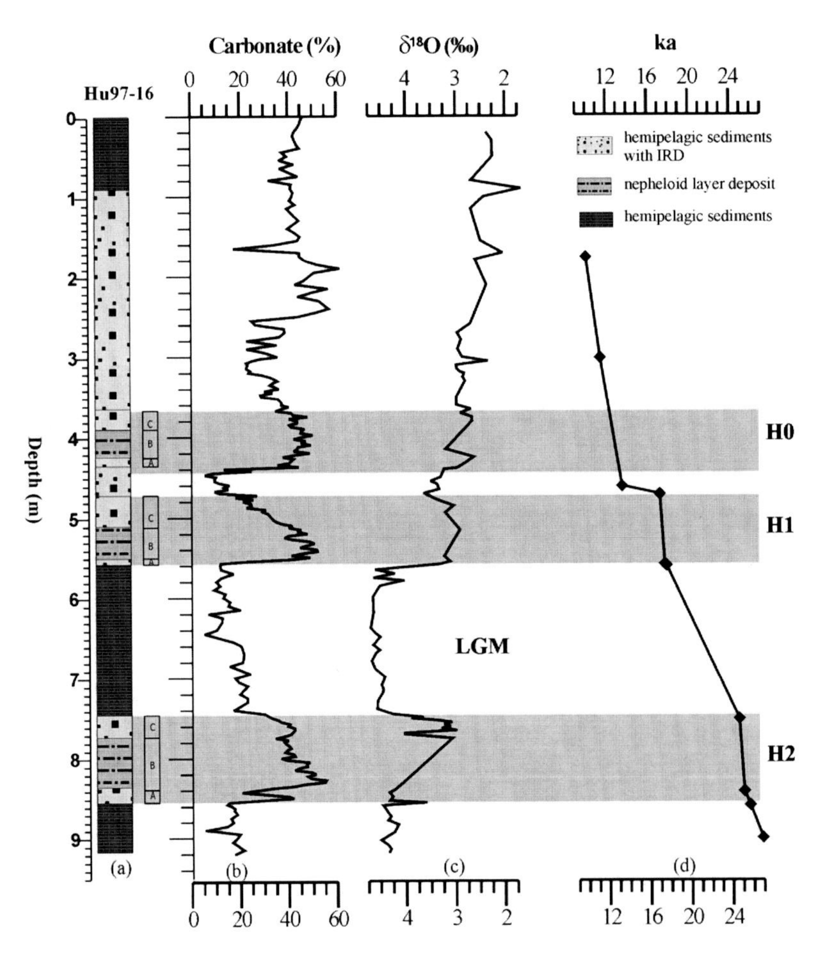

Figure 2. Core Hu97-16 from the slope seaward of Hudson Strait, with down-core plots of (a) sediment facies, (b) carbonate (%), (c) δ^{18}O in *Neogloboquadrina pachyderma* (s), and (d) ^{14}C-AMS (ka) versus depth (m). It was collected from 1085 m water depth. The ^{14}C-AMS dates were used to constrain the age model and were converted to calendar years (B.P.) using the CALIB 6.0.1 program [*Reimer et al.*, 2009] (see text for details). Horizontal gray bars represent H0, H1, and H2 events [*Rashid et al.*, 2003a]. LGM is Last Glacial Maximum.

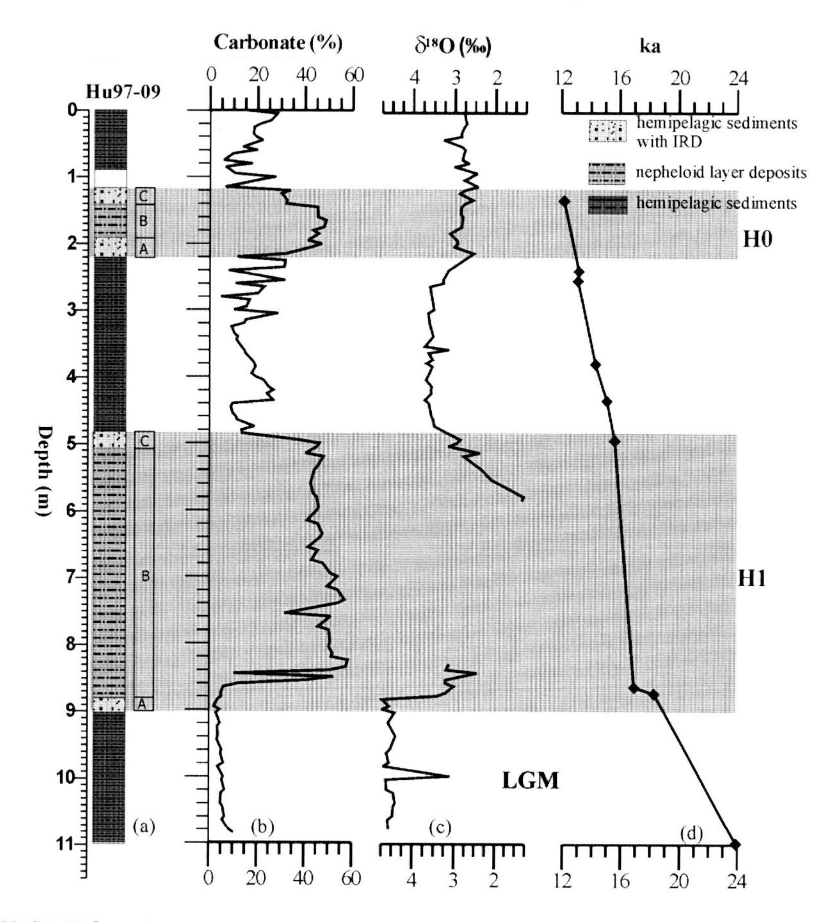

Figure 3. Core Hu97-09 from the slope seaward of Hudson Strait showing (a) sediment facies, (b) carbonate (%), (c) $\delta^{18}O$ in *N. pachyderma* (s), and (d) ^{14}C-AMS dates (ka) versus depth (m). It was collected at 1575 m water depth. Lack of *N. pachyderma* (s) prevented acquiring of $\delta^{18}O$ between 5.8 and 8.4 m. Horizontal gray bars represent H0 and H1 [*Rashid et al.*, 2003a].

show an alternation of thinly laminated IRD and finely laminated detrital carbonate-rich fine-grained sediments with occasional dropstones. The upper part of H0 shows a gradual transition to the overlying hemipelagic sediments. X-ray diffraction on bulk sediments and petrological identification of IRD granules in core Hu97-16 demonstrate the dominance of calcite in H0, as well as in H1 [*Rashid et al.*, 2003a; *Rashid and Piper*, 2007]. The $\delta^{18}O$ in *N. pachyderma* (s) ($\delta^{18}O_{Nps}$) in H0 is 0.5‰ depleted compared with immediately underlying sediment but is little different from overlying Holocene sediment (Figures 2 and 3).

Core Hu97-11, which was retrieved seaward of the central part of the Hatton Basin transverse trough (Figure 4), has no detectable record of an H0 layer, despite a high-resolution record in the trigger weight core. Core Hu97-13 also appears to lack H0 but may have experienced core top loss [*Rashid and Piper*, 2007]. In cores Hu7509-55 and Hu7509-56 at 2440 and 2446 m water depths, respectively, *Kirby* [1998] identified a H0 bed overlying a well-developed H1 and H2, based on a slight increase in carbonate and higher calcite/dolomite ratio (Figures 5e and 5f). In the absence of age control and a clearer carbonate signature, we regard this interpretation of H0 as unconfirmed: the uppermost part of the piston core is commonly disturbed, and the slight carbonate signature might be of Holocene age. Because of the possibility of core top loss, the absence of H0 cannot be confirmed from these cores. H0 in cores Hu82-57 and Hu84-08, which was retrieved at 549 and 580 m water depths from the Resolution Basin, was identified by the high-carbonate intervals (Figures 5a and 5b). The identification of H0 in these cores is constrained by six published dates [*Andrews et al.*, 1995]. On the southeast Baffin Slope, core Hu97048-07, which was retrieved at 938 m water depth, contains a 50 cm thick H0 bed identified by radiocarbon

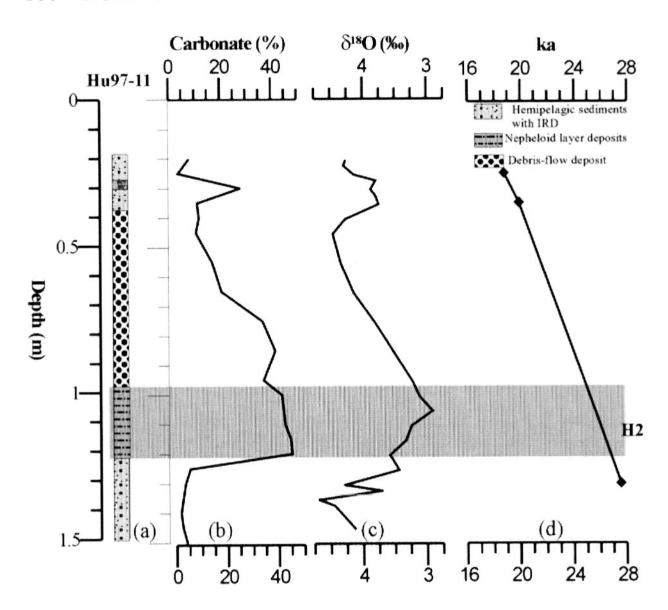

Figure 4. Core Hu97-11twc (trigger weight core) from the slope seaward of Hudson Strait showing (a) sediment facies, (b) carbonate (%), (c) δ^{18}O in *N. pachyderma* (s), and (d) ^{14}C-AMS dates (ka) versus depth (m). It was collected at 2640 m water depth from the rise off Hudson Strait. Horizontal gray bar represents H2.

dates of 12.74 and 12.79 ka (Figure 5c) [*Barber*, 2001]. Nearby core Hu87033-09 has an unusual carbonate and calcite/dolomite profile (Figure 5d), quite unlike other records from this region, which may be the result of one or more debris flow deposits in this core [*Barber*, 2001].

4.2. Distal H0 Records in the Labrador Sea

We briefly reassess cores from the literature in the central and southern Labrador Sea that may contain the H0 layer. Core Hu84-21, seaward of Okak Trough at 2853 m water depth, has previously been reported to contain H0 [*de Vernal and Hillaire-Marcel*, 1987; *Hillaire-Marcel et al.*, 2007]. Five published ^{14}C-AMS dates converted to calibrated ages (Figure 6) were used to constrain the stratigraphy between core top and 180 cm, and the stratigraphy of rest of the interval was obtained by age extrapolation assuming a constant sedimentation rate. The issue is whether a high-carbonate bed between 212 and 236 cm is H0 or H1. Insufficient core sample remains to date this interval. We suggest that the previous interpretation of H0 in this core is not robust, and therefore, its identification remains an open question. However, the transition from the enriched $\delta^{18}O_{Nps}$ to 1.2‰ more depleted across this high-carbonate bed is closer in magnitude to the change in isotopic values at H1, for example, as shown by *Hillaire-Marcel et al.* [1994], *Rashid et al.* [2003a], and *Rashid and Piper* [2007], than the 0.5‰ change at H0. No other well-controlled core from the central or southern Labrador Slope or Labrador Rise contains a H0 carbonate-rich layer [*Rashid et al.*, 2003a; *Rashid and Piper*, 2007].

Core Hu91045-06 (hereinafter Hu91-06) was retrieved at 534 m water depth from Cartwright Saddle on the Labrador Shelf [*Hillaire-Marcel et al.*, 1994, 2007]. It contains a high-carbonate bed from 510 to 540 cm, with an underlying published date from benthic foraminifera of 11.55 ka at 605 cm, indicating that the high-carbonate bed correlates either to the Noble Inlet or the 10.2 ka Gold Cove event [*Miller and Kaufman*, 1990; *Kaufman et al.*, 1993; *Rashid et al.*, 2008]. The bottom of the core is undated, suggesting that either there is no H0 bed or that it was not penetrated. Nearby cores Hu87033-17 and Hu87033-18 [*Andrews et al.*, 1999] also contain a carbonate-rich bed underlain by a date of 10.51 ka from a mollusk shell. The base of core Hu87-18, which appears not to contain a deeper carbonate-rich bed, has an age of 13.80 ka from *N. pachyderma* (s); this date is older than the YD period. In the only other long core on the Labrador Shelf intersecting the YD interval, in Karlsefni Trough [*Hall et al.*, 1999], the YD interval is marked by an unconformity, probably due to advance of continental ice.

On the southern slope of Orphan Basin, carbonate-rich beds in cores Hu2003033-19 and Hu2003033-24 (hereinafter Hu03-19 and Hu03-24) are well constrained by numerous published ^{14}C-AMS dates [*Hillaire-Marcel et al.*, 2007; *Tripsanas and Piper*, 2008a]. In Hu03-19, H1 is constrained by an underlying date of 18.08 ka and is overlain by a distinctive red meltwater plume facies found throughout the Orphan Basin. The overlying carbonate-rich bed from 190 to 210 cm is dated only 10.38 ka at 180 cm. In Hu03-24, H1 is constrained by an underlying age of 18.07 ka and is also immediately overlain by the red meltwater plume facies. The overlying carbonate-rich bed, which stratigraphically correlates with nearby core Hu03-19, is bracketed by ages of 11.29 and 9.54 ka, implying that it correlates with either the Noble Inlet or the 10.2 ka Gold Cove event [*Rashid et al.*, 2008]. This same stratigraphy, with H1, a red meltwater facies bed, and then an early Holocene carbonate-rich bed, is recognized in an additional 12 cores throughout the Orphan Basin [*Tripsanas and Piper*, 2008a]. No H0 carbonate-rich bed was recognized.

Core Hu91-94 off Orphan Knoll (3448 m water depth) contains multiple H layers characterized by high carbonate, a basal peak in coarse fraction, and variation in $\delta^{18}O_{Nps}$ [*Hillaire-Marcel et al.*, 1994]. H0 was convincingly recognized between 167 and 184 cm, where the boundary at the base is constrained by dates of 13.05 and 13.03 ka and the bed is overlain by a date of 10.207 ka (Figure 7).

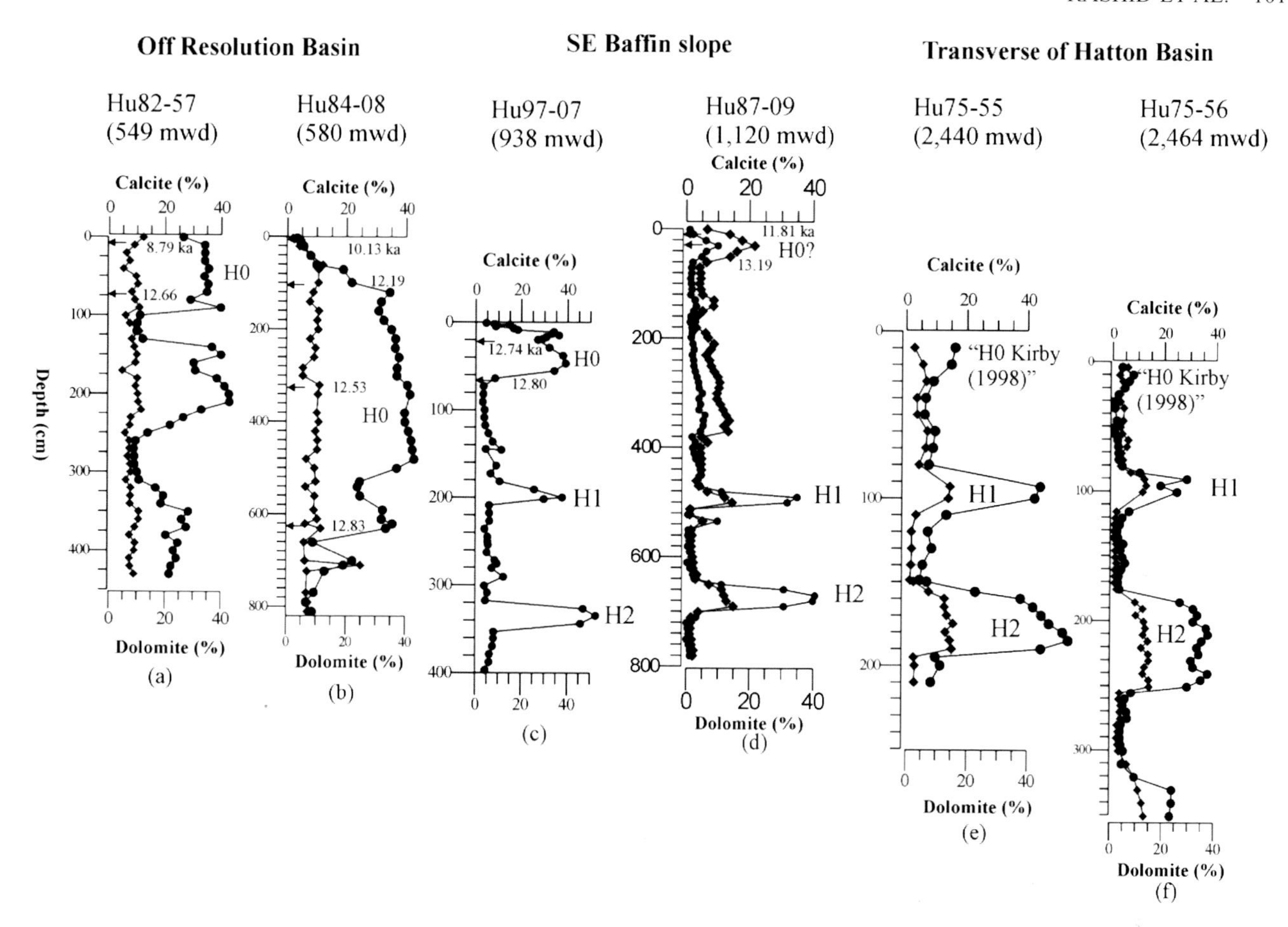

Figure 5. Calcite (%) (solid circles) and dolomite (%) (diamonds) from cores (a) Hu82-57, (b) Hu84-08, (c) Hu97-09 (calcite only), (d) Hu87-09, (e) Hu75-55, and (f) Hu75-56 from the Baffin margin are plotted in depth (cm) scales. H0, H1, and H2 events are constrained by 53 ^{14}C-AMS dates. Calcite and dolomite data of these cores are from the works of *Andrews et al.* [1993, 1995], *Barber* [2001], *Kirby* [1998], and this study. The ^{14}C-AMS dates are converted to calendar years B.P. shown as "ka" (see section 3 for details).

4.3. H0 Records in Sohm Abyssal Plain of the North Atlantic

Fine-grained turbidites with high carbonate content are present on the northeastern SAP in core Hu98-05. These turbidites interbed with terrigenous mud turbidites and very thin hemipelagic intervals (Figure 8). Detrital petrography and the location of the core at the terminus of the NAMOC led *Piper and Hundert* [2002] to identify Hudson Strait ice as the source of the carbonate turbidites, which were transported through the NAMOC. A ~1.2 m thick unit at about 3 m subbottom is characterized by high carbonate, L color, and magnetic susceptibility with prominent laminations picked out by changes in the abundance of silt (Figure 8). The proportion of coarse silt decreases upward through the bed [*Benetti*, 2006]. Two ^{14}C-AMS dates immediately above and below this unit bracket its age between 11.0 and 13.6 ka so that the carbonate interval most likely correlates with the YD.

5. DISCUSSION

5.1. Accuracy of Radiocarbon Dating

Reconstructing stratigraphy close to a former continental ice sheet margin setting is a difficult task, as many of the sediments are devoid of biogenic carbonate required for the ^{14}C-AMS dating. In addition, the transport of older carbon and uncertainty in surface ocean reservoir ages complicate the stratigraphy. We have used a regional reservoir anomaly (ΔR) of 50 years [*Reimer et al.*, 2009] for converting the ^{14}C-AMS dates to calendar years to construct the age models of the cores used in this study. This is probably a minimum estimate of ΔR as discussed in the methods section; this ΔR could be hundreds of years. No accurate estimate of the ΔR is available from the Labrador Sea. However, it is assumed that the ΔR could vary by as much as 400–500 years, as revealed

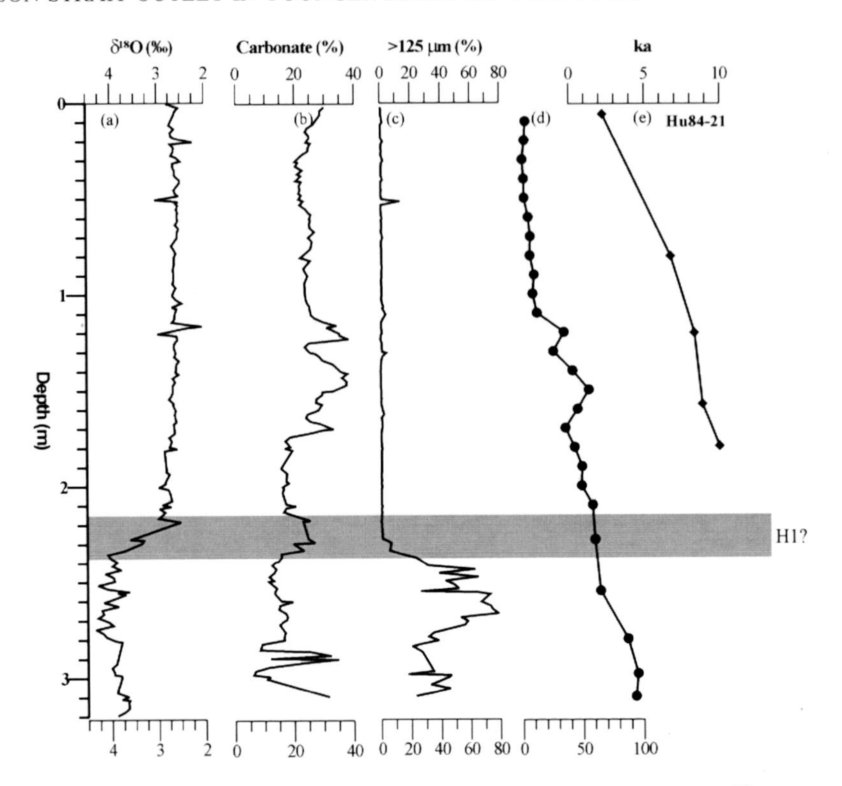

Figure 6. Core Hu84-21 from the lower slope off Okak Trough with down-core plots of (a) $\delta^{18}O$ on *N. pachyderma* (s), (b) carbonate (%), (c) >125 μm (%), (d) dinoflagellate species *Brigantedinium* spp. (%), and (e) ^{14}C-AMS (ka B.P.) versus depth (m). The core was retrieved at 2853 m water depth. Horizontal gray bar denotes H1(?). Two high-carbonate beds between 1.2 and 1.55 m were interpreted as the Agassiz drainage event by *Hillaire-Marcel et al.* [2007], while the high-carbonate event centered at 1.7 m was correlated to the 10.2 ka freshwater forcing event [*Rashid et al.*, 2008].

from marine records of the northeastern Atlantic [*Waelbroeck et al.*, 2002; *Bondevik et al.*, 2006; *Peck et al.*, 2008]. In addition, it is possible that the extension of the ice shelf and sea ice cover in the Labrador Sea may have decreased the ocean-atmospheric exchange further, thus increasing the duration of the ΔR.

Two ^{14}C-AMS dates in core Hu97-09 at 241 and 256 cm are statistically identical (Table 1). Although the 1σ uncertainty of 160 years is high for such a young date, given the small (119 years) age reversal, we do not have any strong reason to discard it. The 14,870 ± 160 ^{14}C year at 461 cm between two dates of 13,130 ± 40 and 13,600 ± 110 ^{14}C year (Table 1 and Figure 3) shows stratigraphic reversal, and we suspect that it is contaminated by older carbon [*Rashid et al.*, 2003b]. Taking the rest of the ^{14}C-AMS dates at face value, it shows that the sedimentation rates in cores Hu97-16 and Hu97-09 abruptly increased during both H0 and H1, resulting in beds 50 to 100 and 90 to 380 cm in thickness, respectively. In contrast, equivalent thicknesses of H0 and H1 vary from merely 5 to 20 cm in the open North Atlantic [*Grousset et al.*, 1993; *Rashid and Boyle*, 2007]. On the other hand, the sedimentation rates in between the H events, such as during the Last Glacial Maximum or in the intervals from 460 to 470 and 865 to 875 cm in cores Hu97-16 and Hu97-09, respectively, are extremely low, consistent with the previous findings of *Hillaire-Marcel et al.* [1994]. In summary, the high sedimentation rate cores from the northwest Labrador Sea allowed us unambiguous identification of H1 and H0.

Some ^{14}C-AMS dates of H layers were obtained at 5–10 cm below the bottom boundary to avoid contamination by reworked sediments, and as a result, our basal dates are always older than the accepted dates for the onset of H layers. This assertion is based on the recently revised Greenland Ice Core Chronology 2005 (GICC05) [*Rasmussen et al.*, 2006; *Svensson et al.*, 2008], where the positions of the stadials/interstadials and hence the H events are clearly constrained. Other ages determined within the H layers may have a minor influence of old carbon. For these reasons, and because of the uncertainties in ΔR, our dates cannot be used to estimate the duration of H events.

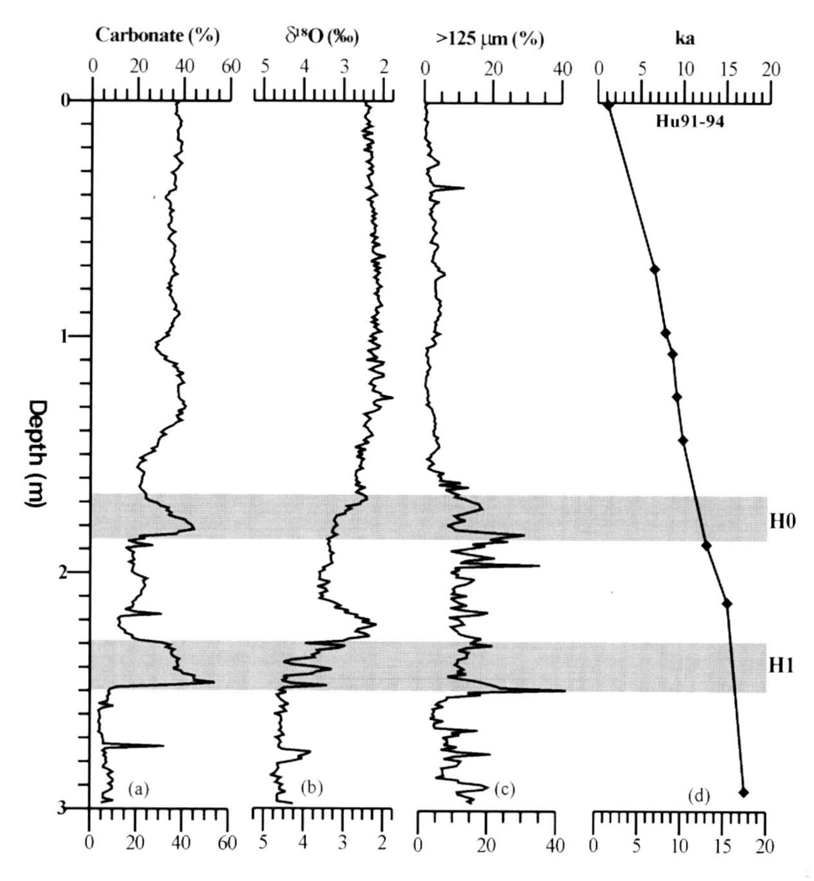

Figure 7. Down-core (a) carbonate (%), (b) $\delta^{18}O$ in *N. pachyderma* (s), (c) >120 μm (%), and (d) ^{14}C-AMS (ka) versus depth (m) are plotted from core Hu91-94 [*Hillaire-Marcel et al.*, 1994]. It was retrieved at 3448 m water depth from the flank of Orphan Knoll. Horizontal gray bars indicate H0 and H1 [*de Vernal et al.*, 2001]. Note that a minor reversal of 21 years between two ^{14}C-AMS dates at 1.88 and 1.93 m was observed; according to *Hillaire-Marcel et al.* [2007], this difference was statistically insignificant and could be discarded.

5.2. Reliability of Identification of the YD in the Northwest Labrador Sea Cores

In both cores Hu97-09 and 97-16, H1 has been identified with confidence on the basis of radiocarbon dating and in Hu97-16 is further supported by the identification of H2 (Figure 2). Dates bounding the H layers are consistent with dates in intervening hemipelagic sediment, indicating that any old carbon effect is minor. Even if the uncertainty in ΔR values was as great as 1000 years, our identification of H1 and H2 would not change. The bounding dates of 13.02 and 13.25 ka below H0 and 11.99 and 11.14 ka above H0 in the two cores mean that the bed identified as H0 can neither be reasonably correlated with the 10.1–10.4 ka Gold Cove event [*Kaufman et al.*, 1993; *Rashid et al.*, 2008] nor the 15.6–16.9 ka H1 event [*Severinghaus et al.*, 2009; *Wang et al.*, 2008], even if the uncertainty in ΔR is hundreds of years. *Bakke et al.* [2009] have precisely constrained the duration of the YD period (12.896 to 11.703 ka) by acquiring 86 ^{14}C-AMS dates correlating the Lake Kråkenes chronology with the GICC05 chronology and the Asian monsoon records [*Dykoski et al.*, 2005; *Wang et al.*, 2008], and our revised chronology for the identification of YD/H0 event from the Labrador margin is well within their estimated errors.

5.3. Source of Sediment to the H0 Layer

In all but one core from around the Hudson Strait outlet, the calcite/dolomite ratio is no different in H0 from what it is in H1 and H2 (Figure 5). The record in core Hu87033-09 is clearly abnormal (Figure 5d), probably due to a debris flow deposit being sampled. Furthermore, counting of IRD in H2, H1, and H0 in core Hu97-16 (Figure 2) shows no significant difference in the relative abundance of limestone and dolostone [*Rashid et al.*, 2003a]. This means that the same source discharging through Hudson Strait supplied

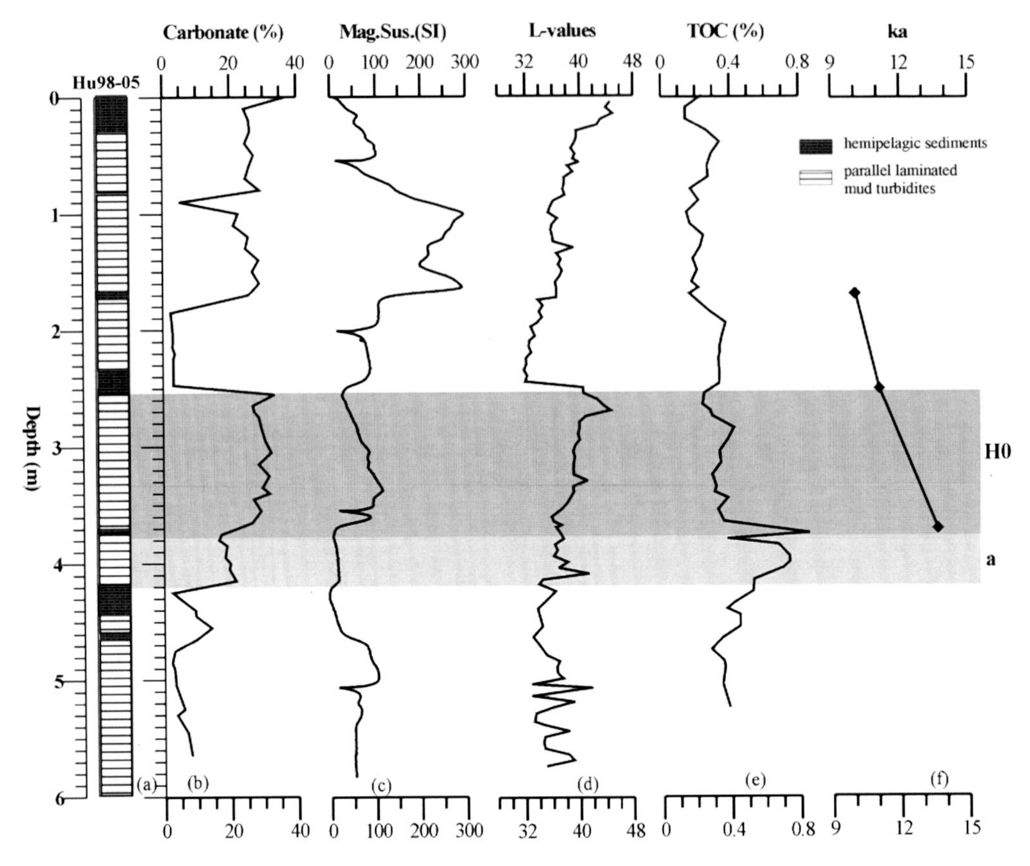

Figure 8. Core Hu98-05 from the Sohm Abyssal Plain (SAP) with down-core plots of (a) sediment facies, (b) carbonate (%), (c) magnetic susceptibility (SI units), (d) L color, (e) total organic carbon (% TOC), and (f) ^{14}C-AMS (ka) versus depth (m). It was retrieved at 5232 m water depth from the northeastern SAP just down flow from the terminus of the NAMOC. Four calibrated ^{14}C-AMS dates in *N. pachyderma* (s) were used to constrain the stratigraphy between 1.7 and 3.71 m. Lack of *N. pachyderma* (s) prevented determination of $\delta^{18}O$ from this core. H0 is denoted by the horizontal gray bar, whereas "a," the lightly shaded bar, represents the underlying carbonate-rich sediments with high TOC (%).

the bulk of the sediment seaward of Hudson Strait during all three H events.

5.4. Distribution of the H0 Detrital Carbonate Bed

The distribution of the high-carbonate H0 bed is quite different from H1 and H2, although all are recognized primarily by the presence of carbonate-rich sediments discharged through Hudson Strait. H0 is virtually absent in the Labrador Sea southeast of core Hu97-16, whereas H1 and H2 are widespread. The variation in their character and distribution would be influenced by several factors: (1) the volume of fresh waters discharged during ice stream retreat, (2) the position of the ice margin, (3) the strength of the Labrador Current, and (4) the extent of sea ice cover in the Labrador Sea.

The similar sedimentological character of the H0 and H1 beds in cores Hu97-16 and Hu97-09 (Figures 2 and 3) suggests similar depositional processes during the two events. Variations in thickness may result from differences in the advection trajectory of the icebergs and surface plume, but the >50% smaller mean thickness of H0 compared with H1 suggests a smaller and/or shorter discharge of sediment. This is consistent with the less depleted $\delta^{18}O_{Nps}$ signature in cores Hu97-09, Hu97-16, and Hu91-94 in H0 and the absence of significant depleted $\delta^{18}O_{Nps}$ signal in many Labrador Sea cores. It thus seems likely that the volume of meltwater discharged during H0 was less than during H1.

Dinoflagellate assemblage data [*Hillaire-Marcel et al.*, 2007; *de Vernal et al.*, 2001] show that sea ice cover varied from 9 to 12 months year^{-1} during both H0 and H1, not only at site Hu84-21 (Figure 6) off Okak Trough but also at a farther offshore site Hu91-94 (Figure 7) at Orphan Knoll.

During H1 and H2, the Hudson Strait ice stream reached at least the middle of the continental shelf and may have briefly extended farther seaward [*Andrews et al.*, 1993; *Rashid and Piper*, 2007]. On the other hand, ice probably reached only the outlet of Hudson Strait and the innermost shelf during H0 [*Kirby and Andrews*, 1999; *Rashid and Piper*, 2007]. The presence of ash zone 1 in the Labrador Sea [*Ruddiman and Glover*, 1987], transported in sea ice from a major eruption at the Katla caldera in southern Iceland during the transition to the YD [*Bond et al.*, 1997], implies that the surface circulation was probably similar to the present. Under these ice margin conditions, the Labrador Current could have restricted the distribution of freshwater plumes and icebergs to a narrow corridor on the outer Labrador Shelf and Labrador Slope. The Labrador Current may have been stronger during H0 than during earlier H events, focusing icebergs in a narrow trajectory zone at the shelf edge, as at present [*Chapman*, 2000]. Glacial ice probably occupied the inner Labrador Shelf, based on core chronology and seismic profiles in inner Karlsefni Trough [*Hall et al.*, 1999]. Arctic outflow was blocked by landfast ice until ~10 ka [*Zreda et al.*, 1999] and the Western Boundary Undercurrent, which underlies the outer portion of the Labrador Current, followed modern trajectories during H0 [*Fillon and Duplessy*, 1980].

5.5. *Sedimentological Processes in the H1 and H0 Events*

Sediment was dispersed during H events by ice rafting, in sediment plumes, and by turbidity currents. During H1, ice rafting carried sediment to the central Labrador Sea [*Dowdeswell et al.*, 1995; *Rashid et al.*, 2003a] and into the open North Atlantic Ocean [*Grousset et al.*, 1993; *Rashid and Boyle*, 2007]. The main sediment plume is recognized along the eastern Canadian continental margin [*Piper and Skene*, 1998]. Carbonate-rich turbidites are present along the NAMOC [*Rashid and Piper*, 2007] and on the SAP [*Piper and Hundert*, 2002].

We have estimated the volume of sediment deposited by estimating the areal extent and thicknesses of H1 and H0 (Table 2). Each bed is subdivided into deposits of

Table 2. Estimates of Sediment Types in H0, H1, and a Major Subglacial Outburst on Laurentian Fan[a]

Region	Mean Thickness (m)	Area ($km^2 \times 10^6$)	Total Volume (km^3)	Proportion of Sediment			Volume of Each Sediment Type (km^3)			References[a]
				IRD	Plume	Turbidite	IRD	Plume	Turbidite	
H0 Event										
Sohm Abyssal Plain (SAP)	1	0.11	110	0	0	1	0	0	110	1, 2
Northwest Atlantic Mid-Ocean Channel (NAMOC) spillover	0.1	0.4	40	0	0	1	0	0	40	1, 3
Plume	0.4	0.09	36	0.4	0.6	0	14.4	21.6	0	1, 4
Total by sediment type[b] (%)							8	12	81	
H1 Event										
Proximal plume	2	0.02	40	0.4	0.6	0	16	24	0	5
Middistance plume	0.1	0.4	40	0.2	0.8	0	8	32	0	5
Distal plume	0.1	0.3	30	0.05	0.95	0	1.5	28.5	0	5, 6
NAMOC proximal turbidites	0.5	0.02	10	0	0	1	0	0	10	3
NAMOC distal mud	0.1	0.4	40	0	0	1	0	0	40	3
SAP	1	0.12	120	0	0	1	0	0	120	2
North Atlantic IRD	0.2	1.6	320	0.8	2	0	256	640	0	7
Total by sediment type[b] (%)							24	62	14	
The 19 ka Laurentian Fan Subglacial Outburst										
Coarse load			150	0	0	1	0	0	150	8
Laurentian Fan mud	5	0.02	100	0	0	1	0	0	100	8
Distal Sohm AP	0.5	0.3	150	0	0	1	0	0	150	8
Scotian Slope mud	2	0.09	180	0.01	0.99	0	1.8	178.2	0	8
Total by sediment type[b] (%)							<1	31	69	

[a]Sources are as follows: 1, data from this study; 2, *Piper and Hundert* [2002]; 3, H. Rashid, unpublished data on Labrador Sea cores; 4, *Tripsanas and Piper* [2008a] and *Tripsanas et al.* [2007]; 5, *Rashid et al.* [2003a]; 6, *Piper and Skene* [1998]; and 7, *Alley and MacAyeal* [1994]; 8, *Tripsanas et al.* [2008].

[b]Estimated uncertainty on volume information is ±20%.

hypopycnal plumes, ice rafting, and deposits from turbidity currents, differentiated based on geographic distribution and sediment facies [*Rashid et al.*, 2003a; *Rashid and Piper*, 2007]. The amount of sediment transported and deposited in H1 was more than four times higher than in H0, with greatest uncertainty in the total amount of sediment transported to the SAP in each event. Further, the estimates suggest that during H0, ~150 km^3 of sediment was transported by sediment gravity flows and 20 km^3 in surface plumes, with a minor amount in icebergs. In contrast, during H1, a similar quantity of sediment was transported by sediment gravity flows, but sediment transported by surface plumes and icebergs is estimated 30 times greater at ~600 km^3. The carbonate turbidite muds immediately followed the deposition of the main phases of H1 and H2, based on the correlation of sedimentological units A–C defined by *Rashid et al.* [2003a] from deepwater H layers, for example, core Hu97-16 [*Rashid*, 2002].

Thus, H0 appears to have been similar to the latest stage of H1 and H2, in which turbidite muds were deposited both proximally in the Labrador Sea and through NAMOC to the SAP. Estimated volumes of turbidites are similar in H1 and H0. The main phase of H1 and H2, resulting in plume and ice-rafted deposits, was 30 times more voluminous than H0 (Table 2).

During H1, the Hudson Strait ice stream extended at least to the middle of the continental shelf and perhaps briefly to the shelf break (Figure 9a), as reviewed above. The duration of ice retreat and iceberg supply has been estimated as many hundreds of years [*Hillaire-Marcel et al.*, 1994; *Dowdeswell et al.*, 1995; *Thomson et al.*, 1995], implying that supply of fresh water, icebergs, and sediment to form H1 and the

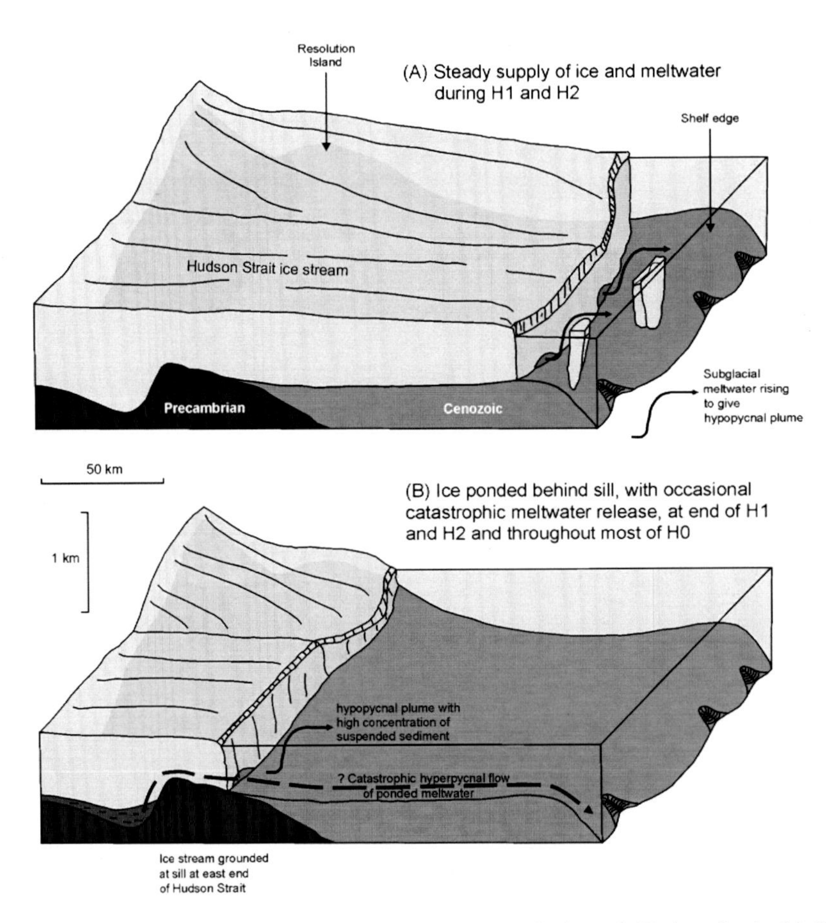

Figure 9. Schematic cartoons showing ice conditions and sediment supply through Hudson Strait. (a) Conditions during main phase of H1 and H2, with the Hudson Strait ice stream extending across much of the continental shelf. Principal sediment transport mechanism in icebergs and in hypopycnal sediment plumes with low concentrations of sediment is shown. (b) Conditions during H0, similar to the terminal phase of H1 and H2, when the ice stream was pinned at the shallow sill at the eastern end of Hudson Strait. Buildup of meltwater behind the sill led to catastrophic subglacial outburst with high sediment concentrations, leading to hyperpycnal flow across the shelf and into deep water.

corresponding retreat of the ice margin was continuous rather than catastrophic.

At the time of H0, ice-contact deposits were formed on the inner shelf, seaward of the sill at the eastern end of Hudson Strait [*Andrews and MacLean*, 2003; *Rashid and Piper*, 2007]. Ice then retreated to behind the sill (Figure 9b). The shallow depth of the sill and the narrowness of the outlet (Figure 1b) would have greatly reduced the supply of large icebergs. However, as suggested by *Johnson and Lauritzen* [1995], unstable ice near the sill would be susceptible to catastrophic freshwater outbursts, as a result of freshwater ponding subglacially or against the northward moving Ungava ice stream, which was active during H0 as well as during the younger Gold Cove and Noble Inlet advances [*Andrews and MacLean*, 2003]. A higher proportion of fresh water might also result from glaciological processes at the end of H events. Studies of sediment supply through the Trinity Trough ice stream to Orphan Basin, offshore Newfoundland, show that supply of subglacial detritus to glacigenic debris flows on the slope was on at least three occasions followed by larger amounts of fresh water that flowed hyperpycnally [*Tripsanas and Piper*, 2008b]. The data are not available to distinguish between these two hypotheses, and indeed, both processes may have occurred concurrently. We suggest that such enhanced freshwater supply at the end of H0 and H1 accounts for the high proportion of gravity flow deposits in H0 and at the end of H1. Freshwater outbursts would have had the velocity to erode and transport previously deposited carbonate rock flour. These outbursts had a much higher instantaneous discharge of water and sediment than that which occurred during the main H event in H1 or H2 when ice extended across the shelf. As a result, the duration of surface plume supply was much shorter, but the power of the discharge to transport sediment across the shelf to deep water was much greater. This is why the distribution of sediment types during H0 more closely resembles that from a classic freshwater outburst, the 19 ka Laurentian Channel outburst (Table 2) [*Piper et al.*, 2007], than a typical H event, with much greater duration and smaller peak freshwater flows.

We can only speculate about the nature of sediment transport process across the shelf because of the lack of core penetration on the shelf to the H0 horizon. The hydraulic head from the sill to the deep continental shelf might have allowed sediment-laden outbursts to flow hyperpycnally as a sediment gravity flow across the continental shelf [cf. *Russell and Arnott*, 2003; *Addington et al.*, 2007]. Because of the Coriolis effect, the flow would tend to occupy the southern part of the transverse trough [*Chapman*, 2000]. On reaching the shelf edge, it would have accelerated down the slope [cf. *Piper and Normark*, 2009], depositing mud turbidites at the foot of slope. Alternatively, freshwater outbursts may have supplied surface freshwater plumes with sediment in much greater concentration than during the main H event. Such energetic flows would have flowed as a jet from the sill outlet across the shelf, producing nepheloid-flow layer deposits [*Rashid et al.*, 2003a] or midwater plumes [*Hesse et al.*, 2004].

In summary, we propose that the contrast between the sedimentological signatures of H1 and H0 in the Labrador Sea results from freshwater outbursts. During H0, the source of most carbonate-rich glacial sediment input into the Labrador Sea was the Hudson Strait, which was kept by the sill before the outburst. A similar process took place at the end of H1 when ice had retreated to the sill; in contrast, during the main depositional phase of H1, when ice extended farther seaward, sediment supply was of lower concentration and spread over a longer time, resulting in a much longer duration of distal plume deposition. The distal plume deposits were thus thicker and hence recognizable over a much larger area.

5.6. Where Is the H0 Freshwater Signal?

The $\delta^{18}O$ signal in Labrador Sea cores is in part a proxy for the supply of fresh water from the former LIS outlet through Hudson Strait. The $\delta^{18}O_{Nps}$ shows a varied signature in H0, H1, and H2 in cores Hu97-09 and Hu97-16, with depleted $\delta^{18}O$ values in H2 (Figures 2 and 3). There was a regional depletion of 1.5‰ sometime just before the beginning of H1 (Figures 2 and 3), which might be attributed to the deglacial changes of the Northern Hemisphere. Since the carbonate and $\delta^{18}O$ were measured in the same sample, there are no artificial leads or lags arising from the tuning process. This advantageous method of data acquisition allowed us to suggest that the changes in $\delta^{18}O_{Nps}$ precede the carbonate increase by ~350–600 years (see also *Stoner et al.* [2000] in MD95-2024). Despite the enormous flux of icebergs released through the strait during H1 [*Dowdeswell et al.*, 1995] and the evidence of a freshwater plume as far south as the Nova Scotian margin [*Piper and Skene*, 1998], the $\delta^{18}O_{Nps}$ during H1 shows only ~1.3‰ depleted isotopic signature in core Hu97-16 and ~2‰ in core Hu97-09. (Lack of adequate numbers of *N. pachyderma* (s) between 840 and 580 cm prevented a full assessment of the magnitude of fresh waters in H1.) The $\delta^{18}O_{Nps}$ perturbation during H0 is only <0.3‰ in core Hu-97-16. Farther downstream, the only other high-resolution record that reveals a minor perturbation in $\delta^{18}O_{Nps}$ at the YD horizon is at site Hu91-94 in the southern Labrador Sea. In this core, $\delta^{18}O_{Nps}$ is ~0.3‰ depleted in the YD compared to ~2‰ $\delta^{18}O_{Nps}$ depleted values during H1 [*Hillaire-Marcel et al.*, 1994]. We cannot evaluate the relative importance of total freshwater discharge and the duration of freshwater discharge in H0 compared to H1 in controlling the isotopic signature of surface water. The other complication with the $\delta^{18}O_{Nps}$ signal is the competing effect of temperature, especially during YD, when the ice margin configuration and the temperature were different than during H1, which may have minimized the depleted $\delta^{18}O_{Nps}$ values resulting from the freshwater input. One of the vexing problems with records like Hu97-09 and Hu97-16 is the absence of enough planktonic foraminifera, preventing paired Mg/Ca and $\delta^{18}O$ measurements in the *N. pachyderma* (s), which could shed light on the relative contribution from temperature and fresh water [*Rashid and Boyle*, 2004]. It is possible that the newer generation of ICP-MS or newer method requiring smaller samples might allow isolation of the seawater $\delta^{18}O$ signal from the $\delta^{18}O_{Nps}$, and this possibility needs to be explored in future studies.

6. CONCLUSIONS

The H0 carbonate-rich sediment layer in the northeastern Labrador Sea corresponds to YD supply of sediment and fresh water through Hudson Strait. H0 is ~1 m thick on the northern Labrador Slope, implying a large supply of sediment and concomitant freshwater discharge. The sedimentology of this layer closely resembles that of H1 and H2, but its distribution is quite different. The H0 layer is restricted to a small area in the northwestern Labrador Sea, and correlative carbonate-rich turbidites are recognized locally in the northeastern SAP. H0 sediment is generally absent in the deep Labrador Sea. Turbidity currents may have flowed hyperpycnally or resulted from plume fallout or slumping of slope sediments from sediment-rich freshwater outbursts that occurred as ice was blocked by the sill at the mouth of Hudson Strait. A sediment plume was transported southward along the Labrador Shelf edge, with deposits recognized in the Orphan Basin, suggesting that the Labrador Current flow may have been similar to the present. The duration of this plume flow was probably less than in H1, resulting in a much thinner distal deposit. Lack of a significant depleted oxygen isotopic signal in the planktonic foraminifera during the H0 could be in part due to the competing temperature and cold freshwater effects, which could be pursued in future studies. The differences in sediment distribution between H0 and earlier H events are a consequence of the ice margin not extending as far seaward during H0. As a result, sedimentation and freshwater supply during H0 were dominated by freshwater outbursts from behind the Hudson Strait sill, whereas in H1, this process was important only during the late stages of the event. Rapid ice retreat and meltwater supply from the same ice stream may thus have a very different impact on the world ocean as a result of differences in ice stream margin extent.

Acknowledgments. We would like to thank L. Polyak for his comments on the earlier version of the manuscript. H. Rashid was partially supported by the Byrd Polar Research Fellowship during the study. This is Geological Survey of Canada contribution 20090094.

REFERENCES

Addington, L. D., S. A. Kuehl, and J. E. McNinch (2007), Contrasting modes of shelf sediment dispersal off a high-yield river: Waiapu River, New Zealand, *Mar. Geol.*, *243*, 18–30.

Alley, R. B., and P. U. Clark (1999), The deglaciation of the Northern Hemisphere: A global perspective, *Annu. Rev. Earth Planet. Sci.*, *27*, 149–182.

Alley, R. B., and D. R. MacAyeal (1994), Ice-rafted debris associated with binge/purge oscillations of the Laurentide Ice Sheet, *Paleoceanography*, *9*, 503–511.

Andrews, J. T., and B. MacLean (2003), Hudson Strait ice streams: A review of stratigraphy, chronology and links with North Atlantic Heinrich events, *Boreas*, *32*, 4–17.

Andrews, J. T., K. Tedesco, and A. E. Jennings (1993), Heinrich events: Chronology and processes, east-central Laurentide Ice Sheet and NW Labrador Sea, in *Ice in the Climatic System*, *NATO ASI Ser.*, vol. 1, edited by R. Peltier, pp. 167–186, Springer, New York.

Andrews, J. T., A. E. Jennings, M. Kerwin, M. Kirby, W. Manley, G. H. Miller, G. Bond, and B. MacLean (1995), A Heinrich-like event, H-0 (DC-0): Source(s) for detrital carbonate in the North Atlantic during the Younger Dryas chronozone, *Paleoceanography*, *10*, 943–952.

Andrews, J. T., L. D. Keigwin, F. Hall, and A. E. Jennings (1999), Abrupt deglaciation events and Holocene palaeoceanography from high-resolution cores, Cartwright Saddle, Labrador Shelf, Canada, *J. Quat. Sci.*, *14*, 383–397.

Bakke, J., Ø. Lie, E. Heegaard, T. Dokken, G. H. Haug, H. Birks, H. P. Dulski, and T. Nilsen (2009), Rapid oceanic and atmospheric changes during the Younger Dryas cold period, *Nat. Geosci.*, *12*, 202–205.

Barber, D. C. (2001), Laurentide Ice Sheet dynamics from 35 to 7 ka: Sr-Nd-Pb isotopic provenance of NW North Atlantic margin sediments, Ph.D. thesis, Univ. of Colo., Boulder.

Barber, D. C., et al. (1999), Forcing of the cold event of 8,200 years ago by catastrophic drainage of Laurentide lakes, *Nature*, *400*, 344–348.

Benetti, S. (2006), Late Quaternary sedimentary processes along the western North Atlantic margin, Ph.D. thesis, Natl. Oceanogr. Cent., Southampton, U. K.

Bond, G. C., W. Showers, M. Cheseby, R. Lotti, P. Almasi, P. deMenocal, P. Priore, H. Cullen, I. Hajdas, and G. Bonani (1997), A pervasive millennial-scale cycle in North Atlantic Holocene and glacial climates, *Science*, *278*, 1257–1266.

Bondevik, S., J. Mangerud, H. H. Birks, S. Gulliksen, and P. J. Reimer (2006), Changes in North Atlantic radiocarbon reservoir ages during the Allerod and Younger Dryas, *Science*, *312*, 1514–1517.

Boyle, E. A., and L. D. Keigwin (1987), North Atlantic thermohaline circulation during the past 20,000 years linked to high-latitude surface temperature, *Nature*, *350*, 35–40.

Chapman, D. C. (2000), Boundary layer control of buoyant coastal currents and the establishment of a shelf break front, *J. Phys. Oceanogr.*, *30*, 2941–2955.

de Vernal, A., and C. Hillaire-Marcel (1987), Marginal paleoenvironments of the eastern Laurentide Ice Sheet and timing of the last ice maximum and retreat, *Geogr. Phys. Quat.*, *41*, 265–277.

de Vernal, A., et al. (2001), Dinoflagellate cyst assemblages as tracers of sea-surface conditions in the northern North Atlantic, Arctic and sub-Arctic seas: The new 'n = 677' data base and its application for quantitative paleoceanographic reconstruction, *J. Quat. Sci.*, *16*, 681–698.

Dowdeswell, J., M. Maslin, J. T. Andrews, and I. McCave (1995), Iceberg production, debris rafting, and the extent and thickness of Heinrich layers (H-1, H-2) in North Atlantic sediments, *Geology*, *23*, 301–304.

Dykoski, C. A., R. L. Edwards, H. Cheng, D. Yuan, Y. Cai, M. Zhang, Y. Lin, J. Qing, Z. An, and J. Revenaugh (2005), A high-resolution, absolute-dated Holocene and deglacial Asian monsoon record from Dongge Cave, China, *Earth Planet. Sci. Lett.*, *233*, 71–86.

Fillon, R. H., and J.-C. Duplessy (1980), Labrador Sea bio-, tephro-, oxygen isotopic stratigraphy and Late Quaternary paleoceanographic trends, *Can. J. Earth Sci.*, *17*, 831–854.

Grachev, A. M., and J. P. Severinghaus (2005), A revised 10±4°C magnitude of the abrupt change in Greenland temperature at the Younger Dryas termination using published GISP2 gas isotope data and air thermal diffusion constants, *Quat. Sci. Rev.*, *24*, 513–519.

Grousset, F. E., L. Labeyrie, J. A. Sinko, M. Cremer, G. Bond, J. Duprat, E. Cortijo, and S. Huon (1993), Patterns of ice-rafted detritus in the glacial North Atlantic (40–55°N), *Paleoceanography*, *8*, 175–192.

Hall, F. R., J. T. Andrews, A. E. Jennings, G. Vilks, and K. Moran (1999), Late Quaternary sediments and chronology of the northeast Labrador Shelf (Karlsefni Trough, Saglek Bank): Link to glacial history, *Geol. Soc. Am. Bull.*, *111*, 1700–1713.

Hesse, R., I. Klaucke, W. B. F. Ryan, M. B. Edwards, and D. J. W. Piperthe NAMOC Study Group (1996), Imaging Laurentide Ice Sheet drainage into the deep sea: Impact on sediments and bottom water, *GSA Today*, *6*(9), 3–9.

Hesse, R., H. Rashid, and S. Khodabakhsh (2004), Fine-grained sediment lofting from meltwater-generated turbidity currents during Heinrich events, *Geology*, *32*, 449–452.

Hillaire-Marcel, C., A. de Vernal, and G. Bilodeau (1994), Isotope stratigraphy, sedimentation rates, deep circulation, and carbonate vent in the Labrador Sea during the last 200 kyr, *Can. J. Earth Sci.*, *31*, 63–89.

Hillaire-Marcel, C., A. de Vernal, and D. J. W. Piper (2007), Lake Agassiz Final drainage event in the northwest North Atlantic, *Geophys. Res. Lett.*, *34*, L15601, doi:10.1029/2007GL030396.

Isarin, R. F. B., H. Renssen, and J. Vandenberghe (1998), The impact of the North Atlantic Ocean on the Younger Dryas climate in northwestern and central Europe, *J. Quat. Sci.*, *13*, 447–453.

Johnson, R. G., and S. E. Lauritzen (1995), Hudson Bay-Hudson Strait jökulhlaups and Heinrich events; a hypothesis, *Palaeogeogr. Palaeoclimatol. Palaeoecol.*, *117*, 123–137.

Kaufman, D. S., G. H. Miller, J. A. Stravers, and J. T. Andrews (1993), Abrupt early Holocene (9.9-9.6 ka) ice-stream advance at the mouth of Hudson Strait, Arctic Canada, *Geology*, *21*, 1063–1066.

Kirby, M. E. (1998), Heinrich event-0 (DC-0) in sediment cores from the northwest Labrador Sea: Recording events in Cumberland Sound?, *Can. J. Earth Sci.*, *35*, 510–519.

Kirby, M. E., and J. T. Andrews (1999), Mid-Wisconsin Laurentide Ice Sheet growth and decay: Implications for Heinrich events 3 and 4, *Paleoceanography*, *14*, 211–223.

Lewis, C. F. M., E. Levac, D. J. W. Piper, and G. V. Sonnichsen (2009), Correlating Lake Agassiz floods to the onset of the 8.2 cal ka cold event, paper presented at Chapman Conference on Abrupt Climate Change, AGU, Columbus, Ohio.

Lippold, J., J. Grützner, D. Winter, Y. Lahaye, A. Mangini, and M. Christl (2009), Does sedimentary $^{231}Pa/^{230}Th$ from the Bermuda Rise monitor past Atlantic Meridional Overturning Circulation?, *Geophys. Res. Lett.*, *36*, L12601, doi:10.1029/2009GL038068.

Miller, G. H., and D. S. Kaufman (1990), Rapid fluctuations of the Laurentide Ice Sheet at the mouth of Hudson Strait: New evidence for ocean/ice-sheet interactions as a control on the Younger Dryas, *Paleoceanography*, *5*, 907–919.

Peck, V. L., I. R. Hall, R. Zahn, and H. Elderfield (2008), Millennial-scale surface and subsurface paleothermometry from the northeast Atlantic, 55–8 ka BP, *Paleoceanography*, *23*, PA3221, doi:10.1029/2008PA001631.

Piper, D. J. W., and T. Hundert (2002), Provenance of distal Sohm Abyssal Plain sediments: History of supply from the Wisconsinan glaciation in eastern Canada, *Geo-Mar. Lett.*, *22*, 75–85.

Piper, D. J. W., and W. R. Normark (2009), The processes that initiate turbidity currents and their influence on turbidites: A marine geology perspective, *J. Sediment. Res.*, *79*, 347–362.

Piper, D. J. W., and K. I. Skene (1998), Latest Pleistocene ice-rafting events on the Scotian margin (eastern Canada) and their relationship to Heinrich events, *Paleoceanography*, *13*, 205–214.

Piper, D. J. W., J. Shaw, and K. I. Skene (2007), Stratigraphic and sedimentological evidence for late Wisconsinan sub-glacial outburst floods to Laurentian Fan, *Palaeogeogr. Palaeoclimatol. Palaeoecol.*, *246*, 101–119.

Rashid, H. (2002), The deep-sea record of rapid Late Pleistocene paleoclimate change and ice-sheet dynamic from the Labrador Sea sediments, Ph.D. thesis, McGill Univ., Montreal, Que., Canada.

Rashid, H., and E. A. Boyle (2004), A ~100 kyr SST record based on the Mg/Ca of *Neogloboquadrina pachyderma* (s) from the Labrador Sea, *Eos Trans. AGU*, *85*(47), Fall Meet. Suppl., Abstract PP23B-1407.

Rashid, H., and E. A. Boyle (2007), Mixed layer deepening during Heinrich Events: A multi-planktonic foraminiferal $\delta^{18}O$ approach, *Science*, *318*, 439–441.

Rashid, H., and E. Grosjean (2006), Detecting the source of Heinrich layers: An organic geochemical study, *Paleoceanography*, *21*, PA3014, doi:10.1029/2005PA001240.

Rashid, H., and D. J. W. Piper (2007), The extent of ice on the continental shelf off Hudson Strait during Heinrich events 1-3, *Can. J. Earth Sci.*, *44*, 1537–1549.

Rashid, H., R. Hesse, and D. J. W. Piper (2003a), Origin of unusually thick ice-proximal Heinrich layers H1 to H3 in the northwest Labrador Sea, *Earth Planet. Sci. Lett.*, *208*, 319–336.

Rashid, H., R. Hesse, and D. J. W. Piper (2003b), Evidence for an additional Heinrich event between H5 and H6 in the Labrador Sea, *Paleoceanography*, *18*(4), 1077, doi:10.1029/2003PA000913.

Rashid, H., D. J. W. Piper, L. Polyak, C. Mansfield, and B. P. Flower (2008), Laurentide Ice Sheet discharge linked to North Atlantic deep water reduction 10,200 years BP, *Eos Trans. AGU*, *89*(53), Fall Meet. Suppl, Abstract PP21C-1444.

Rasmussen, S. O., et al. (2006), A new Greenland ice core chronology for the last glacial termination, *J. Geophys. Res.*, *111*, D06102, doi:10.1029/2005JD006079.

Reimer, P. J., et al. (2009), IntCal09 and Marine09 radiocarbon age calibration curves, 0-50,000 years cal BP, *Radiocarbon*, *51*(4), 1111–1150.

Ruddiman, W. F., and L. K. Glover (1987), Vertical mixing of ice-rafted volcanic ash in North Atlantic sediments, *Geol. Soc. Am. Bull.*, *83*, 2817–2836.

Ruddiman, W. F., and A. McIntyre (1981), The North Atlantic Ocean during the last deglaciation, *Palaeogeogr. Palaeoclimatol. Palaeoecol.*, *35*, 145–214.

Russell, H. A. J., and R. W. C. Arnott (2003), Hydraulic-jump and hyperconcentrated-flow deposits of a glacigenic subaqueous fan: Oak Ridges moraine, southern Ontario, Canada, *J. Sediment. Res.*, *73*, 887–905.

Sarnthein, M., U. Pflaumann, and M. Weinelt (2003), Past extent of sea ice in the northern North Atlantic inferred from foraminiferal paleotemperature estimates, *Paleoceanography*, *18*(2), 1047, doi:10.1029/2002PA000771.

Severinghaus, J. P., R. Beaudette, M. A. Headly, K. Taylor, and E. J. Brook (2009), Oxygen-18 of O_2 Records the impact of abrupt climate change on the terrestrial biosphere, *Science*, *324*, 1431–1434.

Steffensen, J. P., et al. (2008), High-resolution Greenland ice core data show abrupt climate change happens in few years, *Science*, *321*, 608–684, doi:10.1126/science.1157707.

Stoner, J. S., J. E. T. Channell, and C. Hillaire-Marcel (1996), The magnetic signature of rapidly deposited detrital layers from the deep Labrador Sea: Relationship to North Atlantic Heinrich layers, *Paleoceanography*, *11*, 309–325.

Stoner, J. S., J. E. T. Channell, C. Hillaire-Marcel, and C. Kissel (2000), Geomagnetic paleointensity and environmental record from Labrador Sea core MD95-2024: Global marine sediment and ice core chronostratigraphy for the last 110 kyr, *Earth Planet. Sci. Lett.*, *183*, 161–177.

Svensson, A., et al. (2008), A 60,000 year Greenland stratigraphic ice core chronology, *Clim. Past*, *4*, 47–57.

Tarasov, L., and W. R. Peltier (2005), Arctic freshwater forcing of the Younger Dryas cold reversal, *Nature*, *435*, 662–665.

Thomas, E. R., E. W. Wolff, R. Mulvaney, J. P. Steffensen, S. J. Johnsen, C. Arrowsmith, J. W. C. White, B. Vaughn, and T. Popp (2007), The 8.2 ka event from Greenland ice cores, *Quat. Sci. Rev.*, *26*, 70–81.

Thomson, J., N. C. Higgs, and T. Clayton (1995), A geochemical criterion for the recognition of Heinrich events and estimation of their depositional fluxes by the $(^{230}Th_{excess})_0$ profiling method, *Earth Planet. Sci. Lett.*, *135*, 41–56.

Tripsanas, E. K., and D. J. W. Piper (2008a), Late Quaternary stratigraphy and sedimentology of Orphan Basin: Implications for meltwater dispersal in the southern Labrador Sea, *Palaeogeogr. Palaeoclimatol. Palaeoecol.*, *260*, 521–539.

Tripsanas, E. K., and D. J. W. Piper (2008b), Late Quaternary glacigenic debris flows in Orphan Basin, *J. Sediment. Res.*, *78*, 713–723.

Tripsanas, E. K., D. J. W. Piper, and K. A. Jarrett (2007), Logs of piston cores and interpreted ultra-high-resolution seismic profiles, Orphan Basin, *Open File Rep. 5299*, 339 pp., Geol. Surv. Can., Ottawa.

Tripsanas, E., D. J. W. Piper, and D. C. Campbell (2008), Evolution and depositional structure of earthquake-induced mass movements and gravity flows, southwest Orphan Basin, Labrador Sea, *Mar. Pet. Geol.*, *25*, 645–662.

Utting, D. J., J. C. Gosse, D. A. Hodgson, M. S. Trommelen, K. J. Vickers, S. E. Kelley, B. Ward (2007), Report on ice-flow history, deglacial chronology, and surficial geology, Foxe Peninsula, southwest Baffin Island, Nunavut, *Curr. Res. Geol. Surv. Can.*, *2007-C2*, 1–13.

Waelbroeck, C., J.-C. Duplessy, E. Michel, L. Labeyrie, D. Paillard, and J. Duprat (2002), The timing of the last deglaciation in North Atlantic climate records, *Nature*, *412*, 724–727.

Wang, Y., H. Cheng, R. L. Edwards, X.-G. Kong, X. Shao, S. Chen, J.-Y. Wu, X.-Y. Jiang, X.-F. Wang, and Z.-S. An (2008), Millennial- and orbital-scale changes in the East Asian monsoon over the past 224,000 years, *Nature*, *451*, 1090–1093.

Zreda, M., J. England, F. Phillips, D. Elmore, and P. Sharma (1999), Unblocking of the Nares Strait by Greenland and Ellesmere ice-sheet retreat 10,000 years, *Nature*, *398*, 139–142.

B. P. Flower, College of Marine Science, University of South Florida, St. Petersburg, FL 33701, USA.

D. J. W. Piper, Geological Survey of Canada (Atlantic), Bedford Institute of Oceanography, Dartmouth, NS B2Y 4A2, Canada.

H. Rashid, Byrd Polar Research Center, Ohio State University, Columbus, OH 43210, USA. (rashid.29@osu.edu)

Challenges in the Use of Cosmogenic Exposure Dating of Moraine Boulders to Trace the Geographic Extents of Abrupt Climate Changes: The Younger Dryas Example

Patrick J. Applegate[1] and Richard B. Alley

Department of Geosciences, Pennsylvania State University, University Park, Pennsylvania, USA

Cosmogenic exposure dating has sometimes been used to identify moraines associated with short-lived climatic events, such as the Younger Dryas (12.9–11.7 ka). Here we point out two remaining challenges in using exposure dating to identify moraines produced by abrupt climate changes. Specifically, (1) a commonly applied sampling criterion likely yields incorrect exposure dates at some sites, and (2) geomorphic processes may introduce bias into presently accepted nuclide production rate estimates. We fit a geomorphic process model that treats both moraine degradation and boulder erosion to collections of exposure dates from two moraines that were deposited within a few thousand years of the Younger Dryas. Subsampling of the modeled distributions shows that choosing boulders for exposure dating based on surface freshness yields exposure dates that underestimate the true age of the moraine by up to several thousand years. This conclusion applies only where boulders do not erode while buried but do erode after exhumation. Moreover, one of our fitted data sets is part of the global nuclide production rate database. Our fit of the moraine degradation model to this data set suggests that nuclide production rates at that site are several percent higher than previously thought. Potential errors associated with sampling strategies and production rate estimates are large enough to interfere with exposure dating of moraines, especially when the moraines are associated with abrupt climate changes. We suggest sampling strategies that may help minimize these problems, including a guide for determining the minimum number of samples that must be collected to answer particular paleoclimate questions.

1. INTRODUCTION

Cosmogenic exposure dating of moraines is an attractive method for tracing the geographic extents of former abrupt climate changes. Glaciers grow and shrink in response to climate changes [*Lowell*, 2000; *Oerlemans*, 2005; *Denton et al.*, 2005], and they deposit ridges called moraines at their margins [*Gibbons et al.*, 1984]. Thus, for abrupt climate changes that propagated over long distances quickly, we expect to find moraines of about the same age in mountain ranges within the area affected by the change. The crests of

[1]Now at Department of Physical Geography and Quaternary Geology, Stockholm University, Stockholm, Sweden.

Abrupt Climate Change: Mechanisms, Patterns, and Impacts
Geophysical Monograph Series 193

10.1029/2010GM001029

moraines are often studded with large boulders that can be sampled for cosmogenic exposure dating [e.g., *Phillips et al.*, 1990; *Gosse et al.*, 1995a, 1995b]. In principle, cosmogenic exposure dating yields direct estimates of moraine ages; other Quaternary dating methods give only maximum or minimum age estimates, except in rare cases.

The Younger Dryas (12.9–11.7 ka [*Alley et al.*, 1993; *Barrows et al.*, 2007; *Walker et al.*, 2008]) is an example of an abrupt climate change whose geographic extent has been traced partly with cosmogenic exposure dating of moraines. Various proxy records [e.g., *Mangerud et al.*, 1974] show that the Younger Dryas produced strong cooling around the North Atlantic, with weaker negative temperature anomalies elsewhere in the Northern Hemisphere and warming in the Southern Hemisphere [e.g., *Broecker et al.*, 1989; *Denton et al.*, 2005; *Alley*, 2007; see also *Chiang and Bitz*, 2005; *Lowell et al.*, 2005; *Broecker*, 2006]. Climate modeling studies with North Atlantic freshening simulate temperature anomaly patterns consistent with the data [e.g., *Vellinga and Wood*, 2002].

After the first successful exposure dating studies of moraines [*Phillips et al.*, 1990; *Gosse et al.*, 1995a, 1995b], glacial geomorphologists used this new tool to look for the Younger Dryas signal. Moraines dating to the Younger Dryas were identified in the Alps [*Ivy-Ochs et al.*, 1999, 2006, 2007; cf. *Kelly et al.*, 2004] but also far from the North Atlantic (e.g., western North America and New Zealand [*Gosse et al.*, 1995b; *Ivy-Ochs et al.*, 1999]). Given the likely distribution of Younger Dryas cooling, the age assignments for these additional sites are suspect.

Past workers have taken exposure dates falling at any time within the Younger Dryas interval as evidence for Younger Dryas cooling in the region [e.g., *Gosse et al.*, 1995b; *Ivy-Ochs et al.*, 1999; see also *Denton and Hendy*, 1994]. This criterion reflects uncertainties in the exposure dating method and lingering doubt about when moraines should have been deposited during the Younger Dryas.

First-order glaciological considerations suggest that the ages of true Younger Dryas moraines should cluster around the end of the Younger Dryas, but recent modeling studies [*Vacco et al.*, 2009] complicate this simple picture. Changes in glacier margin positions lag temperature change, with a response time and equilibrium length change that vary among glaciers [*Oerlemans*, 2005]. Thus, there was likely a delay between the initial Younger Dryas cooling and glacier advances. Glaciers with short response times reached equilibrium with the new temperature quickly, whereas glaciers with long response times perhaps never did. Until the glaciers reached their maximal positions, any moraines deposited at their margins would not be preserved [*Gibbons et al.*, 1984]. In any case, exposure dating records the time of landform stabilization, assuming no inheritance or long-term landform degradation. This stabilization would not have happened until the glaciers retreated from their moraines at the end of the Younger Dryas. However, a glacier model forced by Greenland ice core temperature records produce a closely spaced complex of moraines distributed over the Younger Dryas interval [*Vacco et al.*, 2009; *Alley et al.*, 2010]. Farther from Greenland, where the event was relatively small, these moraines should be compressed into a single moraine dating to the end of the event.

The analytical precision of cosmogenic exposure dating with beryllium-10 is often very good, suggesting that the method can identify moraines associated with abrupt climate changes such as the Younger Dryas. Confident identification of Younger Dryas moraines probably requires an uncertainty of 10% of the event's duration or about 100 years. Measurements of beryllium-10 concentrations often have uncertainties of ~3% [e.g., *Gosse et al.*, 1995a, 1995b; *Owen et al.*, 2003; *Kelly et al.*, 2008]. Thus, the 1σ analytical uncertainties of beryllium-10 exposure dates from Younger Dryas moraines should be about 360 years (3% of 12.0 kyr). If measurement error were the only source of uncertainty in exposure dating, we would need about 13 samples from a single moraine to reduce this 360 year uncertainty to 100 years [*Bevington and Robinson*, 2003, equation 4-19, Figure 1]. The uncertainty of the weighted mean is appropriate only where the dates are normally distributed and have a scatter consistent with the measurement uncertainties of the dates; more exposure dates are needed where these conditions are violated. Production rate estimates also contribute to the overall uncertainty of exposure dating, as we discuss below.

However, the measurement of nuclide concentrations is only one step in the exposure dating process, and all steps contribute to the total uncertainty of exposure dating. These steps are as follows: (1) collecting the samples [*Gosse and Phillips*, 2001; *Briner*, 2009]; (2) processing the samples [*Kohl and Nishiizumi*, 1992; *Bierman*, 1994]; (3) measuring the nuclide concentrations in the processed samples using accelerator mass spectrometry [*Muzikar et al.*, [2003] (this step also yields the analytical uncertainties of the dates); (4) calculating the apparent exposure times of the samples from the nuclide concentrations [*Lal*, 1991; *Gosse and Phillips*, 2001; *Balco et al.*, 2008]; and (5) estimating the age of the moraine from the exposure dates [*Applegate et al.*, 2008, 2010]. Determining the climatic significance of moraines once their ages are known is also a crucial part of the exposure dating process, but this additional step is beyond the scope of this chapter.

Here we indicate potential problems in the selection of samples for exposure dating and the calculation of exposure dates from nuclide concentrations (steps 1 and 4, above). Briefly, geomorphic process modeling suggests that sampling boulders with minimal surface weathering will yield too young exposure dates on moraines that have lost material from their crests over time. Moreover, geomorphic processes likely introduce errors into the calibration of nuclide production rates. These problems limit our ability to confidently identify moraines associated with abrupt climate changes.

This work is a proof of concept, which we hope will stimulate discussion within the exposure dating community. Practitioners of the exposure dating method are undoubtedly aware of these issues, but we have not seen them discussed in print. Nonpractitioners incorporate the results of exposure dating studies into paleoclimate syntheses (e.g., G. Schmidt, Younger Dry-as dust?, available at http://www.realclimate.org/index.php/archives/2007/10/younger-dry-as-dust/, 2007, accessed 28 July 2010). Thus, an explicit discussion of these uncertainties may be valuable.

2. PRIOR WORK

Much past work indicates that the best simple method of estimating moraine ages varies among moraines (step 5 above). Where measurement error produces all the scatter among exposure dates, the mean is the best estimator of moraine age. The maximum exposure date in a data set is the best estimator of moraine age, where moraine degradation and/or boulder erosion are the dominant processes, and the minimum exposure date provides the best estimate of moraine age, where inheritance is responsible for most of the scatter [*Phillips et al.*, 1990; *Briner et al.*, 2005; *Benson et al.*, 2005].

We have developed models of two processes, moraine degradation and inheritance, that likely increase the scatter among exposure dates from moraines and cause the statistical distributions of these dates to be nonnormal. *Applegate et al.* [2010] provide detailed descriptions of the models, with computer code [see also *Applegate*, 2009; *Zreda et al.*, 1994; *Hallet and Putkonen*, 1994; *Putkonen and Swanson*, 2003; *Benson et al.*, 2005].

This modeling work shows that the statistical distributions of exposure dates provide clues to the geomorphic processes acting at individual field sites [*Applegate et al.*, 2010]. In ideal cases, where geomorphic processes do not affect exposure dating, the exposure dates will be normally distributed. The statistical distributions of exposure dates from moraines should be left-skewed where moraine degradation is predominantly responsible for the scatter among exposure dates and should be right-skewed where inheritance is the dominant process.

However, it is difficult to determine which simple method to apply to a given data set [*Applegate et al.*, 2010]. Given the differences in the statistical distributions produced by different processes, one might choose which method to apply to any given data set based on the skewness of the dates. This method tends to yield results that are close to the correct answer for the parent distributions we have tested, but it sometimes fails spectacularly; the numbers of samples that are typically collected from moraines (20 or fewer per moraine) do not allow us to confidently determine the skewness of the parent distribution. Thus, we sometimes choose the wrong method for estimating moraine ages, leading to errors of thousands of years.

To address this problem, we developed methods for inverting our process models against observations [*Applegate*, 2009]. These methods match the modeled distributions of exposure dates to the observed dates [*Price et al.*, 2005; A. Clauset et al., Power-law distributions in empirical data, 2007, available at http://arxiv.org/abs/0706.1062v1, accessed 30 January 2009, hereinafter referred to as Clauset et al., data set, 2007]. Besides yielding explicit estimates of moraine age, our inverse methods also give estimates of the rates and magnitudes of the geomorphic processes described by the forward models. These inversions require a fairly large number of observations per moraine ($n \approx 10$ or greater) to achieve a good fit.

The models are not appropriate for all field situations because they make certain assumptions that are not met everywhere. We provide a detailed discussion of our models' assumptions elsewhere [*Applegate et al.*, 2010]. In particular, the degradation model assumes that moraines evolve diffusively from initially triangular cross sections [*Hallet and Putkonen*, 1994]. Boulders are distributed uniformly through the removed soil column in the model. Further, the model assumes that the boulders do not erode while buried and erode at a constant rate after exhumation. The model does not apply to moraines with very low surface slopes (e.g., the terminal moraines of the Laurentide Ice Sheet in North America), clast-supported moraines (e.g., the Ledyard moraine in southeastern New England [*Balco and Schaefer*, 2006], or moraines where the boulders are concentrated on the moraine's upper surface (e.g., the outermost Pinedale moraine at Fremont Lake, Wyoming [*Gosse et al.*, 1995a; E. Evenson, personal communication, 2009].

Despite these limitations, the models are useful because their assumptions are transparent. Traditional methods of interpreting exposure dates rely on expert judgment, in which the assumptions underlying a particular interpretation may not be explicitly stated.

3. SELECTED DATA SETS

We have attempted to identify moraines that were deposited at about the time of the Younger Dryas and have a

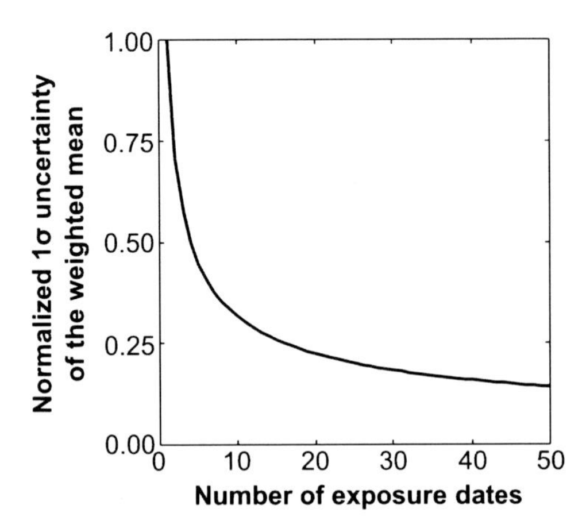

Figure 1. Uncertainty of the weighted mean as a function of the number of exposure dates available from a single moraine [*Bevington and Robinson*, 2003, equation 4.19], normalized by the measurement uncertainty of one exposure date. Figure 1 indicates the minimum number of samples that should be taken from a moraine in order to achieve a desired overall uncertainty. Because geomorphic processes also contribute to the scatter among exposure dates on most moraines [*Putkonen and Swanson*, 2003; *Balco and Schaefer*, 2006; *Ivy-Ochs et al.*, 2007], more samples will generally be needed.

sufficient number of published, independent beryllium-10 exposure dates that we can have reasonable confidence in our age assignments (Figure 1).

Such moraines are rare. The Egesen stade moraines in the Alps seem firmly tied to the Younger Dryas [*Ivy-Ochs et al.*, 2006; cf. *Kelly et al.*, 2004], but we know of no Egesen moraine with more than four independent, published beryllium-10 exposure dates. Because the Egesen I and II moraines are geomorphically distinct in many Alpine valleys [*Ivy-Ochs et al.*, 2006], the age difference between moraine crests within the Egesen stade is likely to be substantial. This conclusion is supported by the exposure dates from Julier Pass, Switzerland, where the difference in the mean of the beryllium-10 exposure dates from the outer and inner Egesen moraines is about 860 years [*Ivy-Ochs et al.*, 2007, Figure 4]. Hence, we do not combine exposure dates from the Egesen I and II moraines to increase the number of exposure dates.

Thus, we have chosen to use the beryllium-10 exposure dates (Figure 2) from the inner Titcomb Lakes moraine (Wind River Range, Wyoming [*Gosse et al.*, 1995b]) and from the Waiho Loop moraine in New Zealand [*Barrows et al.*, 2007, 2008; *Applegate et al.*, 2008]. Each of these moraines has some independent age control. The inner Titcomb Lakes moraine is correlated to the Temple Lake moraine of *Zielinski and Davis* [1987], elsewhere in the Wind River Range. The Temple Lake moraine is bracketed by radiocarbon dates that indicate an age of about 12.0 calendar ka [*Balco et al.*, 2008; cf. *Zielinski and Davis*, 1987; *Gosse and Klein*, 1996]. The Waiho Loop moraine has a large collection of radiocarbon dates that provide a minimum age estimate [*Denton and Hendy*, 1994], although the significance of these dates is debated [*Broecker*, 2003; *Barrows et al.*, 2007; *Turney et al.*, 2007].

For our purposes, it is unimportant whether or not these moraines belong to the Younger Dryas. It is sufficient that they are late glacial or early Holocene in age (10–15 ka) and have a comparatively large number of beryllium-10 exposure dates (n = 10 for the inner Titcomb Lakes moraine; n = 8 for the Waiho Loop). Both of these moraines also have exposure dates determined with other cosmogenic nuclides [*Barrows et al.*, 2007; *Balco et al.*, 2008], but we neglect these other measurements here; our model is most appropriate for evaluating the distributions of beryllium-10 exposure dates.

For consistency, we have recalculated the exposure dates from the inner Titcomb Lakes moraine following *Barrows et al.* [2007]; see Table 1. We also recalculated the Waiho Loop

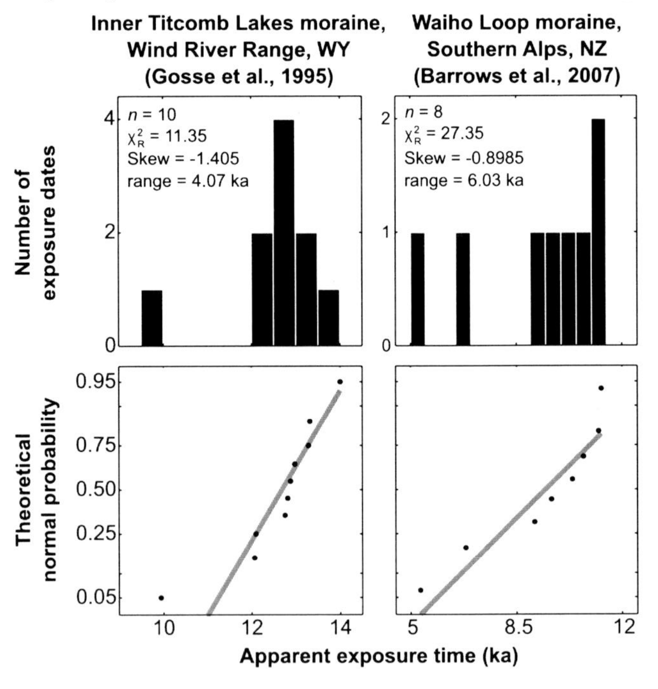

Figure 2. (top) Beryllium-10 exposure dates from the inner Titcomb Lakes moraine [*Gosse et al.*, 1995b] and the Waiho Loop moraine [*Barrows et al.*, 2007]. Both these data sets are likely influenced by geomorphic processes. (bottom) We cannot rule out the hypothesis that these data sets are drawn from normal distributions because the dates fall fairly close to a line when displayed on normal probability plots [*Chambers et al.*, 1983]. However, the scatter in these data sets is much larger than we would predict from the measurement uncertainties of the individual dates; their reduced chi-squared (χ^2_R) [*Bevington and Robinson* [2003] scores are much larger than 1.

Table 1. Beryllium-10 Exposure Dates Recalculated Following *Barrows et al.* [2007][a]

Sample ID	Boulder Height (m)	Nucleon ^{10}Be Production Rate (atoms g^{-1} yr^{-1})	Muon ^{10}Be Production Rate (atoms g^{-1} yr^{-1})	Apparent Age (years)	1σ Uncertainty (years)
		Waiho Loop Moraine, Western New Zealand [Barrows et al., 2007]			
WH-01B		5.554	0.157	1.24	0.42
WH-02		5.580	0.157	10.33	0.70
WH-03		5.576	0.157	10.70	0.38
WH-04B		5.575	0.157	9.65	0.47
WH-05		5.472	0.156	9.10	0.41
WH-08A		5.473	0.156	6.85	0.91
WH-09		5.473	0.156	5.30	0.33
WH-10		5.475	0.156	11.33	1.55
		Inner Titcomb Lakes Moraine, Wind River Range [Gosse et al., 1995a]			
WY-92-138	1.0	51.892	0.568	12.80	0.38
WY-92-139	1.0	51.892	0.568	12.06	0.36
WY-92-140	1.0	51.892	0.568	9.93	0.30
WY-93-333	1.0	51.892	0.568	12.08	0.36
WY-93-334	1.5	51.892	0.568	14.00	0.42
WY-93-335	0.6	51.892	0.568	12.97	0.39
WY-93-336	2.0	51.892	0.568	13.30	0.40
WY-93-337	1.5	51.892	0.568	13.26	0.40
WY-93-338	0.8	51.892	0.568	12.86	0.39
WY-93-339	0.1	51.892	0.568	12.73	0.38
WY-92-138	1.0	51.892	0.568	12.80	0.38
WY-92-139	1.0	51.892	0.568	12.06	0.36
WY-92-140[b]	1.0	51.892	0.568	9.93	0.30

[a]Production rates and exposure dates are recalculated following *Barrows et al.* [2007], using the scaling model of *Stone* [2000] for both nucleon and muon production. The 1σ uncertainties reflect measurement uncertainty only.

beryllium-10 exposure dates; our recalculated dates agree with those reported by *Barrows et al.* [2007] to within 0.6%, suggesting that our calculation method is consistent with theirs. We did not use the CRONUS online calculator [*Balco et al.*, 2008] because the calibration of the online calculator depends in part on the concentration measurements from the inner Titcomb Lakes moraine (see below). Thus, using the online calculator would introduce circularity into our results. In any case, the choice of scaling model has little influence on the scatter among exposure dates from individual moraines [*Balco et al.*, 2008; *Applegate*, 2009], even at midlatitude sites where the effects of geomagnetic field changes are greatest.

Both of these data sets are likely influenced by geomorphic processes. The reduced χ^2 scores of these data sets are much greater than 1 (Figure 2), indicating that the data sets contain more scatter than can be explained by measurement error alone. We cannot rule out the possibility that these data sets are drawn from normal distributions because the observations fall reasonably close to a line on normal probability plots (Figure 2). However, both data sets have skewnesses less than −0.5, and these skewness values are more consistent with moraine degradation than either measurement error alone or inheritance [*Applegate et al.*, 2010].

Explicit fitting of the degradation model to these data sets also suggests that moraine degradation is responsible for most of the scatter in each data set (Figure 3, top, and Table 2), although the model fit to the inner Titcomb Lakes data set is poor. For the purposes of these fits, we prescribed the initial height of each moraine and the erosion rate of the exposed boulders (1.0 mm kyr^{-1} [*Gosse et al.*, 1995a, 1995b]). The constant-erosion-rate assumption is consistent with prior exposure dating studies that correct for the effects of boulder erosion [e.g., *Gosse et al.*, 1995b; *Kelly et al.*, 2008]. We then used the differential evolution genetic algorithm to search for the minimum value of the Kolmogorov-Smirnov test statistic [*Press et al.*, 2005; Clauset et al., data set, 2007]. The model evaluation with the minimum KS statistic indicates the values of moraine age, initial moraine slope, and topographic diffusivity that are most consistent with each data set. We specified the initial moraine heights for these model inversions because the distributions of cosmogenic exposure dates are insensitive to the initial height of the moraine above some minimum value [*Applegate et al.*, 2010].

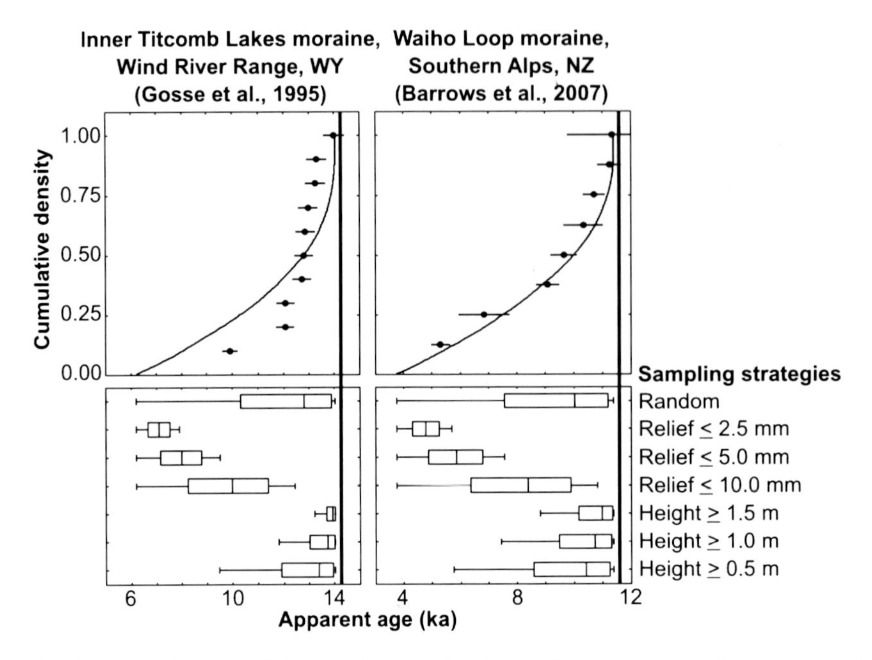

Figure 3. (top) Fits of the moraine degradation model to the beryllium-10 exposure dates from the inner Titcomb Lakes and Waiho Loop moraines [*Gosse et al.*, 1995b; *Barrows et al.*, 2007]. In each graph, the points with error bars represent the exposure dates with their 1σ measurement uncertainties; the curves are the best fit modeled distributions for each data set. These fits presume an erosion rate for exposed boulders of 1.0 mm kyr^{-1}, consistent with earlier exposure dating studies [e.g., *Gosse et al.*, 1995b]. The initial height of the Titcomb Lakes moraine was prescribed at 25 m [cf. *Gosse et al.*, 1995a], and the initial height of the Waiho Loop moraine was prescribed to be 50 m [cf. *Denton and Hendy*, 1994]. (bottom) Effects of different sampling strategies on the resulting distributions of cosmogenic exposure dates. The distributions are shown as box plots [*Chambers et al.*, 1983]. Sampling very tall boulders (height greater than or equal to 1.5 m) produces exposure dates that are within a few thousand years of the true age of the moraine (heavy line); sampling boulders with minimal surface relief (relief less than or equal to 2.5 mm) produces exposure dates that are thousands of years younger than the true age of the moraine.

4. IMPLICATIONS FOR FIELD SAMPLING CRITERIA

After identifying a moraine, the first step in cosmogenic exposure dating is deciding which boulders to sample. Depending on which boulders are sampled, the measured concentrations either will or will not be representative of the moraine's true age. Thus, imperfect boulder selection strategies could interfere with our ability to identify moraines associated with abrupt climate changes.

Table 2. Best Fits of the Degradation Model to the Waiho Loop and Titcomb Lakes Data Sets[a]

Data Set	Moraine Age (ka)	Initial Slope (deg)	Diffusivity (m^2 yr^{-1})	Kolmogorov-Smirnov Statistic
Waiho Loop	11.59	32.77	2.909×10^{-3}	0.1400
Titcomb Lakes	14.27	36.80	6.883×10^{-4}	0.3113

[a]Data sets are from *Barrows et al.* [2007] and *Gosse et al.* [1995b]. The parameter estimates here are probably good to two or three significant figures. Four figures are reported to allow checking of the model fits.

Field workers use a variety of criteria to select samples for cosmogenic nuclide measurements, but the two most common criteria are boulder height and surface freshness [e.g., *Nishiizumi et al.*, 1989; *Phillips et al.*, 1990; *Cerling and Craig*, 1994; *Gosse et al.*, 1995b; *Fabel and Harbor*, 1999; *Licciardi et al.*, 2001; *Laabs et al.*, 2009]. Most field geomorphologists avoid sampling boulders below some minimum height, often 1 m. Fresh boulders retain polish or striations, if they were transported at the glacier bed. Few moraine boulders have polish or striations when they are sampled, so field geomorphologists estimate the thickness of material eroded from each boulder by measuring the relief on the boulders' upper surfaces. For a single boulder, relief is the distance between the lowest point and the highest point on the boulder's upper surface, measured at right angles to the sampled surface. The style of weathering varies with lithology, but the low points on boulder surfaces are often weathering pits, and the high points are often veins of resistant mineralogies such as quartz [e.g., *Barrows et al.*, 2007]. Unless the high points retain polish, boulder surface relief is a minimum estimate of the thickness of material removed from the surfaces of the boulders, again assuming that the boulders were transported subglacially.

These criteria are intended to minimize the chance that the samples have been shielded from cosmic rays during part of their postdepositional history. Tall boulders are less likely to have been covered by sediment or snow; boulders with polished or striated surfaces have not lost their nuclide-rich outer surfaces to erosion.

However, tall and fresh boulders sometimes yield exposure dates that are much younger than shorter and more weathered boulders from the same moraine [e.g., *Briner*, 2009]. There is no correlation between boulder height and apparent exposure time for the samples from the inner Titcomb Lakes moraine (Figure 4) (boulder heights were not reported for the Waiho Loop data set). *Briner* [2009] took both pebble collections and boulder samples from moraines in Colorado and found no relationship between clast size and the apparent exposure time of the sample.

We are unaware of any study that reports on the relationship between surface freshness and apparent exposure time. However, we believe that surface freshness is a poor predictor of the apparent exposure times yielded by individual boulders; that is, fresh boulders are not more likely than weathered boulders to yield exposure dates that are representative of the moraine's actual age. We make this statement on the basis of our modeling results, which we describe below.

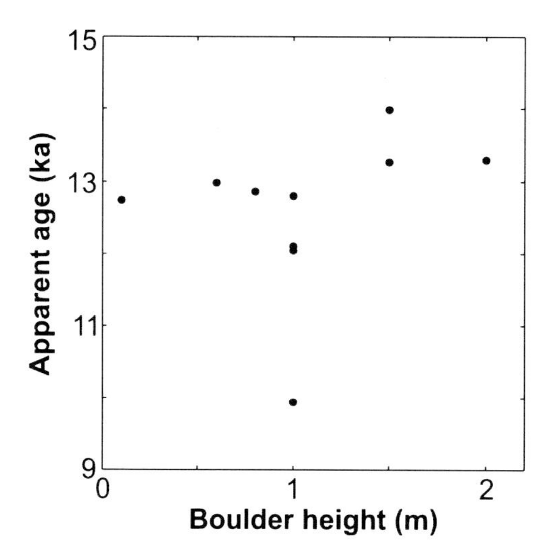

Figure 4. Apparent exposure time as a function of boulder height for the inner Titcomb Lakes moraine [*Gosse et al.* [1995b] (Table 1). For this data set, there appears to be no relationship between boulder height and apparent exposure time. The exposure dates have not been corrected for snow cover or boulder erosion.

Moraine degradation can explain the failure of these sampling criteria to indicate which boulders to sample on moraines (other scenarios are also possible; see Discussion section, below). If moraines lose meters of material from their crests over time, and boulders are distributed throughout the removed soil column, then there will be no correlation between boulder height and apparent exposure time. If boulders do not erode while buried, but do erode at a constant rate after exhumation, then the least eroded boulders are also those that have spent the least amount of time at the surface.

Our model fits and field observations support these assumptions. The scatter among the beryllium-10 exposure dates from the Waiho Loop moraine is best explained by the progressive loss of ~4.2 m of material from the moraine's crest; similarly, the model fit suggests that the inner Titcomb Lakes moraine has lost ~2.6 m of material from its crest. Although these moraine height changes seem extreme, the average rates of crest lowering that they imply are reasonable; 4.2 m of removal over 11.6 kyr (Waiho Loop moraine, Table 2) is ~0.4 mm yr^{-1}, a geomorphically possible rate for a steep, unconsolidated deposit.

It is more difficult to assess the correctness of our assumptions about boulder erosion from the model fits because the shape of the modeled distributions is insensitive to the boulder erosion rate. However, some support for our statements about boulder erosion comes from field observations made by one of us (P.J.A.) on the Huancané II moraines near the Quelccaya Ice Cap [*Mercer and Palacios*, 1977]. On these moraines, clasts buried less than a meter below the surface are fresh, but surface boulders adjacent to the soil pits are weathered. The weathered boulders have pits and pedestals on their upper surfaces that indicate several centimeters of lost material. Taken together, these observations suggest that boulders in this environment do not erode unless they are exposed at the surface.

Given our model fits (Figure 3, top), we can evaluate the effects of different sampling strategies on the distributions of cosmogenic exposure dates (Figure 3, bottom). The model tracks the heights, surface relief values, and exposure dates for each boulder. Thus, we can identify the modeled boulders that are taller than a certain height or have less than a certain amount of surface relief. By comparing the exposure dates within each subsample to the "true" age of the moraine, we can determine which sampling strategies produce exposure dates that are closest to this true value.

Very tall boulders are most likely to yield exposure dates that are close to the true age of the moraine (Figure 3, bottom). The median exposure date moves closer to the true moraine age as the minimum boulder height becomes greater. However, even fairly large boulders, standing at least 1.5 m above the moraine's crest, can yield exposure dates that are several thousand years younger than the moraine's true age. Small boulders do sometimes yield exposure dates that approximate the true age of the moraine, even where moraine degradation is rapid; these boulders originated near the top of the removed soil column.

Sampling boulders with minimal surface relief always yields exposure dates that are younger than the moraine's true age (Figure 3, bottom). Moreover, as the criterion is made more stringent, the median exposure date moves away from the true age. Thus, the more strictly one adheres to the freshness criterion, the worse the resulting exposure dates become.

This result seems intuitive because till erodes more easily than rock in many settings. The changes in moraine heights implied by our model fits (Figure 3, top) are much greater than the thickness of material that can be removed by erosion from boulder surfaces over an equivalent period. Even if we allow a very rapid boulder erosion rate of 100 mm kyr^{-1} [cf. *Gosse et al.*, 1995a, 1995b], a boulder exposed to surface weathering for 12 kyr would lose only 1.2 m of material from its surface. This thickness is much smaller than the 2–4 m of moraine height change implied by our model fits. Thus, till cover has a greater influence on exposure dates than boulder erosion on young, steep, matrix-supported moraines, even allowing for the difference in density between boulders and till [*Hallet and Putkonen*, 1994].

5. IMPLICATIONS FOR DETERMINING NUCLIDE PRODUCTION RATES

Step 5 in determining the age of a moraine using cosmogenic exposure dating involves calculating the apparent exposure time of each sample, using the estimated local production rate of the nuclide [*Balco et al.*, 2008]. Thus, any error in determining the local production rate will translate into errors in the apparent exposure times, reducing our ability to identify moraines associated with short-lived climate events.

The production rates of cosmogenic nuclides are not known a priori. Instead, the concentrations of cosmogenic nuclides are measured in rock surfaces whose exposure ages are known independently from other chronologic methods, usually radiocarbon dating. Not all the calibration samples come from moraine boulders, but some do [*Balco et al.*, 2008]. Ideally, the nuclide concentration in a single rock surface, divided by the independently determined exposure age, yields the local time-averaged production rate after correcting for nuclear decay. In practice, the nuclide concentrations at the calibration sites are highly scattered [*Balco et al.*, 2008]. Consequently, there is uncertainty about the representative nuclide concentration at each calibration site.

Recent efforts to determine a reference production rate from the calibration database [*Balco et al.*, 2008] average the measured nuclide concentrations to determine a representative nuclide concentration for each site. This procedure is reasonable, but it ignores the effects of geomorphic processes on the nuclide concentrations, as well as problems with the independent age constraints. These potential problems were fully acknowledged by *Balco et al.* [2008].

The time-averaged production rate of beryllium-10 at the Titcomb Lakes site may be about 7% larger than previously believed. The measured beryllium-10 concentrations from the inner Titcomb Lakes moraine are part of the global nuclide production rate calibration database [*Gosse and Klein*, 1996; *Balco et al.*, 2008]. The degree of scatter in the exposure dates from the inner Titcomb Lakes moraine is probably larger than can be explained by measurement error, and the skewness of the data set suggests that the maximum exposure date is the best estimator of the moraine's age (Figure 2) [*Applegate et al.*, 2010; *Applegate*, 2009]. Thus, the representative nuclide concentration at the Titcomb Lakes study site is probably the maximum measured concentration. Prior studies that used the Titcomb Lakes concentration measurements to estimate the production rate of beryllium-10 took the average of the nine largest observed concentrations as the representative nuclide concentration for this site, treating the smallest concentration as a statistical outlier [*Gosse and Klein*, 1996; *Balco et al.*, 2008]. The largest concentration is about 7% greater than the mean of the nine largest concentrations.

A potential error of a few percent in estimating nuclide production rates has serious implications for our ability to identify moraines associated with abrupt, short-lived climate changes such as the Younger Dryas. Even a 5% error in estimating nuclide production rates translates into an error in apparent exposure time of ~600 years (5% of 12 kyr), about half the length of the Younger Dryas.

This example suggests that the calibration of beryllium-10 production rates should be reevaluated, taking the effects of geomorphology on the calibration measurements into account. This recalibration might help reduce the extreme mismatch between some of the calibration samples and the best fit of the scaling models to the calibration data set [*Balco et al.*, 2008, Figure 5].

6. DISCUSSION

In this chapter, we have identified two challenges in the use of cosmogenic exposure dating to date moraines associated with abrupt climate changes such as the Younger Dryas. First, pristine boulders will tend to yield exposure dates that are younger than the true age of the moraine, if moraines lose significant thicknesses of material from their crests over time, and boulders erode once exhumed. Second, geomorphic processes likely impart an additional uncertainty of at least several percent in our present estimates of reference cosmogenic nuclide production rates. For moraines of late glacial to early Holocene age, these effects bias exposure dates by hundreds to thousands of years. Thus, these challenges limit our confidence in the ability of cosmogenic exposure dating to identify moraines associated with abrupt climate changes, which have time scales of a few years to a few hundred years.

Moreover, our results confirm prior suggestions that preferentially sampling tall boulders is a good strategy [*Phillips et al.*, 1990; *Gosse et al.*, 1995a, 1995b]. However, even exposure dates from tall boulders may underestimate the age of a moraine by thousands of years, depending on boulder height and the thickness of material lost from the moraine's crest.

Our conclusions do not hold where geomorphic processes other than moraine degradation and boulder erosion affect exposure dates. In particular, inheritance might cause the exposure dates from tall boulders to overestimate the ages of moraines; boulders that fall onto the glacier from oversteepened valley walls will not be eroded in transport and thus may be larger than other clasts that have had most of their inherited nuclides stripped away by subglacial transport.

Moreover, we assume that the erosion rate for exposed boulders is the same for all the rocks on a moraine. However, we expect that boulders on real moraines will weather at different rates, depending on their lithology, position in the landscape, and size. Our models also assume slow, grain-by-grain erosion of moraine boulders, but the rapid loss of several centimeters of rock is also possible [*Zimmerman et al.*, 1995]. In the future, we plan to update our models to represent this style of erosion, perhaps following *Muzikar* [2009].

Here we explain the lack of correlation between boulder height and apparent exposure time on the inner Titcomb Lakes moraine (Figure 4) with moraine degradation. Other explanations are also possible. In cases where inheritance dominates the scatter among exposure dates, there may be no relationship between boulder size and apparent exposure time. If all the boulders on a moraine are taller than the thickness of snow cover or the thickness of sediment that has been removed from the surface of a moraine, there will also be no correlation between boulder height and apparent exposure time. We believe that the geomorphic processes responsible for the scatter among exposure dates vary between moraines, and so the degree and sign of the correlation between boulder height and apparent exposure time probably also vary.

Our results imply that our ability to invert our models against collections of exposure dates depends on how the samples were chosen. The inverse methods can account for the heights of boulders from which samples were taken, as long as these heights are reported with the exposure dates. However, the inverse methods cannot account for use of the surface pristinity criterion. Preferentially sampling fresh boulders produces distributions of exposure dates that are all younger than the true age of the moraine and emphasize the tail of the modeled distribution at the expense of the mode (Figure 3) if both moraine degradation and boulder erosion are active on a particular moraine. Given a data set collected in this way, our methods would likely be unable to determine whether the excess scatter in the data was caused by inheritance or moraine degradation. It is also possible that such a data set would have a deceptively small scatter. Any age estimates made from such a data set would be far too young.

For future field campaigns, and in evaluating data sets from the literature, we recommend determining what overall precision would be necessary to answer the paleoclimate question at hand. For example, confident determination of whether a moraine dates to the Younger Dryas probably requires an overall 1σ uncertainty of about 100 years or about 10% of the event's length.

With that information, the minimum number of samples required to answer the question can be estimated from the expected measurement uncertainty of the exposure dates and Figure 1. Figure 1 represents the uncertainty of the weighted mean [*Bevington and Robinson*, 2003, equation 4-19]. This quantity is the uncertainty of repeated exposure dating on a single moraine in the special case where geomorphology has no effect on the exposure dates. In practice, the uncertainty of exposure dating will be larger than Figure 1 implies, so more samples will be required. Note that repeated sampling will not reduce uncertainty associated with external factors such as production rate estimation errors [*Balco et al.*, 2008].

The number of samples required to achieve a desired precision will sometimes be unrealistically large. The curve in Figure 1 reaches a point of diminishing returns around 10–15 samples, where the uncertainty of the weighted mean is 25%–30% of the uncertainty of one date. The number of samples required to achieve a precision better than 25%–30% of the uncertainty of one date will often be larger than can be produced with available resources. For example, if our measurement uncertainty is 5%, rather than 3%, we will need 36 samples to achieve a 100 year overall uncertainty for a Younger Dryas moraine (Figure 1). We are unaware of any moraine with 36 published exposure dates. Thus, some questions can be answered only approximately with cosmogenic exposure dating.

In choosing which boulders to sample, we recommend sampling all the tallest boulders on the moraine. If resources for further sampling are available, additional boulders should be chosen randomly. All the sampled boulders should be on the crest of the moraine. The samples from the tall boulders have a good chance of producing at least one exposure date that correctly estimates the moraine's age, assuming inheritance is not important. The randomly chosen samples will help identify the source of any geomorphic bias, if the exposure dates from the tall boulders are widely scattered. Moreover, the surface relief of sampled boulders should be recorded in the field and given in papers that report new exposure dates.

Acknowledgments. This work was partly supported by the National Science Foundation (grants 0531211, 0539578, and 0424589), the Comer Science and Education Foundation, and the Geological Society of America (grant 8736-08). Discussions with Tom Lowell, David Pollard, Kaitlin Walsh, Fred Phillips, Meredith Kelly, Aaron Putnam, Roseanne Schwartz, and Ed Evenson sharpened our thinking on these issues. Comments from two anonymous researchers also improved the manuscript.

REFERENCES

Alley, R. B. (2007), Wally was right: Predictive ability of the North Atlantic "conveyor belt" hypothesis for abrupt climate change, *Annu. Rev. Earth Planet. Sci.*, *35*, 241–272, doi:10.1146/annurev.earth.35.081006.131524.

Alley, R. B., et al. (1993), Abrupt increase in Greenland snow accumulation at the end of the Younger Dryas event, *Nature*, *362*, 527–529.

Alley, R. B., et al. (2010), History of the Greenland Ice Sheet: Paleoclimatic insights, *Quat. Sci. Rev.*, *29*, 1728–1756, doi:10.1016/j.quascirev.2010.02.007.

Applegate, P. J. (2009), Estimating the ages of glacial landforms from the statistical distributions of cosmogenic exposure dates, Ph.D. thesis, Pa. State Univ., University Park. (Available at http://etda.libraries.psu.edu/)

Applegate, P. J., T. V. Lowell, and R. B. Alley (2008), Comment on "Absence of cooling in New Zealand and the adjacent ocean during the Younger Dryas chronozone," *Science*, *320*, 746, doi:10.1126/science.1152098.

Applegate, P. J., N. M. Urban, K. Keller, and R. B. Alley (2010), Modeling the statistical distributions of cosmogenic exposure dates from moraines, *Geosci. Model Dev.*, *3*, 293–307.

Balco, G., and J. M. Schaefer (2006), Cosmogenic-nuclide and varve chronologies for the deglaciation of southern New England, *Quat. Geochronol.*, *1*, 15–28, doi:10.1016/j.quageo.2006.06.014.

Balco, G., J. O. Stone, N. A. Lifton, and T. J. Dunai (2008), A complete and easily accessible means of calculating surface exposure ages or erosion rates from 10Be and 26Al measurements, *Quat. Geochronol.*, *3*, 174–195, doi:10.1016/j.quageo.2007.12.001.

Barrows, T. T., S. J. Lehman, L. K. Fifield, and P. de Deckker (2007), Absence of cooling in New Zealand and the adjacent ocean during the Younger Dryas chronozone, *Science*, *318*, 86–89, doi:10.1126/science.1145873.

Barrows, T. T., S. J. Lehman, L. K. Fifield, and P. de Deckker (2008), Response to comment on "Absence of cooling in New Zealand and the adjacent ocean during the Younger Dryas chronozone," *Science*, *320*, 746edoi:10.1126/science.1152216.

Benson, L., R. Madole, G. Landis, and J. Gosse (2005), New data for Late Pleistocene Pinedale alpine glaciation from southwestern Colorado, *Quat. Sci. Rev.*, *24*, 49–65, doi:10.1016/j.quascirev.2004.07.018.

Bevington, P. R., and D. K. Robinson (2003), *Data Reduction and Error Analysis for the Physical Sciences*, 336 pp., McGraw-Hill, New York.

Bierman, P. R. (1994), Using in situ produced cosmogenic isotopes to estimate rates of landscape evolution: A review from the geomorphic perspective, *J. Geophys. Res.*, *99*, 13,885–13,896.

Briner, J. P. (2009), Moraine pebbles and boulders yield indistinguishable ^{10}Be ages: A case study from Colorado, USA, *Quat. Geochronol.*, *4*, 299–305, doi:10.1016/j.quageo.2009.02.010.

Briner, J. P., D. S. Kaufman, W. F. Manley, R. C. Finkel, and M. W. Caffee (2005), Cosmogenic exposure dating of late Pleistocene moraine stabilization in Alaska, *Geol. Soc. Am. Bull.*, *117*, 1108–1120, doi:10.1130/B25649.1.

Broecker, W. S. (2003), Does the trigger for abrupt climate change reside in the ocean or in the atmosphere?, *Science*, *300*, 1519–1522, doi:10.1126/science.1083797.

Broecker, W. S. (2006), Abrupt climate change revisited, *Global Planet. Change*, *54*, 211–215, doi:10.1016/j.gloplacha.2006.06.019.

Broecker, W. S., J. P. Kennett, B. P. Flower, J. T. Teller, S. Trumbore, G. Bonani, and W. Wolfli (1989), Routing of meltwater from the Laurentide Ice Sheet during the Younger Dryas cold episode, *Nature*, *341*, 318–321.

Cerling, T. E., and H. Craig (1994), Geomorphology and in-situ cosmogenic isotopes, *Annu. Rev. Earth Planet. Sci.*, *22*, 273–317.

Chambers, J. M., W. S. Cleveland, P. A. Tukey, and B. Kleiner (1983), *Graphical Methods for Data Analysis*, 395 pp., Duxbury, London.

Chiang, J. C. H., and C. M. Bitz (2005), Influence of high latitude ice cover on the marine Intertropical Convergence Zone, *Clim. Dyn.*, *25*, 477–496, doi:10.1007/s00382-005-0040-5.

Denton, G. H., and C. H. Hendy (1994), Younger Dryas age advance of Franz Josef Glacier in the Southern Hemisphere, New Zealand, *Science*, *264*, 1434–1437.

Denton, G. H., R. B. Alley, G. C. Comer, and W. S. Broecker (2005), The role of seasonality in abrupt climate change, *Quat. Sci. Rev.*, *24*, 1159–1182, doi:10.1016/j.quascirev.2004.12.002.

Fabel, D., and J. Harbor (1999), The use of in-situ produced cosmogenic radionuclides in glaciology and glacial geomorphology, *Ann. Glaciol.*, *28*, 103–110.

Gibbons, A. B., J. D. Megeath, and K. L. Pierce (1984), Probability of moraine survival in a succession of glacial advances, *Geology*, *12*, 327–330.

Gosse, J. C., and J. Klein (1996), Production rate of in-situ cosmogenic ^{10}Be in quartz at high altitude and mid latitude, *Radiocarbon*, *38*, 154–155.

Gosse, J. C., and F. M. Phillips (2001), Terrestrial in situ cosmogenic nuclides: Theory and application, *Quat. Sci. Rev.*, *20*, 1475–1560.

Gosse, J. C., J. Klein, E. B. Evenson, B. Lawn, and R. Middleton (1995a), Beryllium-10 dating of the duration and retreat of the last Pinedale glacial sequence, *Science*, *268*, 1329–1333.

Gosse, J. C., E. B. Evenson, J. Klein, B. Lawn, and R. Middleton (1995b), Precise cosmogenic ^{10}Be measurements in western North America: Support for a global Younger Dryas cooling event, *Geology*, *23*, 877–880.

Hallet, B., and J. Putkonen (1994), Surface dating of dynamic landforms: Young boulders on aging moraines, *Science*, *265*, 937–940.

Ivy-Ochs, S., C. Schlüchter, P. W. Kubik, and G. H. Denton (1999), Moraine exposure dates imply synchronous Younger Dryas glacier advances in the European Alps and in the Southern Alps of New Zealand, *Geogr. Ann.*, *81*, 313–323.

Ivy-Ochs, S., H. Kerschner, A. Reuther, M. Maisch, R. Sailer, J. Schaefer, P. W. Kubik, H. Synal, and C. Schlüchter (2006), The timing of glacier advances in the northern European Alps based on surface exposure dating with cosmogenic ^{10}Be, ^{26}Al, ^{36}Cl, and ^{21}Ne, in *In Situ-Produced Cosmogenic Nuclides and Quantification of Geological Processes*, edited by L. L. Siame, D. L. Bourlès, and E. T. Brown, *Spec. Pap. Geol. Soc. Am.*, *415*, 43–60.

Ivy-Ochs, S., H. Kerschner, and C. Schlüchter (2007), Cosmogenic nuclides and the dating of Lateglacial and early Holocene glacier variations: The Alpine perspective, *Quat. Int.*, *164–165*, 53–63, doi:10.1016/j.quaint.2006.12.008.

Kelly, M. A., P. W. Kubik, F. von Blanckenburg, and C. Schlüchter (2004), Surface exposure dating of the Great Aletsch Glacier Egesen moraine system, western Swiss Alps, using the cosmogenic nuclide ^{10}Be, *J. Quat. Sci.*, *19*(5), 431–441, doi:10.1002/jqs.854.

Kelly, M. A., T. V. Lowell, B. L. Hall, J. M. Schaefer, R. C. Finkel, B. M. Goehring, R. B. Alley, and G. H. Denton (2008), A ^{10}Be chronology of late glacial and Holocene mountain glaciation in the Scoresby Sund region, east Greenland: Implications for seasonality during late glacial time, *Quat. Sci. Rev.*, *27*, 2273–2282, doi:10.1016/j.quascirev.2008.08.004.

Kohl, C. P., and K. Nishiizumi (1992), Chemical isolation of quartz for measurement of in situ-produced cosmogenic nuclides, *Geochim. Cosmochim. Acta*, *56*, 3583–3587.

Laabs, B. J. C., K. A. Refsnider, J. S. Munroe, D. M. Mickelson, P. J. Applegate, B. S. Singer, and M. W. Caffeee (2009), Latest Pleistocene glacial chronology of the Uinta Mountains: Support for moisture-driven asynchrony of the last deglaciation, *Quat. Sci. Rev.*, *28*, 1171–1187, doi:10.1016/j.quascirev.2008.12.012.

Lal, D. (1991), Cosmic ray labeling of erosion surfaces: In situ nuclide production rates and erosion models, *Earth Planet. Sci. Lett.*, *104*, 424–439.

Licciardi, J. M., P. U. Clark, E. J. Brook, K. L. Pierce, M. D. Kurz, D. Elmore, and P. Sharma (2001), Cosmogenic ^{3}He and ^{10}Be chronologies of the late Pinedale northern Yellowstone ice cap, Montana, USA, *Geology*, *29*, 1095–1098.

Lowell, T. V. (2000), As climate changes, so do glaciers, *Proc. Natl. Acad. Sci. U. S. A.*, *97*, 1351–1354.

Lowell, T. V., et al. (2005), Testing the Lake Agassiz meltwater trigger for the Younger Dryas, *Eos Trans. AGU*, *86*(40), 365–372.

Mangerud, J., D. T. Anderson, B. E. Berglund, and J. J. Donner (1974), Quaternary stratigraphy of Norden, a proposal for terminology and classification, *Boreas*, *3*, 109–127.

Mercer, J. H., and M. O. Palacios (1977), Radiocarbon dating of the last glaciation in Peru, *Geology*, *5*, 600–604.

Muzikar, P. (2009), General models for episodic surface denudation and its measurement by cosmogenic nuclides, *Quat. Geochronol.*, *4*, 50–55, doi:10.1016/j.quageo.2008.06.004.

Muzikar, P., D. Elmore, and D. E. Granger (2003), Accelerator mass spectrometry in geological research, *Geol. Soc. Am. Bull.*, *115*, 643–654.

Nishiizumi, K., E. L. Winterer, C. P. Kohl, J. Klein, R. Middleton, D. Lal, and J. R. Arnold (1989), Cosmic ray production rates of ^{10}Be and ^{26}Al in quartz from glacially polished rocks, *J. Geophys. Res.*, *94*(B12), 17,907–17,915.

Oerlemans, J. (2005), Extracting a climate signal from 169 glacier records, *Science*, *308*, 675–677, doi:10.1126/science.1107046.

Owen, L. A., R. C. Finkel, R. A. Minnich, and A. E. Perez (2003), Extreme southwestern margin of late Quaternary glaciation in North America: Timing and controls, *Geology*, *31*, 729–732.

Phillips, F. M., M. G. Zreda, S. S. Smith, D. Elmore, P. W. Kubik, and P. Sharma (1990), Cosmogenic chlorine-36 chronology for glacial deposits at Bloody Canyon, eastern Sierra Nevada, *Science*, *248*, 1529–1532.

Price, K., R. M. Storn, and J. A. Lampinen (2005), *Differential Evolution: A Practical Approach to Global Optimization*, 558 pp., Springer, Berlin.

Putkonen, J., and T. Swanson (2003), Accuracy of cosmogenic ages for moraines, *Quat. Res.*, *59*, 255–261, doi:10.1016/S0033-5894(03)00006-1.

Stone, J. O. (2000), Air pressure and cosmogenic isotope production, *J. Geophys. Res.*, *105*, 23,753–23,759.

Turney, C. S. M., R. G. Roberts, N. de Jonge, C. Prior, J. M. Wilmshurst, M. S. McGlone, and J. Cooper (2007), Redating the advance of the New Zealand Franz Josef Glacier during the last termination: Evidence for asynchronous climate change, *Quat. Sci. Rev.*, *26*, 3037–3042, doi:10.1016/j.quascirev.2007.09.014.

Vacco, D. A., R. B. Alley, and D. Pollard (2009), Modeling dependence of moraine deposition on climate history: The effect of seasonality, *Quat. Sci. Rev.*, *28*, 639–646, doi:10.1016/j.quascirev.2008.04.018.

Vellinga, M., and R. A. Wood (2002), Global climatic impacts of a collapse of the Atlantic thermohaline circulation, *Clim. Change*, *54*, 251–267.

Walker, M., et al. (2008), Formal definition and dating of the GSSP (Global Stratotype Section and Point) for the base of the Holocene using the Greenland NGRIP ice core, and selected auxiliary records, *J. Quat. Sci.*, *24*(1), 3–17, doi:10.1002/jqs.1227.

Zielinski, G. A., and P. T. Davis (1987), Late Pleistocene age of the type Temple Lake moraine, Wind River Range, Wyoming, U. S. A, *Geogr. Phys. Quat.*, *41*, 397–401.

Zimmerman, S. G., E. B. Evenson, J. C. Gosse, and C. P. Erskine (1995), Extensive boulder erosion resulting from a range fire on the type-Pinedale moraines, Fremont Lake, Wyoming, *Quat. Res.*, *42*, 255–265.

Zreda, M. G., F. M. Phillips, and D. Elmore (1994), Cosmogenic ^{36}Cl accumulation in unstable landforms: 2. Simulations and measurements on eroding moraines, *Water Resour. Res.*, *30*(11), 3127–3136.

R. B. Alley, Department of Geosciences, Pennsylvania State University, University Park, PA 16802, USA.

P. J. Applegate, Department of Physical Geography and Quaternary Geology, Stockholm University, SE-106 91 Stockholm, Sweden. (patrick.applegate@natgeo.su.se)

Hypothesized Link Between Glacial/Interglacial Atmospheric CO_2 Cycles and Storage/Release of CO_2-Rich Fluids From Deep-Sea Sediments

Lowell Stott

Department of Earth Sciences, University of Southern California, Los Angeles, California, USA

Axel Timmermann

International Pacific Research Center, SOEST, University of Hawai'i at Mānoa, Honolulu, Hawaii, USA

During the last glacial termination, the rise in atmospheric pCO_2 was accompanied by a precipitous drop in surface ocean $\Delta^{14}C$ that cannot be explained by changes in ^{14}C production alone and therefore appears to require a flux of ^{14}C-depleted carbon into the surface ocean. The magnitude of this $\Delta^{14}C$ excursion is hard to reconcile with an ocean-only mechanism of CO_2 regulation. Here we explore the possibility that hydrothermal sources of CO_2 contributed to glacial/interglacial CO_2 variability and to the $\Delta^{14}C$ variations during the last deglaciation. We hypothesize that as the ocean cooled during glaciations CO_2-hydrate stability expanded upward to shallower depths and over a broader region of the seafloor, reducing the flux of ^{14}C-depleted CO_2 into the ocean from sediment reservoirs that blanket active vents throughout the ocean. Conversely, as the oceans warmed during deglaciation, the CO_2-hydrate stability horizon deepened and caused a transient release of ^{14}C-depleted CO_2 from the sediment reservoirs. Using a transient glacial-interglacial simulation conducted with the Earth system model of LOVECLIM, we estimate that ~3°C temperature increase at intermediate water depths in the Pacific during the last deglaciation would have been large enough to lower the hydrate stability horizon by several hundred meters and significantly reduce the areal extent of seafloor where hydrate was stable. This hypothesis would explain why ^{14}C ages of abyssal water masses were not anomalously old during the last glacial and why there was a large radiocarbon activity ($\Delta^{14}C$) anomaly during the last deglaciation at intermediate (500–700 m) water depths.

1. INTRODUCTION

Among of the most striking aspects of Earth's natural long-term climate behavior are the abrupt terminations of ice ages and their close temporal relationship to rising concentrations of atmospheric CO_2 (Plate 1). The last ice age ended between 19 and 10 kyr B.P., accompanied by an ~80 ppm rise in

Abrupt Climate Change: Mechanisms, Patterns, and Impacts
Geophysical Monograph Series 193

10.1029/2010GM001052

atmospheric CO_2 (Plate 1) [*Jouzel et al.*, 2007; *Luthi et al.*, 2008]. Indeed, each glacial/interglacial climate transition during the past 500 kyr was associated with similar atmospheric CO_2 change. The remarkable similarity in atmospheric CO_2 variability during the late Pleistocene glacial-interglacial cycles points to a regulatory mechanism that systematically limits the range of atmospheric CO_2 change between a glacial and an interglacial climate state. Given the close temporal relationship between atmospheric CO_2 changes and temperature variability at high southern latitudes, it appears the "center of action" for regulation of atmospheric CO_2 lies at high southern latitudes. But is the Southern Ocean itself responsible for the uptake and release of CO_2 to and from the atmosphere, or was the Southern Ocean simply a conduit through which climate change was communicated to the deep ocean where CO_2 was stored during glacials and released during deglaciations?

We set forth a hypothesis wherein the high latitude oceans act as communicator of orbitally induced climate change (temperature) to the deep sea. This temperature change stimulates a thermodynamic response at sites in the ocean where seawater flows through hydrothermal conduits and generates CO_2 and CO_2-rich fluids that accumulate in sediments that flank active volcanic vents. Hydrothermal systems in the Pacific act as both a source and sink for carbon. In the western Pacific, subduction of oceanic crust results in decarbonation of the carbonate sediments that produces CO_2-rich fluids and liquid CO_2 [*Chivas et al.*, 1987; *de Ronde et al.*, 2007; *Embley et al.*, 2006; *Inagaki et al.*, 2006; *Lupton et al.*, 2006, 2008; *Massoth et al.*, 2007; *Resing et al.*, 2004, 2009]. Recent surveys of these vents highlight accumulations of liquid CO_2 and CO_2-rich fluids in sediments that blanket the flanks of active vent sites [*Inagaki et al.*, 2006]. The flux of CO_2 and CO_2-rich fluids from sediments at active vent sites is regulated in part by CO_2 hydrate that can form at the sediment/water interface where warm, buoyant CO_2-rich fluids come in contact with cold seawater. The flux of CO_2 into the overlying ocean is regulated by diffusion and dissolution of CO_2 at the hydrate-sediment/water interface. Today, the temperature and pressure-dependent hydrate stability horizon occurs at intermediate water depths below ~700 m where temperatures are less than 9°C (Plate 2). Many of the active hydrothermal vents in the Pacific occur at intermediate depths [*Cheminée et al.*, 1991; *Chivas et al.*, 1987; *Hilton et al.*, 1998; *Inagaki et al.*, 2006; *Lupton et al.*, 2008; *McMurtry et al.*, 1993].

Coinciding with the discovery of liquid CO_2 accumulations at active submarine arcs in the Pacific, there have also been significant advances in the estimates of the flow of seawater and heat exchange at active vents along the axis and off axis of spreading centers in the eastern Pacific [*Fisher et al.*, 2003; *Hutnak et al.*, 2008; *Stein and Stein*, 1994]. These studies provide insight about the flux of seawater through the hydrothermal conduits and how this flux is influenced by temperature differences between cold seawater that enters a hydrothermal system and the warmer fluids that emanate at active vents [*Fisher et al.*, 2003; *Hutnak et al.*, 2008]. If shown to be representative of a wider distribution of active submarine vent systems throughout the Pacific, these estimates from the northeast Pacific imply very high volume flows, as much as 50 L s^{-1} at some sites [*Walker et al.*, 2008]. Using heat flow estimates and measurements of dissolved inorganic carbon (DIC) together with the isotopic composition ($\delta^{13}C$ and $\Delta^{14}C$) of DIC from the cold inflow water and warm outflow water, *Walker et al.* [2008] made a compelling case that hydrothermal systems may sequester significant volumes of carbon, both as precipitated carbonate in shallow crustal rocks and presumably as CO_2-rich subsurface fluids.

We are drawn to these observations as possible influences on glacial/interglacial CO_2 cycles. We begin by reviewing some of the previous hypotheses that have been considered in efforts to explain atmospheric CO_2 variability during the Pleistocene. In reviewing the previous hypotheses, it is evident that no single hypothesis can explain all of the phenomena that accompanied glacial/interglacial CO_2 variability, particularly the marked shifts in surface ocean $\Delta^{14}C$ during the most recent glacial termination. We then set forth our reasoning and a physical basis for considering an alternative mechanism for CO_2 variability.

2. PREVIOUS HYPOTHESES

During the last glacial termination, the rise in atmospheric CO_2 began at ~18 kyr B.P. Between 17.5 and 14.5 kyr B.P., the concentration of atmospheric CO_2 rose by about 50 ppm. The source of carbon released to the atmosphere must account for a 190‰ drop in atmospheric and surface ocean $\Delta^{14}C$ between ~18 and 14 kyr B.P. (Figure 1) because it appears that there was no change in the production of cosmogenic isotopes at that time [*Broecker and Clark*, 2010; *Finkel and Nishiizumi*, 1997]. *Broecker and Clark* [2010] estimated that ~5000 Gt of radiocarbon-dead carbon would be required to shift the entire ocean and the atmosphere carbon reservoirs by −190‰. With this constraint in mind, the most likely reservoir for sequestration of CO_2 would seem to be the abyssal ocean [*Broecker and Clark*, 2010]. Calling upon such a reservoir requires there to be a distinctive isotopic fingerprint in biogenic carbonate that came in contact with this water mass. Yet after more than 20 years of research, there is no clear evidence that an abyssal water mass was isolated from the atmosphere for anomalously longer periods of time during the last glacial [*Broecker and*

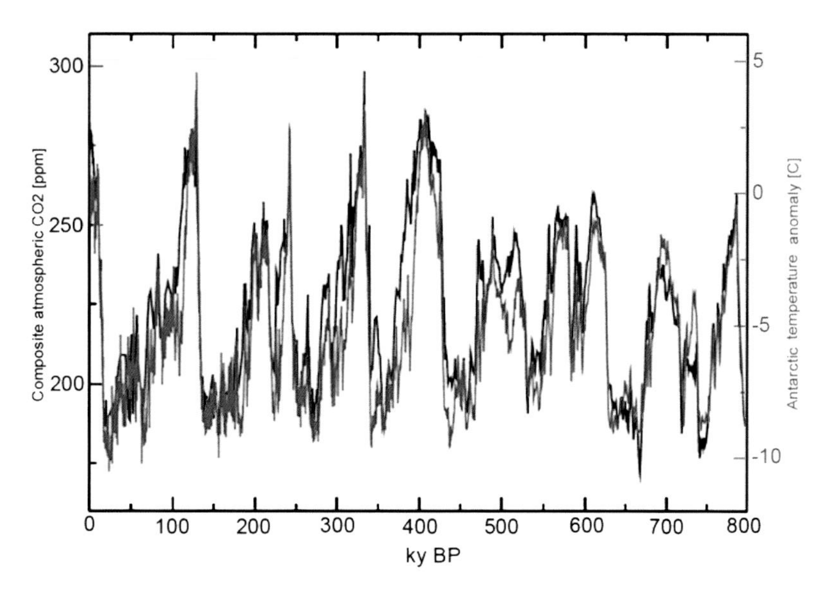

Plate 1. Graph of CO_2 (red) obtained from a composite of different ice core CO_2 measurements [*Jouzel et al.*, 2007; *Luthi et al.*, 2008] and reconstructed temperature (back), measured in the EPICA Dome C ice core [*Jouzel et al.*, 2007].

Clark, 2010]. There have been suggestions that some regions of the abyssal ocean contained anomalously old carbon [*Sikes et al.*, 2000; *Skinner et al.*, 2010]. However, interpretations of available data from the deep sea are not unequivocal. For example, two studies that used ^{14}C age differences between planktonic (*Globigerina bulloides*) and benthic foraminifera from abyssal cores from the South Atlantic (TN057-5, 41°S, 10°E, ~4700 m by C. Charles (unpublished results, 2010) and TN057-21, 41°S, 7°E, ~4900 m by *Barker et al.* [2010]) found no clear evidence of anomalously old abyssal water mass ages during the last glacial, whereas *Skinner et al.* [2010] made ^{14}C age measurements for planktonic and benthic foraminifera from another South Atlantic core located at 44°4.46'S, 14°12.47'W, 3.8 km (MD07-3076 CQ) that suggest ventilation ages were higher during the last glacial. The differences between these records are enigmatic and underscore the difficulty in using planktonic-benthic ^{14}C age differences as estimator of water mass age. As Skinner et al. point out, there are significant uncertainties associated with reservoir age corrections for planktonic age estimates, the species used to estimate surface ages and inherent errors due to contamination and reworking. Given the magnitude of $\Delta^{14}C$ change during the last glacial termination, it would seem most likely that abyssal sites in the Atlantic and the Pacific should record anomalously old ^{14}C ages. Yet in reviewing the ^{14}C data from sites at 41°S in the Atlantic [*Barker et al.*, 2010] and abyssal sites within the Pacific, *Broecker and Clark* [2010] conclude there is no convincing evidence to support the existence of an anomalously old water mass at abyssal depths during the last glacial.

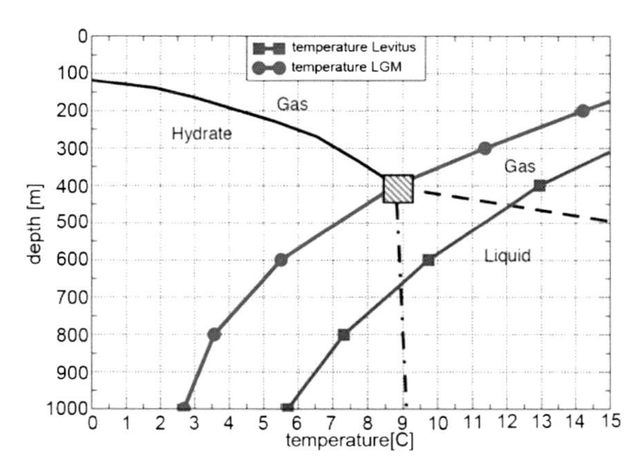

Plate 2. Phase diagram for CO_2 as a function of temperature and pressure showing how cooling during glaciations would shift the depth of hydrate stability upward. The modern temperature gradient through the water column [*Levitus*, 1994] is shown in red for a site located at 164°E, 30.5°S. An estimate of temperature gradient at this site for glacial conditions (blue), obtained by adding the difference of the Last Glacial Maximum (LGM)-present-day anomaly simulated by the LOVECLIM LGM simulation [*Timmermann et al.*, 2009] to the Levitus [*Levitus*, 1994] present-day climatology, is shown. Note that today, CO_2 hydrate is stable below 700 m, but in the glacial ocean, it expanded upward more than 300 m at this location.

Identifying an appropriate oceanographic mechanism to explain enhanced sequestration of atmospheric CO_2 during glaciations has also been problematic. *Sigman and Boyle* [2000] pointed out, for example, that simple ocean mechanisms for atmospheric CO_2 regulation, such as enhanced

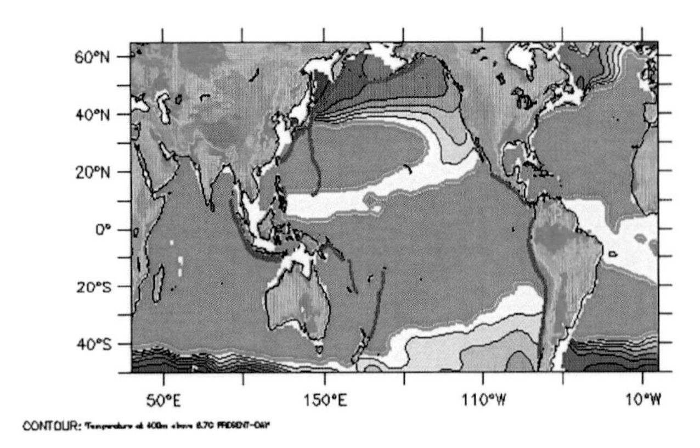

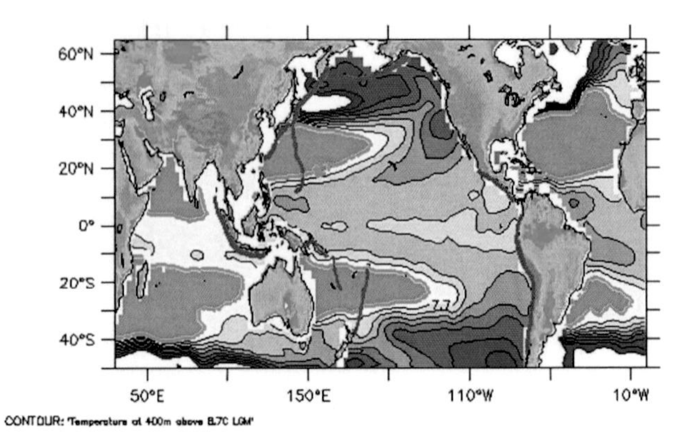

Plate 3. (left) Present-day temperature at 400 m: blue shading (orange coloring) characterizes temperatures below (above) CO_2 triple-point temperature indicating CO_2 clathrate stability (instability). The temperature data are taken from the work of *Levitus* [1994]; cyan contours indicate the triple-point temperature at 400 m depth. (right) LGM temperature at 400 m; coloring same as left. The LGM temperature data are obtained by subtracting an LGM temperature anomaly estimate from the *Levitus* [1994] temperature climatology. The LGM temperature anomaly estimate is obtained by averaging the difference between LGM and present-day simulations obtained from two climate models: (1) the LOVECLIM climate model [*Timmermann et al.*, 2009] and (2) the CCSM3 model [*Otto-Bliesner et al.*, 2006].

solubility into a colder glacial ocean would account for only a small, ~30 ppm, portion of the total 80 ppm decrease in glacial atmospheric concentrations. Higher salinities in the glacial ocean, as well would have offset the effects of enhanced solubility. Carbonate chemistry and shifting carbonate deposition have also been considered as influences on the atmospheric CO_2 variability during glacials [*Ainsworth et al.*, 2004; *Archer and Maier-Reimer*, 1994; *Brigault et al.*, 1998; *Broecker*, 1982, 2009b; *Broecker and Clark*, 2003; *Marchitto et al.*, 2005]. For this mechanism to explain the recurrent ~80–100 ppm drops in atmospheric CO_2 during glacials, decreased carbonate deposition in the oceans would have to be invoked. The deep-sea carbonate record itself does not provide a clear indication that this factor was the primary influence on glacial/interglacial CO_2 variability, although it very likely contributed to the variability [*Broecker and Clark*, 2001; *Farrell and Prell*, 1989]. Furthermore, while shallow water carbonate deposition was reduced during glacials, the deglacial increase in deposition on the continental shelves and at shallow carbonate reefs did not occur until after sea level had risen and, therefore, well after atmospheric CO_2 had begun to increase.

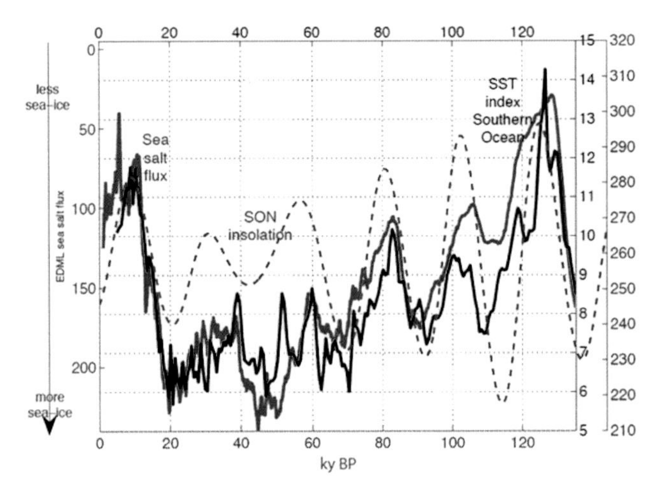

Plate 4. Timing of Southern Ocean temperature (black) and sea ice (red) change during the last glacial termination in relation to the austral spring insolation at 70°S (W m^{-2}, blue). The Southern Ocean SST index is obtained by averaging over the spline-interpolated SST records from MD88-770 MD97-2120, RC11-120; the sea ice proxy is derived from the sea salt sodium flux in the EPICA Dronning Maud Land (EDML) ice core and plotted on common EDML1/EDC3 time scale.

Considerable research has focused on mechanisms to enhance carbon transfer into the ocean's interior during glacials via biological processes termed the "biological pump." The Southern Ocean has received most attention because macronutrients are not fully utilized there. If biological sequestration of carbon was increased within the Southern Ocean during glacials, it could have enhanced transfer of CO_2 from the atmosphere into the ocean interior [*Adkins et al.*, 2002; *Ito and Follows*, 2005; *Keeling and Stephens*, 2001a, 2001b; *Masson-Delmotte et al.*, 2005; *Rozanski et al.*, 1992; *Sarmiento and Toggweiler*, 1984; *Siegenthaler and Wenk*, 1984; *Sigman et al.*, 2010; *Stephens and Keeling*, 2000; *Toggweiler and Sarmiento*, 1985]. It is also possible, although debated, that weakening or equatorward displacement of westerly winds over the Southern Ocean during glacials reduced Ekman pumping and upwelling of carbon-rich waters to the surface during glaciation

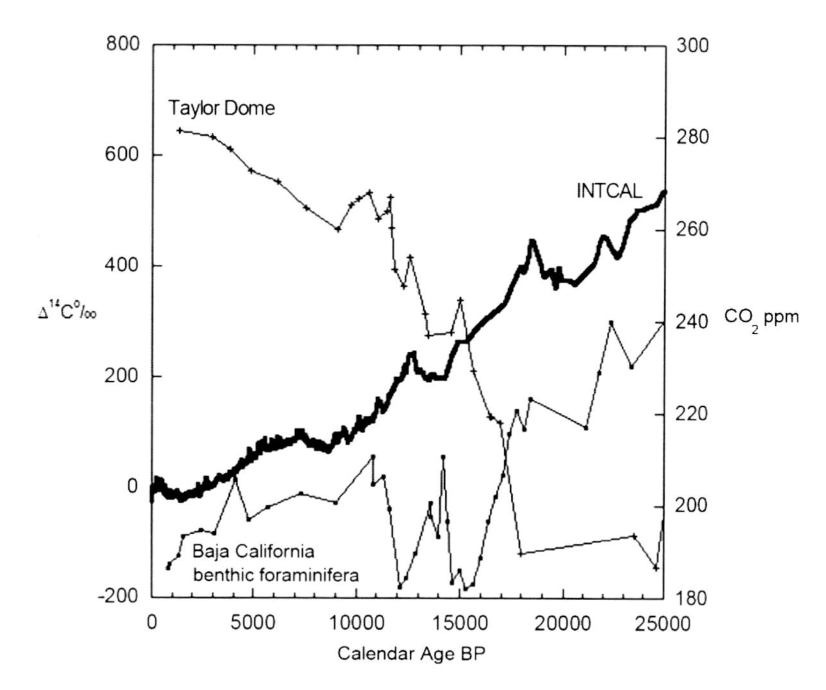

Figure 1. Taylor Dome record of atmospheric CO_2 over the most recent glacial termination [*Smith et al.*, 1999] and the INTCAL [*Reimer et al.*, 2004] reconstruction of atmospheric $\Delta^{14}C$. The $\Delta^{14}C$ excursion record from the Baja margin by *Marchittoet al.* [2007] occurs during the rise in deglacial atmospheric CO_2 and the decline in atmospheric $\Delta^{14}C$.

[*Toggweiler et al.*, 2006]. Depending on the ratio of upwelling to downward carbon export, this could reduce the flux of metabolic CO_2 to the glacial atmosphere. However, recent estimates suggest that ventilation of CO_2 through the Southern Ocean today is roughly balanced by the uptake of CO_2 by marine primary productivity and export of carbon to the deep ocean [*Gruber et al.*, 2009]. For the Southern Ocean to have played a primary role in regulating atmospheric CO_2 changes on glacial/interglacial time scales, its biological system must have worked in tandem with dynamical changes in ocean ventilation to reduce the net flux of metabolic CO_2 to the atmosphere [*Gruber et al.*, 2009; *Sigman et al.*, 2010].

If Southern Ocean biological and physical processes were responsible for glacial/interglacial CO_2 variations, there should be evidence to support this idea in proxy records collected from the Southern Ocean. *Martin* [1990] first suggested atmospheric CO_2 would decrease during glacials due to increased biological export production in the Southern Ocean that would result from increased wind-blown iron. The flux of soluble iron to the Southern Ocean did increase during glacials [*Gaspari et al.*, 2006; *Martínez-Garcia et al.*, 2009; *Wolff et al.*, 2006]. However, proxy data indicating marine carbon export south of the Polar Front was not substantially higher during the last glacial maxima when atmospheric CO_2 was at a minimum [*Downes et al.*, 1999; *Masson-Delmotte et al.*, 2005], although export production may have increased over some parts of the Southern Ocean [*Abelmann et al.*, 2006]. Instead, biogenic opal accumulation rates, a proxy for export production, was highest during the last deglaciation when atmospheric CO_2 was rising [*Anderson et al.*, 2009]. Ice core records of NH_4^+ and SO_4^{+2} further indicate there was no substantial change in marine productivity within the Southern Ocean during glacials [*Wolff et al.*, 2006]. In fact, ice core records show that dustiness and transport of iron to the Southern Ocean began to decrease ~4–8 kyr before the beginning of the rise in atmospheric CO_2 during the last glacial termination [*Wolff et al.*, 2006]. Consequently, the available records do not provide clear evidence that there was major changes in the biological sequestration of metabolic carbon in the Southern Ocean during the last glacial. The extent to which iron affected carbon export in the Southern Ocean and influenced atmospheric CO_2 is therefore highly uncertain [*Archer et al.*, 2000; *Boyd and Trull*, 2007; *Boyd et al.*, 2000].

Carbon cycle models that attempt to simulate a drawdown of atmospheric CO_2 via a Southern Ocean mechanism depend significantly on various parameterizations employed, particularly a biological response to Fe availability. Recent modeling results suggest that Fe fertilization would have a smaller overall influence on CO_2 sequestration than early estimates indicated [*Fischer et al.*, 2010]. Furthermore, in experiments using a dynamical atmosphere-ocean model coupled to an interactive carbon cycle model, a weakening of the westerly winds and reduced equatorward Ekman transport during glacials lowers the concentration of dissolved carbon in surface waters. A concomitant reduction in the upwelling of bioactive nutrients further reduces carbon export production. Combined, these processes would produce only small decreases in atmospheric CO_2 during glacials. Model simulations of CO_2 cycling in the Southern Ocean under glacial conditions must also simulate the net meridional transport or "residual circulation," which is the sum of both the Ekman transport and the opposing eddy transport [*Augustin et al.*, 2004; *Karsten and Marshall*, 2002]. The development of realistic simulations of these dynamical processes and obtaining validations from proxies remain an important challenge to hypotheses that call upon Southern Ocean mechanisms to explain glacial/interglacial CO_2 cycles [*Fischer et al.*, 2010].

3. THE $\Delta^{14}C$ RECORD AND THE TIMING OF ATMOSPHERIC CO_2 CHANGE

The record of surface ocean $\Delta^{14}C$ change during the last glacial termination provides an important constraint to any hypothesis that attempts to explain glacial/interglacial atmospheric CO_2 variability via an ocean-only mechanism. The rise in atmospheric CO_2 during the last deglaciation coincided with a long-term decrease in atmospheric radiocarbon ($\Delta^{14}C$) [*Beck et al.*, 2001; *Chiu et al.*, 2007; *Fairbanks et al.*, 2005; *Hughen et al.*, 2004; *Muscheler et al.*, 2005; *Voelker*, 2000] (Figure 1). The magnitude and duration of the atmospheric $\Delta^{14}C$ change during the last deglaciation implies either a change in production of ^{14}C in the atmosphere or large redistribution of carbon between the surface ocean and a ^{14}C-depleted reservoir. Reconstructions of surface ocean $\Delta^{14}C$ reveal several shorter-term excursions during the past 30 kyr that were not associated with cosmogenic isotope events, and thus, the excursions cannot be explained by changes in the production rate of ^{14}C [*Finkel and Nishiizumi*, 1997; *Muscheler et al.*, 2005].

The largest $\Delta^{14}C$ excursion was a 190‰ decrease between 17.5 and 14.5 kyr B.P. (Figure 1) at the beginning of the last glacial termination. This excursion accompanied a 40 ppm rise in atmospheric CO_2 documented in ice core records from Antarctica [*Monnin et al.*, 2001]. *Denton et al.* [2006] and *Broecker and Barker* [2007] referred to the interval between 17.5 and 14.5 kyr B.P. as a "Mystery Interval" (MI) because of several widespread ocean and atmospheric changes that coincided with the enigmatic $\Delta^{14}C$ excursion. The MI was a time of massive iceberg discharge into the North Atlantic [*Bond et al.*, 1993; *Bond and Lotti*, 1995; *Maslin et al.*, 1995] and weakened North Atlantic Deep Water overturning circulation [*McManus et al.*, 2004]. A reduction in North Atlantic overturning itself would not explain such a large decrease in atmospheric $\Delta^{14}C$, and thus, there must be another cause.

Two recent discoveries shed additional light on the glacial CO_2 mystery and prompt us to consider a new hypothesis to explain events during MI. First, *Marchitto et al.* [2007] presented evidence for a two-part excursion in the $\Delta^{14}C$ at intermediate water depths (705 m) within the northeastern Pacific during last glacial termination. By measuring the ^{14}C ages of benthic foraminiferal calcite taken from a sediment core, these authors documented a ~200‰ drop in $\Delta^{14}C$ from ~17 to 15 kyr B.P. and a second equivalently large negative excursion between ~14 and 11 kyr B.P. (Figure 1). *Marchitto et al.* [2007] argued that these $\Delta^{14}C$ excursions resulted from renewed ventilation of a deepwater mass that had been isolated from the atmosphere during the last glacial. They speculated that retreating sea ice in the Southern Ocean led to enhanced ventilation of CO_2 from the deep sea and to the transport of a low-$\Delta^{14}C$ signal through the ocean's interior via Antarctic Intermediate Water (AAIW)/Subantarctic Mode Water (SAMW). In another study, *Stott et al.* [2009] reported a much larger $\Delta^{14}C$ excursion during the MI from a marine core collected in the eastern tropical Pacific (VM21-30, 607 m) from the Galapagos margin. In the Galapagos core, the benthic to planktonic ^{14}C age differences increased by more than 6000 years during the MI (Figure 2). With no indication of diagenesis and with several replicated ^{14}C ages for different benthic and planktonic species from the same samples, it appears that this $\Delta^{14}C$ excursion in the eastern equatorial Pacific is the same event documented by Marchitto et al. from the Baja margin. However, the planktonic-benthic ^{14}C age differences at the Galapagos site are 5000 to 6000 years. If the source of ^{14}C-depleted carbon was a formerly isolated water mass, it would have contained excess metabolic CO_2 that accumulated over thousands of years and therefore would have distinctly lower $^{13}C/^{12}C$ composition as well. The metabolic CO_2 would have made that water highly corrosive to carbonate and highly depleted in dissolved oxygen. In the VM21-30 core, there is neither evidence of increased carbonate dissolution through the deglacial section [*Stott et al.*, 2009], nor is there any evidence that the benthic fauna were affected by a decrease in oxygen availability that should have resulted from such an episode, and there is no large $\delta^{13}C$ anomaly in the benthic foraminifera.

Additional ^{14}C results have now been obtained from other intermediate depth sites, including one site located on the southern Peru margin, which is bathed by the Subantarctic Intermediate Water. At this site, *De Pol-Holz et al.* [2010] found no anomalously old benthic foraminiferal ^{14}C ages during the last deglaciation. In the northeastern Pacific, the Santa Barbara Basin should record the same deglacial $\Delta^{14}C$ excursion as documented on the Baja margin by Marchitto

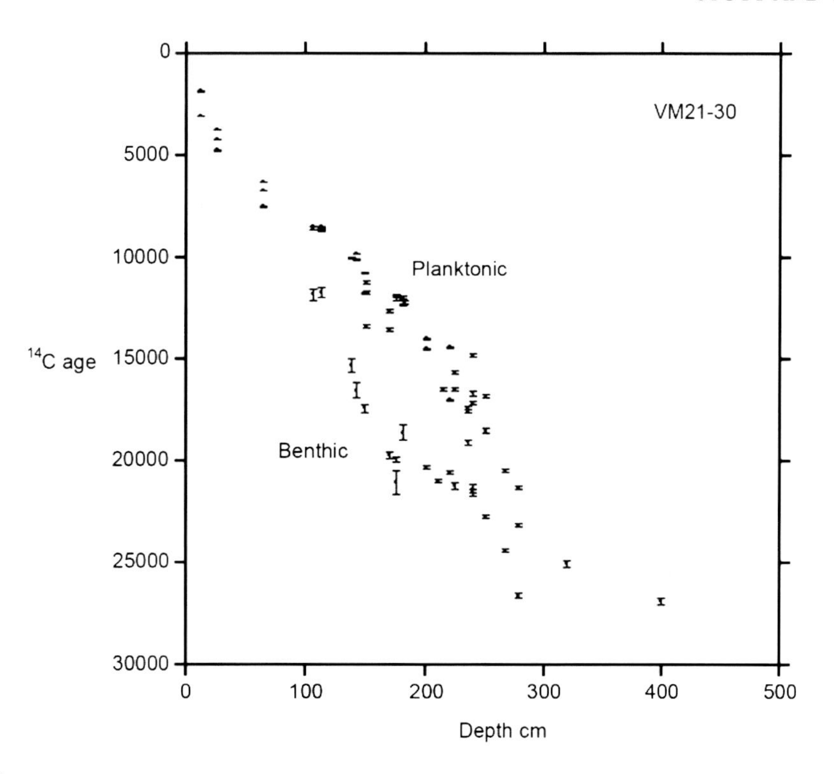

Figure 2. The ^{14}C ages of fossil planktonic and benthic foraminifera from core VM21-30 as depicted by *Stott et al.* [2009].

et al. However, in a study by *Magana et al.* [2010], the benthic foraminiferal ^{14}C ages from a core collected in the Santa Barbara Basin are not anomalously old during the last glacial termination. Yet in a study of a core collected at intermediate depth in the Arabian Sea, *Bryan et al.* [2010] report a deglacial $\Delta^{14}C$ anomaly that appears to match the Baja margin record, arguing that it is evidence of a formally isolated deepwater mass that ventilated through the Southern Ocean during the deglaciation. On the other hand, *Rose et al.* [2010] reconstructed ^{14}C ages of planktonic and benthic foraminifera from two cores collected at intermediate water depths in the southwestern Pacific and Southern Ocean and found no $\Delta^{14}C$ excursion and hence no evidence that SAMW core RR0503-JPC64) or AAIW (core MD97-2120) were anomalously old during the last deglaciation. Rose et al. suggest that perhaps the $\Delta^{14}C$ excursion found on the Baja margin (they do not mention the Galapagos record) is indicative of a North Pacific source. But this would not be consistent with the Santa Barbara Basin or the Arabian Sea record.

4. ARE THERE OTHER SOURCES OF $\Delta^{14}C$-DEPLETED CARBON IN THE OCEAN?

During the past decade, surveys of active submarine volcanic arcs in the Pacific and hydrothermal vents in the northeastern and tropical eastern Pacific have documented CO_2-rich fluids venting [*Chivas et al.*, 1987; *Embley et al.*, 2006; *Lupton et al.*, 2008; *Lupton et al.*, 2006; *Resing et al.*, 2009] at intermediate water depths (~1000 m). Estimates of the CO_2 flux at these sites are sparse, and the areal extent of active vents that are emitting a separate CO_2 gas or liquid phase is not known. Therefore, our knowledge of how these vent systems contribute to an overall global carbon budget is changing as new information becomes available. Nonetheless, the initial results from the surveys that have documented liquid CO_2 or CO_2-rich fluids venting in the Pacific appear to greatly exceed estimates of CO_2 flux based on observations at Mid Ocean Ridge (MOR) systems. Hence, submarine vents in the Pacific may represent a greater source of carbon to the global carbon budget than previously estimated [*Lupton et al.*, 2008; *Resing et al.*, 2009]. For example, at just one vent site on the Mariana Trench, *Lupton et al.* [2006] estimated the carbon flux to be 8×10^8 mol CO_2 yr^{-1}, which is about 0.1% of the global MOR carbon flux. Remarkably, at Eifuku on the Mariana system, approximately 98% of the CO_2 flux is liquid CO_2 droplets. Most importantly, liquid CO_2 is not only venting from these sites, it also accumulates within the sediments on the margins of the active volcanoes and is being stored beneath hydrate caps that regulate the flux of CO_2 from the sediments.

The total volume of CO_2 that is stored in sediments blanketing the Pacific trench and active seamount systems is unknown. At present, liquid CO_2 accumulations have only been observed at sites in the western Pacific. However, CO_2-rich fluids have been identified at other active vents sites throughout the Pacific. The distribution of sites where there is liquid CO_2 includes the Okinawa Trough [*Inagaki et al.*, 2006], the volcanoes along the Mariana and the Tonga-Kermadec arcs [*Lupton et al.*, 2008]. These and other occurrences of CO_2-rich vent fluids [*Cheminée et al.*, 1991; *Hilton et al.*, 1998; *McMurtry et al.*, 1993; *Wheat et al.*, 2000] hint at a potentially far greater distribution of sites that emit liquid CO_2 and CO_2-rich fluids. We are drawn to these recent observations as a way to explain the enigmatic nature of the glacial/interglacial CO_2 changes summarized above, and particularly the large $\Delta^{14}C$ anomaly during the deglaciation, while recognizing that hypothesis presented here will require a more quantitative estimate of carbon fluxes and CO_2 storage at active vent sites.

The CO_2-rich fluids produced at volcanic arcs in the Pacific have $\delta^{13}C$ values similar to that of marine carbonate [*Lupton et al.*, 2006]. In fact, the elevated CO_2:^3He and carbon isotope chemistry indicate a slab origin for the CO_2. *Lupton et al.* [2006] argue that the CO_2 is derived from decarbonation of marine carbonate that is carried down the trench. As a result, the CO_2 that vents into the water column at this vent site would have virtually no influence on the $^{13}C/^{12}C$ of dissolved carbon in the ocean. The alkalinity of vent fluids also varies significantly among various vent sites, and this appears to reflect the stage of magmatic development [*McMurtry et al.*, 1993; *Resing et al.*, 2009]. This is a critical observation because at the Kasuga vent, for example, *McMurtry et al.* [1993] report titration alkalinities that are 13 times higher than ambient seawater. The dissolved CO_2 content is estimated to be 230 mmol kg^{-1} indicating fluids that are highly supersaturated with CO_2, principally as bicarbonate.

5. AN OCEAN CO_2 CAPACITOR?

The hypothesis we set forth here calls upon expansion of the CO_2 hydrate stability horizon in the ocean during glacials as temperatures cooled. Expansion and shoaling of the hydrate stability horizon during glaciations would increase the volume of sediment within which CO_2 accumulated and lowered the net flux of CO_2 to the water column. This hypothesis makes two explicit but testable assumptions: (1) sediments that blanket volcanic centers are capable of storing large volumes of CO_2 as hydrate and CO_2-rich fluids and (2) present-day estimates of the steady state flux of CO_2 from volcanic centers is significantly underestimated. If either of these assumptions proves incorrect, our hypothesis is nullified.

Today, submarine volcanoes that emit a separate liquid and gas phase of CO_2 occur at shallow to intermediate water depths. *Lupton et al.* [2008] point out that it is at these depths and at the elevated temperatures that circulating fluids boil and degassing occurs. The CO_2 and CO_2-rich fluids released in this process vent directly into the water column or pass through sediments blanketing the margin of volcanic centers. Our hypothesis calls for increased storage and reduced net flux of CO_2 as liquid CO_2 and CO_2-rich fluids from sediments at volcanic centers located at intermediate depth as the hydrate stability zone expanded vertically and horizontally during glacial cooling (Plate 3). At the water depths where hydrate is stable today (Plate 2), the flux of CO_2 from active volcanic centers is governed by the CO_2 production rate. But the flux of CO_2 from the sediments through which much of the venting fluids pass also depends on the rate of diffusion of CO_2 across the hydrate-water interface and the rate of formation and subsequent dissolution of CO_2 at the bottom and top of the hydrate layer (respectively) [*Rehder et al.*, 2004]. Whereas the rate of diffusion between seawater and pure CO_2 hydrate has been determined empirically [*Rehder et al.*, 2004], the in situ diffusional flux across a sediment hydrate has not. Boundary layer constraints will influence the diffusion flux [*Rehder et al.*, 2004]. Furthermore, because the formation of hydrate leads to salt rejection, high salinity, CO_2-saturated pore waters would further reduce the rate of exchange of CO_2 with the overlying seawater.

The critical depth for hydrate stability in the ocean today occurs at 8.5°C and ~400 m (Plate 3, left). Plate 3 illustrates the temperature (left) at ~400 m water depth [*Levitus*, 1994] where in the modern ocean, the critical point of hydrate-liquid-gaseous CO_2 occurs and how hydrate stability would have changed in the glacial ocean (right) in response to cooling. Orange shading highlights the area for which CO_2 hydrate becomes unstable at 400 m. Glacial subsurface temperatures were estimated by adding the averaged glacial-interglacial subsurface temperature anomalies simulated by the CCSM3-coupled general circulation model and the LOVECLIM Earth system model [*Otto-Bliesner et al.*, 2006; *Timmermann et al.*, 2009] to the observed Levitus temperatures at 400 m.

During the Last Glacial Maximum, most of the active volcanic centers within the Pacific would have fallen within the hydrate stability zone ($\leq$8.5°C, 400 m) as the 8.5°C isotherm shoaled by more than 100 m. As a result, the overall flux of CO_2 from active volcanic centers would have been reduced due to a transition from buoyant liquid/gas CO_2 to hydrate CO_2. The Mariana and the Kermadec volcanic arcs are located in the region where the hydrate stability zone expanded during the last glacial. There are also extensive regions throughout the Pacific, particularly in the eastern

Pacific where the hydrate stability horizon would have shoaled by several hundred meters (Plate 3).

The deglacial warming at intermediate water depths within the Pacific would have increased the steady state flux of CO_2 from those parts of the ocean where liquid and hydrate CO_2 accumulated during the glacial. Sea level rise during deglaciation would influence the stability horizon in the opposite direction. However, as much of the MI CO_2 rise occurred between 18 and 15 kyr B.P., prior to the rapid sea level rise, we assume that pressure changes due to sea level rise were not a major influence on the hydrate stability. The release of this volcanic CO_2 would have injected carbon that was ^{14}C dead into intermediate waters but would not have produced a negative $\delta^{13}C$ excursion in DIC if the bulk of the CO_2 was from sites such as those along the Mariana [*Lupton et al.*, 2006]. Such a release would therefore explain why there is no evidence for a $\delta^{13}C$ excursion in benthic foraminifera in association with the deglacial $\Delta^{14}C$ excursion [*Gonfiantini et al.*, 1997]. It would, however, explain the small negative $\delta^{13}C$ excursion in atmospheric CO_2 [*Smith et al.*, 1999; *Spero and Lea*, 2002; *Stott et al.*, 2009]. The $\delta^{13}C$ of atmospheric CO_2 decreased from ~−6.5‰ to −6.8‰ between 18 and 12 kyr B.P. in association with the deglacial $\Delta^{14}C$ excursion. Taking the fractionation between atmospheric CO_2-dissolved CO_2 in surface waters to be −8‰, the deglacial $\delta^{13}C$ excursion documented in the ice core and planktonic foraminifera in the tropical ocean implies the CO_2 added to the ocean/atmosphere system had a $\delta^{13}C$ composition close to −1‰, the same as the CO_2 venting today from submarine volcanoes in the Pacific [*Lupton et al.*, 2006].

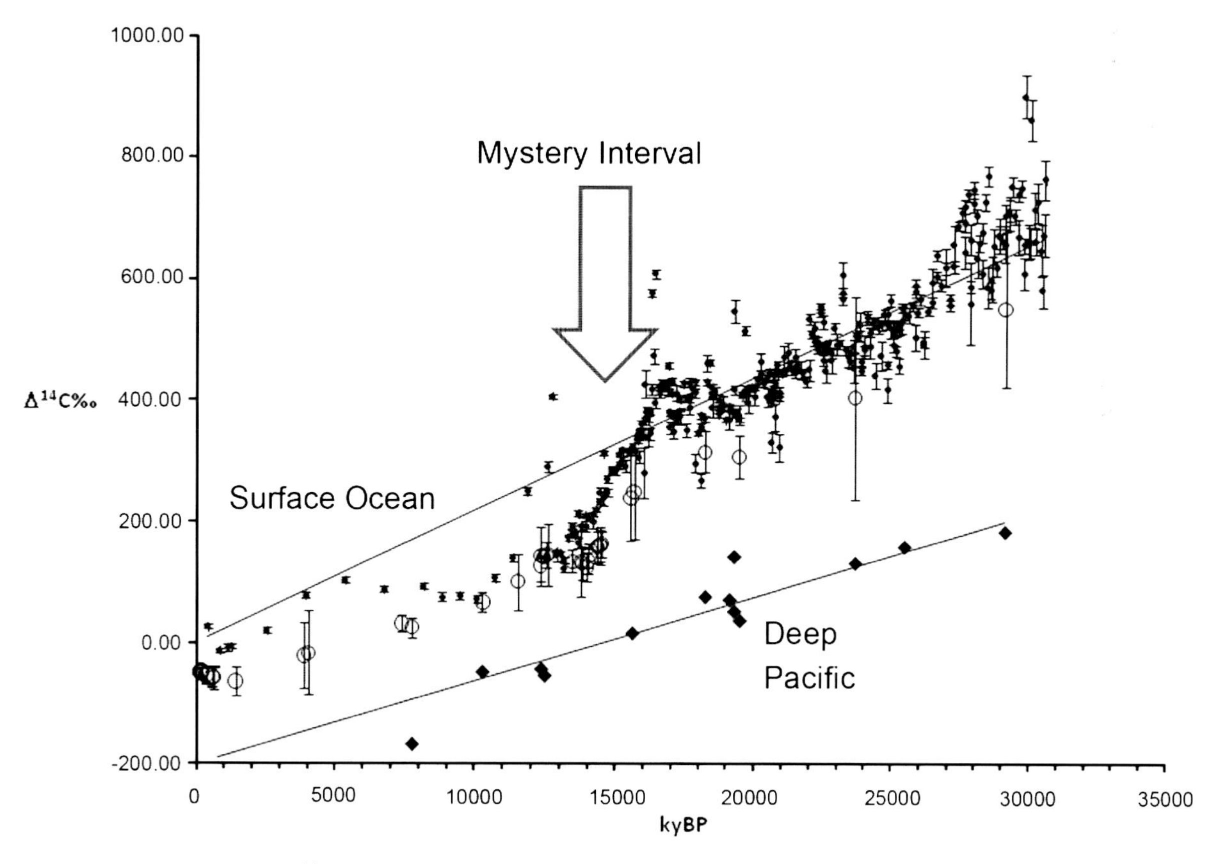

Figure 3. Surface ocean $\Delta^{14}C$ (solid circles) from Cariaco Basin [*Hughen et al.*, 2006] compared to planktonic (open circles) and benthic $\Delta^{14}C$ estimates (diamonds) for the western tropical Pacific core MD98-2181 [*Broecker et al.*, 2004; *Stott et al.*, 2007]. Planktonic $\Delta^{14}C$ values from MD9821081 track the surface ocean record, including the marked decline during the Mystery Interval. However, benthic $\Delta^{14}C$ values (open circles) at 2.1 km in the western tropical Pacific do not exhibit the large −190‰ excursion between 17.5 and 14 kyr B.P. Linear regressions through the surface and deepwater values are forced through approximate modern values for surface waters in the tropical Atlantic and Pacific Deep waters at 2.1 km in the western Pacific [*Key et al.*, 2004]. Calendar ages for MD98-2181 were obtained with the CALIB-4 program (M. Stuiver et al., CALIB Radiocarbon Calibration, 2005, http://calib.qub.ac.uk/calib). Error bars for MD98-2181 planktonic $\Delta^{14}C$ are the standard deviation of $\Delta^{14}C$ estimates based on the 1σ calendar age ranges.

Our hypothesis requires a significant increase in CO_2 storage (and reduced CO_2 fluxes) during glacials and a production rate of CO_2 sufficient to replenish the CO_2 released during deglaciations. This requirement has an important implication for current estimates of volcanic-derived CO_2 in the global carbon cycle budget. If correct, this hypothesis requires an upward adjustment in the relative contribution of CO_2 from submarine volcanism [*Lupton et al.*, 2006]. We use a one-dimensional, three-box carbon cycle box model to estimate the amount of CO_2 required to explain a 50 ppm rise in atmospheric CO_2 over 5 kyr during the MI. The box model simulates exchange of carbon between the atmosphere and the shallow, intermediate, and deep ocean reservoirs [*Tyrell et al.*, 2007]. This model simulates a release of carbon directly to the atmosphere and subsequent equilibration with the entire ocean. The model simulates a scenario similar to that which we propose in which carbon is released from shallow to intermediate depths (<1000 m) and ventilates more or less directly to the atmosphere. The model does not simulate carbon exchange with terrestrial reservoirs and parameterizes carbon, phosphorous, and alkalinity fluxes. For our simulation, we hold phosphorous and alkalinity fluxes constant and impose a transient flux of carbon over 5000 years.

Between 19 and 14 kyr B.P., the concentration of atmospheric CO_2 increased by approximately 50 ppm. In the box model, a transient 50 ppm increase in atmospheric CO_2 over 5000 years requires a release of 600 Gt of carbon to the atmosphere. As the carbon enters the atmosphere, CO_2 re-equilibrates with the ocean, a process that takes several thousand years. The transient release of 600 Gt of C to the atmosphere increases atmospheric pCO_2 by 45–50 ppm for roughly 7–10 kyr B.P., the length of an entire interglacial, and then gradually decays back to lower concentrations as the CO_2 equilibrates with the deep ocean reservoir. The 600 Gt of carbon would be equivalent to ~2200 Gt of CO_2.

The box model simulates a rise in atmospheric CO_2 that is consistent with ice core records. However, 600 Gt of ^{14}C-free carbon released to the atmosphere would not shift $\Delta^{14}C$ of carbon throughout the deep ocean by −190‰ [*Broecker*, 2009a]. In order to shift the deep ocean reservoir of carbon by −190‰ along with the atmosphere and surface ocean would require nearly 5000 Gt of ^{14}C-free carbon. On the other hand, there are not many well-calibrated, high-resolution $\Delta^{14}C$ records for the deep ocean that span the MI. Hence, we do not know if the entire deep ocean carbon reservoir also decreased by −190‰ between 17.5 and 14 kyr. Taking planktonic and benthic foraminiferal ^{14}C ages from sediment core MD9821-81, (6°N, 126°E, 2.1 km water) as estimators of surface and deepwater $\Delta^{14}C$ values through the last glacial termination suggests the −190‰ decline between 18 and 14 kyr B.P. was not recorded in Pacific Deep Water (Figure 3). The benthic $\Delta^{14}C$ record from MD9821-81 implies ^{14}C-depleted carbon entering the surface ocean during the MI was highly diluted by the time it mixed with Pacific Deep Water. A mass balance that does not take into account the differences in $\Delta^{14}C$ between surface and deep water during the MI would greatly overestimate the amount of carbon that entered the ocean. Thus far, a $\Delta^{14}C$ excursion during the MI has only been documented in surface ocean records and from benthic foraminifera that inhabited water depths between ~500 and 700 m. We therefore conclude that the amount of ^{14}C-dead carbon released during the MI was substantially less than 5000 Gt if it was released from shallow water depths and entered the atmosphere before equilibrating with the entire deep ocean.

A transient release of CO_2 to the atmosphere during the MI as we hypothesize would have caused a simultaneous adjustment in the carbonate ion concentration of surface waters. We have constructed a preliminary Li/Ca record through the MI from two planktonic foraminifer species taken from the Galapagos core VM21-30, *Globigerinoides*

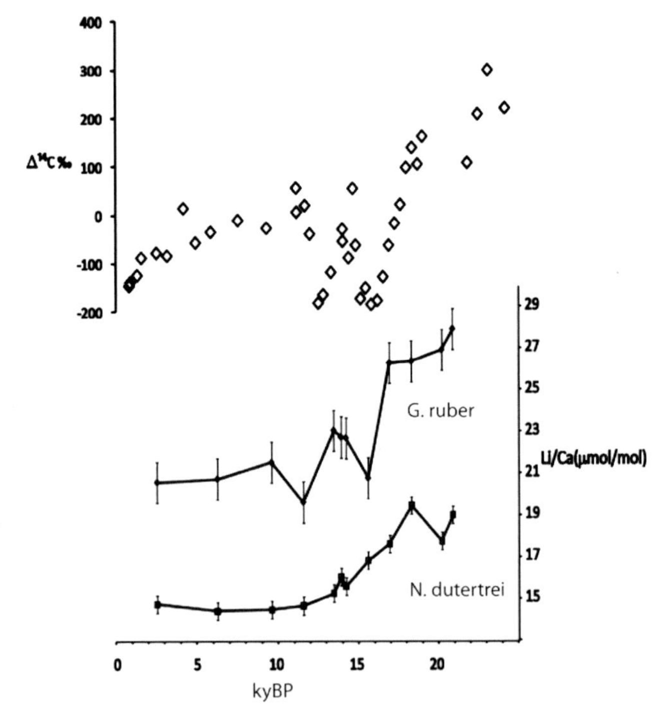

Figure 4. (top) $\Delta^{14}C$ for benthic foraminifera from Baja margin core MC19/GC31/PC08 [*Marchitto et al.*, 2007] and (bottom) Li/Ca for two species of planktonic foraminifera, *Globigerinoides ruber*, and *Neoglobquadrina dutertrei* from Galapagos margin core VM21-30 (K. Harazin and L. Stott, manuscript in preparation, 2011). Error bars on Li/Ca are standard deviations of replicate measurements of a sample.

ruber and *Neogloboquadrina dutertrei*, (a mixed layer and thermocline dweller, respectively). There is a marked decrease in Li/Ca of both species during the MI that is consistent with lower carbonate ion concentrations (Figure 4). A precise transfer function for ΔLi/Ca to $[CO_3^{=}]$ is complicated by temperature changes that would have occurred during deglaciation [*Hall and Chan*, 2004; *Lear and Rosenthal*, 2006]. Nonetheless, the two lowest Li/Ca values for *G. ruber* correspond in timing with excursions in $\Delta^{14}C$ documented by *Marchittoet al.* [2007] (Figure 4). The magnitude of Li/Ca change is also larger than could be accounted for by temperature change alone [*Marriott et al.*, 2004], particularly at the VM21-30 site where the glacial to interglacial SST change was only on the order of 1.5°C [*Koutavas and Sachs*, 2008]. In fact, virtually all of the glacial to interglacial SST warming at this location occurred after the decrease in Li/Ca (Figure 5). We therefore interpret the decrease in Li/Ca at this location to a decrease in $[CO_3^{=}]$ during the MI in response to a release of CO_2 to the atmosphere and subsequent equilibration with the ocean.

6. STORAGE AND PRODUCTION OF CO_2 DURING GLACIATIONS

The density of liquid CO_2 at intermediate water depths (~1200–500 m) is close to that of seawater. Hence, approximately 2200 km^3 of storage space is required to accommodate the accumulation of excess volcanogenic CO_2 during glaciations. The only reservoir where CO_2 and CO_2-rich fluids could be stored is in sediments that accumulate on flanks of volcanic centers. Typically, these sediments have porosities between 50% and 80% [*Dadey and Klaus*, 1992], which implies a needed storage capacity of ~3000–5000 km^3. The thickest deposits of volcaniclastic sediments today are in waters deeper than 1000 m where hydrate would have remained stable on glacial/interglacial time scales. On the other hand, liquid CO_2 and CO_2-rich fluids are mobile and can migrate upward from deeper horizons through hydrothermal conduits [*Fisher and Wheat*, 2010]. The removal of a hydrate cap at shallower depths during deglaciation would open up additional conduits for upward migration CO_2-rich fluids. The volume of sediment required to accommodate 2200 Gt of CO_2 storage during glaciations may therefore include sediments at deeper water depths. Further modeling of thermodynamic constraints on subsurface flow and the storage capacity are also required to better constrain this aspect of the hypothesis.

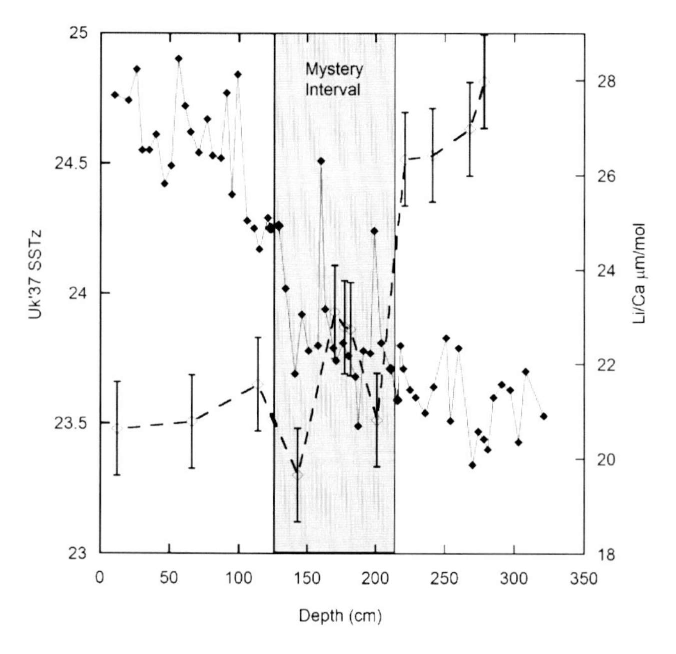

Figure 5. Li/Ca (µmol mol^{-1}) of *G. ruber* from VM21-30 (open diamonds) and the UK'37-based SST reconstruction for VM21-30 by *Koutavas and Sachs* [2008] (solid diamonds).

In addition to the large storage volume required by this hypothesis, the production rate of CO_2 at subducting margins must also accommodate the amount of CO_2 stored and released on the 100 kyr glacial/interglacial timescale. This calculation implies a production of ~2400 Gt of excess CO_2 during a 100 kyr glacial cycle. Taking average spreading rates and average carbonate content on subducting slabs provides an approximate estimate of the CO_2 production potential. Using this approach does not negate this hypothesis. However, the production rate of CO_2 emitted at subducting margins is not necessarily in steady state on geologic time scales. Indeed, there are several lines of evidence indicating that CO_2 production at submarine arcs is not a steady state function of the long-term spreading rates [*Leeman and Davidson*, 2005; *Nkrintra Singhrattna et al.*, 2005; *Resing et al.*, 2009; *Smith and Price*, 2006]. *Resing et al.* [2009] point out that there is a wide range of CO_2 fluxes between active arcs in the western Pacific, and the large diversity of CO_2 fluxes reflects a diverse history of magmatic development and subsequent arc aging. These authors proposed a simple model for magmatic development and evolution that would account for a wide range of CO_2 fluxes over Earth's recent geologic past that involves progressive magmatic development, followed by aging and geochemical evolution. Over time, the distance between the magmatic source and the surface of the arc volcano increases. During this aging process, acid is ultimately consumed, and the circulating fluids become increasingly enriched in CO_2. As the magmatism evolves further, the distance between the magmatic source and the volcano's surface becomes larger, and all of the magmatic SO_2 is consumed, causing pH to rise. The subsequent weathering reactions produce fluids that are further elevated in both alkalinity and CO_2. We surmise it is this CO_2 production cycle that is the key to understanding variable fluxes of CO_2-rich fluids during the late Pleistocene along the arc margins.

7. A MECHANISM FOR CO_2 RELEASE DURING DEGLACIATION

The onset of the warming in the Southern Ocean during glacial terminations preceded the initial rise in atmospheric CO_2 [*Blunier et al.*, 1998; *Stott et al.*, 2007] (although the precise lead-lag timing is not yet established) and coincided with a substantial increase in austral spring insolation [*Stott et al.*, 2007; *Timmermann et al.*, 2009] (Plate 4) or the related length of the austral summer season [*Huybers*, 2009]. The spring season forcing arises from changes of Earth's orbital precession. Between 19 and 10 austral spring shortwave radiation rose by ~40 m^{-2}, which led to a decrease of sea ice extent and an increase of temperatures in the Southern Ocean and Antarctica [*Timmermann et al.*, 2009] (Plate 4). Deep Pacific temperatures warmed by ~2°C between 19 and 17 kyr B.P. [*Stott et al.*, 2007]. Warmer water was injected into the formation zones of SAMW and AIWs, leading to an overall warming of the Pacific and Atlantic oceans at depths between several hundred meters to more than 1500 m [*Herguera et al.*, 1991; *Lynch-Stieglitz and Fairbanks*, 1994; *Mashiotta et al.*, 1999; *Matsumoto and Lynch-Stieglitz*, 1999; *Stott et al.*, 2007]. Here we argue that this warming triggered release of radiocarbon-dead CO_2, further accelerating deglacial warming. As deglaciation progressed, sea surface temperatures in the northwestern Pacific varied in concert with climatic changes in the North Atlantic region, and this influenced thermal properties of mode water and deep water in the North Pacific [*Okazaki et al.*, 2010]. The phasing of south to north warming during the deglaciation may have influenced subsequent warming and a further release of CO_2 that would account for the millennial-scale oscillations observed in oceanic and atmospheric radiocarbon during termination I [*Broecker*, 2009b; *Hughen et al.*, 2006; *Marchitto et al.*, 2007].

8. SUMMARY

Our hypothesis does not necessarily account for the entire glacial/interglacial CO_2 change. However, it would reconcile the lack of evidence for an isolated deepwater mass during glacials that is otherwise required to explain the large $\Delta^{14}C$ excursion during the last deglaciation. It would also explain why atmospheric CO_2 began to rise very soon after the Southern Ocean began to warm. The time required to propagate a surface warming in the Southern Ocean through the Pacific to sites where hydrate occurs determines the relative phasing of high-latitude temperature change and the rise of atmospheric CO_2. The deepwater temperature changes in the Pacific during glacial/interglacial cycles were large enough to have shifted the hydrate stability zone up and down by as much as several hundred meters and thereby affected the areal extent of CO_2 hydrate stability.

A comprehensive test of our CO_2 hypothesis requires a more thorough assessment of the present-day CO_2 flux at sites of active magmatism and the extent of liquid and hydrate CO_2 accumulations. There is also need to trace the flow of ^{14}C-depleted waters during the last deglaciation to determine where it was ventilated to the atmosphere. In the meantime, the hypothesis presented here offers an important opportunity to reexamine that causes of glacial/interglacial CO_2 variations and the sequence of events that punctuated the last deglaciation.

Acknowledgments. We wish to express our gratitude to John Southon for his generous support and insightful comments and suggestions. Miguel Rincon was instrumental in developing the data presented for cores VM21-30 and MD98-2181. We are grateful to numerous people who have provided critical and constructive advice as this paper developed. Kathleen Harazin provided unpublished Li/Ca data. Support for this research was provided by the National Science Foundation to both Stott and Timmermann.

REFERENCES

Abelmann, A., R. Gersonde, G. Cortese, G. Kuhn, and V. Smetacek (2006), Extensive phytoplankton blooms in the Atlantic sector of the glacial Southern Ocean, *Paleoceanography*, *21*, PA1013, doi:10.1029/2005PA001199.

Adkins, J. F., K. McIntyre, and D. P. Schrag (2002), The salinity, temperature, and $\delta^{18}O$ of the glacial deep ocean, *Science*, *298*(5599), 1769–1773.

Ainsworth, E. A., A. Rogers, R. Nelson, and S. P. Long (2004), Testing the source-sink hypothesis of down-regulation of photosynthesis in elevated [CO_2] in the field with single gene substitutions in *Glycine max*, *Agric. For. Meteorol.*, *122*(1–2), 85–94.

Anderson, R. F., S. Ali, L. I. Bradtmiller, S. H. H. Nielsen, M. Q. Fleisher, B. E. Anderson, and L. H. Burckle (2009), Wind-driven upwelling in the Southern Ocean and the deglacial rise in atmospheric CO_2, *Science*, *323*(5920), 1443–1448.

Archer, D., and E. Maier-Reimer (1994), Effect of deep-sea sedimentary calcite preservation on atmospheric CO_2 concentration, *Nature*, *367*, 260–263.

Archer, D., A. Winguth, D. Lea, and N. Mahowald (2000), What caused the glacial/interglacial atmospheric pCO_2 cycles?, *Rev. Geophys.*, *38*, 159–189.

Augustin, L., et al. (2004), Eight glacial cycles from an Antarctic ice core, *Nature*, *429*(6992), 623–628.

Barker, S., G. Knorr, M. J. Vautravers, P. Diz, and L. C. Skinner (2010), Extreme deepening of the Atlantic overturning circulation during deglaciation, *Nat. Geosci.*, *3*(8), 567–571.

Beck, J. W., et al. (2001), Extremely large variations of atmospheric ^{14}C concentration during the last glacial period, *Science*, *292*(5526), 2453–2458.

Blunier, T., et al. (1998), Asynchrony of Antarctic and Greenland climate change during the last glacial period, *Nature*, *394*(6695), 739–743.

Bond, G. C., and R. Lotti (1995), Iceberg discharges into the North Atlantic on millennial time scales during the last glaciation, *Science*, *267*(5200), 1005–1010.

Bond, G. C., W. Broecker, S. Johnsen, J. McManus, L. Labeyrie, J. Jouzel, and G. Bonani (1993), Correlations between climate records from North-Atlantic sediments and Greenland ice, *Nature*, *365*(6442), 143–147.

Boyd, P. W., and T. W. Trull (2007), Understanding the export of biogenic particles in oceanic waters: Is there consensus?, *Prog. Oceanogr.*, *72*(4), 276–312.

Boyd, P. W., et al. (2000), A mesoscale phytoplankton bloom in the polar Southern Ocean stimulated by iron fertilization, *Nature*, *407*(6805), 695–702.

Brigault, S., E. Sacchi, R. Gonfiantini, and G. M. Zuppi (1998), Functioning of a French hydroelectric reservoir. Isotopic approach, *C. R. Acad. Sci., Ser. IIa Sci. Terre Planetes*, *327*(6), 397–403.

Broecker, W. S. (1982), Ocean chemistry during glacial time, *Geochim. Cosmochim. Acta*, *46*, 1698–1705.

Broecker, W. S. (2009a), The mysterious ^{14}C decline, *Radiocarbon*, *51*(1), 109–119.

Broecker, W. S. (2009b), Wally's quest to understand the ocean's $CaCO_3$ cycle, *Annu. Rev. Mar. Sci.*, *1*(1), 1–18.

Broecker, W. S., and S. Barker (2007), A 190‰ drop in atmosphere's Δ14C during the "Mystery Interval" (17.5 to 14.5 kyr), *Earth Planet. Sci. Lett.*, *256*(1–2), 90–99.

Broecker, W. S., and E. Clark (2001), Glacial-to-Holocene redistribution of carbonate ion in the deep sea, *Science*, *294*(5549), 2152–2155.

Broecker, W. S., and E. Clark (2003), Glacial-age deep sea carbonate ion concentrations, *Geochem. Geophys. Geosyst.*, *4*(6), 1047, doi:10.1029/2003GC000506.

Broecker, W. S., and E. Clark (2010), Search for a glacial-age ^{14}C-depleted ocean reservoir, *Geophys. Res. Lett.*, *37*, L13606, doi:10.1029/2010GL043969.

Broecker, W. S., S. Barker, E. Clark, I. Hajdas, G. Bonani, and L. Stott (2004), Ventilation of the glacial deep Pacific Ocean, *Science*, *306*, 1169–1172.

Bryan, S. P., T. M. Marchitto, and S. J. Lehman (2010), The release of ^{14}C-depleted carbon from the deep ocean during the last deglaciation: Evidence from the Arabian Sea, *Earth Planet. Sci. Lett.*, *298*(1–2), 244–254.

Cheminée, J. L., P. Stoffers, G. McMurtry, H. Richnow, D. Puteanus, and P. Sedwick (1991), Gas-rich submarine exhalations during the 1989 eruption of Macdonald Seamount, *Earth Planet. Sci. Lett.*, *107*(2), 318–327.

Chiu, T.-C., R. G. Fairbanks, L. Cao, and R. A. Mortlock (2007), Analysis of the atmospheric ^{14}C record spanning the past 50,000 years derived from high-precision $^{230}Th/^{234}U/^{238}U$, $^{231}Pa/^{235}U$ and ^{14}C dates on fossil corals, *Quat. Sci. Rev.*, *26*(1–2), 18–36.

Chivas, A. R., I. Barnes, W. C. Evans, J. E. Lupton, and J. O. Stone (1987), Liquid carbon dioxide of magmatic origin and its role in volcanic eruptions, *Nature*, *326*(6113), 587–589.

Dadey, K. A., and A. Klaus (1992), Physical properties of volcaniclastic sediments in the Izu-Bonin area, *Proc. Ocean Drill. Program Sci. Results*, *126*, 543–550.

Denton, G., W. Broecker, and R. Alley (2006), The Mystery Interval 17.5 to 14.5 kyrs, *Pages News*, *13*(2), 14–16.

De Pol-Holz, R., L. Keigwin, J. Southon, D. Hebbeln, and M. Mohtadi (2010), No signature of abyssal carbon in intermediate waters off Chile during deglaciation, *Nat. Geosci.*, *3*(3), 192–195.

de Ronde, C. E. J., et al. (2007), Submarine hydrothermal activity along the mid-Kermadec Arc, New Zealand: Large-scale effects on venting, *Geochem. Geophys. Geosyst.*, *8*, Q07007, doi:10.1029/2006GC001495.

Downes, D., et al. (1999), Proposed identification of Hubble Deep Field submillimeter source HDF 850.1, *Astron. Astrophys.*, *347*(3), 809–820.

Embley, R. W., et al. (2006), Long-term eruptive activity at a submarine arc volcano, *Nature*, *441*(7092), 494–497.

Fairbanks, R. G., R. A. Mortlock, T.-C. Chiu, L. Cao, A. Kaplan, T. P. Guilderson, T. W. Fairbanks, A. L. Bloom, P. M. Grootes, and M.-J. Nadeau (2005), Radiocarbon calibration curve spanning 0 to 50,000 years BP based on paired $^{230}Th/^{234}U/^{238}U$ and ^{14}C dates on pristine corals, *Quat. Sci. Rev.*, *24*(16–17), 1781–1796.

Farrell, J. W., and W. L. Prell (1989), Climatic change and $CaCO_3$ preservation: An 800,000 year bathymetric Reconstruction from the central equatorial Pacific Ocean, *Paleoceanography*, *4*(4), 447–466.

Finkel, R. C., and K. Nishiizumi (1997), Beryllium 10 concentrations in the Greenland Ice Sheet Project 2 ice core from 3–40 ka, *J. Geophys. Res.*, *102*(C12), 26,699–26,706.

Fischer, H., et al. (2010), The role of Southern Ocean processes in orbital and millennial CO_2 variations – A synthesis, *Quat. Sci. Rev.*, *29*(1–2), 193–205.

Fisher, A. T., and C. G. Wheat (2010), Seamounts as conduits for massive fluid, heat, and solute fluxes on ridge flanks, *Oceanography*, *23*(1), 74–87.

Fisher, A. T., et al. (2003), Hydrothermal recharge and discharge across 50 km guided by seamounts on a young ridge flank, *Nature*, *421*(6923), 618–621.

Gaspari, V., C. Barbante, G. Cozzi, P. Cescon, C. F. Boutron, P. Gabrielli, G. Capodaglio, C. Ferrari, J. R. Petit, and B. Delmonte (2006), Atmospheric iron fluxes over the last deglaciation: Climatic implications, *Geophys. Res. Lett.*, *33*, L03704, doi:10.1029/2005GL024352.

Gonfiantini, R., S. Valkiers, P. D. P. Taylor, and P. DeBievre (1997), Adsorption in gas mass spectrometry. 1. Effects on the measurement of individual isotopic species, *Int. J. Mass Spectrom. Ion Processes*, *163*(3), 207–219.

Gruber, N., et al. (2009), Oceanic sources, sinks, and transport of atmospheric CO_2, *Global Biogeochem. Cycles*, *23*, GB1005, doi:10.1029/2008GB003349.

Hall, J. M., and L. H. Chan (2004), Li/Ca in multiple species of benthic and planktonic foraminifera: Thermocline, latitudinal, and glacial-interglacial variation, *Geochim. Cosmochim. Acta*, *68*(3), 529–545.

Herguera, J. C., L. D. Stott, and W. H. Berger (1991), Glacial deep-water properties in the west-equatorial Pacific: Bathyal thermocline near a depth of 2000 m, *Mar. Geol.*, *100*(1–4), 201–206.

Hilton, D. R., G. M. McMurtry, and F. Goff (1998), Large variations in vent fluid $CO_2/^3He$ ratios signal rapid changes in magma chemistry at Loihi seamount, Hawaii, *Nature*, *396*(6709), 359–362.

Hughen, K., S. Lehman, J. Southon, J. Overpeck, O. Marchal, C. Herring, and J. Turnbull (2004), ^{14}C activity and global carbon cycle changes over the past 50,000 years, *Science*, *303*(5655), 202–207.

Hughen, K., J. Southon, S. Lehman, C. Bertrand, and J. Turnbull (2006), Marine-derived ^{14}C calibration and activity record for the past 50,000 years updated from the Cariaco Basin, *Quat. Sci. Rev.*, *25*(23–24), 3216–3227.

Hutnak, M., A. T. Fisher, R. Harris, C. Stein, K. Wang, G. Spinelli, M. Schindler, H. Villinger, and E. Silver (2008), Large heat and fluid fluxes driven through mid-plate outcrops on ocean crust, *Nat. Geosci.*, *1*(9), 611–614.

Huybers, P. (2009), Antarctica's orbital beat, *Science*, *325*(5944), 1085–1086.

Inagaki, F., et al. (2006), Microbial community in a sediment-hosted CO_2 lake of the southern Okinawa Trough hydrothermal system, *Proc. Natl. Acad. Sci. U. S. A.*, *103*(38), 14,164–14,169.

Ito, T., and M. J. Follows (2005), Preformed phosphate, soft tissue pump and atmospheric CO_2, *J. Mar. Res.*, *63*, 813–839.

Jouzel, J., et al. (2007), Orbital and millennial Antarctic climate variability over the past 800,000 years, *Science*, *317*(5839), 793–796.

Karsten, R. H., and J. Marshall (2002), Constructing the residual circulation of the ACC from observations, *J. Phys. Oceanogr.*, *32*(12), 3315–3327.

Keeling, R. F., and B. B. Stephens (2001a), Antarctic sea ice and the control of Pleistocene climate instability, *Paleoceanography*, *16*(1), 112–131.

Keeling, R. F., and B. B. Stephens (2001b), Correction to "Antarctic sea ice and the control of Pleistocene climate instability" by Ralph F. Keeling and Britton B. Stephens, *Paleoceanography*, *16*(3), 330–334.

Key, R. M., A. Kozyr, C. L. Sabine, K. Lee, R. Wanninkhof, J. L. Bullister, R. A. Feely, F. J. Millero, C. Mordy, and T.-H. Peng (2004), A global ocean carbon climatology: Results from Global Data Analysis Project (GLODAP), *Global Biogeochem. Cycles*, *18*, GB4031, doi:10.1029/2004GB002247.

Koutavas, A., and J. P. Sachs (2008), Northern timing of deglaciation in the eastern equatorial Pacific from alkenone paleothermometry, *Paleoceanography*, *23*, PA4205, doi:10.1029/2008PA001593.

Lear, C. H., and Y. Rosenthal (2006), Benthic foraminiferal Li/Ca: Insights into Cenozoic seawater carbonate saturation state, *Geology*, *34*(11), 985–988.

Leeman, W. P., and J. P. Davidson (2005), Energy and mass fluxes in volcanic arcs: A preface to the special SOTA issue, *J. Volcanol. Geotherm. Res.*, *140*(1–3), vii–x.

Levitus, S., and T. P. Boyer (1994), *World Ocean Atlas 1994*, vol. 4, *Temperature, NOAA Atlas NESDIS*, vol. 4, 129 pp., NOAA, Silver Spring, Md.

Lupton, J., et al. (2006), Submarine venting of liquid carbon dioxide on a Mariana Arc volcano, *Geochem. Geophys. Geosyst.*, *7*, Q08007, doi:10.1029/2005GC001152.

Lupton, J., M. Lilley, D. Butterfield, L. Evans, R. Embley, G. Massoth, B. Christenson, K. Nakamura, and M. Schmidt (2008), Venting of a separate CO_2-rich gas phase from submarine arc volcanoes: Examples from the Mariana and Tonga-Kermadec arcs, *J. Geophys. Res.*, *113*, B08S12, doi:10.1029/2007JB005467.

Luthi, D., et al. (2008), High-resolution carbon dioxide concentration record 650,000-800,000 years before present, *Nature*, *453*(7193), 379–382.

Lynch-Stieglitz, J., and R. G. Fairbanks (1994), A conservative tracer for glacial ocean circulation from carbon isotope and palaeo-nutrient measurements in benthic foraminifera, *Nature*, *369*(6478), 308–310.

Magana, A. L., J. R. Southon, J. P. Kennett, E. B. Roark, M. Sarnthein, and L. D. Stott (2010), Resolving the cause of large differences between deglacial benthic foraminifera radiocarbon measurements in Santa Barbara Basin, *Paleoceanography*, *115*, PA4102, doi:10.1029/2010PA002011.

Marchitto, T. M., J. Lynch-Stieglitz, and S. R. Hemming (2005), Deep Pacific $CaCO_3$ compensation and glacial-interglacial atmospheric CO_2, *Earth Planet. Sci. Lett.*, *231*(3–4), 317–336.

Marchitto, T. M., S. J. Lehman, J. D. Ortiz, J. Fluckiger, and A. van Geen (2007), Marine radiocarbon evidence for the mechanism of deglacial atmospheric CO_2 rise, *Science*, *316*(5830), 1456–1459.

Marriott, C. S., G. M. Henderson, R. Crompton, M. Staubwasser, and S. Shaw (2004), Effect of mineralogy, salinity, and temperature on Li/Ca and Li isotope composition of calcium carbonate, *Chem. Geol.*, *212*(1–2), 5–15.

Martin, J. H. (1990), Glacial-interglacial CO_2 change: The iron hypothesis, *Paleoceanography*, *5*(1), 1–13, doi:10.1029/PA005i001p00001.

Martínez-Garcia, A., A. Rosell-Melé, W. Geibert, R. Gersonde, P. Masqué, V. Gaspari, and C. Barbante (2009), Links between iron supply, marine productivity, sea surface temperature, and CO_2 over the last 1.1 Ma, *Paleoceanography*, *24*, PA1207, doi:10.1029/2008PA001657.

Mashiotta, T. A., D. W. Lea, and H. J. Spero (1999), Glacial-interglacial changes in subantarctic sea surface temperature and $\delta^{18}O$-water using foraminiferal Mg, *Earth Planet. Sci. Lett.*, *170*(4), 417–432.

Maslin, M. A., N. J. Shackleton, and U. Pflaumann (1995), Surface-water temperature, salinity and density changes in the northeast

Atlantic during the last 45,000 years: Heinrich events, deep-water formation and climate rebounds, *Paleoceanography*, *10*(3), 527–544.

Masson-Delmotte, V., et al. (2005), Changes in European precipitation seasonality and in drought frequencies revealed by a four-century-long tree-ring isotopic record from Brittany, western France, *Clim. Dyn.*, *24*(1), 57–69.

Massoth, G., et al. (2007), Multiple hydrothermal sources along the south Tonga Arc and Valu Fa Ridge, *Geochem. Geophys. Geosyst.*, *8*, Q11008, doi:10.1029/2007GC001675.

Matsumoto, K., and J. Lynch-Stieglitz (1999), Similar glacial and Holocene deep water circulation inferred from southeast Pacific benthic foraminiferal carbon isotope composition, *Paleoceanography*, *14*(2), 149–163.

McManus, J. F., R. Francois, J.-M. Gherardi, L. D. Keigwin, and S. Brown-Leger (2004), Collapse and rapid resumption of Atlantic meridional circulation linked to deglacial climate changes, *Nature*, *428*, 834–837.

McMurtry, G. M., P. N. Sedwick, P. Fryer, D. L. VonderHaar, and H. W. Yeh (1993), Unusual geochemistry of hydrothermal vents on submarine arc volcanoes: Kasuga Seamounts, northern Mariana Arc, *Earth Planet. Sci. Lett.*, *114*(4), 517–528.

Monnin, E., A. Indermuhle, A. Dallenbach, J. Fluckiger, B. Stauffer, T. F. Stocker, D. Raynaud, and J. M. Barnola (2001), Atmospheric CO_2 concentrations over the last glacial termination, *Science*, *291*(5501), 112–114.

Muscheler, R., J. Beer, P. W. Kubik, and H. A. Synal (2005), Geomagnetic field intensity during the last 60,000 years based on ^{10}Be and ^{36}Cl from the Summit ice cores and ^{14}C, *Quat. Sci. Rev.*, *24*(16–17), 1849–1860.

Nkrintra Singhrattna, B. R., M. Clark, and K. Krishna Kumar (2005), Seasonal forecasting of Thailand summer monsoon rainfall, *Int. J. Climatol.*, *25*(5), 649–664.

Okazaki, Y., A. Timmermann, L. Menviel, N. Harada, A. Abe-Ouchi, M. O. Chikamoto, A. Mouchet, and H. Asahi (2010), Deepwater formation in the North Pacific during the last glacial termination, *Science*, *329*(5988), 200–204.

Otto-Bliesner, B. L., E. C. Brady, G. Clauzet, R. Tomas, S. Levis, and Z. Kothavala (2006), Last Glacial Maximum and Holocene climate in CCSM3, *J. Clim.*, *19*(11), 2526–2544.

Rehder, G., S. H. Kirby, W. B. Durham, L. A. Stern, E. T. Peltzer, J. Pinkston, and P. G. Brewer (2004), Dissolution rates of pure methane hydrate and carbon-dioxide hydrate in undersaturated seawater at 1000-m depth, *Geochim. Cosmochim. Acta*, *68*(2), 285–292.

Reimer, P. J., et al. (2004), IntCal04 terrestrial radiocarbon age calibration, 0–26 cal kyr BP, *Radiocarbon*, *46*(3), 1029–1058.

Resing, J. A., J. E. Lupton, R. A. Feely, and M. D. Lilley (2004), CO_2 and ^{3}He in hydrothermal plumes: Implications for mid-ocean ridge CO_2 flux, *Earth Planet. Sci. Lett.*, *226*(3–4), 449–464.

Resing, J. A., E. T. Baker, J. E. Lupton, S. L. Walker, D. A. Butterfield, G. J. Massoth, and K.-I. Nakamura (2009), Chemistry of hydrothermal plumes above submarine volcanoes of the Mariana Arc, *Geochem. Geophys. Geosyst.*, *10*, Q02009, doi:10.1029/2008GC002141.

Rose, K. A., E. L. Sikes, T. P. Guilderson, P. Shane, T. M. Hill, R. Zahn, and H. J. Spero (2010), Upper-ocean-to-atmosphere radiocarbon offsets imply fast deglacial carbon dioxide release, *Nature*, *466*(7310), 1093–1097.

Rozanski, K., L. Araguasaraguas, and R. Gonfiantini (1992), Relation between long-term trends of oxygen-18 isotope composition of precipitation and climate, *Science*, *258*(5084), 981–985.

Sarmiento, J. L., and J. R. Toggweiler (1984), A new model for the role of the oceans in determining atmospheric pCO_2, *Nature*, *308*(5960), 621–624.

Siegenthaler, U., and T. Wenk (1984), Rapid atmospheric CO_2 variations and ocean circulation, *Nature*, *308*, 624–625.

Sigman, D. M., and E. A. Boyle (2000), Glacial/interglacial variations in atmospheric carbon dioxide, *Nature*, *407*, 859–869.

Sigman, D. M., M. P. Hain, and G. H. Haug (2010), The polar ocean and glacial cycles in atmospheric CO_2 concentration, *Nature*, *466*(7302), 47–55.

Sikes, E. L., C. R. Samson, T. P. Guilderson, and W. R. Howard (2000), Old radiocarbon ages in the southwest Pacific Ocean during the last glacial period and deglaciation, *Nature*, *405*(6786), 555–559.

Skinner, L. C., S. Fallon, C. Waelbroeck, E. Michel, and S. Barker (2010), Ventilation of the deep Southern Ocean and deglacial CO_2 rise, *Science*, *328*(5982), 1147–1151.

Smith, H. J., H. Fischer, M. Wahlen, D. Mastroianni, and B. Deck (1999), Dual modes of the carbon cycle since the Last Glacial Maximum, *Nature*, *400*(6741), 248–250.

Smith, I. E. M., and R. C. Price (2006), The Tonga-Kermadec Arc and Havre-Lau back-arc system: Their role in the development of tectonic and magmatic models for the western Pacific, *J. Volcanol. Geotherm. Res.*, *156*(3–4), 315–331.

Spero, H. J., and D. W. Lea (2002), The cause of carbon isotope minimum events on glacial terminations, *Geochim. Cosmochim. Acta*, *66*(15A), A731.

Stein, C., and S. Stein (1994), Constraints on hydrothermal heat flux through the oceanic lithosphere from global heat flow, *J. Geophys. Res.*, *99*, 3081–3095.

Stephens, B. B., and R. F. Keeling (2000), The influence of Antarctic sea ice on glacial-interglacial CO_2 variations, *Nature*, *404*(6774), 171–174.

Stott, L., A. Timmermann, and R. Thunell (2007), Southern Hemisphere and deep-sea warming led deglacial atmospheric CO_2 rise and tropical warming, *Science*, *318*(5849), 435–438.

Stott, L., J. Southon, A. Timmermann, and A. Koutavas (2009), Radiocarbon age anomaly at intermediate water depth in the Pacific Ocean during the last deglaciation, *Paleoceanography*, *24*, PA2223, doi:10.1029/2008PA001690.

Timmermann, A., O. Timm, L. Stott, and L. Menviel (2009), The roles of CO_2 and orbital forcing in driving Southern Hemispheric temperature variations during the last 21 000 yrs, *J. Clim.*, *22*(7), 1626–1640.

Toggweiler, J. R., and J. L. Sarmiento (1985), Glacial to interglacial changes in atmospheric carbon dioxide: The critical role of ocean surface water in high latitudes, in *The Carbon Cycle and Atmospheric CO_2: Natural Variations Archean to Present*, *Geophys. Monogr. Ser.*, vol. 32, edited by E. T. Sundquist and W. S. Broecker, pp. 163–184, AGU, Washington, D. C.

Toggweiler, J. R., J. L. Russell, and S. R. Carson (2006), Midlatitude westerlies, atmospheric CO_2, and climate change during the ice ages, *Paleoceanography*, *21*, PA2005, doi:10.1029/2005PA001154.

Tyrell, T., J. G. Shepherd, and S. Castle (2007), The long-term legacy of fossil fuels, *Tellus, Ser. B*, *59*(4), 664–672.

Voelker, A. H. L., P. M. Grootes, M.-J. Nadeau, and M. Sarnthein (2000), ^{14}C levels in the Iceland Sea from 25–53 kyr and their link to the Earth's magnetic field intensity, *Radiocarbon*, *42*, 437–452.

Walker, B. D., M. D. McCarthy, A. T. Fisher, and T. P. Guilderson (2008), Dissolved inorganic carbon isotopic composition of low-temperature axial and ridge-flank hydrothermal fluids of the Juan de Fuca Ridge, *Mar. Chem.*, *108*(1–2), 123–136.

Wheat, C. G., H. W. Jannasch, J. N. Plant, C. L. Moyer, F. J. Sansone, and G. M. McMurtry (2000), Continuous sampling of hydrothermal fluids from Loihi Seamount after the 1996 event, *J. Geophys. Res.*, *105*(B8), 19,353–19,367.

Wolff, E. W., et al. (2006), Southern Ocean sea-ice extent, productivity and iron flux over the past eight glacial cycles, *Nature*, *440*(7083), 491–496.

L. Stott, Department of Earth Sciences, University of Southern California, 3651 Trousdale Pkwy., Los Angeles, CA 90089, USA. (stott@usc.edu)

A. Timmermann, International Pacific Research Center, SOEST, University of Hawai'i at Mānoa, 2525 Correa Rd., Honolulu, HI 96822, USA. (axel@hawaii.edu)

The Impact of the Final Lake Agassiz Flood Recorded in Northeast Newfoundland and Northern Scotian Shelves Based on Century-Scale Palynological Data

Elisabeth Levac

Environmental Studies and Geography, Bishop's University, Sherbrooke, Quebec, Canada

C. F. M. Lewis

Geological Survey of Canada-Atlantic, Natural Resources Canada, Dartmouth, Nova Scotia, Canada

A. A. L. Miller

marine g.e.o.s., Wolfville, Nova Scotia, Canada

Two high-resolution century-scale palynological records from the eastern Canadian margin were analyzed to estimate the impact of Lake Agassiz's final drainage at circa 8.3 ka on sea surface conditions and to track the path of the meltwater plume. Core HU87033-19 from Notre Dame Channel on Northeast Newfoundland Shelf contains four distinct detrital carbonate (DC) beds, known to be sediment transported from Hudson Strait and Hudson Bay, and one layer is coeval with the drainage of Lake Agassiz. Within that DC layer, significant changes in dinoflagellate cyst assemblages indicate lower sea surface temperatures and salinity. The drop in salinity is a doublet, suggesting two episodes of meltwater drainage. Core HU84011-12, from St. Anne's Basin, on the northern Scotian Shelf contains similar changes in dinoflagellate cyst assemblages at the time of the drainage, indicating sea surface cooling accompanied by a slight decrease in salinity. The impact of the meltwater was greater in the Notre Dame Channel. This suggests that most of the meltwater from the final drainages of Lake Agassiz flowed south over the Labrador and Northeast Newfoundland shelves and was not dispersed directly into the Labrador Sea. Instead, it was possibly dispersed into the slope water system and subsequently into the North Atlantic after flowing initially over the continental shelf. This is the first paper describing paleoecological data indicating the presence of the Agassiz meltwater along the eastern Canadian margin.

Abrupt Climate Change: Mechanisms, Patterns, and Impacts
Geophysical Monograph Series 193

10.1029/2010GM001051

1. INTRODUCTION

The abrupt cooling event known as the 8.2 ka event, first detected in the Greenland Ice Sheet oxygen isotope records [*Johnsen et al.*, 1992; *Dansgaard*, 1993; *Alley et al.*, 1997] was imputed to have resulted from a slowing in Atlantic meridional overturning circulation (AMOC) [*Alley et al.*,

1997]. *Barber et al.* [1999] proposed that the final meltwater drainage from combined glacial lakes Agassiz and Ojibway (hereafter Lake Agassiz), flowed directly into the Labrador Sea (Figure 1), perturbed the North Atlantic Ocean overturning circulation, and suppressed the northward transport of heat, inducing the cold event. A large volume of meltwater impounded by the Laurentide Ice Sheet was released in a flood lasting about 0.5 year at peak transport rate of approximately 5 Sv (1 Sv = 10^6 m^3 s^{-1}) as the ice dam failed in Hudson Bay [*Teller et al.*, 2002; *Clarke et al.*, 2004]. Some computer models have shown that this volume of fresh water would have been sufficient to affect the overturning circulation in the North Atlantic [*Wiersma and Renssen*, 2006].

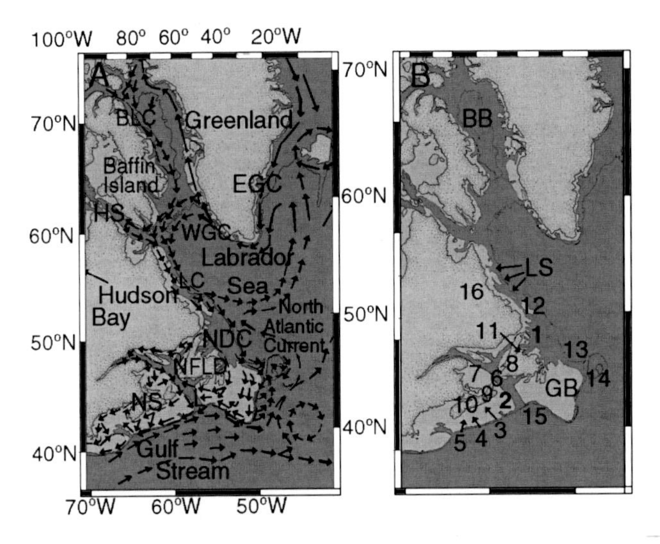

Figure 1. Location maps and surface currents along the eastern Canadian margin. (a) Surface currents in the northwest North Atlantic, Labrador Sea, and Baffin Bay, based on the works of *Greenberg and Petrie* [1988], *Han et al.* [1999], and *Shore et al.* [2000]. The acronyms refer to the following surface currents: BLC, Baffin Land Current; EGC, East Greenland Current; WGC, West Greenland Current; and LC, Labrador Current. Other acronyms refer to the following locations: NDC, Notre Dame Channel; NFLD, Newfoundland; and NS, Nova Scotia. The areas with a water depth between 0 and 200 m are shaded light gray; the 200 and 1000 m depth contours are shown. (b) Location of core sites and other important sites mentioned in the text. Abbreviations are LS, Labrador Shelf; BB, Baffin Bay; and GB, Grand Banks. The locations of the two cores published in this chapter are shown by bold numbers: 1, Notre Dame Channel core 87033-19 and 2, St. Anne's Basin core 84011-12. Other marine sites are 3, Canso Basin; 4, Emerald Basin; 5, La Have Basin; 6, Cabot Strait; 7, Gulf of St. Lawrence; 8, Bay of Islands; 12, Cartwright Saddle; 13, Orphan Knoll; 14, Flemish Pass; and 15, Laurentian Fan. Terrestrial sites are 9, Everitt Lake; 10, Silver Lake; 11, Compass Pond; and 16, Labrador region.

1.1. Evidence for the Lake Agassiz Outburst Flood

Lajeunesse and St-Onge [2008] mapped large sand waves and iceberg scours on the Hudson Bay seafloor, which they attributed to a catastrophic subglacial release. *Kerwin* [1996] recognized a marker red bed in sediments from Hudson Strait, believed to be the result of settling of suspended glacial sediment transported from NW Hudson Bay by meltwater floods [*Barber et al.*, 1999]. The red bed was subsequently found to be a double set of graded sediments [*Lajeunesse and St-Onge*, 2008]. Paleohydraulic modeling by *Clarke et al.* [2004] suggested there were two episodes of drainage, caused by the ice dam resealing to the seafloor after the first drainage, allowing Lake Agassiz to refill and undergo a second flood 13–48 years later. Radiocarbon dates in this chapter are expressed in the form of 8.2 ka, for example, meaning 8200 calibrated years before present (A.D. 1950).

Many North Atlantic records, which show evidence for slowing of the AMOC at the time of the 8.2 ka event [*Oppo et al.*, 2003; *Hall et al.*, 2004], were deemed inconclusive by some [*Alley and Agustsdottir*, 2005; *Rohling and Pälike*, 2005]. However, *Ellison et al.* [2006] present evidence of a cooling trend in sea surface temperature (SST) and a reduction in deep water flow speed in the eastern North Atlantic that are coincident with the Lake Agassiz discharge, all of which culminated in a cold climate event 200 years later at 8.2 ka. *Kleiven et al.* [2007] have recently presented evidence of reduced North Atlantic Deep Water flow in the northwest Atlantic for a period of about 100 years beginning at about 8.38 ka, coeval with the catastrophic drainage of Lake Agassiz (dated 8.47 ka with an error range of 8.16–8.74 ka by *Barber et al.* [1999]). Based on high-resolution chemistry and isotope records from Greenland ice cores, *Thomas et al.* [2007] estimated the duration of the event to be about 160 years, between 8.21 and 8.14 ka. Because of the relatively short duration of the 8.2 ka event, high-resolution records are needed if its relationship to oceanic change is to be detected and understood.

Though Lake Agassiz's drainage was large enough to suggest transport volumes of approximately 5 Sv through Hudson Strait [*Clarke et al.*, 2004], *Hillaire-Marcel et al.* [1994] and *MacLean et al.* [2001] found no evidence in the delta ^{18}O records indicating a freshening of surface waters in Hudson Strait, the Labrador Sea, or on the Labrador Shelf. However, *Andrews et al.* [1999] observed a weak signal at Cartwright Saddle, and one has been found spanning the early Holocene at Orphan Knoll [*de Vernal and Hillaire-Marcel*, 2000, 2006]. In fact, the lack of a widespread shift in isotopic values has been noted [*Andrews et al.*, 1999; *Keigwin et al.*, 2005; *Hillaire-Marcel et al.*, 2007]. *Hillaire-Marcel et al.* [2007, 2008] suggested that freshwater mixing and dispersal occurred quickly and that the observed shift would be very small (less than 1 practical salinity unit or psu), which would account for the lack of signal. However, a distinct $\delta^{18}O$ minimum is recognized at this time on the Laurentian Fan (Figure 1) and as far south as Cape Hatteras [*Keigwin et al.*, 2005]. *Keigwin et al.* [2005] proposed that the meltwater might have flowed south as a coastal current, hence the lack of isotopic signal along the Labrador Slope or in the Labrador Sea, but this does not adequately explain why they find a signal off the Scotian margin and as far south as the Middle Atlantic Bight. The paucity of isotopic evidence north of Grand Banks means that other avenues of investigation are needed to determine the routing and the impact of the outburst [*Lewis and Miller*, 2005; *Miller et al.*, 2006; *Lewis et al.*, 2009] on the surface waters along the eastern Canadian margin; therefore we propose to use sedimentological evidence (e.g., detrital carbonate (DC)) and palynological data to estimate the magnitude of changes in sea surface conditions.

Sediment layers with enhanced DC contents are a proxy for sediment transport out of Hudson Bay [*Andrews and Tedesco*, 1992; *Hillaire-Marcel et al.*, 2007]. These carbonates originate from the entrainment and transport of particles of glacial sediment derived from the limestone and dolomite bedrock in Hudson Bay and Strait [*Josenhans et al.*, 1986]. Holocene-age carbonate-rich sediment layers have been recognized in cores from several locations on the Labrador Shelf and Newfoundland shelves and slopes [*Piper et al.*, 1978; *Andrews et al.*, 1999; *Miller et al.*, 2006; *Hillaire-Marcel et al.*, 2007; *Tripsanas and Piper*, 2008; *Lewis et al.*, 2009], and one has been found to be coeval with the final Lake Agassiz discharge [*Lewis and Miller*, 2005; *Hillaire-Marcel et al.*, 2007]. The DC was likely transported southward via the Labrador Current (Figure 1) along the shelf and upper slope [*Lewis and Miller*, 2005; *Lewis et al.*, 2009]. Although carbonate bedrock also outcrops in the northern Gulf of St. Lawrence region, it is unlikely that this bedrock contributed early Holocene DC erosion products as the region was largely deglaciated by 10.85 ka (9.5 ^{14}C ka) [*Dyke et al.*, 2003].

Based on the distribution of Agassiz DC beds, *Lewis and Miller* [2005] and *Lewis et al.* [2009] proposed that the outer portion of the meltwater plume would have dispersed into the northern branch of the North Atlantic Current (NAC) along the eastern Grand Banks slope, and the remainder would have continued to flow southward inshore of the Gulf Stream. Along Cape Hatteras, the temperature is warmer and salinity higher, and hence, the contrast with the isotopic composition of the meltwater influx would be pronounced and could account for the almost 1‰ isotopic shift observed by *Keigwin et al.* [2005].

Although bottom water changes at 400–500 m depth, concurrent with the Agassiz discharge have been recognized on Northeast Newfoundland Shelf [*Miller et al.*, 2001, 2006], the sea surface impacts of the outflow along the eastern Canadian margin have not been studied. Here we investigate the outflow's impact on sea surface conditions, using new high-resolution palynological records from Notre Dame Channel on the Northeast Newfoundland Shelf and St. Anne's Basin on the northern Scotian Shelf. Basinal shelf deposits have higher sedimentation rates and thus contain expanded sequences allowing high-resolution analyses. We selected the Notre Dame Channel core because it contains DC beds deposited from discharge out of Hudson Strait [*Miller et al.*, 2006], one of which is coeval with the drainage of Lake Agassiz. St. Anne's Basin was selected because we want to show that although most of the meltwater probably was incorporated into the NAC, the impact of the meltwater drainage was also felt farther downstream, although with reduced amplitude of change in the sea surface conditions. St. Anne's Basin provides a link between the DC proxy evidence of the Agassiz discharge found off Labrador and Newfoundland [*Hillaire-Marcel et al.*, 2007; *Lewis et al.*, 2009] and the $\delta^{18}O$ minimum observed off Cape Hatteras [*Keigwin et al.*, 2005].

1.2. Chronology for the Lake Agassiz Outburst Floods

The final Lake Agassiz outburst floods through Hudson Bay and Hudson Strait were radiocarbon dated to 8.47 ka (7.7 ^{14}C ka) with an error range of 8.16–8.74 ka by *Barber et al.* [1999] using marine biogenic carbonate (e.g., mollusc shells) from sediment levels above and below a flood-generated red bed. This age was considered conservative (too old) by *Barber et al.* [1999] because the ^{14}C dates were not corrected (reduced) for the expected longer-than-present annual sea ice cover during the early Holocene, which would have limited the transfer of atmospheric ^{14}C in CO_2 to oceanic surface waters [*Bard et al.*, 1994; *Mangerud et al.*, 2006]. The effect of present seasonal sea ice cover is accounted for in the normal practice of dating pre-A.D. 1950 museum specimens of mollusc shells to establish local ^{14}C reservoir ages (and corrections for conventional ^{14}C dates). These ages consist of a surface ocean model age (about 400 years in the Northern Hemisphere) plus a ΔR term specific to the locality of interest [*Stuiver and Braziunas*, 1993]. The ΔR term accounts for local factors, which increase the reservoir age; it is the difference between the model ocean age and the known age of a museum specimen (elapsed time since collection). The mean and standard deviation of present-day ΔR values as determined by *McNeely et al.* [2006] are as follows: southern Hudson Bay 275±40 years, Hudson Strait 125±40 years, Labrador and Northeast Newfoundland shelves 144±38 years, and Scotian Shelf 83±14 years.

Recently, an added correction of −200±10 years was established for the region using a palynological transfer function estimate of early Holocene annual sea ice duration on the central Labrador Shelf (about 11 months versus 5–6 months at present) [*Lewis et al.*, 2009] in conjunction with a model of annual sea ice duration versus increase in marine ^{14}C reservoir age [*Bard et al.*, 1994; *Mangerud et al.*, 2006]. This correction of −200 years, when added to the conventional radiocarbon ages for constraints on the Agassiz floods and calibrated to calendar years, resulted in an adjusted age of the final Lake Agassiz drainage of 8.33 ka (7.5 ^{14}C ka) with an error range of 8.15–8.48 ka [*Lewis et al.*, 2009]. Beds of elevated DC with bracketing ages that correlate with this age range (e.g., HU87033-19P: 295–380 cm, this chapter) are attributed to deposition from iceberg detritus and plumes of sediment transported from Hudson Bay via the Lake Agassiz floodwaters in the Labrador Current [*Lewis et al.*, 2009].

2. REGIONAL SETTING

2.1. Location of Study Areas

Notre Dame Channel is an intrashelf basin located on the Northeast Newfoundland Shelf (Figure 1). Intrashelf basins and outer shelf banks are common on the eastern Canadian margin and are probably the result of glacial erosion and deposition, respectively [*Grant and McAlpine*, 1990; *Piper et al.*, 1990]. Notre Dame Channel is oriented approximately SW-NE and is separated from the Northeast Newfoundland Slope by a 330 m deep shelf-edge sill. Sediment deposition is uniform and continuous, without hiatus [*Dale and Haworth*, 1979; *Miller*, 1999]. Core HU87033-19 (hereafter referred to as core 19) was recovered at 50°54.51′N and 53° 15.63′W at a water depth of 453 m [*Vilks and Powell*, 1987].

St. Anne's Basin is located on the northeast part of the Scotian Shelf, 50 km from the edge of the Laurentian Channel (Figure 1). The basin has a length of 7 km and a width of 3 km. It lies on the inner shelf, about 40 km south-southeast of Cape Breton Island. Core HU84011-12 (hereafter referred to as core 12) was taken near the center of the basin at 45° 46.72′N; 58°39.16′W, at a water depth of 270 m.

2.2. Oceanographic Conditions of the Area

Water over the Northeast Newfoundland Shelf is heavily influenced by the Labrador Current, which flows southeastward and parallel to the Labrador coast (Figure 1). The main branch of the Labrador Current is less than 100 km wide and flows over the upper slope. It combines cold, low-salinity water from the southward flowing Baffin Island Current (originating in the Arctic) with warmer more saline waters from the West Greenland Current moving westward across Davis Strait. The cold, low-salinity inner branch of the Labrador Current derives mainly from the Baffin Island Current and water out of the Hudson Strait and is seasonally fed by sea ice melt [*Loder et al.*, 1998]. The Labrador Current flows southward from the Labrador Shelf onto the Northeast Newfoundland Shelf, and then the greater part turns east and north along the eastern margin of the Grand Banks and runs parallel with the NAC before joining it north of Flemish Cap (Figure 1). A smaller part of the Labrador Current continues south and west to follow the southern margin of the Grand Bank, to the Laurentian Fan, and along the Scotian Shelf margin. There is an outflow of slightly fresher water from the Gulf of St. Lawrence on the western side of Cabot Strait [*Han et al.*, 1999].

Water on the Northeast Newfoundland Shelf is characterized by low temperatures and low salinity. Over the core site, the SST in August is between 9°C and 10°C, and sea surface salinity (SSS) is 32.5 ‰ (in psu scale) [*Levitus*, 1982; *Loder et al.*, 1998]. The seasonal duration of sea ice cover is 4 months [*Markham*, 1988]. In winter, the minimum SST is −1.5° C, and the SSS is between 33.0‰ and 33.5‰ [*Loder et al.*, 1998].

St. Anne's Basin is located near the Laurentian Channel, which is a major exchange pathway between the Atlantic Ocean and the Gulf of St. Lawrence waters and also a confluence zone for continental runoff from the St. Lawrence River, the Labrador Current, and offshore slope waters [*Loder et al.*, 1997].

The sea surface conditions over the northeast Scotian Shelf clearly reflect the input from the Gulf of St. Lawrence through Cabot Strait. Salinity is lower along the coast of Cape Breton and on the inner Scotian Shelf and gradually increases offshore [*Loder et al.*, 1997; *Petrie et al.*, 1996]. Over St. Anne's Basin, the mean SST is 0°C in February and 16°C in August. SSS is higher in February (31.5‰) than in August (30 psu) [*Levitus and Boyer*, 1994]. The duration of the sea ice cover is variable around Cape Breton. While the average duration is 80–100 days on the west side of the island, it is only 40 days along the southern coast and over St. Anne's Basin [*Drinkwater et al.*, 1999].

2.3. Modern Vegetation Distribution and Associated Pollen Record

2.3.1. Vegetation in Northern Newfoundland. Northern Newfoundland forests lie within the spruce-fir subdivision of the boreal forest [*Rowe*, 1977]. The most abundant tree in the boreal forest is *Picea mariana*, accompanied by *Abies balsamea*, *Betula papyrifera*, *Populus tremuloides*, and *Populus balsamifera*. *Picea glauca*, *Pinus strobus*, and *Betula alleghaniensis* are also found [*Rowe*, 1977]. The boreal forest is replaced by small bands of spruce-fir-heath open woodland along the coastline because of wind exposure. This open woodland is composed of spruce (*P. mariana*) woodland interspersed with heath barrens (Ericaceae), bogs, and muskeg [*Rowe*, 1977].

2.3.2. Vegetation in Nova Scotia and Cape Breton. Most of Nova Scotia, with the exception of the coastal areas and Cape Breton, belongs to the Atlantic Uplands and Central Lowlands subdivision of the Acadian forest region of *Rowe* [1977]. The forest of the Atlantic Uplands subregion is composed mainly of *Picea rubens*, with *Tsuga canadensis*, *P. mariana*, *B. papyrifera*, *Acer rubrum*, *A. balsamea*, *P. strobus*, and *Pinus resinosa*. In addition to these species, the vegetation of the Central Lowlands also include *Acer saccharum*, *Fagus grandifolia*, *Quercus* spp, *Ulmus* spp, *Fraxinus nigra*, *Fraxinus americana*, and *Tilia americana*. Dense lowstands of *A. balsamea*, *P. mariana*, and *P. glauca* are found along the coast in the Eastern Atlantic Shore subregion [*Rowe*, 1977]. *Sphagnum* bogs occupy areas with poor drainage [*Rowe*, 1977; *Ecoregions Working Group*, 1989].

Over Cape Breton (northern part of Nova Scotia), *P. glauca* and *A. balsamea* are the main trees, accompanied by *P. balsamifera*, *F. americana*, and *Ulmus americana* [*Rowe*, 1977]. On hilly lands, *A. rubrum*, *B. papyrifera*, and *B. alleghaniensis*, can also be found. *T. canadensis* and *F. grandifolia* are locally abundant on slopes and ravines. In areas with poor drainage, *P. mariana* is common [*Rowe*, 1977]. The forest becomes predominantly deciduous on the slopes of the highlands and the main tree species are *A. saccharum*, *F. grandifolia*, *B. alleghaniensis*, and *A. rubrum*. Conifers become more prominent on the Cape Breton Plateau, with *A. balsamea* as the dominant species, with *P. mariana* and *P. glauca* [*Rowe*, 1977]. Shrubs (*Salix* spp, Ericaceae, *Alnus* spp) are locally abundant throughout all vegetation zones described here.

2.4. Taphonomic Issues Affecting Pollen Assemblages in Marine Sediments

Modern pollen assemblages in marine sediments from the eastern Canadian margin reflect the vegetation of adjacent lands; however, representation of the principal taxa varies with distance from the shoreline [*Mudie*, 1982]. Pollen transport tends to be selective. Bisaccate pollen (mainly *Pinus* spp and *Picea* spp) is preferentially transported by wind, and their proportions gradually increase with distance from shore. Other taxa (*Betula*, *Alnus*, *Tsuga*, and *Quercus*, to list only a few) see their proportions rapidly decrease across the Scotian Shelf [*Mudie*, 1982; *Mudie and McCarthy*, 1994]. In addition to these selective transport mechanisms, taxa representation in pollen assemblages is not proportional to their abundance in vegetation. Certain trees such as *A. balsamea* and *Acer* spp generally tend to be underrepresented in pollen diagrams, while big pollen producers such as *Pinus* are overrepresented [*Davis and Webb*, 1975].

Pollen assemblages in both locations reflect these selective transport mechanisms and differences in pollen production. Pollen diagrams of both cores 19 and 12 display greater proportions of bisaccate pollen (mainly *Pinus* and *Picea*)

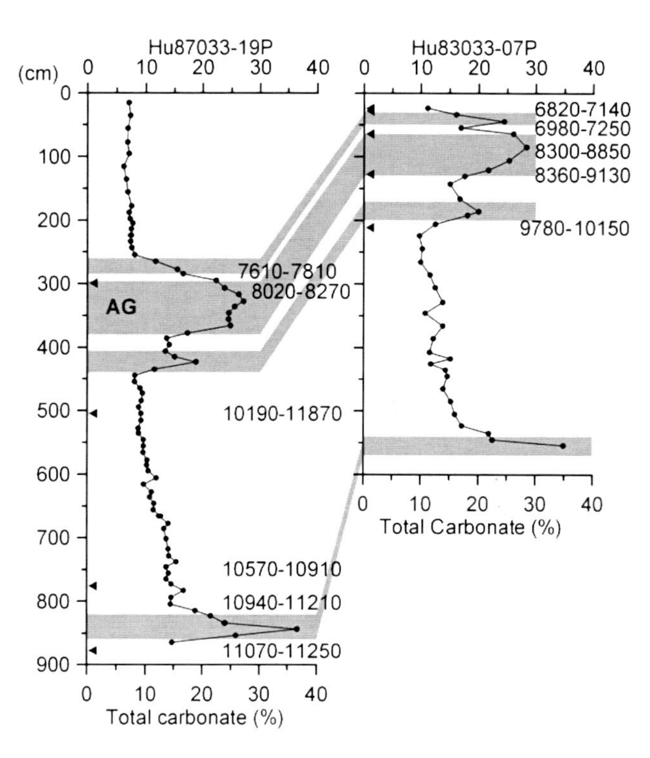

Figure 2. Detrital carbonate (DC) peaks in cores HU87033-19P and HU83033-07P from Notre Dame Channel on the Northeast Newfoundland Shelf. The thick beds of DC labeled AG in these two cores correlate to the Agassiz drainage as shown by the bracketing ^{14}C dates on foraminifera. Total carbonate is the sum of percent calcite and dolomite on a dry weight basis in the sediment, measured by the Chittick method using acid dissolution and CO_2 evolution [*Dreimanis*, 1962]. Constraints on the Lake Agassiz floods were recently reassessed as 8.33 kyr with an error range of 8.15–8.48 kyr [*Lewis et al.*, 2009]. The calibrated 2σ range of the dates is shown here for levels marked by the triangles in each core. Dates were obtained on the planktic foraminifer *Neogloboquadrina pachyderma* s (sinistral) coiling. Where two dates are posted for a single level, the upper date is on *N. pachyderma* s, and the lower date is on the benthic foraminifer *Nonionellina labradorica*.

than their onshore counterparts. In core 12, the proportion of *Quercus* pollen is also markedly lower compared to onshore pollen records.

Despite the bias introduced by the selective nature of long-distance pollen transport, it is possible to recognize the zonations of terrestrial pollen records in the marine pollen records [*Mudie*, 1982; *Mudie and McCarthy*, 1994], and these zonations will be presented in the results section. However, since the marine pollen assemblages potentially integrate the pollen signal from a much larger surface area, reconstructing a climatic signal using them is ruled out.

3. METHODS

Sediments were analyzed for carbonate content and for palynological analysis. Calcite and dolomite were measured by the Chittick method using acid dissolution and CO_2 evolution [*Dreimanis*, 1962]. The total carbonate shown on Figure 2 is the sum of percent calcite and dolomite on a dry weight basis in the sediment.

The dated levels in core 19 were determined either directly by ^{14}C dating of monospecific foraminiferal samples or mollusc shells or indirectly by correlating to ^{14}C-dated DC beds in nearby core HU83033-07 (50°53.2′N 53°18.0′W, 457 m water depth; hereafter core 07) and then by correlating core 19 to ^{14}C-dated onshore lake sediments from lakes along the Northeast Newfoundland coast (Table 1) using the similarity of pollen profile features or pollen zone boundaries (see section 5). The DC beds contained reduced concentrations of biogenic carbonate; thus, it was not possible to date the maximum concentrations of DC; instead, foraminifera were separated from sediments above and below each interval and dated in order to bracket the age of the DC beds.

Conventional ^{14}C dates on foraminifera from the early Holocene sections of cores 19 and 12 (Table 2) were calibrated with the CALIB 6.0.1 online software program [*Stuiver and Reimer*, 1993] using the MARINE09 calibration data set [*Reimer et al.*, 2009] with a 10 year smoothing function. A sea ice correction (200 years for core 19 and 35 years for 12) to account for longer-than-present annual sea ice cover duration in the early Holocene was added to the present-day ΔR values of 144±38 years for the Labrador and Northeast Newfoundland shelves and 83±14 years for the Scotian Shelf [*McNeely et al.*, 2006] during the calibration process. This procedure is equivalent to subtracting the sea ice correction from the conventional radiocarbon age before calibration, as done by *Lewis et al.* [2009]. Terrestrial conventional ^{14}C dates on lake organic sediment were calibrated with the CALIB 6.0.1 program [*Stuiver and Reimer*, 1993] using the INTCAL09 data set [*Reimer et al.*, 2009] with a 10 year smoothing function. A single calibrated age for a radiocarbon date, as used in this chapter, is the mean of the most probable calibrated age range in the CALIB 6.0.1 calibration program output (commonly the mean of the two-sigma calibrated age range). The age is expressed as *x.y* ka meaning *x.y* thousands of calibrated years before present where present is A.D. 1950.

Sampling resolution for palynological analysis varies with depth in the cores and inferred sedimentation rates. Samples for palynological analyses were taken every 10 or 20 cm, which represents a sample every 100 or 200 years, on average. In the lower part of core 12, including the interval representing the Agassiz flood, the sampling resolution is a sample every 40 years. In core 19, this interval is between 100 and 200 years duration.

Sediments were wet sieved at 120 μm to remove the coarsest particles and at 10 μm to eliminate the fine silt and clay. The 10–120 μm fraction was processed with cold hydrochloric acid to remove carbonates and with cold hydrofluoric acid (50% concentration) to remove silicates. *Lycopodium* markers were used to estimate concentrations of

Table 1. Location of Core Sites Studied for This Study and Core Sites Providing Mass Spectrometer Dates Used in This Study

Location	Core Number	Depth/Elevation (m)	Latitude (°N)	Longitude (°W)	Reference
			Marine Sites		
Notre Dame Channel	HU87033-19P	453	50°54.51′	53°15.63′	*Lewis et al.* [2009], *Smith and Licht* [2000], this study
Notre Dame Channel	HU83033-07P	457	50°53.2′	53°18′	*Miller* [1999], *Lewis et al.* [2009], this study
St. Anne's Basin	HU84011-12P	270	45°46.72′	58°39.16′	*Levac* [2002]
			Terrestrial Sites		
Compass Pond		236 asl[a]	50°02.05′	56 °11.78′	*Dyer* [1986], *Macpherson* [1995]
Everitt Lake		80 asl[a]	44°27′	65°52′	*Green* [1987]
Silver Lake		61 asl[a]	44°34′	63°38.5′	*Livingstone* [1968], *Ogden* [1987]

[a]Here asl indicates above sea level.

Table 2. Radiocarbon Dates Used in This Study From Cores Listed in Table 1

Location, Core	Down-Core Depth (cm)	Corresponding Depths in Core	Material Dated	Lab Number	Conventional ^{14}C Age (years B.P.)	Sea Ice Correction	ΔR	Calibrated Age (1σ range calBP)	Calibrated Age (2σ range calBP)	Stratigraphic or Biostratigraphic Markers
HU87033-19P	298–300		*Neogloboquadrina pachyderma s*	UCIAMS74194	7585±20	200	144±38	7650 to 7760	7610 to 7810	
HU87033-19P	298–300		*N. labradorica*	UCIAMS74195	8025±20	200	144±38	8070 to 8190	8020 to 8270	
HU87033-19P	503–505		*N. labradorica*	AA21750	10300±300[a]	200	144±38	10500 to 11250	10190 to 11870	
HU87033-19P	775–780		*N. pachyderma s*	UCIAMS45220	10175±25	200	144±38	10590 to 10760	10570 to 10910	
HU87033-19P	775–780		*N. labradorica*	UCIAMS45221	10450±25	200	144±38	11100 to 11170	10940 to 11210	
HU87033-19P	875–877		*N. pachyderma s*	UCIAMS 45372	10500±40	200	144±38	11130 to 11200	11070 to 11250	
HU83033-07P	24	254[b]	*N. labradorica*	Beta-191949	6810±40	200	144±38	6880 to 7040	6820 to 7140	top DC bed above AG
HU83033-07P	28	260[b]	*N. labradorica*	Beta-191950	6950±40	200	144±38	7060 to 7210	6980 to 7250	base DC bed above AG
HU83033-07P	63–65	295[b]	*N. pachyderma s*	Beta-194865	8390±100	200	144±38	8380 to 8630	8300 to 8850	top of AG DC bed
HU83033-07P	125–130	378[b]	*N. pachyderma s*	TO1353	8550±150	200	144±38	8550 to 8950	8360 to 9130	base of AG DC bed
HU83033-07P	206–216	448[b]	*N. pachyderma s*	UCIAMS20521	9520±35	200	144±38	9900 to 10100	9780 to 10150	base DC bed below AG
Compass Pond	195–200	143[b]	bulk sediments	GSC-3906	3050±140			3040 to 3430	2870 to 3560	top of zone CP3-ii
Compass Pond	395–400	230[b]	bulk sediments	GSC-3902	6280±120			7010 to 7320	6900 to 7430	top of zone CP3-I
Compass Pond	435–440	450[b]	bulk sediments	GSC-3992	8310±140			9140 to 9460	8820 to 9550	top of zone CP2-ii
HU84011-12P	65–67		foraminifera	TO6232	3100±60	35	83±14	2700 to 2820	2610 to 2920	
HU84011-12P	640–650		foraminifera	TO6233	7930±80	35	83±14	8190 to 8354	8090 to 8440	
Everitt Lake	150–160	100[c]	bulk sediments	DAL209	3340±175			3390 to 3830	3160 to 4080	top of zone C3a
Silver Lake	245–255	250[c]	bulk sediments	GSC-791	4540±140			4970 to 5450	4860 to 5580	end *Tsuga* peak
Silver Lake	345–355	430[c]	bulk sediments	GSC-792	7140±140			7830 to 8160	7680 to 8290	start *Tsuga* peak

[a]The date at 503 cm in core 19 is from *Smith and Licht* [2000].

[b]Corresponding to core 19.

[c]Corresponding to core 12.

palynomorphs in the sediments [*Stockmarr*, 1971]. Palynomorphs were examined, identified, and counted under a light transmission microscope (Nikon Eclipse) at a magnification of 400 times. At least 300 pollen grains and 300 dinocysts were counted.

Zonation (for pollen as well as dinoflagellate cyst data) is based on cluster analysis of the data, using the PSIMPOLL and PSCOMB software [*Bennett*, 1996] and visual inspection. The ecological affinities of dinoflagellate cysts species were also considered when determining the zonations (see regional overview).

Sea surface conditions were reconstructed by the best analog technique, using the GEOTOP dinoflagellate cyst database [*de Vernal et al.*, 2001; *Guiot and de Vernal*, 2007]. Owing to the strong regionalism in the distribution of the dinoflagellate cyst taxa [*Taoufik and de Vernal*, 2008; *Marret and Zonneveld*, 2003], we selected sites from the North Atlantic Ocean and adjacent basins (Gulf of St. Lawrence, Gulf of Maine, Hudson Bay, Labrador Sea, Norwegian Sea) and the Arctic Ocean. Surface dinoflagellate cyst assemblages from another 58 sites from the Scotian Shelf and Slope were added to provide additional analogs for core 12 [*Levac*, 2002]. A number of unpublished surface sediment samples from the Northern Newfoundland Shelf were added to fill a void in that area of the database [see *Levac*, 2002, 2003]. In total, 601 reference sites were used.

For the reconstructions, each fossil assemblage is compared with all the modern spectra in the reference database, and the 10 most similar spectra are selected, using the same weight for each taxon. The degree of similarity (the weight) between the fossil assemblage and each of the modern analogs is determined by the Euclidian distance between them. The set of analogs provides the paleoenvironmental data from which a weighted average is calculated for each sea surface parameter (SST, SSS, seasonal duration of sea ice cover). The minimum and maximum environmental values in the set of analogs are used to define the confidence interval [*de Vernal et al.*, 2001; *Guiot and de Vernal*, 2007]. The accuracy of the method, with the data set used, is defined as the standard deviation of reconstructed values. For August SST, the accuracy is ±1.35°C, for SSS, it is ±2‰, and for the seasonal duration of the sea ice cover, it is ±1.28 months year^{-1}.

4. CORE STRATIGRAPHY

Core 19 is composed of clay or silty clay, with colors ranging from gray, to olive gray, or dark gray (Figure 3). Some intervals were initially observed to be calcareous, which is confirmed by carbonate analysis (Figure 2).

Background values in total carbonate are low (less than 10%) in core 19. However, four intervals enriched in DC were evident where total carbonate values rise to between 15% and 40% (Figure 2). Similar DC beds are found at other locations along the eastern Canadian margin [*Lewis et al.*, 2009]. In core 19, the four DC beds occur at 265–285, 295–380, 405–440, and 820–860 cm depth with the thickest bed at 295–380 cm identified as coeval with the Agassiz drainage (Figure 2). A sympathetic variation of dolomite ($CaMg(CO_3)_2$) (not shown) and total carbonate (Figure 2) indicates these beds are mainly of detrital origin and are not derived from biogenic sources, which would produce mainly calcite or aragonite ($CaCO_3$). Nearby core 07 (50°53.2′N 53° 18.0′W, 457 m water depth) also has radiocarbon chronological control [*Miller*, 1999; *Miller et al.*, 2006; *Lewis et al.*, 2009]. This control was transferred to core 19 using the correlation of DC layers between the two cores (Figure 2). In addition to the marine radiocarbon dates, terrestrial radiocarbon ages in a radiocarbon-dated pollen record from a Northeastern Newfoundland coastal site (Compass Pond) [*Dyer*, 1986] were correlated to core 19 using similar features and zones in their pollen records. More details are given in the pollen zonation section below. The same procedure was done for core 12.

Core 12 is composed mainly of olive to olive gray clay or mud [*Freeman*, 1986]. Carbonate analysis was not done on core 12. Two radiocarbon ages were obtained on foraminifera from piston core 12 (Table 2). As for core 19, we supplemented our ^{14}C radiocarbon chronology by comparing the pollen record from core 12 to onshore pollen records from Cape Breton and mainland Nova Scotia [*Livingstone*, 1968; *Green*, 1987; *Ogden*, 1987].

The DC beds were found to contain reduced concentrations of palynomorphs and foraminifera (see section 5)

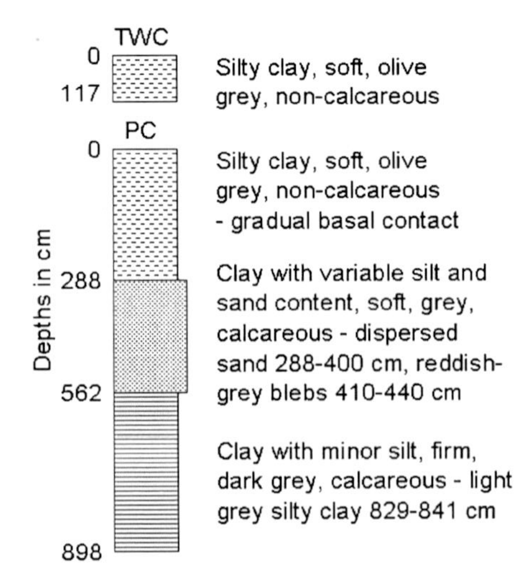

Figure 3. Sedimentology of core HU87033-19, 0–898 cm, Notre Dame Channel, on the Northeast Newfoundland Shelf. TWC means trigger weigh core; PC is piston core.

compared with underlying and overlying sediments, indicating that the DC bed sediment accumulated rapidly. This result is consistent with the inferred 1 year or less duration of the floods, as suggested by paleohydraulic and glaciological modeling of the Lake Agassiz discharge [*Clarke et al.*, 2004]. The zones of elevated DC content contained reduced concentrations of biogenic carbonate; thus, it was not possible to date the peaks of the DC intervals; instead, foraminifera were separated from sediments above and below each interval and dated in order to bracket the age of the DC beds.

The age model adopted for core 19 (Figure 4) assumes linear sediment accumulation between most dated levels, with the beds of elevated DC interpreted to have been laid down rapidly (see section 5.1). Sedimentation rates below 440 cm in core 19 were nonlinear, progressively diminishing upward from extremely high rates at 880 cm. The dated levels in core 19 were determined, either directly by AMS ^{14}C dating of monospecific foraminiferal samples or indirectly by correlating ^{14}C-dated DC beds in the adjacent core 07P, and then by comparing core 19 to ^{14}C-dated onshore pollen records from a lake (Compass Pond) along the Northeast Newfoundland coast using the similarity of pollen profile features or pollen zone boundaries. A similar methodology was used to develop an age model for core 12 (Figure 5).

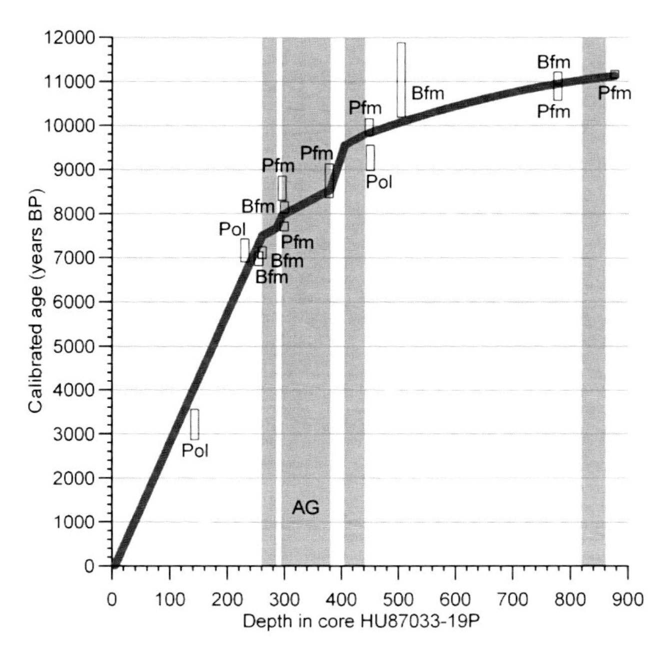

Figure 4. Age-depth model for core HU87033-19P (dark gray line). The gray bands represent beds of DC settled from plumes of suspended sediment arising from energetic outflows from Hudson Strait. The thick bed of DC-labeled AG correlates to the Agassiz floods. Note that the age-depth interpretation generally shows higher sedimentation rates (steeper curves) in these zones than in adjacent sediment layers. Age data from Table 2 are plotted as rectangles. The horizontal dimension of 14 cm (±7 cm) represents uncertainty due to bioturbation that was measured in La Have Basin on the Scotian Shelf using ^{210}Pb activity dating (J. Smith, Bedford Institute of Oceanography, personal communication, 2000). The vertical dimensions are the calibrated 2σ age ranges for ^{14}C dates. The source material for dates is indicated by the rectangle labels: Pfm, planktic foraminifera; Bfm, benthic foraminifera; and Pol, correlation by pollen from a dated lake sediment record.

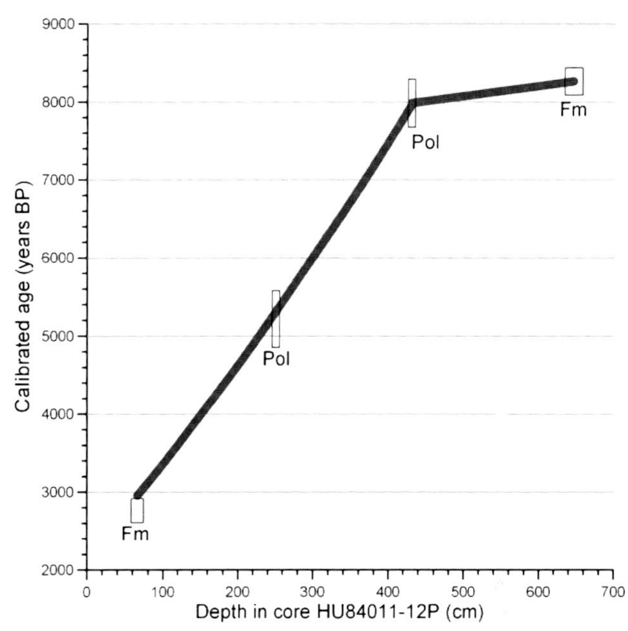

Figure 5. Age-depth model for core HU84011-12P (dark gray line). Age data from Table 2 are plotted as rectangles. The horizontal dimensions of 14 cm (±7 cm) represent uncertainty due to bioturbation which was measured in La Have Basin on the Scotian Shelf using ^{210}Pb activity dating (J. Smith, Bedford Institute of Oceanography, personal communication, 2000). The vertical dimensions are the calibrated 2σ age ranges for ^{14}C dates. The source material for dates is indicated by the rectangle labels: Fm, foraminifera; and Pol, correlation by pollen from a dated lake sediment record.

5. RESULTS

5.1. Palynomorph Concentrations

Concentrations of most categories of palynomorphs in core 19 (Figure 6) seem to reflect dilution by greater sedimentation rates within the DC beds (especially the Agassiz (AG) DC bed and the DC bed below). The presence of low-growing tundra vegetation and subsequently low shrub vegetation explains why pollen concentrations were low below the 5 m level in core 19. Pollen concentrations began to increase

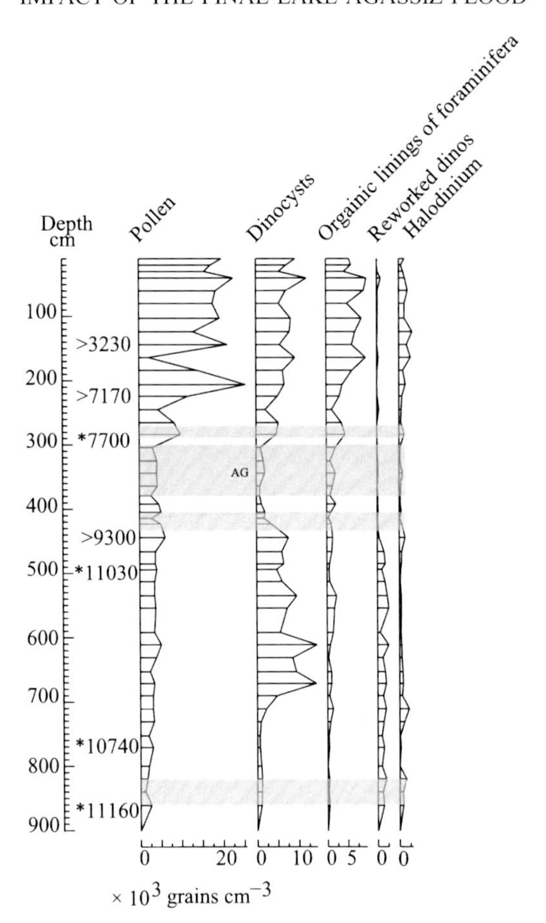

Figure 6. Palynomorph concentrations per cm^3 of sediments (times 1000) in core HU87033-19P, Notre Dame Channel. Numbers on the left are calibrated ages from foraminifera and pollen zonations. Calibrated ages based on foraminifera are indicated by the asterisks on the *x* axis (only ages from planktic foraminifera are shown here). The pollen-correlated ages are indicated by the greater than symbols. The gray shading indicates layers with high DC. AG is Lake Agassiz.

slightly in the Holocene, at the time forests became established in North Central Newfoundland (forest vegetation produces more pollen), sometime around 9.9 ka (around 450 cm), but they are lower in the DC beds, probably due to dilution by higher sedimentation rates. Pollen concentrations increase only above 305 cm, i.e., not until after the AG episode, indicating that the deposition of the AG and the next DC bed below occurred during episodes of more rapid sedimentation. Dinocysts show a sharp upcore increase around 7 m (circa 10.3 ka) but then drop to late glacial values in the AG interval.

Halodinium concentrations exhibit a different trend compared to the other categories of palynomorphs, with high concentrations in the early Holocene but very low concentrations in the last 7000 years. *Halodinium* is an acritarch of unknown biological affinities that is abundant in Baffin Bay fjord sediments [*Mudie and Short*, 1985]. The greater concentrations of *Halodinium* could represent greater influence from Arctic and/or meltwater until about 7.6 ka. The increased concentrations of organic foraminiferal test linings in the upper 3 m of core 19 could reflect either greater productivity or increased carbonate dissolution.

Concentrations of reworked palynomorphs are consistent throughout core 19, but the presence of reworked Cretaceous pollen and spores within the AG and the lower carbonate layers should be noted as well as the presence of reworked dinocysts of Cretaceous and Mesozoic age. Trilete *Cicatricosisporites* sp. spores of Upper Jurassic-Cretaceous age [*Batten*, 1996] are found at 324 and 343 cm. Mesozoic *Tsugaepollinites* [*Burden and Holloway*, 1985] are found at the following depths: 324, 343, and 391 to 443 cm. Reworked dinocysts include *Dapsilidinium pseudocolligerum*

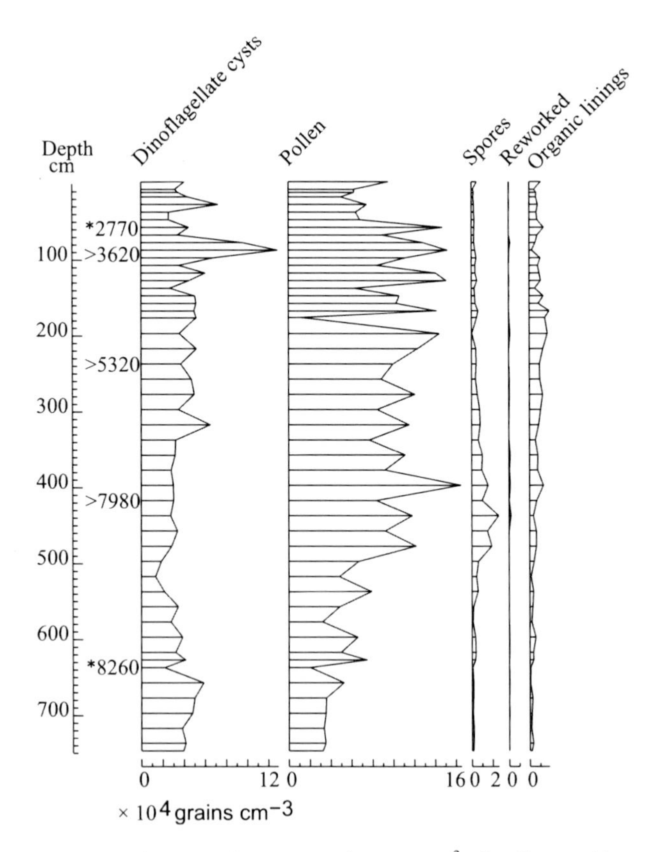

Figure 7. Palynomorph concentrations per cm^3 of sediments (times 10000) in core HU84011-12, St. Anne's Basin. Numbers on the left are calibrated ages from foraminifera and pollen zonations. Ages based on foraminifera are indicated by the asterisks on the *x* axis (only ages from planktic foraminifera are shown here). The pollen-correlated ages are indicated by the greater than symbols.

(Mesozoic-Tertiary) [*Stover et al.*, 1996] at 465 cm and Cretaceous *Surculosphaeridinium longifurcatum* at 413 cm [*Burden and Holloway*, 1985].

Concentrations of various palynomorph categories are generally constant in piston core 12 (Figure 7). The only notable variations are the peak amounts in spores and reworked Mesozoic palynomorphs between 4 and 5 m, which probably corresponds to the end of the Lake Agassiz flood. Concentrations of organic lining of foraminifera show an increase toward the late Holocene, which could suggest greater carbonate dissolution or low bottom oxygen.

5.2. Pollen Zonation

5.2.1. Pollen Zones in Core 19, Notre Dame Channel. The pollen record from core 19 (Figure 8) was compared with three pollen records from Newfoundland lakes: Leading Tickles [*Macpherson*, 1995], Small Scrape Pond [*Dyer*, 1986; *Macpherson*, 1995], and Compass Pond [*Dyer*, 1986; *Macpherson*, 1995]. The succession of pollen assemblages at Compass Pond is most similar to that of core 19, probably because Compass Pond is at a higher location than the other sites and is surrounded by forest vegetation, pollen from trees being transported more readily by wind than pollen from low-growing plants.

Below 900 cm (not illustrated on Figure 9), pollen assemblages are similar to Compass Pond (CP) pollen zones CP1 and CP2-i. Pollen zone CP1 is characterized by a dominance of grass pollen (Poaceae), shrub (Cyperaceae, *Salix*, etc.), and tundra taxa (*Rumex*, *Artemisia*). Pollen zone CP2-i represents the beginning of the Holocene in Newfoundland [*Dyer*, 1986; *Macpherson*, 1995].

Pollen assemblages between 450 and 900 cm are characterized by high total *Betula* percentages peaking around 600 cm, high percentages of *Salix*, and a gradual increase in the proportions of tree pollen. This interval was correlated with Compass Pond zone CP2-ii. The top of zone CP2-ii is dated at 9.3 ka. The ^{14}C dates for Compass Pond (and other lakes mentioned herein) are based on total organic carbon from bulk sediments and are reported in Table 2.

The interval between 230 and 450 cm corresponds with zone CP3-i based on increasing percentages of *Betula* pollen and the initial peak in *Abies* pollen. There is a date of 7.17 ka for the top of zone CP3-i at Compass Pond. The second peak

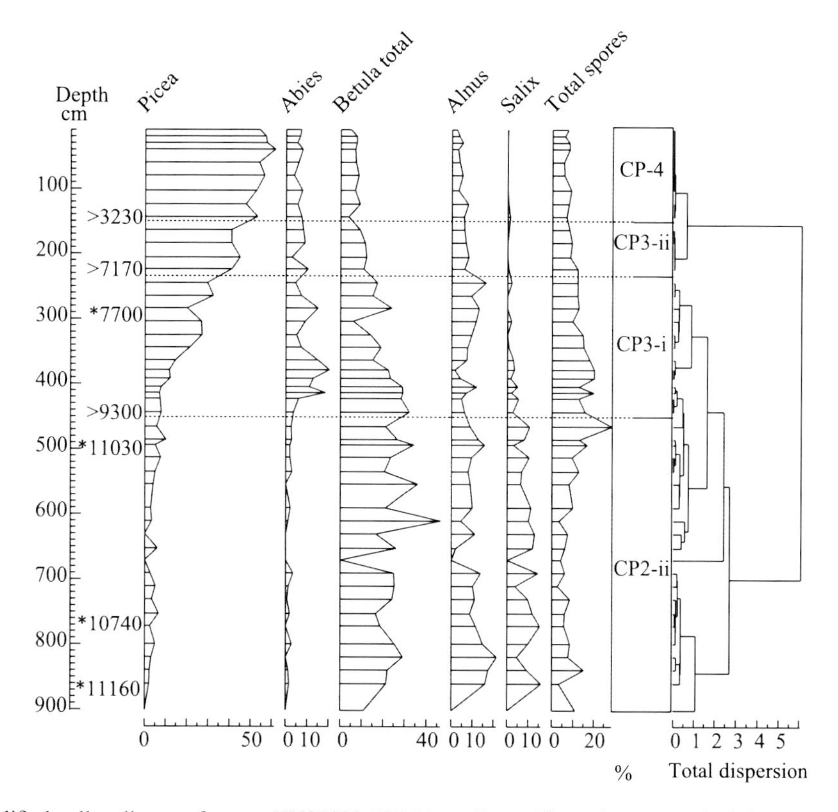

Figure 8. Simplified pollen diagram for core HU87033-19P, Notre Dame Channel. Ages on the left are calibrated ages from planktic foraminifera (asterisks) and dates from the Compass Pond pollen record with which core 19 was correlated (greater than symbols). The dates are for the boundaries between pollen zones. Lines on the right are the results of the cluster analysis.

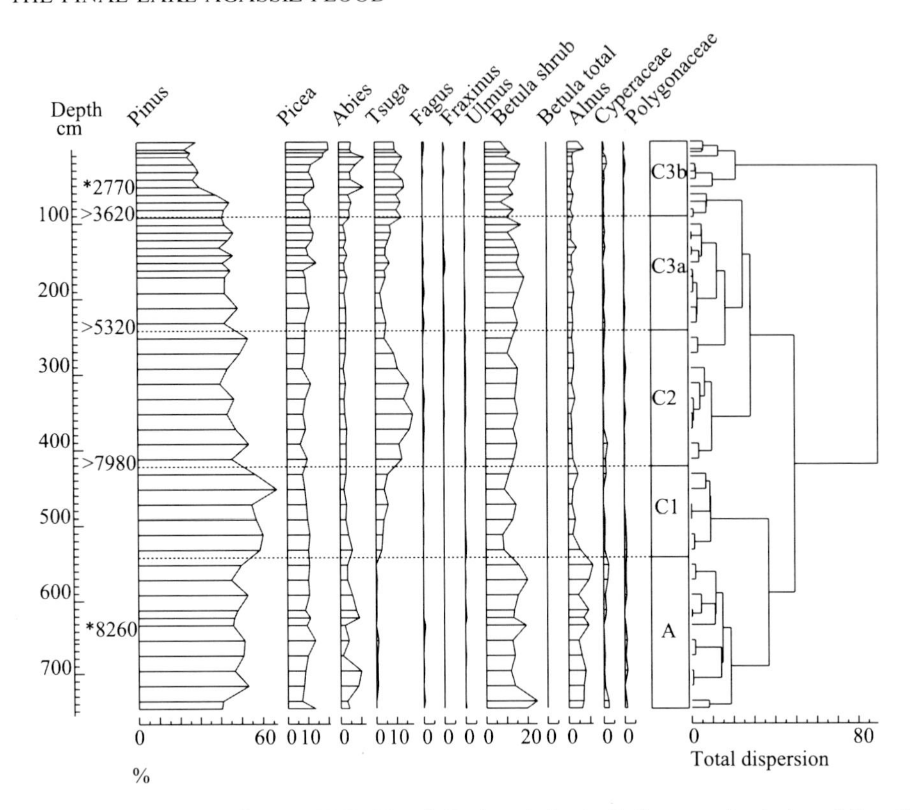

Figure 9. Simplified pollen diagram for core HU84011-12, St. Anne's Basin. Pollen zonation is that of Cape Breton. Ages on the left side are calibrated dates based on foraminifera (asterisks) and from Cape Breton pollen records with which core 12 was correlated (greater than symbols). Lines on the right are the results of the cluster analysis.

in *Betula* pollen and the peak in *Sphagnum* percentages between 143 and 280 cm have been correlated with zone CP3-ii, and the top of this zone is dated at 3.23 ka. The upper part of core 19 (0–150 cm) is characterized by decreasing percentages of *Betula* pollen and increasing percentages of *Picea* pollen, similar to zone CP4.

5.2.2. Pollen Zones in Core 12, St. Anne's Basin. Pollen assemblages from core 12 (Figure 9) are dominated by pollen of conifers (mainly *Pinus*, *Picea*, *Abies*, and *Tsuga*) and deciduous shrubs (*Betula* and *Alnus*) with some deciduous tree pollen (*Betula*, *Acer*, *Quercus*, *Fagus*, *Fraxinus*, and *Ulmus*) and herbs (Cyperaceae). Because of the similarity between the pollen assemblages from this core and those of onshore lake pollen records, the pollen zonation from Nova Scotia lakes [*Livingstone*, 1968] is used.

Pollen assemblage zone A (745–550 cm) is characterized by relatively low pollen concentrations, high percentages of *Pinus* pollen, and the maximum percentages of *Abies*, *Alnus*, and shrub *Betula* (*Betula* pollen grains smaller than 20 μm are considered to represent shrub varieties [*Dyer*, 1986]). *Salix* and herbs (Cyperaceae and Polygonaceae) are also relatively important.

Pollen assemblage zone C1 (540–430 cm) is defined by the maximum percentages of *Pinus*, the increase in *Tsuga*, and by decreased proportions of shrub (*Betula* and *Alnus*) and *Abies* pollen. It is also the zone where the proportion of total *Betula* is the lowest.

The peak of *Tsuga* pollen that defines pollen assemblage zone C2 (430–250 cm) suggests a warmer climate and has been traditionally used to define the hypsithermal in Nova Scotia [*Anderson*, 1985]. The beginning of the *Tsuga* peak is dated at 7.98 ka in Silver Lake [*Livingstone*, 1968; *Ogden*, 1987]. Percentages of tree *Betula* and deciduous hardwoods gradually increase in this zone, and *Fraxinus* first appears in the middle of the zone.

Pollen assemblage zone C3 (upper 250 cm of core 12) starts with the decline of *Tsuga* and is characterized by the highest percentages of tree *Betula* (also total *Betula*) and hardwoods. The decline of *Tsuga* is dated at 5.32 ka in Silver Lake [*Livingstone*, 1968; *Ogden*, 1987]. A similar decline was observed in pollen records from all over North America around circa 5.45 ka [*Davis*, 1979]. Zone C3b records the decline of *Pinus* accompanied by an increase of *Tsuga* and *Abies* percentages. The base of zone C3b at 100 cm is dated at 3.22 ka at Everitt Lake [*Green*, 1987].

5.3. Dinoflagellate Cyst Zonation

5.3.1. Dinoflagellate Cyst Zones in Core 19. The dinoflagellate cyst assemblage in core 19 zone 1 (500–900 cm) (Figure 10) is dominated by *Islandinium minutum*, with *Brigantedinium* spp. Other cysts (*Pentapharsodinium dalei*, *Operculodinium centrocarpum*, and *Nematosphaeropsis labyrinthus*) occur sporadically in small proportions.

I. minutum and *Brigantedinium* spp are abundant in arctic-subacrtic neritic environments (i.e., Newfoundland and Labrador shelves) [*Rochon et al.*, 1999] and in the Canadian Arctic [*Mudie et al.*, 2001; *de Vernal et al.*, 2001]. *P. dalei* is found in small percentages in assemblages of the eastern Canadian margin, including our study region, and it is found in high percentages only in the Gulf of St. Lawrence [*Rochon et al.*, 1999; *de Vernal et al.*, 2001]. This assemblage probably reflects cold postglacial sea surface conditions.

In dinoflagellate cyst assemblage zone 2 (400–500 cm), assemblages are dominated by both *I. minutum* and *Brigantedinium* spp, with 10%–20% of *Spiniferites elongatus*. Low percentages of *O. centrocarpum* are present at the top of the zone. At high latitudes, high percentages of *Brigantedinium* spp are typical of ice-edge environments [*Mudie et al.*, 2001]. Proportions of *S. elongatus* greater than 20% are only seen in Baffin Bay and in the Iceland and Norwegian Seas [*Rochon et al.*, 1999; *de Vernal et al.*, 2001]. The presence of *S. elongatus* indicates greater northern influence, and could signal an increase in meltwater flow southward as well as colder SST.

Assemblages are once again dominated by *I. minutum* in dinoflagellate cyst assemblage zone 3 (350–400 cm). Proportions of *Brigantedinium* spp decrease throughout the zone. *O. centrocarpum* and *P. dalei* are consistently present but in small proportions. *S. elongatus* is also present, but its proportion decreases upward. The appearance of *Impagidinium pallidum* and *Alexandrium tamarense* type cysts in this zone (in the DC layers) is noted.

I. pallidum is typically an Arctic taxon [*Rochon et al.*, 1999; *de Vernal et al.*, 2001] and today is not found south of the northern Labrador Shelf. Its presence suggests cold sea surface conditions as it is rare where temperatures are higher than 10°C. *A. tamarense* is a taxon responsible for harmful algal blooms, which are favored by increased water stratification [*Mudie et al.*, 2002]. Its presence could be an

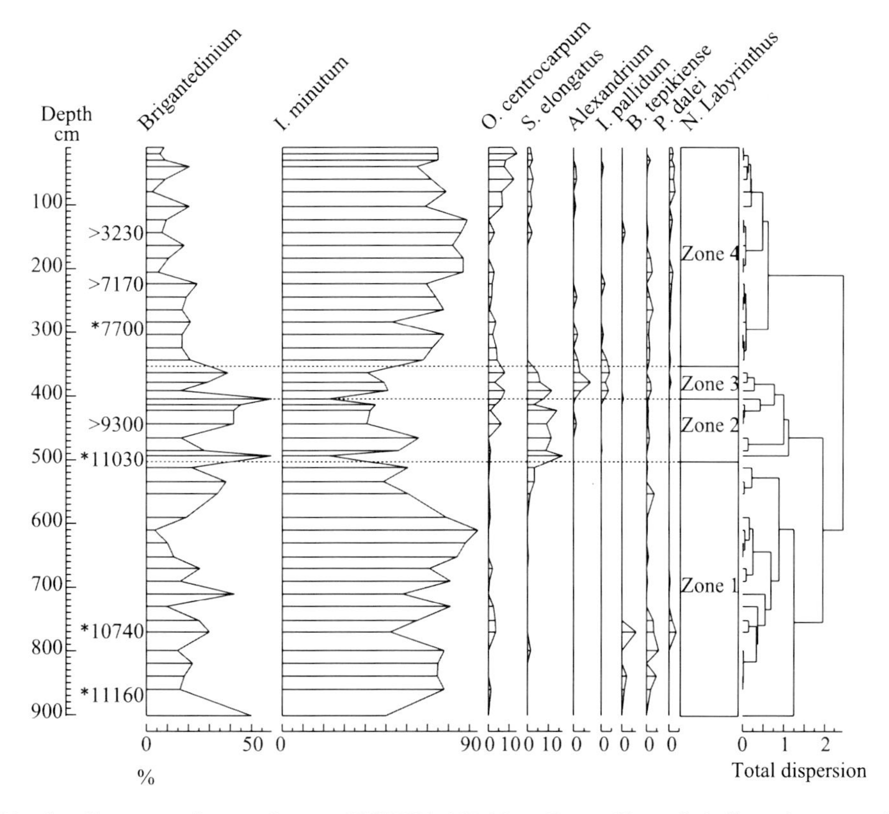

Figure 10. Dinoflagellate cysts diagram for core HU87033-19P, Notre Dame Channel. Calibrated ages are the same as in Figures 6 and 8. Lines on the right are the results of the cluster analysis.

indication of lower SSS. Increased nutrient inputs can also trigger dinoflagellate blooms, which suggest increased meltwater flow to the area during that interval. Similar increases in *A. tamarense* were observed in early Holocene assemblages from the Scotian Shelf [*Levac*, 2001].

The lower part of dinocyst zone 4 (200–350 cm) is dominated by *I. minutum* and *Brigantdinium* spp, with small proportions of *P. dalei*, *O. centrocarpum* and *N. labyrinthus*. The proportions of *A. tamarense*, and *I. pallidum* are notably lower, and *S. elongatus* is absent. High percentages of *I. minutum* indicate that sea surface conditions are still cold, but the near disappearance of *I. pallidum* and *A. tamarense* suggests that the meltwater influence was reduced.

In the upper part of dinocyst zone 4 (0–200 cm), the proportion of most minor taxa increases. *O. centrocarpum* reaches 10%–15%, *S. elongatus* reappears, and *N. labyrinthus* is present in every sample. This assemblage is similar to those of the neritic portion of the Northeast Newfoundland Shelf today [*Levac*, 2002].

5.3.2. Dinoflagellate Cysts in Core 12. Dinoflagellate cyst assemblages throughout core 12 (Figure 11) are dominated by *Brigantedinium* spp, *O. centrocarpum*, and *P. dalei*. Minor species are *I. minutum*, *Selenopemphix quanta*, *Bitectatodinium tepikiense*, *N. labyrinthus*, *Spiniferites* spp, *Ataxiodinium choanum*, and *A. tamarense* type cyst. Similar assemblages occur in Canadian Arctic channel core tops [*Mudie and Rochon*, 2001].

Dinoflagellate cyst assemblage zone SA1 (650–750 cm) is dominated by *O. centrocarpum* (50–60%), *P. dalei* and *Brigantedinium* spp, with small percentages of *Spiniferites* spp (mainly *S. ramosus* and *S. elongatus*).

Assemblages in dinocyst zone SA2 (500–650 cm) are markedly different from the rest of the core. This zone is defined by the maximum relative abundance of *N. labyrinthus* and of other minor species, including *B. tepikiense* and *S. quanta*. It is also characterized by a peak in the abundance of *I. minutum*. *S. quanta* is more abundant in the middle of zone SA2. This depth interval contains the Lake Agassiz flood deposits, and assemblages (especially the peak values of *I. minutum*) indicate lower SST. In the upper part of the zone, the presence of two warm indicators, *A. choanum* and *S. mirabilis*, as well as decreased proportions of *I. minutum*, indicate the cold episode had ended.

O. centrocarpum and *Brigantedinium* spp dominate the assemblages of dinocyst zone SA3 (180–500 cm). Among the accompanying taxa, *Spiniferites* spp (mainly *S. ramosus* and *S. elongatus*) and *P. dalei* are the most abundant. There is

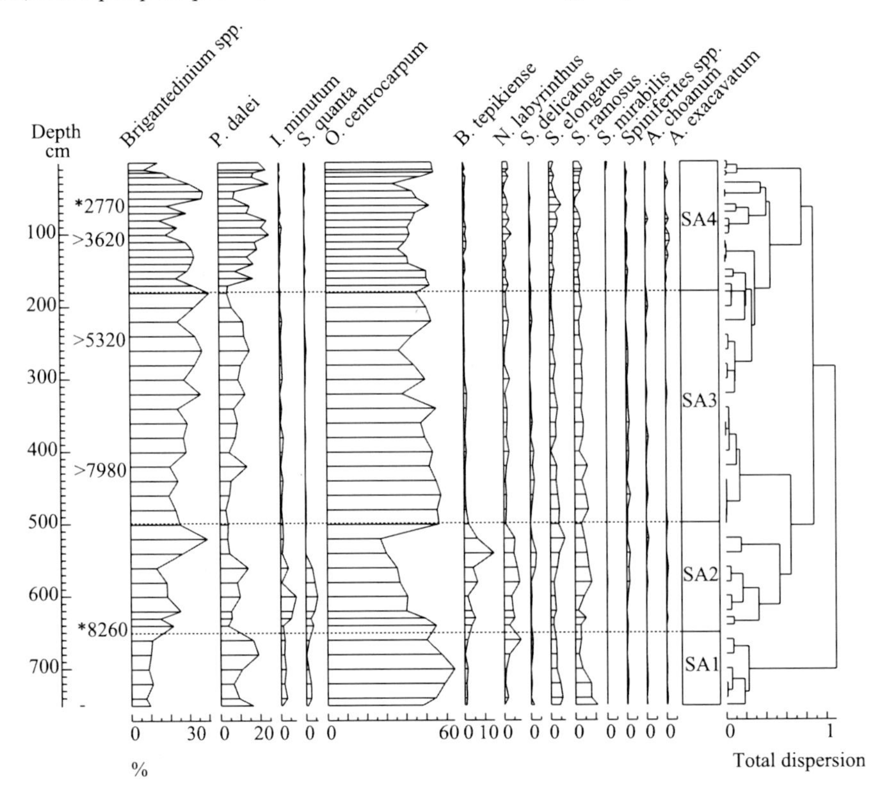

Figure 11. Dinoflagellate cysts diagram for core HU84011-12, St. Anne's Basin. Ages are the same as in Figures 7 and 9. Lines on the right are the results of the cluster analysis.

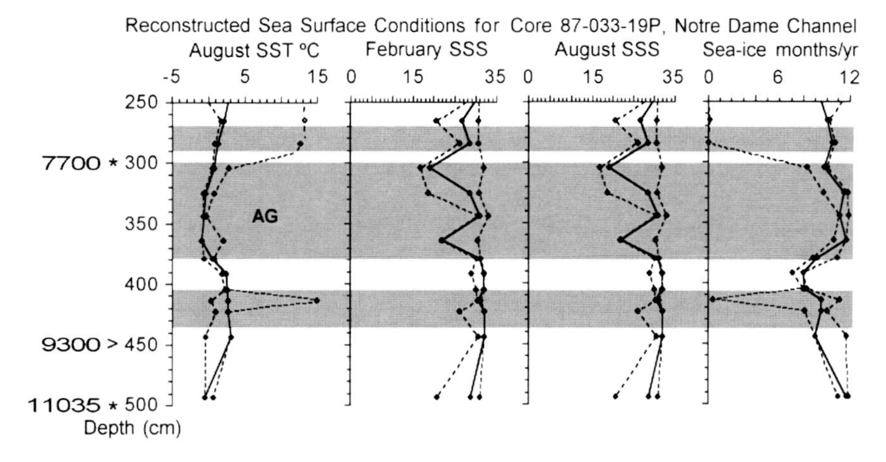

Figure 12. Reconstruction of sea surface conditions, core HU87033-19P, from Notre Dame Channel, Northeast Newfoundland Shelf. Calibrated ages are the same as in Figures 6, 8, and 10. The solid line represents the best estimate, while the space between the dashed lines is the confidence interval.

a sharp decrease in the proportion of *B. tepikiense* and *N. labyrinthus* relative to zone SA2. Maximum percentages of *S. mirabilis*, whose occurrence on the North American margin does not extend north of the Gulf of Maine today, occur at the base of the zone, suggesting warmer conditions than present.

Dinocyst zone SA4 0–180 cm (approximately the last 4000 years) is dominated by *O. centrocarpum*, *Brigantedinium* spp, and *P. dalei*, which reaches its maximum relative abundance in this zone. A peak in total concentrations between 80 and 100 cm (3.0–3.22 ka) coincides with lower percentages of *Brigantedinium* spp, higher percentages of *P. dalei*, and the occurrence of *A. choanum*. This zone is similar to the present-day assemblages in surface samples for the area [*Levac*, 2002].

5.4. Reconstructions of Sea Surface Conditions

Sea surface conditions were reconstructed for the core 19 250–500 cm interval (Figure 12), which records the Lake Agassiz outbursts. SSS shows a significant decrease to 25‰ within the DC layer corresponding to the drainage of Lake Agassiz. Two drops in SSS are centered around 300 and 365 cm. The seasonal duration of sea ice cover also shows

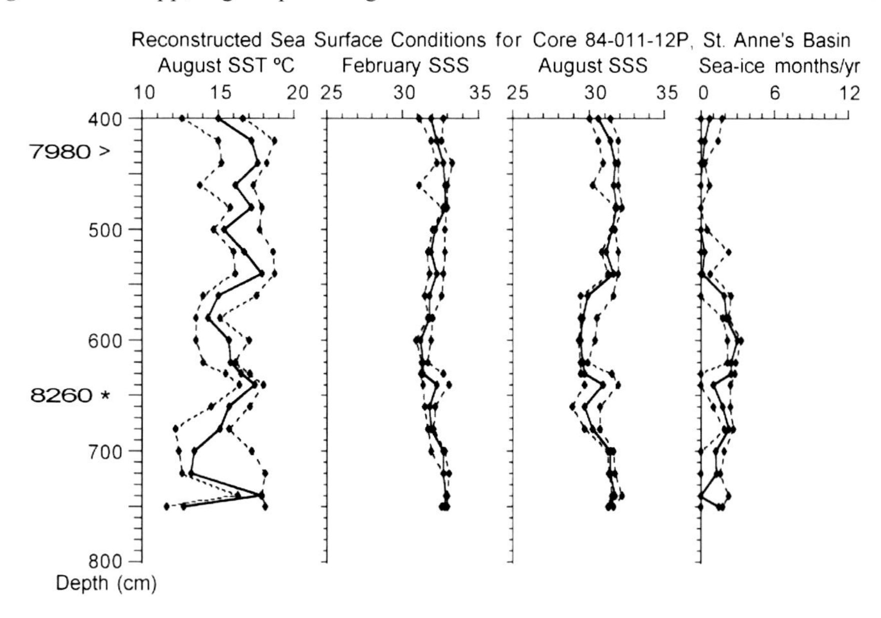

Figure 13. Reconstruction of sea surface conditions for core HU84011-12, from St. Anne's Basin, Northeast Scotian Shelf. Ages are the same as in Figures 7, 9, and 11.

an increase to 11 months year^{-1} within that interval. The impact of the meltwater flow on SST appears to be rather small with a 3°C difference.

Reconstructions for core 12 (Figure 13) show a drop in temperature between 680 and 720 cm, just before the level dated at 8.26 ka and another decline immediately after that (560–640 cm). These drops in SST are between 2°C and 3°C. August SSS also shows small drops of 1‰–2‰ in nearby intervals. The sea ice cover duration would have been longer by 1–2 months during the interval with lowered temperature and salinity.

6. DISCUSSION

The recognition of an AG-aged DC layer in core 19 [*Miller et al.*, 2006] and coincident changes in dinoflagellate cyst assemblages in both cores 12 and 19, indicate shifts to colder, fresher, surface waters. Combined with previously published paleoceanographic data from other sites in the area, there is strong evidence that the Lake Agassiz meltwater outburst transited via the Labrador Current [*Lewis and Miller*, 2005; *Lewis et al.*, 2009] and had a significant impact on sea surface conditions along the eastern Canadian margin.

6.1. Impact of Fresh Water on Dinoflagellate Cyst Assemblages

The significant changes in dinoflagellate cyst assemblages within DC layers of core 19, especially those occurring within the DC layer coeval with the Agassiz outburst, clearly indicate that the meltwater flow over Notre Dame Channel impacted sea surface conditions. Increases in the percent occurrences of Arctic taxa (e.g., *I. pallidum*) and of other taxa indicative of greater water stratification (in this case *A. tamarense*) correspond to an approximately 10‰ lowering of SSS. Farther south, on the Scotian Shelf, Arctic taxa such as *I. pallidum* are far less abundant, and cold conditions are indicated instead by high percentages of *I. minutum*. In core 12, a peak in *I. minutum* is centered at 640 cm, dated 8.26 ka, that corresponds to a 2°C–3°C drop in SST and a 1‰–2‰ drop in salinity, as determined by the transfer functions.

Changes in dinoflagellate cyst assemblages around the time of the 8.2 ka event, consistent with lower SSS and stronger stratification, are also observed in other palynological records on the southeast Canadian margin. Increased percentages of Arctic taxon *I. pallidum*, as seen in core 19, are recorded in core HU03033-03 on the Scotian Slope (E. Levac, unpublished data, 2009) and in Bay of Islands core MD99-2225 [*Levac*, 2003]. In La Have Basin (core HU95030-24 [*Levac*, 2001]), Emerald Basin (CSS Dawson core DA77002-20 [*Scott et al.*, 1984]), and Canso Basin (core DA80004-33 [*Scott et al.*, 1984]), maximum *I. minutum* values similar to those of core 12 at the time of the 8.2 ka event are observed. In Cabot Strait core 111 [*de Vernal et al.*, 1993], *I. minutum* is present, but adequate chronology is lacking, and it is not possible to correlate it to the Agassiz drainage. The change in dinocyst assemblages corresponds to a 1‰–2‰ drop in August SSS around 8.2 ka in La Have Basin and in Bay of Islands [*Levac*, 2001, 2003]. In La Have Basin, the Agassiz flood also marks the end of the Holocene optimum in SST, during which SST was at least 5°C higher than present [*Levac*, 2001]. This was possibly the result of a more northerly path of the Gulf Stream in the early Holocene [*Levac*, 2001] as also proposed by *Keigwin et al.* [2005].

6.2. Evidence From Foraminiferal and Isotopic Records

The meltwater drainage has been reported in other offshore planktic foraminiferal and δ^{18}O isotope records. Low isotopic values, coincident with the drainage of Lake Agassiz, are recorded in cores from the Laurentian Fan, Nova Scotia rise, and as far south as Cape Hatteras [*Keigwin and Jones*, 1995, *Keigwin et al.*, 2005]. Isotopic records from Cartwright Saddle on the Labrador Shelf show slight shifts that may indicate the AG meltwater pulse, but there is no evidence of it in records from the Labrador Sea [*Andrews et al.*, 1999; *Hillaire-Marcel et al.*, 2007]. Similar records are not available for the Scotian Shelf because planktic foraminifera are affected by dissolution in that region [*Scott et al.*, 1984].

On the Northeast Newfoundland Shelf, benthic foraminiferal evidence from core 07 indicates bottom water mass changes and increased stratification in outer Notre Dame Channel at the time of both the AG-aged meltwater influxes and at the time of deposition of the underlying DC bed [*Miller*, 1999; *Miller et al.*, 2006]. A unique fauna characterized by *Melonis barleeanum*, *Cassidulina carinata*, and *Pullenia* spp (*Pullenia osloensis*, *Pullenia quinqueloba*, and *Pullenia bulloides*) suddenly appear (and disappear) twice in intervals that correlate closely to the AG DC bed and the DC bed below. Enhanced flux of the Labrador Current moving south over the upper slope could have laterally overflowed the shelf-edge sill and carried the indigenous slope fauna with it into the deep shelf basin. With subsequent flux reduction, freshened waters were isolated in the basins, which then mixed with overlying waters and decreased shelf water salinities. The upper slope benthic fauna was replaced with an assemblage characterized by hyposaline-tolerant species, notably *Islandiella* spp (*Islandiella helenae* and *Islandiella norcrossi*) and increases in *Stainforthia* spp (*Stainforthia complexa*, *Stainforthia concave*, and *Stainforthia pauciloculata*) [*Miller*, 1999; *Miller et al.*, 2001]. Bottom water salinities gradually returned to normal ranges, and shelf faunas reappeared [*Miller*, 1999; *Miller et al.*, 2006].

On the Scotian Shelf, the impact of the meltwater plume might have been restricted to the upper part of the water column, as benthic foraminiferal evidence of the influx is absent in Emerald Basin (core DA77002-20 [*Scott et al.*, 1984]), Canso Basin (core DA80004-33 [*Scott et al.*, 1984]), and St. Anne's Basin (core 12 [*Freeman*, 1986]). There is no evidence of a freshwater excursion in benthic foraminiferal isotopic records from Emerald and Canso basins [*Scott et al.*, 1989].

6.3. Geographical Distribution of Paleoecological and Isotopic Evidence for Meltwater Presence

The Agassiz outburst event is recorded on the Northeast Newfoundland Shelf, Scotian Shelf, Laurentian Fan, and off Cape Hatteras, with a latitudinal decrease in the amplitude of the SSS changes associated with the floods. We suggest that the Agassiz fresh water traveled southward via the Labrador Current and gradually started to disperse into the NAC along the eastern margin of the Grand Banks. A smaller portion continued flowing south toward the Scotian Shelf, and along the Scotian margin, and some of this water may have mixed with the Gulf Stream. This is consistent with *Keigwin et al.* [2005] who proposed that the meltwater could have initially followed a narrow path along the coast and then dispersed into the shelf/upper slope water farther south.

While the DC proxy for floods out of Hudson Strait is distinctive and continuous along the Labrador and Newfoundland shelves and upper slopes [*Lewis et al.*, 2009], the event is not recorded by foraminiferal proxies in records closer to the meltwater source or in Hudson Strait and the Labrador Sea [*Andrews et al.*, 1999; *Keigwin et al.*, 2005; *Hillaire-Marcel et al.*, 2007]. The lack of a $\delta^{18}O$ signal in the Labrador Sea is not surprising if, as *Keigwin et al.* [2005] and *Hillaire-Marcel et al.* [2007] suggested, the freshwater transport was along the shelf with some overflow downslope, a conclusion also reached by *Lewis et al.* [2009] after analyses at additional sites along the margin. The water that would have been transported farther offshore would have dispersed quickly, too quickly to leave an isotopic signature. The absence of the signal in Hudson Strait and along the margin can be explained by the small difference between the isotopic composition of Lake Agassiz waters and the present-day freshwater discharge in the NW North Atlantic, a difference of only about 4‰ [*Hillaire-Marcel et al.*, 2007, 2008], which corresponds to less than a 0.1‰ shift in isotopic composition in *Neogloboquadrina pachyderma* within the salinity range of its habitat [*Hillaire-Marcel et al.*, 2007, 2008].

The discontinuous geographical distribution of micropaleontological evidence for a freshwater presence could stem from the monospecific nature of assemblages (dinoflagellate cysts as well as planktic foraminifera). Extreme conditions can lead to monospecific or low-diversity assemblages, making it difficult to interpret palaeoenvironmental changes. Another explanation is that because the Labrador Shelf region remained under a meltwater "regime" until the end of the melting [*Levac*, 2001, 2002], the Labrador Shelf was already affected by relatively frequent freshwater pulses, so that microfossil assemblages were relatively insensitive to yet another (even if much larger) meltwater event. Lowered SSS characterized the North Atlantic margins in the early Holocene [*de Vernal and Hillaire-Marcel*, 2000]. Analysis of early Holocene summer-melt layers in ice caps from the Canadian Arctic [*Koerner and Fisher*, 1990] also suggest that pulses of meltwater from the Arctic might have continued to affect the Labrador Shelf episodically in the early Holocene. This idea is supported by the fact that dinoflagellate cyst assemblages similar to the modern ones are found in most cores from the eastern Canadian margin sometime after about 7 ka, once the influence from the melting ice sheet had decreased sufficiently [*Levac*, 2001, 2002; *Levac et al.*, 2001].

7. CONCLUSIONS

A DC layer in Notre Dame Channel core 19 (the Northeast Newfoundland Shelf) has been radiocarbon dated to the age of the Lake Agassiz final drainage [*Lewis et al.*, 2009] and contains dinoflagellate cyst assemblages indicative of lower SST, SSS, and of higher water stratification. Reconstructions reveal two pulses of lower SSS within the Agassiz DC bed (Figure 12). Core 12 from St. Anne's Basin (Scotian Shelf) records a smaller-amplitude drop in SSS at that same time (Figure 13). The event is also recorded at other sites on the Scotian Shelf (e.g., La Have Basin, Emerald Basin) [*Levac*, 2001, 2003; *Scott et al.*, 1984].

Based on this new evidence, we suggest that the Lake Agassiz meltwater flux traveled southward over the Northeast Newfoundland Shelf, and some was dispersed in the NAC along the eastern Grand Banks slope. Some water continued to flow over the Scotian Shelf/upper slope as evidenced from the signal found in the Laurentian Fan [*Keigwin et al.*, 2005] and on the Scotian Shelf [*Levac*, 2001; this chapter]. Some of this water may have then dispersed into the Gulf Stream. The remainder flowed south, inshore of the Gulf Stream and was recorded off Cape Hatteras [*Keigwin et al.*, 2005].

The absence of micropaleontological evidence for the event in cores from the northern Labrador Shelf could be explained by the low diversity of the microfossil assemblages that characterized the early Holocene and the fact that these assemblages, being closer to the meltwater sources, were already reflecting baseline and episodic, even though smaller in volume, freshwater pulses from the melting ice sheet. These assemblages then were insensitive to additional meltwater and could not resolve the Lake Agassiz flood at 8.33 ka, whereas more diverse dinoflagellate cyst assemblages from sites farther south, where the meltwater pulse was less pronounced, were able to do so.

Acknowledgments. E. L. thanks A. de Vernal for use of the GEOTOP facilities, Maryse Henri, research assistants at Bishop's, and T. Neulieb for preparing illustrations. We acknowledge and thank the scientific personnel, officers, and crew of the CCGS *Hudson*, the staff at GSC-A (I. Hardy and K. Jarrett) for core collection and curation, I. Girard (GSC-Ottawa) for the Chittick analyses, and P. Mudie and D. Piper (GSC-A) and two anonymous reviewers for insightful reviews of the manuscript. We would like to honor Gus Vilks (deceased), Chief Scientist of cruise 87-033, who collected core 19 at the specific request of A. M. Funding from NSERC (discovery grant and CHSD scholarship to E. L.), SRC grants from Bishop's University and support of the Natural Resources Canada Earth Science Sector Offshore Geoscience program are acknowledged. This is contribution 20100120 of the Earth Sciences Sector of Natural Resources Canada.

REFERENCES

Alley, R. B., and A. M. Agustsdöttir (2005), The 8 k event: Cause and consequences of a major Holocene abrupt climate change, *Quat. Sci. Rev.*, *24*(10–11), 1123–1149, doi:10.1016/j.quascirev.2004.12.004.

Alley, R. B., P. A. Mayewski, T. Sowers, M. Stuiver, K. C. Taylor, and P. U. Clark (1997), Holocene climatic instability: A prominent widespread event 8200 years ago, *Geology*, *25*, 483–486.

Anderson, T. W. (1985), Late Quaternary pollen records from Eastern Ontario, Quebec, and Atlantic Canada, in *Pollen Records of Late-Quaternary North American Sediments*, edited by V. M. Bryant and R. G. Holloway, pp. 281–326, AASP Found., Tulsa, Okla.

Andrews, J. T., and K. Tedesco (1992), Detrital carbonate-rich sediments, northwestern Labrador Sea: Implications for ice-sheet dynamics and iceberg rafting (Heinrich) events in the North Atlantic, *Geology*, *20*, 1087–1090.

Andrews, J. T., L. Keigwin, F. Hall, and A. E. Jennings (1999), Abrupt deglaciation events and Holocene palaeoceanography from high-resolution cores, Cartwright Saddle, Labrador Shelf, Canada, *J. Quat. Sci.*, *14*, 383–397.

Barber, D. C., et al. (1999), Forcing of the cold event of 8200 years ago by catastrophic drainage of Laurentide lakes, *Nature*, *400*, 344–348.

Bard, E., M. Arnold, J. Mangerud, M. Paterne, L. Labeyrie, J. Duprat, J. Sønstegaard, and J. P. Duplessy (1994), The North Atlantic atmosphere-sea surface ^{14}C gradient during the Younger Dryas climatic event, *Earth Planet. Sci. Lett.*, *126*, 275–287.

Batten, D. J. (1996), Chapter 20E: Upper Jurassic and Cretaceous miospores, in *Palynology: Principles and Applications*, edited by J. Jansonius and D. C. McGregor, pp. 807–830, AASP Found., Tulsa, Okla.

Bennett, K. D. (1996), Determination of the number of zones in a biostratigraphical sequence, *New Phytol.*, *132*, 155–170.

Burden, E., and D. Holloway (1985), Palynology and age of the Scott Inlet Inliers of Baffin Island (Northwest Territories), *Can. J. Earth Sci.*, *22*, 1542–1545.

Clarke, G. K. C., D. W. Leverington, J. T. Teller, and A. S. Dyke (2004), Paleohydraulics of the last outburst flood from glacial Lake Agassiz and the 8200 BP cold event, *Quat. Sci. Rev.*, *23*, 389–407, doi:10.1016/j.quascirev.2003.06.004.

Dale, C. T., R. T. Haworth (1979), High resolution reflection seismology studies on late Quaternary sediments of the northeast Newfoundland continental shelf, *Curr. Res. Geol. Surv. Can., Pap. 79-1B*, 57–364.

Dansgaard, W. (1993), Evidence for general instability of past climate from a 250 kyr ice core record, *Nature*, *364*, 218–220.

Davis, M. B. (1979), The Quaternary record of forest pathogens, in *Proceedings of the 10th Annual Meeting, Palynology*, vol. *3*, p. 283, AASP Found., Tulsa, Okla.

Davis, R. B., and T. Webb (1975), The contemporary distribution of pollen in eastern North America: A comparison with the vegetation, *Quat. Res.*, *5*, 395–434.

de Vernal, A., and C. Hillaire-Marcel (2000), Sea-ice cover, sea-surface salinity, and halo -/thermohaline structure of the northwest Atlantic: Modern versus full glacial conditions, *Quat. Sci. Rev.*, *19*, 65–85.

de Vernal, A., and C. Hillaire-Marcel (2006), Provincialism in trends and high frequency changes in the northwest North Atlantic during the Holocene, *Global Planet. Change*, *54*, 263–290, doi:10.1016/j.gloplacha.2006.06.023.

de Vernal, A., J.-L. Turon, and J. Guiot (1993), Late and postglacial paleoenvironments of the Gulf of St. Lawrence: Marine and terrestrial palynological evidence, *Geogr. Phys. Quat.*, *47*, 167–180.

de Vernal, A., et al. (2001), Dinoflagellate cyst assemblages as tracers of sea-surface conditions in the northern North Atlantic, Arctic and sub-Arctic seas: The new 'n = 677' data base and its application for quantitative paleoceanographic reconstructions, *J. Quat. Sci.*, *16*, 681–698.

Dreimanis, A. (1962), Quantitative gasometric determinations of calcite and dolomite by using Chittick apparatus, *J. Sediment. Res.*, *32*, 520–529.

Drinkwater, K. F., R. G. Pettipas, G. L. Bugden, and P. Languille (1999), Climatic variations in northern North America (6000 BP to present) reconstructed from pollen and tree-ring data, *Arct. Antarct. Alp. Res.*, *21*, 45–59.

Dyer, A. K. (1986), A palynological investigation of the late Quaternary vegetational history of the Baie Verte Peninsula, north-central Newfoundland, M.S. thesis, Dep. of Geogr., Mem. Univ., St. John's, Newfoundland, Canada.

Dyke, A. S., A. Moore, and L. Robertson (2003), Deglaciation of North America, *Open File Rep. 1574*, Geo. Surv. of Can., Ottawa, Ont.

Ellison, C. R. W., M. R. Chapman, and I. R. Hall (2006), Surface and deep ocean interactions during the cold climate event 8200 years ago, *Science*, *312*(5782), 1929–1932, doi:10.1126/science.1127213.

Ecoregions Working Group (1989), *Ecoclimatic Regions of Canada*, *Ecol. Land Classif. Ser.*, vol. 23, Can. Wildl. Serv., Environ. Canada, Ottawa, Ont.

Freeman, J. M. (1986), Paleoceanographic trends on the Northern Scotian Shelf, honours thesis, Dep. of Earth Sci., Dalhousie Univ., Halifax, N. S., Canada.

Grant, A. C., and K. D. McAlpine (1990), The continental margin around Newfoundland, in *Geology of Canada: Geology of the Continental Margin of Eastern Canada*, edited by M. J. Keen and G. L. Williams, chap. 6, pp. 241– 292, Geol. Surv. of Can., Ottawa, Ont.

Green, D. G. (1987), Pollen evidence for the postglacial origin of Nova Scotia's forest, *Can. J. Bot.*, *65*, 1163–1179.

Greenberg, D. A., and B. D. Petrie (1988), The mean barotropic circulation on the Newfoundland Shelf and Slope, *J. Geophys. Res.*, *93*, 15,541–15,550.

Guiot, J., and A. de Vernal (2007), Transfer functions: Methods for quantitative paleoceanography based on microfossils, in *Proxies in Late Cenozoic Paleoceanography*, edited by C. Hillaire-Marcel and A. de Vernal, pp. 523–563, Elsevier, Amsterdam, Netherlands.

Hall, I. R., G. Bianchi, and J. R. Evans (2004), Centennial to millennial scale Holocene climate-deep water linkage in the North Atlantic, *Quat. Sci. Rev.*, *23*, 1529–1536, doi:10.1016/j.quascirev.2004.04.004.

Han, G., J. H. Loder, and P. C. Smith (1999), Seasonal-mean hydrography and circulation in the Gulf of St. Lawrence and on the eastern Scotian and southern Newfoundland shelves, *J. Phys. Oceanogr.*, *29*, 1279–1301.

Hillaire-Marcel, C., A. de Vernal, G. Bilodeau, and G. Wu (1994), Isotope stratigraphy, sedimentation rates, deep circulation, and carbonate events in the Labrador Sea during the last 200 ka, *Can. J. Earth Sci.*, *31*, 63–89.

Hillaire-Marcel, C., A. de Vernal, and D. J. W. Piper (2007), Lake Agassiz final drainage event in the northwest North Atlantic, *Geophys. Res. Lett.*, *34*, L15601, doi:10.1029/2007GL030396.

Hillaire-Marcel, C., J.-F. Helie, J. McKay, and A. de Vernal (2008), Elusive isotopic properties of deglacial meltwater spikes into the North Atlantic: Example of the final drainage of Lake Agassiz, *Can. J. Earth Sci.*, *45*, 1235–1242, doi:10.1139/E08-029.

Johnsen, S. J., H. B. Clausen, W. Dansgaard, K. Fuhrer, N. Gundestrup, C. U. Hammer, P. Iversen, J. Jouzel, B. Stauffer, and J. P. Steffensen (1992), Irregular glacial interstadials recorded in a new Greenland ice core, *Nature*, *359*, 311–313.

Josenhans, H. W., J. Zevenhuizen, and R. A. Klassen (1986), The Quaternary geology of the Labrador Shelf, *Can. J. Earth Sci.*, *23*, 1190–1213.

Keigwin, L. D., and G. A. Jones (1995), The marine record of deglaciation from the continental margin off Nova Scotia, *Paleoceanography*, *10*, 973–985.

Keigwin, L. D., J. P. Sachs, Y. Rosenthal, and E. A. Boyle (2005), The 8200 year event in the slope water system, western subpolar North Atlantic, *Paleoceanography*, *20*, PA2003, doi:10.1029/2004PA001074.

Kerwin, M. W. (1996), A stratigraphic isochron (ca. 8000 ^{14}C yr B.P.) from final deglaciation of Hudson Strait, *Quat. Res.*, *46*, 89–98.

Kleiven, H. F., C. Kissel, C. Laj, U. S. Ninnemann, T. O. Richter, and E. Cortijo (2007), Reduced North Atlantic deep water coeval with the glacial Lake Agassiz freshwater outburst, *Science*, *319*, 60–64, doi:10.1126/science.1148924.

Koerner, R. M., and D. A. Fisher (1990), A record of Holocene summer climate from a Canadian high-Arctic ice core, *Nature*, *343*, 630–631.

Lajeunesse, P., and G. St-Onge (2008), The subglacial origin of the Lake Agassiz–Ojibway final outburst flood, *Nat. Geosci.*, *1*, 184–188, doi:10.1038/ngeo130.

Levac, E. (2001), High resolution Holocene palynological records from the Scotian Shelf, *Mar. Micropaleontol.*, *43*, 179–197.

Levac, E. (2002), High resolution palynological records from Atlantic Canada: Regional Holocene paleoceanographic and paleoclimatic history, Ph.D. thesis, Dep. of Earth Sci., Dalhousie Univ., Halifax, N. S., Canada.

Levac, E. (2003), Palynological records from Bay of Islands, Newfoundland: Direct correlation of Holocene paleoceanographic and climatic changes, *Palynology*, *27*, 135–154.

Levac, E., A. de Vernal, and W. Blake Jr. (2001), Holocene palynology of cores from the North Water Polynya, Baffin Bay, *J. Quat. Sci.*, *16*, 353–363.

Levitus, S. (1982), Climatological atlas of the World Ocean, *NOAA Prof. Pap. 13*, U.S. Gov. Print. Off., Washington, D. C.

Levitus, S., and T. Boyer (1994), *National Oceanography Data Center World Ocean Atlas 1994*, vol. 4, *Temperature*, *NOAA Atlas NESDIS*, vol. 2, NOAA, Silver Spring, Md.

Lewis, C. F. M., and A. A. L. Miller (2005), Glacial dam failure and oceanic routing of the final Lake Agassiz-Ojibway outburst flood, 7.6–7.7 ka: New evidence from the Northeast Newfoundland Shelf, Abstract, paper presented at 30th Joint Annual Meeting, Geol. Assoc. of Can. and Mineral. Assoc. of Can., Halifax, N. S.

Lewis, C. F. M., A. A. L. Miller, E. Levac, D. J. W. Piper, and G. V. Sonnichsen (2009), Tracking Agassiz floodwaters beyond Hudson Strait: Correlation to the onset of the 8.2 cal ka cold event, *Eos Trans. AGU*, *90*(22), Jt. Assem. Suppl, Abstract GA23A-02.

Livingstone, D. A. (1968), Some interstadial and postglacial pollen diagrams from eastern Canada, *Ecol. Monogr.*, *38*, 87–125.

Loder, J. W., G. Han, C. G. Hannah, D. A. Greenberg, and P. C. Smith (1997), Hydrography and baroclinic circulation in the Scotian Shelf region: Winter versus summer, *Can. J. Fish. Aquat. Sci.*, *54*, 40–56.

Loder, J. W., B. Petrie, and G. Gawarkiewicz (1998), The coastal ocean over northeastern North America: A large scale review, in *The Sea*, vol. 10, in *The Global Coastal Ocean: Processes and Methods*, edited by A. R. Robinson and K. H. Brink, pp. 105–133, John Wiley, Hoboken, N. J.

MacLean, B., G. Vilks, I. Hardy, B. Deonarine, A. E. Jennings, and W. F. Manley (2001), Quaternary sediments in Hudson Strait and Ungava Bay, in *Marine Geology of Hudson Strait and Ungava Bay, Eastern Canadian Arctic: Late Quaternary Sediments, Depositional Environments, and Late Glacial-Deglacial History Derived From Marine and Terrestrial Studies*, edited by B. MacLean, *Bull.Geol. Surv. Can.*, *566*, 71–125.

Macpherson, J. B. (1995), A 6 ka BP reconstruction for the Island of Newfoundland from a synthesis of Holocene lake-sediment pollen records, *Geogr. Phys. Quat.*, *49*, 163–182.

Mangerud, J., S. Bondevik, S. Gulliksen, A. K. Hufthammer, and T. Høisaeter (2006), Marine ^{14}C reservoir ages for 19th century whales and molluscs from the North Atlantic, *Quat. Sci. Rev.*, *25*, 3228–3245, doi:10.1016/j.quascirev.2006.03.010.

Markham, W. E. (1988), *Ice Atlas: Hudson Bay and Approaches*, Environ. Can., Ottawa, Ont.

Marret, M., and K. A. F. Zonneveld (2003), Atlas of modern organic-walled dinoflagellate cyst distribution, *Rev. Palaeobot. Palynol.*, *125*, 1–200, doi:10.1016/S0034-6667(02)00229-4.

McNeely, R., A. S. Dyke, and J. R. Southon (2006), Canadian marine reservoir ages, preliminary data assessment, *Open File Rep. 5049*, Geol. Surv. of Can., Ottawa, Ont.

Miller, A. A. L. (1999), The Quaternary sediments and seismostratigraphy of the Grand Banks of Newfoundland and the Northeast Newfoundland Shelf: Foraminiferal refinements and constraints, Ph.D. dissertation, Dep. of Geol., George Washington Univ., Washington, D. C.

Miller, A. A. L., G. J. B. Fader, and K. Moran (2001), Late Wisconsinan ice advances, ice extent, and glacial regimes interpreted from seismic data, sediment physical properties, and foramininfera: Halibut Channel, Grand Banks of Newfoundland, in *Deglacial History and Relative Sea-Level Changes, Northeastern New England and Adjacent Atlantic Canada*, edited by T. K. Weddle and M. J. Ratelle, *Spec. Pap.Geol. Soc. Am.*, *351*, 51–107.

Miller, A. A. L., C. F. M. Lewis, E. Levac, J. B. Macpherson, and I. S. Spooner (2006), The post-glacial evolution of the Labrador Current: Micropaleontological evidence of outburst floods, and climate changes in Atlantic Canada, paper presented at 31st Joint Annual Meeting, Geol. Assoc. of Can. and Mineral. Assoc. of Can., Montreal, Quebec.

Mudie, P. J. (1982), Pollen distribution in recent marine sediments, eastern Canada, *Can. J. Earth Sci.*, *19*, 729–747.

Mudie, P. J., and F. M. G. McCarthy (1994), Late Quaternary pollen transport processes, western North Atlantic: Data from box models, cross-margin and N-S transects, *Mar. Geol.*, *118*, 79–105.

Mudie, P. J., and A. Rochon (2001), Distribution of dinoflagellate cyst in the Canadian Arctic marine region, *J. Quat. Sci.*, *16*, 603–620.

Mudie, P. J., and S. K. Short (1985), Marine palynology of Baffin Bay, in *Studies of Baffin Island, West Greenland and Baffin Bay*, edited by J. T. Andrews, pp. 263–308, Allen and Unwin, London, U. K.

Mudie, P. J., R. Harland, J. Matthiessen, and A. de Vernal (2001), Marine dinoflagellate cysts and high latitude Quaternary paleoenvironmental reconstructions: An introduction, *J. Quat. Sci.*, *16*, 595–602.

Mudie, P. J., A. Rochon, and E. Levac (2002), Palynological records of red tides in Canada: Past trends and implications for the future, *Palaeogeogr. Palaeoclimatol. Palaeoecol.*, *180*, 159–186.

Ogden, J. G. (1987), Vegetational and climatic history of Nova Scotia: Radiocarbon-dated pollen profiles from Halifax, Nova Scotia, *Can. J. Bot.*, *65*, 1482–1490.

Oppo, D. W., J. F. McManus, and J. L. Cullen (2003), Palaeo-oceanography: Deepwater variability in the Holocene epoch, *Nature*, *422*, 277–278, doi:10.1038/422277b.

Petrie, B., K. Drinkwater, D. Gregory, R. Pettipas, and A. Sandstrom (1996), Temperature and salinity Atlas for the Scotian Shelf and the Gulf of Maine, *Can. Tech. Rep. Hydrogr. Ocean Sci. 171*, Dep. Fish. and Oceans, Dartmouth, N. S., Canada.

Piper, D. J. W., P. J. Mudie, A. E. Aksu, and P. R. Hill (1978), Late Quaternary sedimentation, 50°N, North-east Newfoundland Shelf, *Geogr. Phys. Quat.*, *22*, 321–332.

Piper, D. J. W., P. J. Mudie, G. B. J. Fader, H. W. Josenhans, B. MacLean, and G. Vilks (1990), Quaternary geology, in *Geology of the Continental Margin of Eastern Canada*, *Geol. Can.*, vol. 2, edited by M. J. Keen and G. L. Williams, pp. 475–607, Geol. Surv. of Can., Ottawa, Ont.

Reimer, P. J., et al. (2009), IntCal09 and Marine09 radiocarbon age calibration curves, 0–50,000 years cal BP, *Radiocarbon*, *51*, 1111–1150.

Rochon, A., A. de Vernal, J.-L. Turron, J. Matthiessen, and M. J. Head (1999), *Distribution of Recent Dinoflagellate Cysts in Surface Sediment From the North Atlantic Ocean and Adjacent Seas in Relation to Sea-Surface Parameters*, *Contrib. Ser.*, vol. 35, AASP Found., Tulsa, Okla.

Rohling, E. J., and H. Pälike (2005), Centennial-scale climate cooling with a sudden cold event around 8,200 years ago, *Nature*, *434*, 975–979.

Rowe, J. S. (1977), Forest regions of Canada, *Can. For. Serv. Publ. 1300*, Dep. of Fish. and the Environ., Ottawa, Ont., Canada.

Scott, D. B., P. J. Mudie, G. Vilks, and D. C. Younger (1984), Latest Pleistocene-Holocene paleoceanographic trends on the continental margin of eastern Canada: Foraminiferal, dinoflagellate and pollen evidence, *Mar. Micropaleontol.*, *9*, 181–218.

Scott, D. B., V. Baki, and C. D. Younger (1989), Late Pleistocene-Holocene paleoceanographic changes on the eastern Canadian margin: Stable isotopic evidence, *Palaeogeogr. Palaeoclimatol. Palaeoecol.*, *74*, 279–295.

Shore, J. A., C. G. Hannah, and J. W. Loder (2000), Drift pathways on the western Scotian Shelf and its environs, *Can. J. Fish. Aquat. Sci.*, *57*, 2488–2505.

Smith, L. M., and K. J. Licht (2000), Radiocarbon date list IX: Antarctica, Arctic Ocean and the northern North Atlantic, *Occas. Pap. 54*, Inst. of Arct. and Alp. Res., Univ. of Colo., Boulder.

Stockmarr, J. (1971), Tablets with spores used in absolute pollen analysis, *Pollen Spores*, *13*, 615–621.

Stover, L. E., H. Brinkhaus, L. De Verteuil, R. J. Helby, E. Monteil, A. D. Partridge, A. J. Powell, J. B Riding, M. Smelmor, and G. L. Williams (1996), Mesozoic-Tertiary dinoflagellates, acritarchs and prasinophytes, in *Palynology: Principles and Applications*, edited by J. Jansonius and D. C. McGregor, chap. 19, pp. 641– 750, AASP Found., Tulsa, Okla.

Stuiver, M., and T. F. Braziunas (1993), Modeling atmospheric ^{14}C influences and ^{14}C ages of marine samples to 10,000 BC, *Radiocarbon*, *35*, 137–189.

Stuiver, M., and P. J. Reimer (1993), Extended ^{14}C database and revised CALIB radiocarbon calibration program, *Radiocarbon*, *35*, 315–330.

Taoufik, R., and A. deVernal (2008), Dinocysts as tracers of hydrographical conditions and productivity along the ocean margins, *Mar. Micropaleontol.*, *68*, 84–114.

Teller, J. T., D. W. Leverington, and J. D. Mann (2002), Freshwater outbursts to the oceans from glacial Lake Agassiz and their role in climate change during the last deglaciation, *Quat. Sci. Rev.*, *21*, 879–887.

Thomas, E. R., E. W. Wolff, R. Mulvaney, J. P. Steffensen, S. J. Johnsen, C. Arrowsmith, J. W. C. White, B. Vaughn, and T. Popp (2007), The 8.2 ka event from Greenland ice cores, *Quat. Sci. Rev.*, *26*, 71–80.

Tripsanas, E. K., and D. J. W. Piper (2008), Late Quaternary stratigraphy and sedimentology of Orphan Basin: Implications for meltwater dispersal in the southern Labrador Sea, *Palaeogeogr. Palaeoclimatol. Palaeoecol.*, *260*, 521–539.

Vilks, G., and C. Powell (1987), Cruise report and preliminary results, CSS Hudson 87-033, September 18–October 7 1987, *Open File Rep. 1702*, Geol. Surv. of Can., Ottawa, Ont.

Wiersma, A. P., and H. Renssen (2006), Model-data comparison for the 8.2 ka BP event: Confirmation of a forcing mechanism by catastrophic drainage of Laurentide lakes, *Quat. Sci. Rev.*, *25*, 63–83.

E. Levac, Environmental Studies and Geography, Bishop's University, 2600 College, Sherbrooke, QC J1M 1Z7, Canada. (elevac@ubishops.ca)

C. F. M. Lewis, Geological Survey of Canada-Atlantic, Natural Resources Canada, Box 1006, Dartmouth, NS B2Y 4A2, Canada.

A. A. L. Miller, marine g.e.o.s., PO Box 2253, Wolfville, NS B4P 2N5, Canada.

The 1500 Year Quasiperiodicity During the Holocene

A. Ruzmaikin and J. Feynman

Jet Propulsion Laboratory, California Institute of Technology, Pasadena, California, USA

We investigate the quasiperiodic 1500 year oscillation in climate during the Holocene and its relation to the Atlantic meridional overturning circulation (AMOC). Our data analyses of paleodata reveal the nonlinear nature of this oscillation. The data analysis also indicates that the AMOC has two major equilibrium states characterized by a weak and a strong meridional flow in accord with modeling. Based on the results of the data analysis and earlier ideas about the origin of the 1500 year oscillation, we suggest a conceptual model explaining its origin. The model includes two basic equilibrium states. Transitions between the states are driven by noise and the combined action of the ocean centennial variability and the 90 year solar variability. We briefly discuss a relationship of the 1500 year millennium oscillation to the global warming problem.

1. INTRODUCTION

Although the Holocene climate is more stable than climates of preceding glacial-interglacial periods, it displays a full palette of high- and low-frequency variations. Identifying the timing, spatial distribution, and origin of these variations is critical for understanding the mechanisms of climate change. One of the prominent low-frequency variations is the quasiperiodic 1500 year oscillation originally found in ice-rafted debris fluctuations in North Atlantic, which have the quasiperiod 1470 ± 500 years thus amounting to eight periods over the Holocene [*Bond et al.*, 1997]. The extrema of this oscillation are often referred to as the "Bond events" and are considered to be related to the glacial Dansgaard-Oeschger events.

The origin of the 1500 year oscillation, which is even more prominent in the glacial time than in the Holocene, remains debatable. *Bond et al.* [1997] suggested that it is driven by the solar forcing of the same periodicity. However, the proxies for the solar irradiance variations (C^{14} and B^{10}), as well as the sunspot number reconstructed over 11,000 years, do not show this periodicity at the time when it is clearly seen in the North Atlantic paleodata [*Debret et al.*, 2007; *Dima and Lohmann*, 2009]. The oceanic forcing is associated with the Atlantic meridional overturning circulation (AMOC), which transports the salty warm water to high latitudes where it cools, sinks, and returns southward. A combined solar-ocean forcing has been considered in attempts to explain the origin of the 1500 year oscillation in the glacial climate. Thus, *Braun et al.* [2005] showed that an intermediate-complexity model could simulate an apparent 1500 year quasiperiodicity in the glacial climate when forced by periodic inputs of fresh water into the North Atlantic. As this input, they selected the superposition of two solar signals: 90 and 210 year cycles (Gleissberg and de Vries cycles [*Feynman and Fougere*, 1984; *Damon and Sonnet*, 1991; *Peristykh and Damon*, 2003; *Ruzmaikin et al.*, 2006]), which are rational fractions (1/17 and 1/7) of the 1500 year period. This mechanism does not involve the interaction of solar and ocean variabilities. *Dima and Lohmann* [2009] put forward a concept that the 1500 year variation arises due to a threshold response of the AMOC

Abrupt Climate Change: Mechanisms, Patterns, and Impacts
Geophysical Monograph Series 193

Published in 2011 by the American Geophysical Union.
10.1029/2010GM001024

to the low-frequency (>550 years) solar forcing represented by the long-term reconstruction of sunspot number [*Solanki et al.*, 2004].

The model experiments by *Schulz et al.* [2007] show that the AMOC has two states corresponding to a strong and a weak return flow (Figure 1). These experiments take into account that the surface AMOC flow in the North Atlantic consists of the current into the Nordic Seas passing between the subpolar and subtropical gyres from which it draws water. When a continuous freshwater input to the Labrador Sea exceeds a small threshold, the deep convection is suppressed leading to a reduced circulation (the weak state) in the subpolar gyre. The deepwater formation in the Nordic Seas, which is always active, provides a negative feedback returning the system to the strong state. The duration between the transitions from one state to another is found to be random with a mean value close to 1500 years but with a very large standard deviation [*Schulz et al.* 2007]. Follow-up modeling [*Jongma et al.*, 2007] indicates that the transitions are susceptible to a weak periodic solar irradiance forcing via a noise-assisted stochastic resonance. The reconstruction of temperature and salinity of the inflow to the Nordic Seas [*Thornalley et al.*, 2009] provides some support for this mechanism of the Labrador Sea-Nordic Seas interaction. The reconstruction shows a stratified upper ocean flow during the earlier Holocene followed by an abrupt change to well-mixed flow oscillating with the about 1500 year period after about 8.4 ka B.P.

Here we investigate the 1500 year variability during the Holocene using paleodata at three locations of the AMOC and suggest a simple conceptual model simulating a possible cause of this oscillation. The model includes the interaction of solar and ocean centennial variabilities as well as the threshold response of the AMOC to forcing.

The standard spectral (Fourier and integrated wavelet) analysis sums spectral features over the whole time period and assumes that underlying process is stationary, which is not the case as indicated straight away by inspection of the data. To analyze the nonstationary time series of the paleodata, we apply two methods of analysis to the interpolated data. The first is the wavelet method [*Torrence and Compo*, 1998] previously used in studies of a selected set of North Atlantic paleodata by *Debret et al.* [2007]. The wavelet method tracks the nonstationary features well but is not adaptive to data since it is based on the use of given basic functions (the Morlet functions in this case). To complement the wavelet analysis, we apply the empirical mode decomposition (EMD) method, which has no prescriptions and is data adaptive [*Huang and Wu*, 2009]. The EMD represents the data as a sum of a small number of

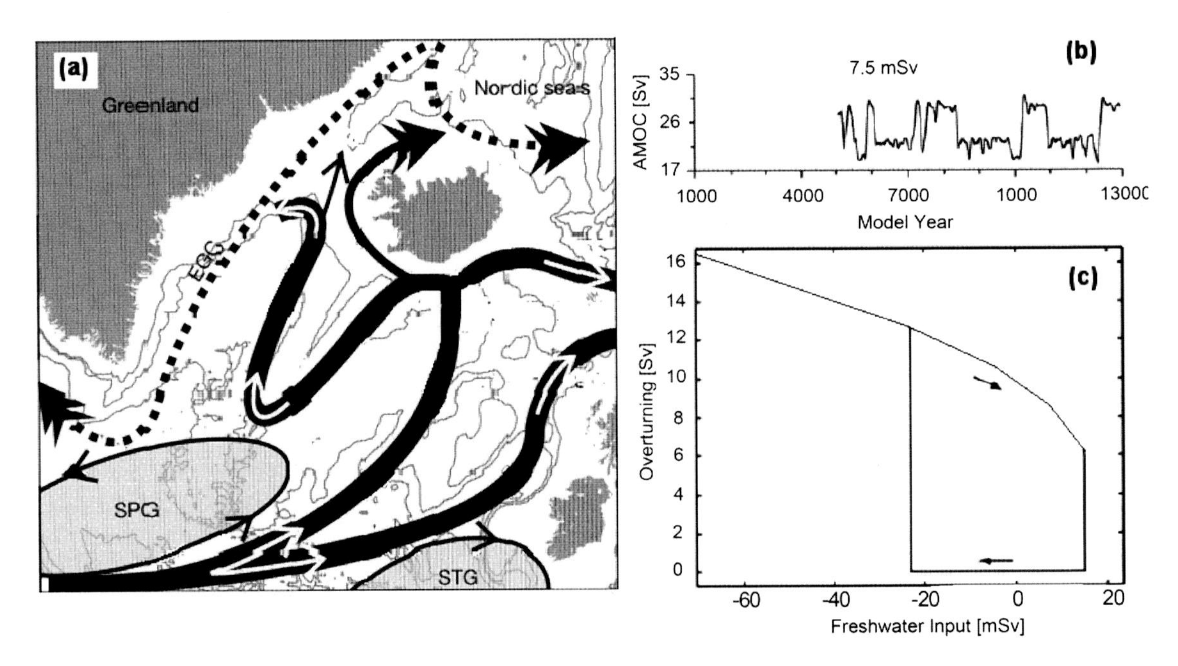

Figure 1. (a) A map of the ocean circulation in the northeast Atlantic. Abbreviations are SPG, subpolar gyre; STG, subtropical gyre; and EGC, East Greenland Current. Adopted from *Thornalley et al.* [2009]. Reprinted by permission from Macmillan Publishers Ltd: *Nature*, copyright 2009. (b) Time series of the AMOC poleward of 30°N below 500 m depth generated for 7.7 mSv freshwater perturbation. (c) Conceptual two-state (hysteresis loop) model. Figures 1b and 1c are adopted from *Schulz et al.* [2007].

quasi-orthogonal empirical modes. The quasi-orthogonality prevents leaks from one mode to another thus overcoming a common problem in the use of the conventional band-pass filtering. The EMD modes have time-variable amplitudes and frequencies. The number of modes depends on the number of data points n as $\log_2 n$; i.e., it is always small. Each mode is equivalent to a filtered signal in an empirically determined (not imposed!) frequency band. A mode has an envelope defined by local maxima and minima so that its mean amplitude is zero everywhere. A mean period of a mode can easily be estimated by counting the number of its maxima. Since data contain noise, which can be subjected to the same decomposition, it is important to know whether each mode represents a true signal or a component of noise. EMD modes of white or colored noise have progressively double mean periods and the log variances of these modes are inversely proportional to the logarithms of the mean periods. The variance of each EMD mode must obey a χ^2 distribution [*Flandrin et al.*, 2004; *Wu and Huang*, 2004]. By determining the number of degrees of freedom of this distribution, they derived the spread function of the variances of the noise modes. This allows a true signal mode to be discriminated from that of white noise at any desired statistical significance level.

2. DATA SELECTION

In our data analyses, we use the following paleodata in the Holocene:

1. The temperature and salinity in the Nordic Sea are reconstructed using measurements of two species of planktonic foraminifera: *Globigerina bulloids*, which occupies 0–50 m seasonal mixed layer below the ocean surface, and *Globorotalia inflata*, which calcifies at the base of seasonal thermocline (100–200 m) during winter. The data are taken from the RAPiD-12-1K core at 62.05°N, 17.49°W using Mg/Ca-δ^{18}O measurements [*Thornalley et al.*, 2009].

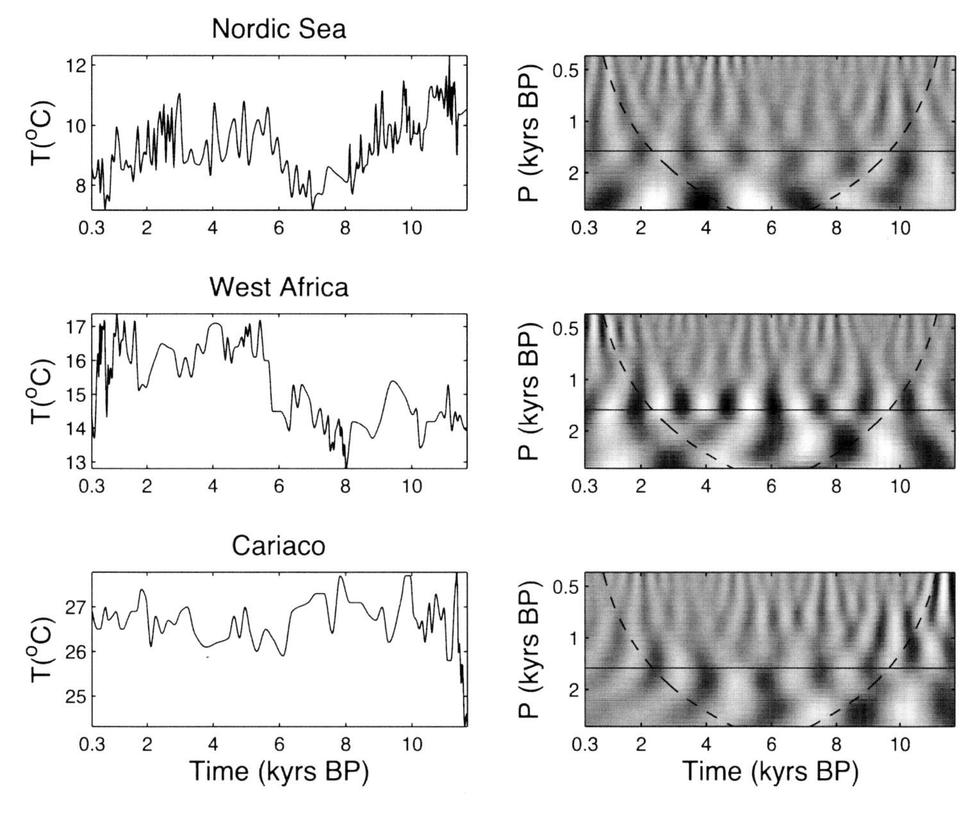

Figure 2. Reconstructed temperatures (top) at the thermocline depth of the ocean at 62.05°N, 17.49°W [*Thornalley et al.*, 2009]; (middle) off the coast of West Africa at 20.75°N, 18.58°W [*deMenocal et al.*, 2000]; and (bottom) Cariaco Basin core PL07-39PC at 10.70°N, 64.94°W, 790 m) [*Lea et al.*, 2003]. (right) Wavelet decompositions of the corresponding data in the range of millennium scales as indicated along the y axis. The solid line marks 1500 year period. The dashed line shows the cone of influence of edges (the results inside the cone are the statistically significant at 90% level of significance), as designed by *Torrence and Compo* [1998].

2. Holocene variations in subtropical Atlantic sea surface temperature (SST) at the west coast of Africa from the World Data Center A for Paleoclimatology [*deMenocal et al.*, 2000] are used. The data were taken at site 658C and were reconstructed using planktonic foraminiferal assemblage data. Counts of 29 species were converted to warm and cold season SSTs using a special transfer function. Average temporal resolution of the SST data ranges between approximately 50 and 100 years. Warm and cold season SST anomalies relative to core top values and the warm-cold season SST seasonality were also computed.

3. Cariaco Basin SSTs were reconstructed based on foraminiferal Mg/Ca measurements in the western tropical Atlantic from the World Data Center for Paleoclimatology [*Lea et al.*, 2003]. The data were taken from Cariaco Basin core PL07-39PC (10.70°N, 64.90°W, 790 m).

4. Greenland Ice Sheet Project 2 (GISP2) ions and temperature data were used. The ion data set was produced by the Glacier Research Group for the GISP2 core from 2 to 3040 m depth in Greenland [*Mayewsky et al.*, 2004]. The temperature data are provided by *Alley* [2004].

5. Alkenone SSTs were reconstructed from the Pacific [*Isono et al.*, 2009]. Analysis of alkenones and calculation of temperature were conducted at tree cores, labeled MD01-2421, KR02-06 St.A GC, and KR02-06 St.A MC sampled at 36°N, 141.5°E, at a water depth of 2224 m off the coast of central Japan in the northwestern Pacific.

6. SSTs were reconstructed for the eastern Mediterranean from a site located near the Nile River Delta [*Castañeda et al.*, 2010]. This reconstruction utilized two organic geochemical proxies, the "$U_{37}^{k'}$" and "TEX_{86}".

7. Sunspot numbers were reconstructed for 11,000 years based on dendrochronologically dated radiocarbon concentration and models relating the radiocarbon to the sunspot number [*Solanki et al.*, 2004].

The paleodata have different intervals of record and different time cadences. For comparison of the data and data analysis, we selected equal record intervals covering the Holocene and interpolated the data to 20 year time steps using the MatLab "pschip" interpolation code similar to a cubic spline.

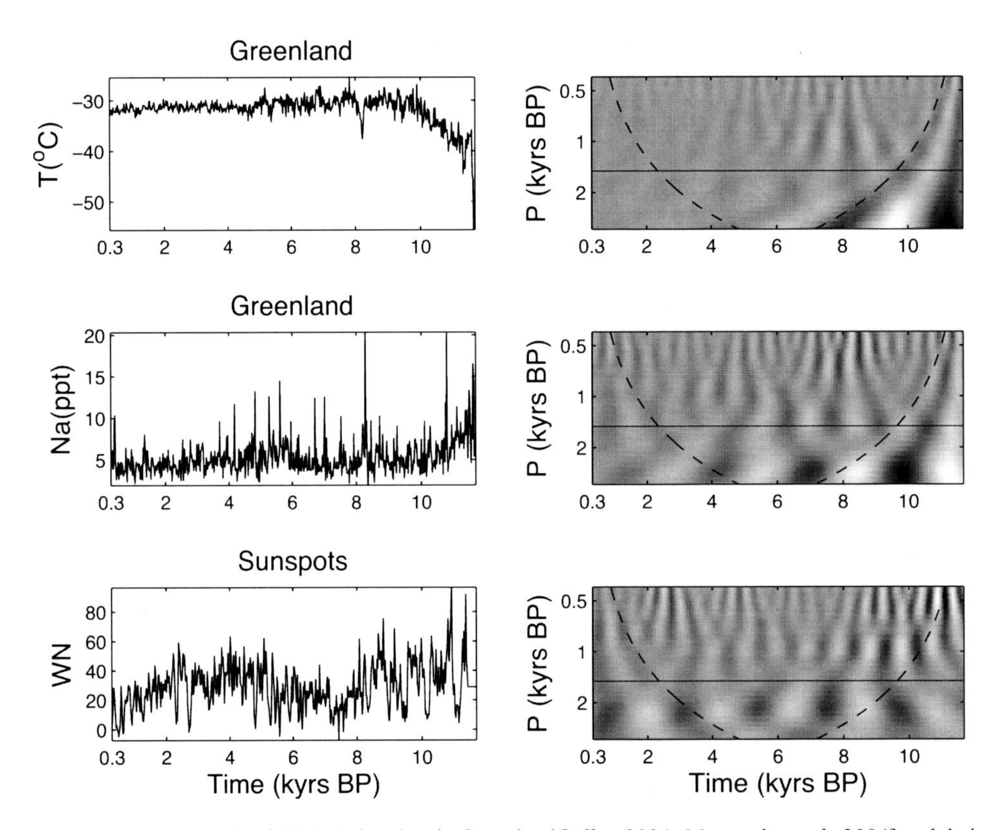

Figure 3. (top) Temperature and (middle) Na ion data in Greenland [*Alley*, 2004; *Mayewsky et al.*, 2004] and their wavelet decompositions. (bottom) Sunspot number reconstruction [*Solanki et al.*, 2004] and its wavelet decomposition. The solid line marks 1500 year period. The dashed line shows the cone of influence of edges. The major solar variations are seen at 2100–2400 year periods.

3. PALEOTEMPERATURE VARIABILITY IN ATLANTIC

First, we investigate the millennium variability at three sites of the Atlantic characterizing the subpolar and subtropical gyres. Figure 2 shows the data and their wavelet decompositions for (1) the deep water of the subpolar gyre in the Nordic Seas, (2) off the coast of West Africa, and (3) in the Cariaco Basin, which characterize the subtropical gyre. We see the presence of the 1500 year signal in both subpolar and subtropical gyres.

However (Figure 3), the 1500 year signal is almost invisible in the Greenland temperature, and it is only slightly expressed in the Greenland Na ion record, which characterizes the wind conditions over Greenland [*Mayewsky et al.*, 2004]. The sunspot reconstruction record (Figure 3, bottom) does not display 1500 year quasiperiodicity at all, thus confirming the finding by *Debret et al.* [2007] and *Dima and Lohmann* [2009]. We conclude that the solar forcing does not "directly" drive the 1500 year climate variation.

4. SUBPOLAR ATLANTIC VARIABILITY OF PALEOTEMPERATURE AND SALINITY

To investigate the variability in the North Atlantic in more detail, we complement the temperature data used in Figure 2 with the salinity data from the same site and with the surface temperature and salinity reconstructions (Figure 4) made by *Thornalley et al.* [2009]. Figure 5 shows the wavelet decomposition of these data. The wavelets of salinity and temperature display a nonstationary 1500 year variability. We see also more powerful variations at about 3000 years.

The EMD method, which is adaptive to the data, permits a better investigation of evolution of the modes with frequency bands in the vicinity of 1/1500 year frequency. Figure 6 shows the time evolution of the modes. We see again that the oscillations with the quasiperiods close to 1500 year are highly nonlinear. These findings imply a nonlinear mechanism of excitation of the oscillations.

A combination of data sets for temperature and salinity allows us to investigate their joint probability distributions.

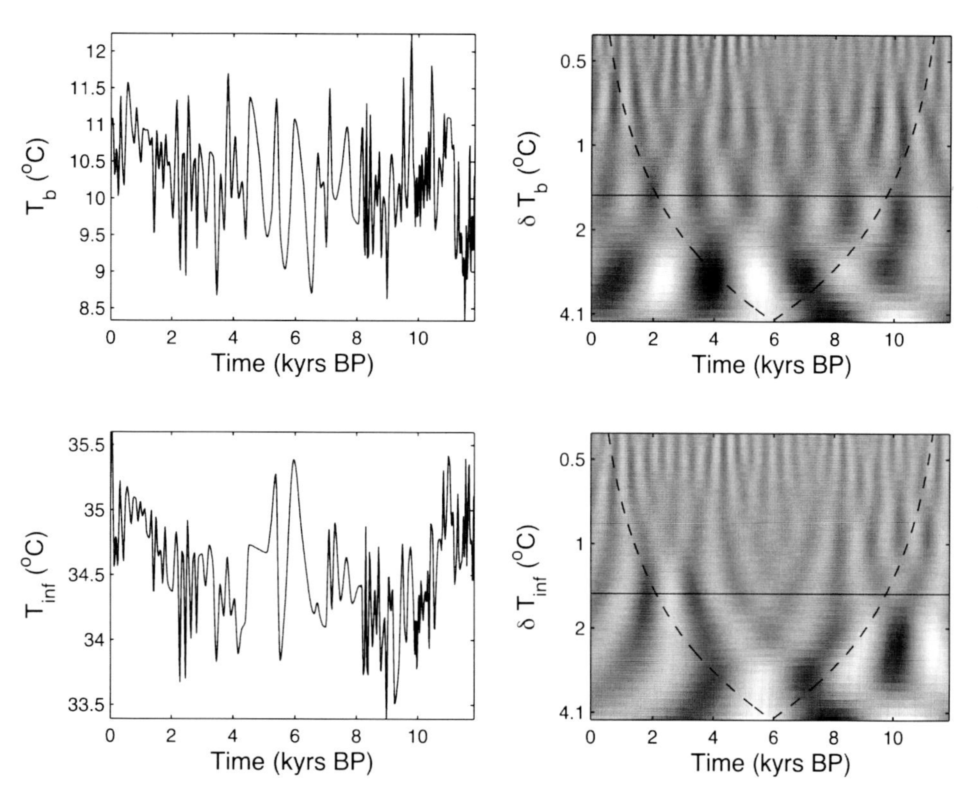

Figure 4. (left) Temperatures reconstructed and interpolated with 20 year steps for the *Globigerina bulloides* (T_b), which characterize the ocean surface conditions, and *Globorotalia inflata* (T_{inf}), which characterize the conditions at the thermocline depth of the ocean at 62.05°N, 17.49°W [*Thornalley et al.*, 2009]. (right) Wavelet decompositions of data shown in Figure 4, left. The solid horizontal line marks the 1500 year period. The dashed line outlines the 90% statistical significant part (inside the cone).

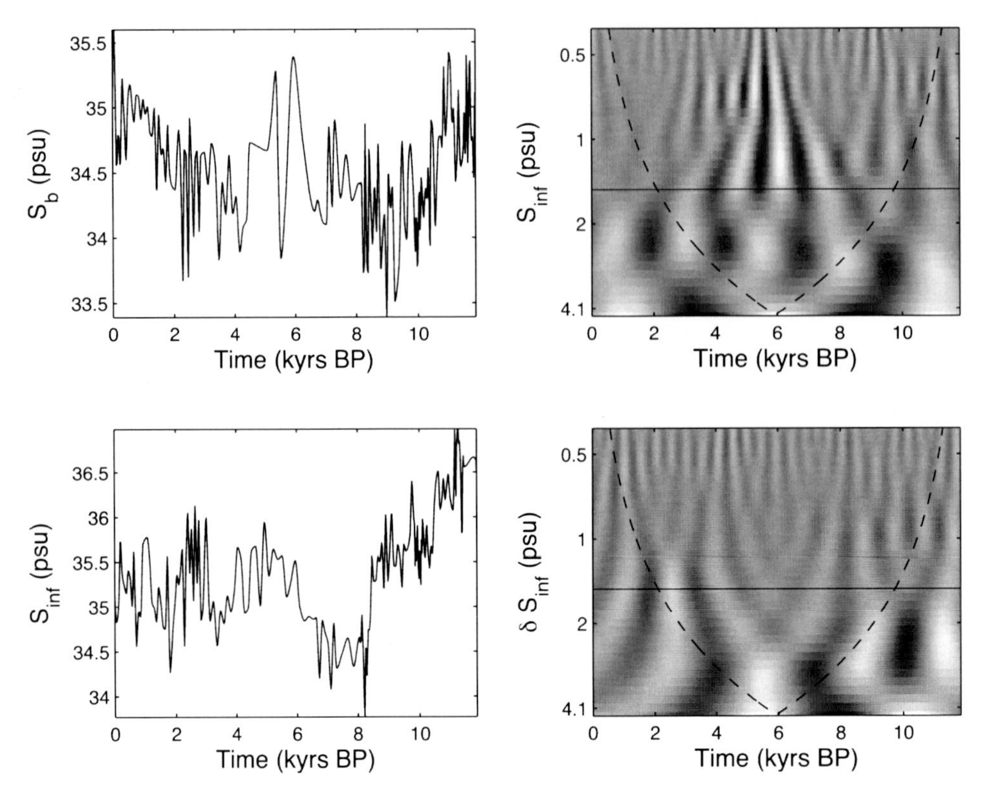

Figure 5. Same as Figure 4 for the near-surface salinity and the salinity at the base of seasonal thermocline.

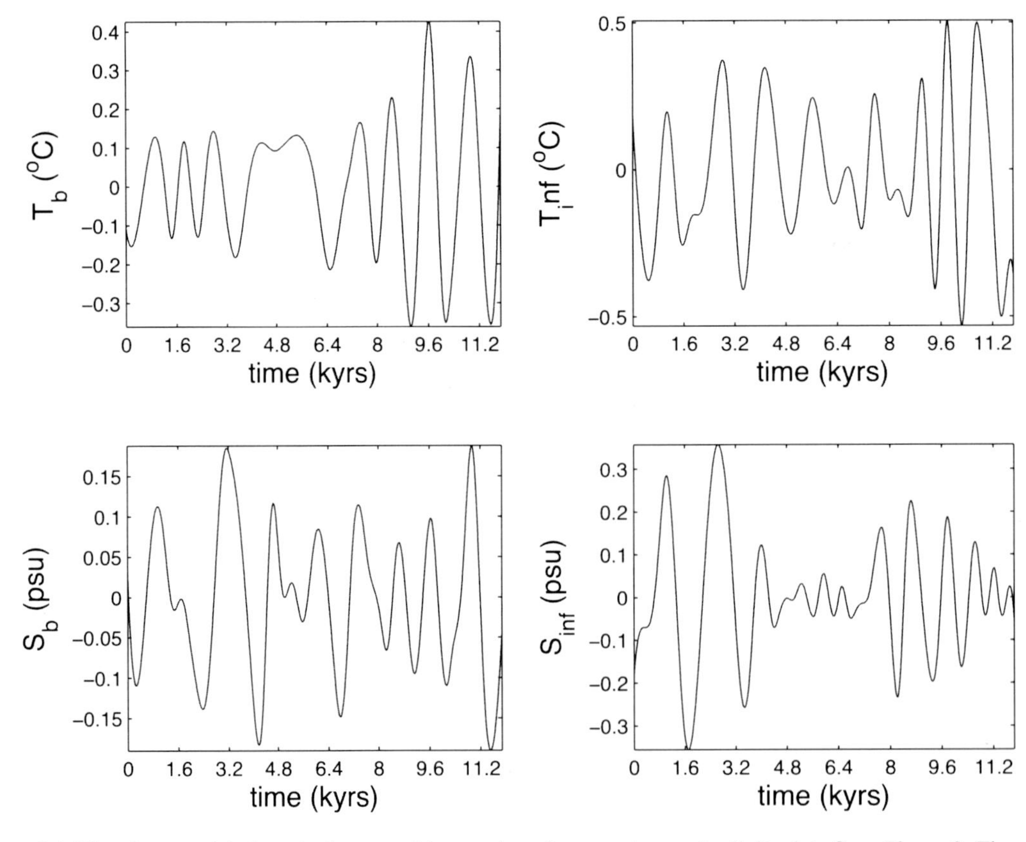

Figure 6. Millennium empirical mode decomposition modes of temperature and salinity data from Figure 3. The modes have frequency bands close to 1500 year periodicity. The varying amplitudes of the modes indicate that the 1500 year periodicity is highly nonlinear.

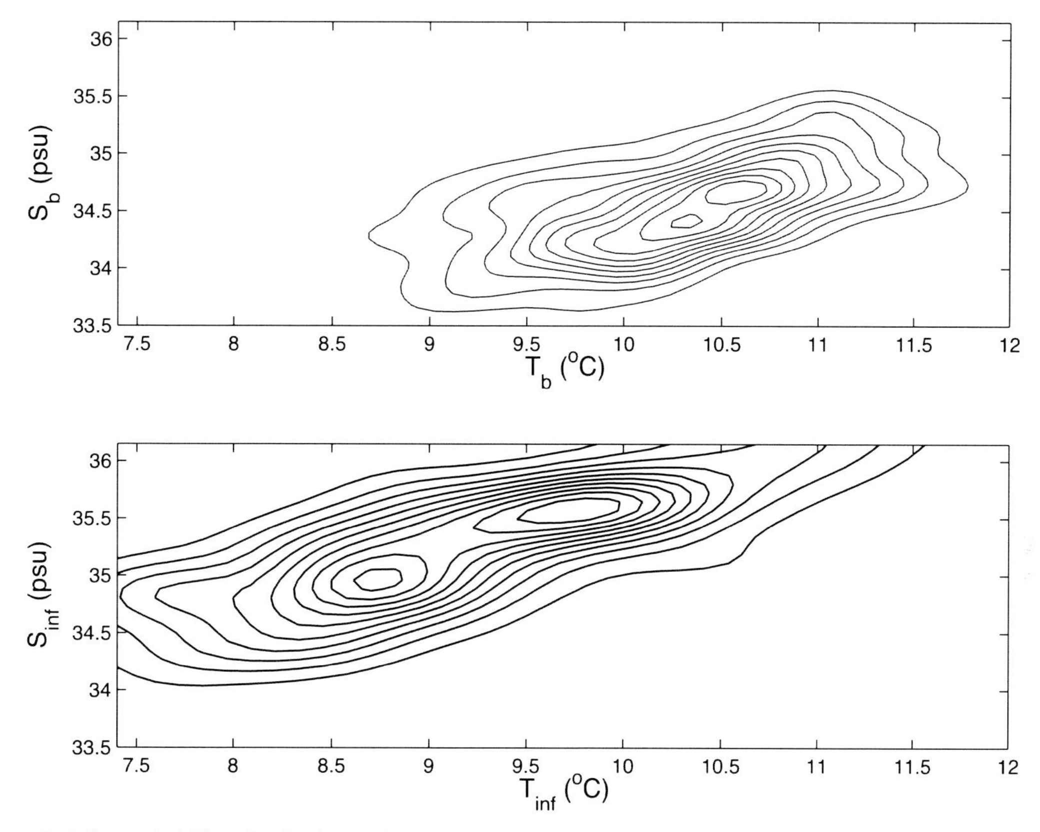

Figure 7. Joint probability distributions of the temperature and salinity (top) near the surface and (bottom) near thermocline for the data shown in Figure 3. For plotting purposes, the distributions are normalized to unity, 10 contour lines shown in both plots are spaced by 0.1 with the maxima located in the centers. Both distributions have two maxima, which we associate with the two states of the AMOC. Using the two-dimensional extension of the method developed by *Marron and Chaudhuri* [1998], we estimated that the near-thermocline maxima are statistically significant at 90% level of significance. The significance level of the surface maxima is lower.

In Figure 7, we show the joint distributions of these variables for the surface and near-thermocline data. The two peaks seen in both distributions may be interpreted as the weak and strong states of the AMOC in accord with the modeling findings by *Schulz et al.* [2007]. The weak state has a lower salinity and slightly lower temperature compared with the strong state.

5. ON THE ORIGIN OF THE 1500 YEAR OSCILLATION

The origin of the 1500 year oscillation remains a matter of debate. What has become clear and is supported by our data analysis is that it could not be caused simply by direct solar forcing as suggested earlier by *Bond et al.* [2001]. Combining the current ideas described in the introduction, we propose here a simple mechanism, which involves the solar and ocean variability as well as a threshold effect in response to the forcing.

The well-established solar activity quasiperiodicities are positioned at 11 year solar cycle (a relatively short periodicity in the paleocontext) and 90 years (the centennial Gleissberg cycle). There is no direct solar irradiance (or other solar variability proxy variations) at 1500 year oscillation as shown by *Debret et al.* [2007], *Dima and Lohmann* [2009], *Usoskin* [2008], and by our data analysis. The proposed mechanism is based on a simple physical idea of the interaction (beats) of the 90 year solar signal with the centennial ocean variability of the AMOC. We note that the 90 year cycle is a nonlinear oscillation not a sinusoid. We also warn that "Gleissberg cycle" is a vague term used by different authors to refer to different variations. It may have a different meaning depending on what kinds of proxies are used to identify it. Thus, *Feynman and Fougere* [1984] identified a narrowband 90 year cycle in 1000 years of aurora data (450–1450 A.D.), which directly represents the solar variability. It is also seen in the solar cycle modulation observed at the beginning of the nineteenth, twentieth, and now the twenty-first centuries. *Ogurtsov et al.* [2005], using the tree ring data, which is an Earth's climate proxy, have found a spectral band of 50–130 years. They divided it into two components: one with 90–100 year period and the second one having 50–60 year periods (see also *Hathaway* [2010]). According to our model, it is the first component of this proxy that beats with the centennial ocean variability.

Coupled ocean-atmosphere modeling indicates that the internal variability of the AMOC is well distinguished at interannual and centennial time scales, with the centennial (≈100 years) mode being dominant [*Vellinga and Wu*, 2004]. This mode, which is shown in Figure 4 of Vellinga and Wu, also indicates an unstable, intermittent nature of this variability. In the model this mode is excited due to the following mechanism advocated by Vellinga and Wu: The AMOC in its strong state produces a weak cross-equatorial SST contrast (of about 0.1°C in the tropics) that increases the strength of the Intertropical Convergence Zone (ITCZ) and shifts the summer ITCZ to a more northerly position. The resulting extra rainfall brings fresh water into the ocean lowering its salinity. This salinity anomaly slowly propagates toward the subpolar North Atlantic where it causes the AMOC flow to slow down to the weak state. In the weak state of the AMOC, the ITCZ returns to its normal location thus closing the cycle, which takes a century long period P_c mostly defined by the transport time of the salinity anomaly. Since this period of natural ocean variability is close to the centennial Gleissberg period P_s = 90 years of solar variability, one may expect beats between the two close periods: $1/P = 1/P_s - 1/P_c$. The resulting modulation period would be equal to P = 1500 years for $P_c \approx 97$ years. Since, in reality, there is no strict periodicity either in ocean or in solar variability, the beats are expected to be unstable and thus in accord with observations, suggestive of a nonlinear nature of oscillations.

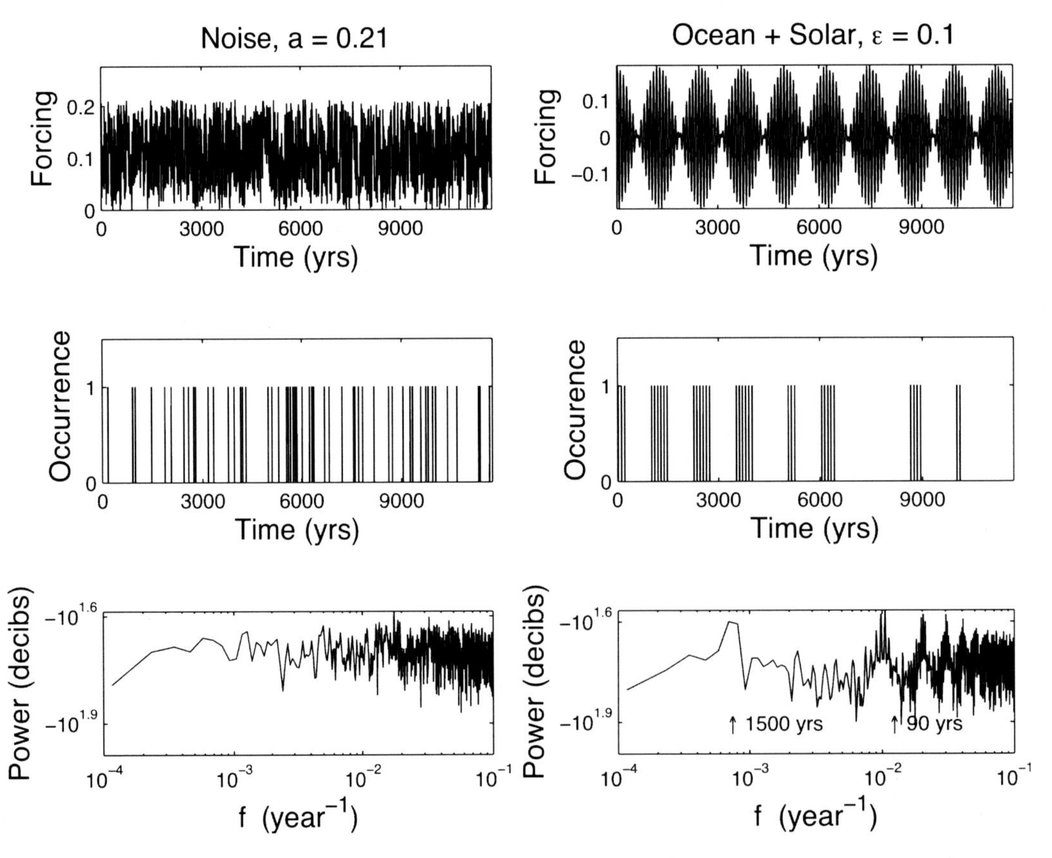

Figure 8. (left) A realization of the white noise forcing causes the system to randomly transit between two states marked (middle) by 0 and 1. The resulting spectrum shows no specific features. (right) A weak periodic forcing by two sinusoids with close frequencies (1/90 year and 1/97 year), which simulate the solar and centennial ocean variabilities, groups the transitions so that the beat frequency 1/1500 year is seen in the spectrum of the system response. The parameters used for this realizations are p = 0.885 (to make the threshold equal to 1), a = 0.21, and b = 1, ε = 0.1.

To convert the idea into a mathematical form, we consider a simple conceptual model described by the equation

$$\frac{d\varphi}{dt} + sin\varphi = p + a \cdot N(t) + \varepsilon \cdot [O(t) + b \cdot S(t)], \quad (1)$$

which is an extension of the equation introduced in the context of stochastic resonance by *Wiesenfeld et al.* [1994]. Here $p < 1$ is a parameter which will define two basic states of the AMOC (the weak and strong), $N(t)$ is a random noise of amplitude a, and ε is a small amplitude of the ocean $O(t) = \sin(2\pi/P_c)$ centennial variability. The amplitude of the solar signal $S(t) = \sin(2\pi/P_s)$ is distinguished by an extra parameter $b \leq 1$. The basic variable φ characterizes the state of the AMOC, i.e., the magnitude of the flow.

Without the noise, ocean centennial variability, and solar signal ($N = 0$, $\varepsilon = 0$), this system has two equilibrium states defined by the condition $\sin \varphi = p$: a stable $\varphi_S = a \sin(p)$ state and an unstable $\varphi_W = \pi - a \sin(p)$ state, which we can associate with the strong and weak states discussed above in the observational context. The stability of these states can easily be checked by a perturbation of the equilibrium or by presenting the terms $-\sin \varphi + p$ in equation (1) via a potential $-\partial/\partial\varphi$ with $U = -\cos \varphi + px$ having a potential well corresponding to the stable state and unstable hump. Equation (1) has been integrated in time using a stiff ODE code from the MATLAB library. The value of ε is chosen such that the ocean and solar signal are smaller than the threshold distance $\varphi - 2 \arcsin(p)$ between the two states, and thus these forcings alone cannot move the system from one state to another. At $\varepsilon = 0$, driven only by the irregular noise kicks, the state point φ jiggles around the stable strong state and, from time to time when the amplitude of an event is large enough, goes over the threshold and moves into the unstable weak state. After about a mean residence time in the unstable state, sec (φ_s), the point returns back to the stable state. The noise forcing alone produces a random sequence of events. However, when $\varepsilon \neq 0$, the added ocean-solar forcing produces different outputs via modulating (grouping) the random transitions between the states so that the 1500 year signal arising from the beats becomes feasible. Figure 8 shows two specific examples of numerical solutions of equation (1): the first one with the white noise input only (left plots) and the second one with the addition of the periodic

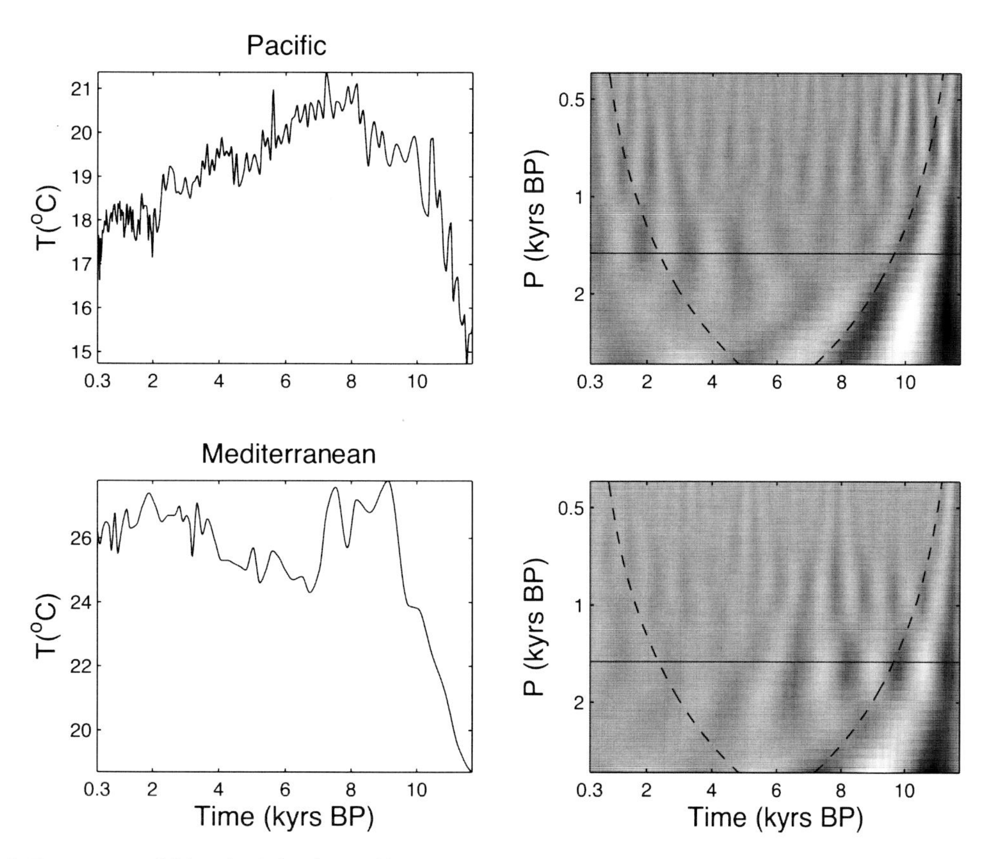

Figure 9. Reconstructed SSTs (top) for the Pacific Ocean at 62.05°N, 17.49°W [*Isono et al.*, 2009] and (bottom) near the Nile delta (10.70°N, 64.94°W at the depth 790 m) [*Castañeda et al.*, 2010].

functions characterizing the ocean and solar variabilities (right plots).

6. DISCUSSION

The progress in the collection of paleodata greatly stimulates the reconstruction of the climate of the past and gives insight into the climate of the future. With a focus on the nature of the millennium variability during the Holocene, we have shown that the 1500 year oscillation is not purely random; thus, forcing is involved. We have suggested a simple model of excitation of this oscillation in a nonlinear dynamical system with two equilibrium states. The transitions between the two states are caused by noise, ocean, and solar variability. This implies that it is the beats between the centennial ocean variability [*Vellinga and Wu*, 2004] and 90 year solar variability that produces the 1500 year oscillation in the noisy system. Note that the frequencies of solar and ocean variabilities are both known to be unstable. This simple model needs to be validated by more sophisticated coupled ocean-atmosphere models. The modeling should start with a GCM simulating the centennial ocean variability and then be forced by the 90 year solar variation.

An interesting question is whether or not the 1500 year variability is limited to the Atlantic. *Isono et al.* [2009] using a multidecadal resolution record of alkenone SST in the northwestern Pacific off central Japan have found the millennial variability with an amplitude of about 1°C throughout the entire Holocene. Spectral analysis carried out by Isono et al. revealed a statistically significant peak with 1470 year periodicity in the SST. They also found that the SST correlated with the variations of ice-rafted hematite-stained grain content in North Atlantic sediments. These findings indicate that the mean latitude of the Kuroshio Extension has varied on a 1500 year cycle and suggest that a climatic link exists between the North Pacific gyre system and the high-latitude North Atlantic thermohaline circulation. Our wavelet analysis of the SST data supports these conclusions (Figure 9). Figure 9 also shows the result of our analysis of a paleotemperature record from the Mediterranean region (the Nile Delta) [*Castañeda et al.*, 2010], indicating traces of the 1500 year oscillation.

Finally, we would like to comment on the efforts relating the 1500 year oscillation in the Holocene to the current global warming. *Avery* [2008, p.153] presents a counterpoint to the Intergovernmental Panel on Climate Change statement that CO_2 and other greenhouse gases are the cause of the warming trend in the current global climate. Taking on historical and paleoclimate records, these authors argue that "the current global warming (is) part of a natural trend with a cycle that results in elevated climatic temperatures every 1,500 years. That means humanity's response must be keyed to adaptation, not prevention." The current warming trend, according to these authors, began about 1850, and "The 1500 year cycle typically produces a temperature change of four to five degrees centigrade from peak to trough. Temperature changes are greater in the Arctic. At the equator, however, the temperatures do not change—the rainfall does. The tropical rain belts at the equator can move hundreds of miles north and south during the cycle."

Our data analysis does not support the basic assessment of these authors since the amplitude of the 1500 year oscillation is smaller than 1°C; see, for example, Figure 5 and paper by *Isono et al.* [2009]. So the rise of 1500 year oscillation could not compete with the predicted temperature rise of 2°C–5°C per century due to the increase in CO_2 and other greenhouse gases. However, the 1500 year variability may be taken into account in historical and possibly future considerations.

Acknowledgements. We have obtained the GISP2, West Africa, and Cariaco paleodata from the NOAA/NGDC Paleoclimatology Program and acknowledge the scientists who provided these data (see references in Section 2). We are very grateful to David Thornalley, Masanobu Yamamoto, and Isla Castañeda for providing their paleodata and helpful communications. This work was supported in part by the Jet Propulsion Laboratory of the California Institute of Technology, under a contract with the National Aeronautics and Space Administration.

REFERENCES

Alley, R. B. (2004), GISP2 Ice Core Temperature and Accumulation Data, IGBP PAGES/World Data Center for Paleoclimatology, *Data Contrib. Ser. N2004-013*, http://www.ncdc.noaa.gov/paleo, NOAA/NCDC, Paleoclimatol. Program, Boulder, Colo.

Avery, D. T., (2008), Global warming every 1,500 years: Implications for an engineering vision, *Leadership Manage. Eng.*, *8*(3), 153–159, doi: 10.1061/(ASCE)1532-6748(2008)8:3(153).

Bond, G., M. Cheseby, R. Lotti, P. Almasi, P. deMenocal, P. Priore, H. Cullen, I. Hajdas, and G. Bonani (1997), A pervasive millennial-scale cycle in North Atlantic Holocene and glacial climates, *Science*, *278*, 1257–1266.

Bond, G., B. Kromer, J. Beer, R. Muscheler, M. Evans, W. Showers, S. Hoffmann, R. Lotti-Bond, I. Hajdas, and G. Bonani (2001), Persistent solar influence on North Atlantic climate during the Holocene, *Science*, *294*, 2130–2136.

Braun, H., M. Christl, S. Rahmstorf, A. Ganopolski, A. Mangini, C. Kubatzki, K. Roth, and B. Kromer (2005), Possible solar origin of the 1,470-year glacial climate cycle demonstrated in a coupled model, *Nature*, *438*, 208–211, doi:10.1038/nature04121.

Castañeda, I. S., E. Schefuß, J. Pätzold, J. S. Sinninghe Damsté, S. Weldeab, and S. Schouten (2010), Millennial-scale sea surface temperature changes in the eastern Mediterranean (Nile River

Delta region) over the last 27,000 years, *Paleoceanography*, *25*, PA1208, doi:10.1029/2009PA001740.

Damon, P. E., and C. P. Sonett (1991), *The Sun in Time*, edited by C. P. Sonett, M. S. Giampapa, and M. S. Matthews, 360 pp., Univ. of Ariz. Press, Tucson.

Debret, M., V. Bout-Roumazeilles, F. Grousset, M. Desmet, J. F. McManus, N. Massei, D. Sebag, J.-R. Petit, Y. Copard, and A. Trentesaux (2007), The origin of the 1500-year climate cycles in Holocene North-Atlantic records, *Clim. Past*, *3*, 569–575.

deMenocal, P. B., J. Ortiz, T. Guilderson, and M. Sarnthein (2000), Coherent high- and low-latitude climate variability during the Holocene Warm Period, *Science*, *288*, 2198–2202.

Dima, M., and G. Lohmann (2009), Conceptual model for millennial climate variability: A possible combined solar-thermohaline circulation origin for the 1500-year cycle, *Clim. Dyn.*, *32*, 301–311.

Feynman, J., and P. F. Fougere (1984), Eighty-eight year periodicity in solar-terrestrial phenomena confirmed, *J. Geophys. Res.*, *89*, 3023–3027.

Flandrin, P., G. Rilling, and P. Goncalves (2004), Empirical mode decomposition as a filter bank, *IEEE Signal Process. Lett.*, *11*, 112–114.

Hathaway, D. (2010), The solar cycle, *Living Rev. Solar Phys.*, *7*, 1.

Huang, N. E., and Z. Wu (2009), A review on Hilbert-Huang transform: Method and its applications to geophysical studies, *Adv. Adapt. Data Anal.*, *1*, 1–23.

Isono, D., M. Yamamoto, T. Irino, T. Oba, M. Murayama, T. Nakamura, and H. Kawahata (2009), The 1500-year climate oscillation in the midlatitude North Pacific during the Holocene, *Geology*, *37*, 591–594.

Jongma, J. I., M. Prange, H. Renssen, and M. Schulz (2007), Amplification of Holocene multicentennial climate forcing by mode transitions in North Atlantic overturning circulation, *Geophys. Res. Lett.*, *34*, L15706, doi:10.1029/2007GL030642.

Lea, D. W., D. K. Pak, L. C. Peterson, and K. A. Hughen (2003), Synchroniety of tropical and high-latitude Atlantic temperatures over the Last Glacial termination, *Science*, *301*, 1361–1366.

Marron, J. S., and P. Chaudhuri (1998), Significance of features via SiZer, in *Statistical Modelling, Proceedings of 13th International Workshop on Statistical Modelling*, edited by B. Marx and H. Friedl, pp. 65–75, Stat. Modell. Soc., Amsterdam.

Mayewski, P. A., et al. (2004), Holocene climate variability, *Quat. Res.*, *62*, 243–255.

Ogurtsov, M. G., G. E. Kocharov, M. Lindholm, J. Merilainen, M. Eronen, and Y. A. Nagovitsyn (2005), Evidence of solar variations in tree-ring-based climate reconstructions, *Sol. Phys.*, *205*(2), 403–417.

Peristykh, A. N., and P. E. Damon (2003), Persistence of the Gleissberg 88-year solar cycle over the last ~12,000 years: Evidence from cosmogenic isotopes, *J. Geophys. Res.*, *108*(A1), 1003, doi:10.1029/2002JA009390.

Ruzmaikin, A., J. Feynman, and Y. L. Yung (2006), Is solar variability reflected in the Nile River?, *J. Geophys. Res.*, *111*, D21114, doi:10.1029/2006JD007462.

Schulz, M., M. Prange, and A. Klocker (2007), Low-frequency oscillations of the Atlantic Ocean meridional overturning circulation in a coupled climate model, *Clim. Past*, *3*, 97–107.

Solanki, S. K., I. G. Usoskin, B. Kromer, M. Schussler, and J. Beer (2004), Unusual activity of the Sun during recent decades compared to the previous 11,000 years, *Nature*, *431*, 1084–1087.

Thornalley, D. J. R., H. Elderfield, and I. N. McCave (2009), Holocene oscillations in temperature and salinity of the surface subpolar North Atlantic, *Nature*, *457*, 711–714, doi:10.1038/nature07717.

Torrence, C., and G. P. Compo (1998), A practical guide to wavelet analysis, *Bull. Am. Meteorol. Soc.*, *79*, 61–78.

Usoskin, I. G. (2008), A history of solar activity over millennia, *Living Rev. Solar Phys.*, *5*, 3.

Vellinga, M., and P. Wu (2004), Low-latitude influence on centennial variability of the Atlantic thermohaline circulation, *J. Clim.*, *17*, 4498–4511.

Wiesenfeld, K., D. Pierson, E. Pantalezelou, C. Dames, and F. Moss (1994), Stochastic resonance on a circle, *Phys. Rev. Lett.*, *72*, 2125–2129.

Wu, Z., and N. E. Huang (2004), A study of the characteristics of white noise using the empirical mode decomposition method, *Proc. R. Soc. London, Ser. A*, *460*, 1597–1611.

J. Feynman and A. Ruzmaikin, Jet Propulsion Laboratory, 4800 Oak Grove Dr., Pasadena, CA 91109, USA. (Alexander.Ruzmaikin@jpl.nasa.gov)

Abrupt Climate Changes During the Holocene Across North America From Pollen and Paleolimnological Records

Konrad Gajewski and Andre E. Viau

Laboratory for Paleoclimatology and Climatology, Department of Geography, University of Ottawa, Ottawa, Ontario, Canada

Databases of ecological and cultural records, especially of pollen diagrams, record climate variability of several time scales during the Holocene and late glacial. Results from lake and wetland ecosystems geographically extend the evidence of rapid climate change obtained from ice cores and ocean sediments. Continental and regional climate curves for North America, based on pollen diagrams from the North American Pollen Database, illustrate abrupt changes on the order of every ~1000 years during the past 12 kyr, and major times of change in North American pollen records are coherent with vegetation changes across Europe. Novel analyses of the database show that even taxa that are widespread and with presumably broad climate tolerances were affected by abrupt climate changes such as the Younger Dryas and illustrate the complexity of ecosystem response to these changes. Reconstructions of freshwater as well as terrestrial ecosystems across northern Canada also show how climate variability affects terrestrial and freshwater ecosystem-level properties such as nutrient cycling. These results can be used to reconstruct the spatial patterns of abrupt climate change, as well as the impacts of climate change on ecosystem and cultures.

1. INTRODUCTION

Although most of the "iconic" paleoclimate time series are from ocean [e.g., *Imbrie et al.*, 1989] or ice cores [e.g., *Johnsen et al.*, 2001; *EPICA Members*, 2006], a considerable amount of information is known about the terrestrial paleoclimates of the past 21 kyr [e.g., *Bryant and Holloway*, 1985; *Wright et al.*, 1993]. Evidence from marine and ice cores suggests abrupt climate changes superimposed on the "Milankovitch-scale" variability that occurred throughout the last million years [e.g., *Bond et al.*, 1997, 2001; *MacManus et al.*, 1999; *O'Brien et al.*, 1995; *Oppo et al.*, 1998; *Raymo et al.*, 1998; *Bianchi and McCave*, 1999; *Heinrich*, 1988; *Dansgaard et al.*, 1993; *Rahmstorf*, 2003]. However, one major uncertainty that arises in paleoclimate research is the lack of spatial corroborating evidence. For example, the Greenland ice core records offer a temporal record of past climate variations on several time scales; however, they remain a single point in space until results can be compared to records found in other regions. This lack of the spatial dimension has plagued large-scale paleoclimate research for many years. The last few decades have seen assembled large databases of paleoclimate information. The availability of large databases of pollen, tree ring, and other data enables syntheses of the terrestrial paleoclimate at several time and space scales.

Considerable progress has been achieved in the past 30 years in producing paleoclimate estimates for the postglacial using pollen as a climate proxy for all of North America and Europe. This has come about due to the availability of a critical mass of fossil and calibration data and methodological developments suitable for large-scale reconstructions,

Abrupt Climate Change: Mechanisms, Patterns, and Impacts
Geophysical Monograph Series 193

10.1029/2010GM001015

for example, the modern analogue technique (MAT) [*Overpeck et al.*, 1985]. This work has been done in the context of an extensive knowledge of the ecology and biogeography of the principal plant species of the region, which lends theoretical justification to the conclusions.

Records of vegetation change preserved in pollen time series have been used to document climate variability on several time and space scales [*Wright et al.*, 1993]. Studies have shown that vegetation responds rapidly to climate change [e.g., *Webb*, 1986; *Gajewski*, 1987], and there is evidence of synchronous vegetation response to abrupt climate changes during the late glacial and glacial period [*Grimm and Jacobson*, 1991; *Shuman et al.*, 2002a, 2002b; *Viau et al.*, 2002; *Williams et al.*, 2002, 2004; *Peros et al.*, 2008] (see also collection of papers in special section "Vegetation Response to Millennial-scale Variability during the Last Glacial" in *Quaternary Science Reviews*, *29*(21–22), 2010) as well as during the Little Ice Age (LIA) and Medieval Warm Period (MWP) of the past 1000 years [*Gajewski*, 1987]. The pollen-based paleoclimate reconstructions are beginning to be supplemented by results using other proxy-climate records found as fossils in lake sediments.

In this chapter, we will discuss Holocene climate variability in North America using a spatially extensive pollen database [*Grimm*, 2010] supplemented with other terrestrial proxy-climate data. Quantitative reconstructions of Holocene climates are made possible due to a large publically available modern pollen database [*Whitmore et al.*, 2005] for quantitative calibration of the fossil paleoecological time series. This research effort provides information of how climate variability affects terrestrial ecosystem structure and function. Although radiocarbon dating of these series is currently a limitation to the resolution that can be obtained, this can be overcome, and high-resolution series of past vegetation and ecosystem dynamics can be obtained.

Using these pollen databases and the modern analogue method, maps have been prepared of the climates of North America at 100 year intervals for the past 12 kyr, and when compared to climate model simulations of 6 kyr, both simulated and reconstructed patterns show important similarities [e.g., *Wright et al.*, 1993; *Gajewski et al.*, 2000; *Sawada et al.*, 2004]. Regional averaging of several pollen sequences is one method to enable high-resolution time series to be obtained [*Viau et al.*, 2006] and is a methodology comparable to the development of tree ring chronologies. Time series of the climate of the Holocene have been developed for North America, Europe, and several regions. Although this chapter will focus on North America, studies have shown that the timing of millennial-scale transitions is coherent in North American and European pollen diagrams, as might be expected [*Gajewski et al.*, 2006]. The pollen-based reconstructions also show coherence with marine and ice core records [*Viau et al.*, 2002, 2006].

Increasingly, studies are using several different proxy-climate indicators in the same core to provide more information about past environments. In cases when different fossil organisms from the same core are used for quantitative paleoclimate reconstructions, the time series based on the different indicators can be compared. Other indicators besides pollen are being used, although these are far from being as well studied or understood. For example, chironomids, the larval form of nonbiting midges (insects) are fossilized in lake sediments and can be identified. Initial work suggested that these organisms provide useful paleoclimate estimates [*Batarbee*, 2000], and although subsequent work has suggested interpretation may be more subtle [*Walker and Cwynar*, 2006], several local studies have been accomplished. Other fossils, such as diatoms, record changes in aquatic production and factors such as pH. To the extent that the lake chemical environment is affected by climate, diatoms may therefore be an indirect recorder of past climates, although in practice, it has proven difficult to quantify past changes due to the huge diversity of species and large differences even in neighboring lakes [*Bouchard et al.*, 2004]. Many other fossil groups have been studied, but extensive enough data sets and knowledge are not yet available for their use in other than local studies. Chemical and physical components of the sediment are also widely used as indices of past carbon dynamics and temperature [*Willemse and Törnqvist*, 1999; *Kaufman*, 2009; *Fortin and Gajewski*, 2009]. These are especially useful as they are easy to obtain, and thus, high-temporal-resolution records can be developed.

2. DATA AND METHODS

Pollen data are collected from sedimentary deposits and, over the past few decades, mostly from lake sediments. Cores from the sediments are sampled and pollen extracted from a series of levels. The cores are dated using radiocarbon and other methods such as ^{210}Pb. A pollen diagram from a sediment core describes the vegetation history of the region, and since plant growth is limited in part by climate, changes in climate can be interpreted. However, pollen data are multivariate time series, and to be useful as paleoclimate indicators, they must be transformed, using some kind of transfer function, to climate series in the units of degrees Celsius or millimeters precipitation.

2.1. The North American Pollen Database

Continental-scale maps and time series of Holocene climate are made possible due to the availability of databases of accumulated pollen data. The North American Pollen Database (NAPD) [*Grimm*, 2010] has incorporated pollen diagrams from across North America, with a relatively high density of sites in the northeastern region (Figure 1). Available at the National Climate Data Center (www.ncdc.noaa.gov/paleo), the data are stored in a uniform and accessible format, so queries can extract all the data needed for a particular project. The database is taxonomically complex, so usage is not straightforward; some examples are reviewed in the work of *Gajewski* [2008].

For quantitative paleoclimate reconstructions, a spatial array of modern pollen samples, along with associated climate data are necessary. Recently, a continental database of modern pollen data has been made publically available in easily accessible form [*Whitmore et al.*, 2005]. This database consists of roughly 4600 sites from across North America, and new data are being periodically added (Figure 1). Analysis of these data illustrates the climatic constraints on the major pollen taxa used in paleoclimate reconstructions [*Williams et al.*, 2006].

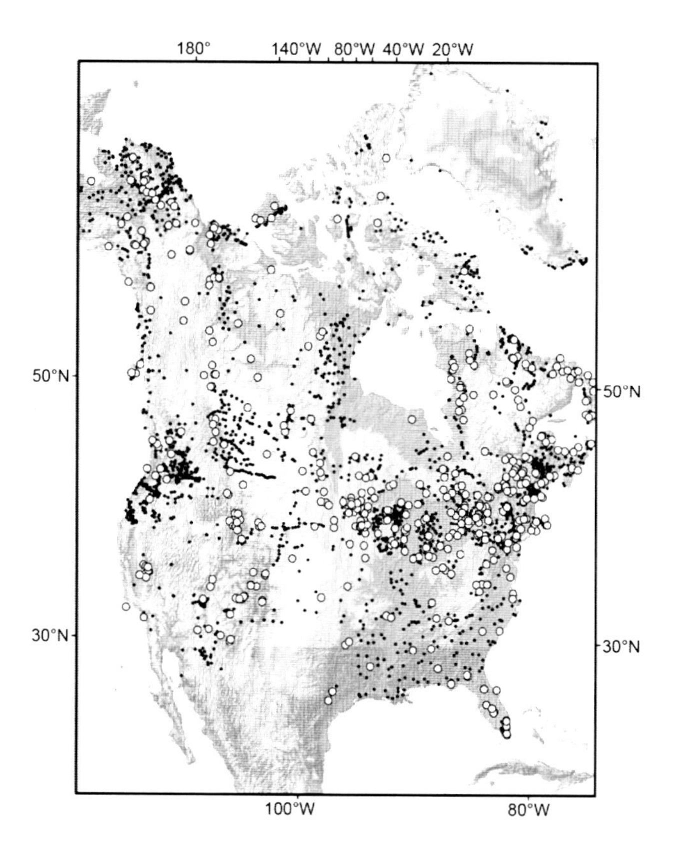

Figure 1. Location of pollen cores (open circles) in the Global Pollen Database [*Grimm*, 2010]. Closed circles are location of modern pollen data, used for calibration of paleoclimate reconstructions [*Whitmore et al.*, 2005].

The large numbers of sites provides opportunities but also challenges. A dense network of sites across an entire continent provides opportunities for regional and continental maps depicting climates at various times in the past [*Wright et al.*, 1993; *Viau et al.*, 2006]. However, when synthesizing the large numbers of pollen diagrams, each slightly different due to local site conditions or disturbance histories, it is easy to get lost in all of the details. Diagrams from different regions of North America can contain dissimilar pollen taxa, so to determine which changes are caused by climate variations and which by some local factor requires knowledge of the local ecological and biogeographic conditions (see below). Many of these diagrams have few radiocarbon dates and low temporal resolution of samples. The spatial replication of the results nevertheless lends confidence to our conclusions, as inspection and multivariate analysis of many pollen diagrams from a region show similarities between sites, suggesting a climate cause to many of the vegetation transitions [*Gajewski*, 1987, 1993]. Studies of the transitions show preferred times of transition in pollen data across North America [*Viau et al.*, 2002] and that these are coherent with changes in Europe [*Gajewski et al.*, 2006].

2.2. Data Suitability and Biogeography

At a descriptive level, paleoecologists have been able to make general conclusions about past climate changes using pollen and plant macrofossil data. Indeed, many of the terms used in discussing Holocene climate variability, such as Younger Dryas, Bolling, and Allerød were originally seen in peat and lake sediments from Scandinavia [*Birks and Birks*, 1980]. The overall Milankovitch-scale climate variability has been successfully reconstructed using pollen data from several regions of the world and compared to climate model output [*Wright et al.*, 1993; *Sawada et al.*, 2004]. Many issues remain, and research is continuing to develop the data needed for this research effort. There are still large areas with few available fossil or modern data, for example, in desert regions, in very cold regions, and in mountain areas.

One reason for the success in using vegetation and pollen records for paleoclimate work is that the ecological tolerances and biogeography of the major tree taxa is well known. Forestry and ecological data and knowledge, accumulated over a century, have resulted in not only an understanding of the site conditions associated with optimal growth for a species but also in maps depicting the distribution in space of most vascular plants. In Europe and eastern North America, the postglacial migration of the major tree taxa has been mapped [*Williams et al.*, 2004; *Huntley and Webb*, 1989]. This basic biological knowledge means that the reasonableness of resulting climate reconstructions can be assessed.

A question that has for a long time affected paleoclimate work using pollen data is the ability of vegetation to respond rapidly to abrupt climate changes and, therefore, the ability of pollen records to be used for the study of abrupt climate changes. Since many trees are long-lived, it is argued that vegetation responds very slowly to climate changes with a long lag time [*Davis*, 1986; *Wright*, 1976]. However, *Webb* [1986] and *Huntley and Webb* [1989] have argued that vegetation changes are in equilibrium with climate changes at the Milankovitch scale [*Webb*, 1986]. Studies have shown that vegetation responds to perturbation at decadal to millennial scales [*Wendland and Bryson*, 1974; *Gajewski*, 1987, 1993, 2008; *Grimm and Jacobson*, 1991; *Shuman et al.*, 2002a, 2002b; *Williams et al.*, 2002; *Jackson and Williams*, 2004; *Peros et al.*, 2008], and tree ring and ecological analysis has extensively documented the response of trees to climate variations at annual to multidecadal scales [*Intergovernmental Panel on Climate Change*, 2007]. In part, the response time depends on the type of vegetation; herbaceous or shrubby vegetation responds more rapidly, and indeed, changes in shrub growth at tree line are being observed in Alaska, presumably in response to global warming [*Tape et al.*, 2006]. However, even in forests, structural changes in the vegetation can occur rapidly, after a fire, for example. In an uneven aged forest, subcanopy trees can quickly start to grow faster following a climatic or local perturbation, causing a change in the relative pollen production of the species in the forest. Finally, it is important to note that the fossil indicator used in sediments is pollen and not plants [*Webb and Bryson*, 1972]. Changes in pollen production of different plant taxa can respond rapidly following rapid climatic transitions (see below) and can be registered in sediment with a high enough sedimentation rate. Thus, abrupt changes can be interpreted from pollen records [*Gajewski*, 1987, 1993; *Shuman et al.*, 2002b; *Gajewski et al.*, 2007].

Another concern when using pollen records is the potential importance of human impact on the landscape, especially as it may influence paleoclimate reconstructions. If human activities, such as agriculture or urbanization affect the landscape, this may make paleoclimate impacts more difficult to discern. *Ruddiman* [2003] has recently argued that human activities have had a significant impact on atmospheric composition, suggesting such a large human impact on the global landscape that it should be discernible in pollen records. In Europe, palynologists have identified the impact of human activities on the landscape during the past several millennia, and these impacts are so extensive that much of the subcontinent is considered a "cultural landscape." However, millennial-scale climate changes still impacted the vegetation, as shown by the coherence of major transitions in pollen diagrams with those occurring in North America [*Gajewski et al.*, 2006]. Although the impact of European settlement on the vegetation of North America is well known, that of Native Americans is under debate. Several local studies have shown indications of Native American agriculture in pollen records from eastern North American, and using pollen and archeological databases, *Munoz and Gajewski* [2010, and references therein] showed this at a regional scale. There is ongoing research into these questions, as geographers, archaeologists, and paleobotanists continue to debate questions such as the size of Native American populations and their impact on the vegetation. But it seems that climate change impacts on the vegetation can still be identified in spite of human impacts on the landscape.

Among other aspects of the paleoenvironmental record that are not well understood include the importance of CO_2 fertilization on plant production and biodiversity, where it has been suggested that this may contribute to the so-called "nonanalogue" communities of the full glacial [*Cowling and Sykes*, 1999; *Williams et al.*, 2000]. Nonanalogue conditions during the late glacial, when CO_2 reached preindustrial levels, are also not entirely explained, although Northern Hemisphere summer insolation changes may be at cause [*Webb*, 1986; *Williams and Jackson*, 2007]. In addition, the impact of changes in seasonality on plants and vegetation needs further study, and there appears to be a fundamental difficulty in reconstructing seasonal climates using the MAT [*Viau et al.*, 2008]. At the moment, these effects seem to be secondary and the broad-scale reconstructions are in general agreement with independent proxy records.

3. RECORDS OF ABRUPT CLIMATE CHANGE IN NORTH AMERICA DURING THE HOLOCENE

3.1. Regional Holocene Climate Reconstructions

In North America, the most studied area is the deciduous forest region of eastern North America. Owing in large part to the concentration of data from the region, a long research program has shown conclusively the importance of climate change in affecting the vegetation of the region [e.g., *Bernabo and Webb*, 1977; *Webb*, 1986; *Williams et al.*, 2004]. This work has provided key data supporting the importance of Milankovitch-scale climate variations in affecting the climate [*COHMAP Members*, 1988; *Wright et al.*, 1993]. More recent work has illustrated the importance of abrupt climate changes [*Gajewski*, 1987; *Shuman et al.*, 2002a, 2002b] on vegetation composition.

In areas with less dense site networks, however, many sites are being studied at high temporal resolution and are showing abrupt climate changes. As an example of a recent multiproxy synthesis from one site in the Canadian Arctic, Lake

KR02 from western Victoria Island, Northwest Territories, Canada, provides a high temporal resolution, multiproxy record of rapid environmental change during the past 10 kyr (Figure 2). Pollen, chironomid, diatom, and sediment parameters were analyzed in the same core [*Podritskie and Gajewski*, 2007; *Peros and Gajewski*, 2008; *Fortin and Gajewski*, 2010b]. Quantitative estimates of July temperature were developed from the pollen and chironomid records using different methods (transfer function and MAT) for both proxy-climate series, and lake water pH changes through the Holocene could be estimated using biogenic Si and organic matter content of the sediment [*Fortin and Gajewski*, 2009].

Results show abrupt changes in July temperatures throughout the Holocene, which can be related to independent records of climate variability such as from the Agassiz Ice Cap ice melt record [*Fisher et al.*, 1995]. Interestingly, estimates of both terrestrial primary and aquatic secondary production (pollen and chironomid influx) are also comparable and show a highly productive ecosystem in the early Holocene, with a decrease in aquatic and terrestrial biological production associated with a cooling over the past several thousand years. The general form of these production records from Lake KR02 parallels the ice melt record from the Agassiz Ice Cap, suggesting aquatic and terrestrial production, as well as ice melt on glaciers are related to summer conditions. The millennial-scale climate changes also affected the lake water pH, although interpretation of this curve is complicated by dissolution of the diatoms. Similar results were obtained in another multiproxy record from nearby Melville Island [*Peros et al.*, 2010] and north central Victoria Island [*Fortin and Gajewski*, 2010a]. Apart from the long-term changes during the Holocene, which clearly indicate a warm early Holocene and cooling in the past millennia [*Kaufman et al.*, 2004; *Gajewski and Atkinson*, 2003], more rapid changes are coherent across the Canadian Arctic and with records such as Greenland and Ellesmere Island ice cores [*Alley*, 2004; *Fisher et al.*, 1995] (Figure 2). Thus, a regional picture is emerging from the Canadian Arctic, where abrupt climate changes seen in records from elsewhere also impacted the ecosystems of the region during the Holocene.

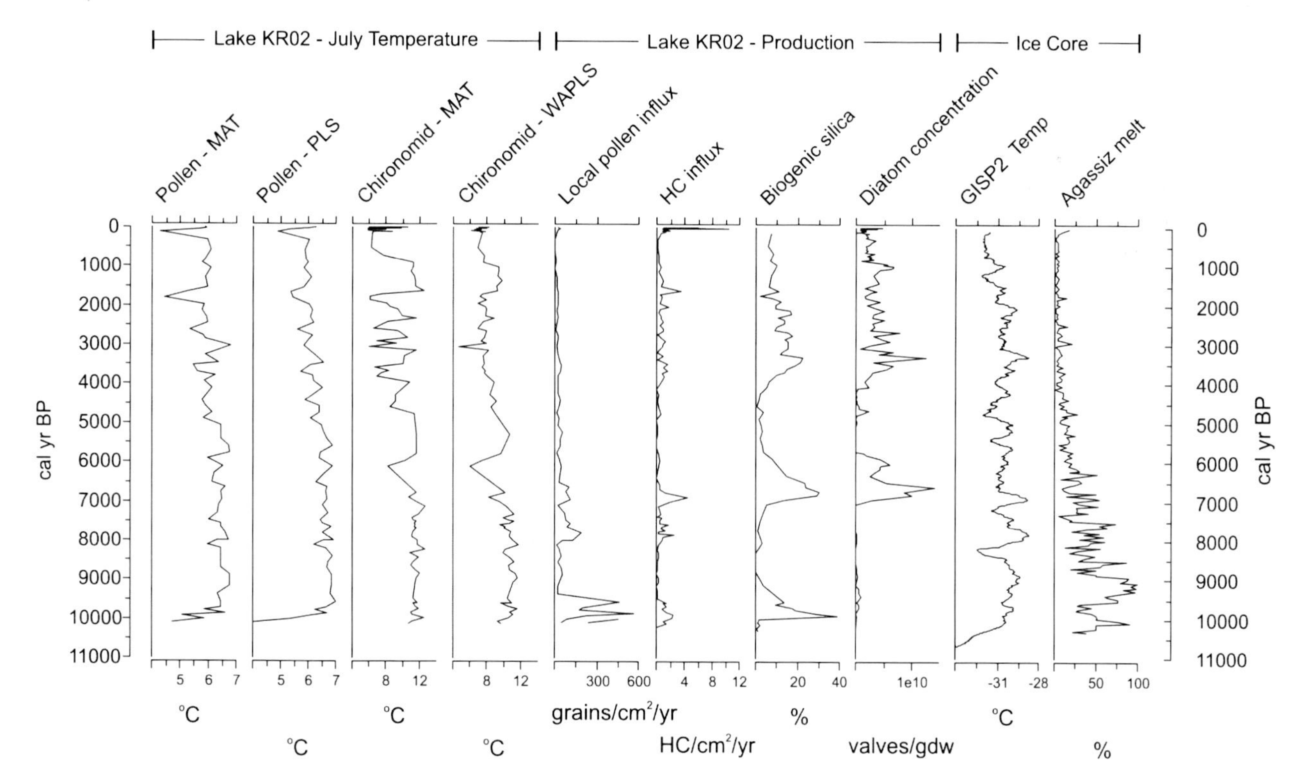

Figure 2. Paleoclimate estimates and indices of biological production in a lake sediment core from western Victoria Island, Lake KR02. See text for references. Temperatures are mean July temperature reconstructed using both pollen and chironomids and the modern analogue technique (MAT) or a transfer function (partial least squares (PLS) and weighted-averaged partial least squares (WAPLS)). Biological production indices are pollen influx and two indices of aquatic primary production. Reconstructed temperature from the Greenland Ice Sheet Project 2 (GISP2) core [*Alley*, 2004] and ice melt percentage from the Agassiz ice core [*Fisher et al.*, 1995] are shown.

Recently, paleoclimate reconstructions using pollen data quantified Holocene climate variability on several timescales across the Canadian Boreal and Alaskan regions [*Viau and Gajewski*, 2009; *Viau et al.*, 2008]. Regional reconstruction of the past 12,000 years show a time-transgressive pattern of climate change on orbital scales where eastern Canada was relatively cool compared to western Canada and Alaska during the early Holocene due to the remnants of the Laurentide Ice Sheet in the east [*Viau and Gajewski*, 2009]. A similar time-transgressive pattern is seen at the millennial scale. Results for the past 2 kyr identified multicentennial-scale climate variability (i.e., MWP and LIA) across the boreal region [*Viau and Gajewski*, 2009], which shows that even relatively brief climate variations cause rapid responses in regional pollen production and vegetation patterns [*Gajewski*, 1987; *Gajewski et al.*, 2007; *Williams et al.*, 2002].

3.2. Continental Paleoclimate Records of Abrupt Climate Changes

Peros et al. [2008] illustrated the impacts of rapid climate change, especially during the Younger Dryas, on the populations of poplar (*Populus*) across North America. They found that poplar pollen increased at both the beginning and end of the Younger Dryas. At the beginning of the Younger Dryas, spruce (*Picea*) pollen decreased in many sites across North America, whereas poplar increased in abundance, probably as the forest opened in many areas, and poplar could tolerate the conditions that were too cold for spruce. Today, poplar is found at or near tree line and forms tree line in a few small areas of North America. Because it is a pioneer species, it could quickly invade the landscape as conditions ameliorated. During the entire period, alder (*Alnus*) pollen changed little.

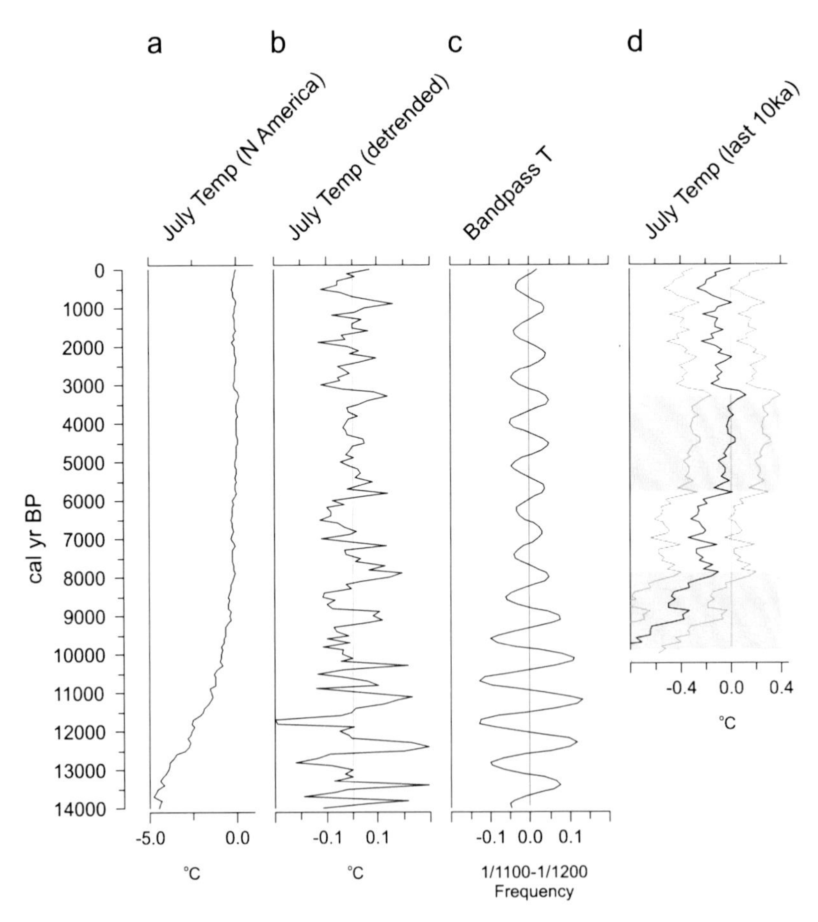

Figure 3. Mean July temperature anomaly for North America estimated using the MAT and the pollen data shown in Figure 1. (a) Mean July temperature anomalies for the past 14 kyr. (b) As in Figure 3a, detrended using 10th-order polynomial. (c) Band-pass-filtered curve of Figure 3b. (d) Curve from Figure 3a with x axis expanded. Gray lines are confidence limits, and gray shading shows general subdivisions of the curve for the Holocene. After the work of *Viau et al.* [2006].

Alder is abundant in the forest-tundra, and thus can probably tolerate more cold conditions than spruce or poplar. However, this example also illustrates another aspect to be considered when interpreting pollen records. Just before 9 ka, poplar populations decreased in abundance and remained low for the remainder of the Holocene, until the European disturbance of the past 400 years. In this case, it was suggested that competition from other trees may have been responsible for the decrease. Analyses such as these indicate the rapid response of tree populations to climate changes. Although the changes may be complex and not always intuitive, they provide good examples of the rapid nature of climate transitions, as well as the response.

Viau et al. [2006] estimated the climate of North America during the past 14 kyr using over 750 pollen records that comprised over 30,000 samples dated by approximately 2500 radiocarbon dates extracted from the NAPD from across North America. After the July temperature was estimated using the MAT for each pollen sample, data were interpolated to uniform time intervals, and a mean temperature curve for North America was computed. This continental-scale reconstruction quantified temperature variations of several timescales during the past 14 kyr (Figure 3). Temperatures increased during the late glacial, and maximum Holocene temperatures occurred between 6000 and 3000 cal years B.P. in North America. Millennial-scale temperature variations, with a range of ±0.2°C and separated by abrupt transitions, are superimposed on the Milankovitch-scale variability. The dominant frequency of the July temperature variability is ~1100 years (as opposed to the ~1500 year cycle found during the last glacial period), with abrupt transitions between climate states. In addition, there was a scale interaction, with the millennial-scale variability being more pronounced during both the late glacial warming and the late Holocene cooling. A comparable study was performed using European pollen data [*Davis et al.*, 2003].

In a related study, *Gajewski et al.*, [2006] showed synchronous abrupt vegetation transitions during the Holocene between both North American and European continents. In this study, two independent data sets from Europe were used to test the robustness of the results. Major vegetation transitions averaging ~1100 years intervals between both continents during the Holocene were also found to be synchronous with North Atlantic ice-rafted detritus (IRD) events [*Bond et al.*, 2001].

3.3. Relating Terrestrial to Ice Core and Marine Records

Studies of pollen in marine cores [e.g., *Heusser and Morley*, 1985; *Lezine and Denefle*, 1997] have been used for large-scale paleoclimate reconstructions. Comparison of pollen records from lakes in Labrador and Quebec with pollen and dinoflagellate records in offshore marine cores showed broad-scale coherence [*Sawada et al.*, 1999]. Pollen have been studied in ice cores [*Bourgeois et al.*, 2000], permitting a direct comparison of the ice core record with vegetation change. Indeed, a high-resolution study of seasonal ice layers in the Agassiz ice core has shown that pollen are transported to the ice cap in the same season as they are liberated and thus faithfully record the phenology of the major plant groups [*Bourgeois*, 2001; *Gajewski*, 2006]. Studies such as this can be used to directly relate climate variability reconstructed from terrestrial and marine or ice caps, as the pollen are a common proxy found in all systems.

When well-dated records of abrupt changes in the past are available, they can be compared and lend insight into potential climate forcing mechanisms. A spectral analysis of the reconstruction of *Viau et al.* [2006] found a major peak around 1100 years (Table 1). A similar peak has been identified in several studies, including the Greenland Ice Core Project ^{18}O record during the Holocene [*Schulz and Paul*, 2002], North Atlantic IRD records [*Bond et al.*, 2001] and North Atlantic circulation patterns [*Chapman and Shackleton*, 2000], among others. A similar periodicity was identified in high-resolution records from central United States [*Overpeck*, 1987], Scotland [*Langdon et al.*, 2003], and Alaska [*Hu et al.*, 2003]. *Bond et al.* [2001] have shown a cross-spectral coherence of their IRD record with cosmogenic nuclide records (^{14}C and ^{10}Be, proxies for solar variability), at the 900 to 1100 year frequency bands for the Holocene.

Table 1. Dominant Peaks in Spectral Analyses of Several Time Series From the North Atlantic Region and North America

Reference	Periodicities	
	10^2 Time Scale	10^3 Time Scale
Cross-spectral ice-rafted detritus (IRD)/^{14}C [*Bond et al.*, 2001]	400–500	900–1100
Cross-spectral IRD/^{10}Be [*Bond et al.*, 2001]	400–500	900–1100
North Atlantic IRD [*Bond et al.*, 1997, 2001]		~1350
GRIP/GISP2 ^{18}O [*Schulz and Paul*, 2002]		900
THC/NADW ^{13}C [*Chapman and Shackleton*, 2000]	550	1000
North America [*Viau et al.*, 2006]		1150
Subarctic Alaska [*Hu et al.*, 2003]		950
Midwest United States [*Overpeck*, 1987]		1100
Southeast Scotland [*Langdon et al.*, 2003]		1100

Comparisons in the frequency domain provide a useful strategy for analyzing possible causes of the variability at any time scale; however, there is interest in direct comparison of individual transitions to determine leads and lags in the climate system as well as to describe the nature of the transition (Figure 4). Comparing the various series, there are similarities but also differences among them, and it is difficult to propose an overall spatial model of Holocene climate variability using these data. For example, the North American mean July temperature anomaly, Santa Barbara, and West African marine records all show major changes at around 8, 6, and 3.2 ka, but these transitions are not as clear in the other records. Higher-frequency changes are not always coherent. This is due in part to the low density of sites with high enough temporal resolution and also due to issues with dating the sediments. As noted by *Viau et al.* [2006] and *Gajewski et al.* [2007], "wiggle-matching" series from across large distances can lead to difficulties. In any geophysical series, any individual climate transition may be missing, due to local factors such as sedimentation hiatus or simply a lack of local response to the climate forcing. Therefore, oscillations may appear shorter or longer in some records but not even be seen in others, and this may explain, in part, the differences between records. The interactions between slow variations and more abrupt changes further complicates the interpretation of Holocene climate records, especially during the mid-Holocene, when scale interactions between the orbital and suborbital scales seems to have reduced the millennial-scale climate variability [*Fisher*, 1982; *Gajewski*, 1983]. An individual paleoclimate series records only the regional climate, but it is the spatial pattern of climate that is the response to forcing [*Gajewski*, 1987; *Viau et al.*, 2006], and a sufficiently dense series of sites is needed to understand changes in the general circulation.

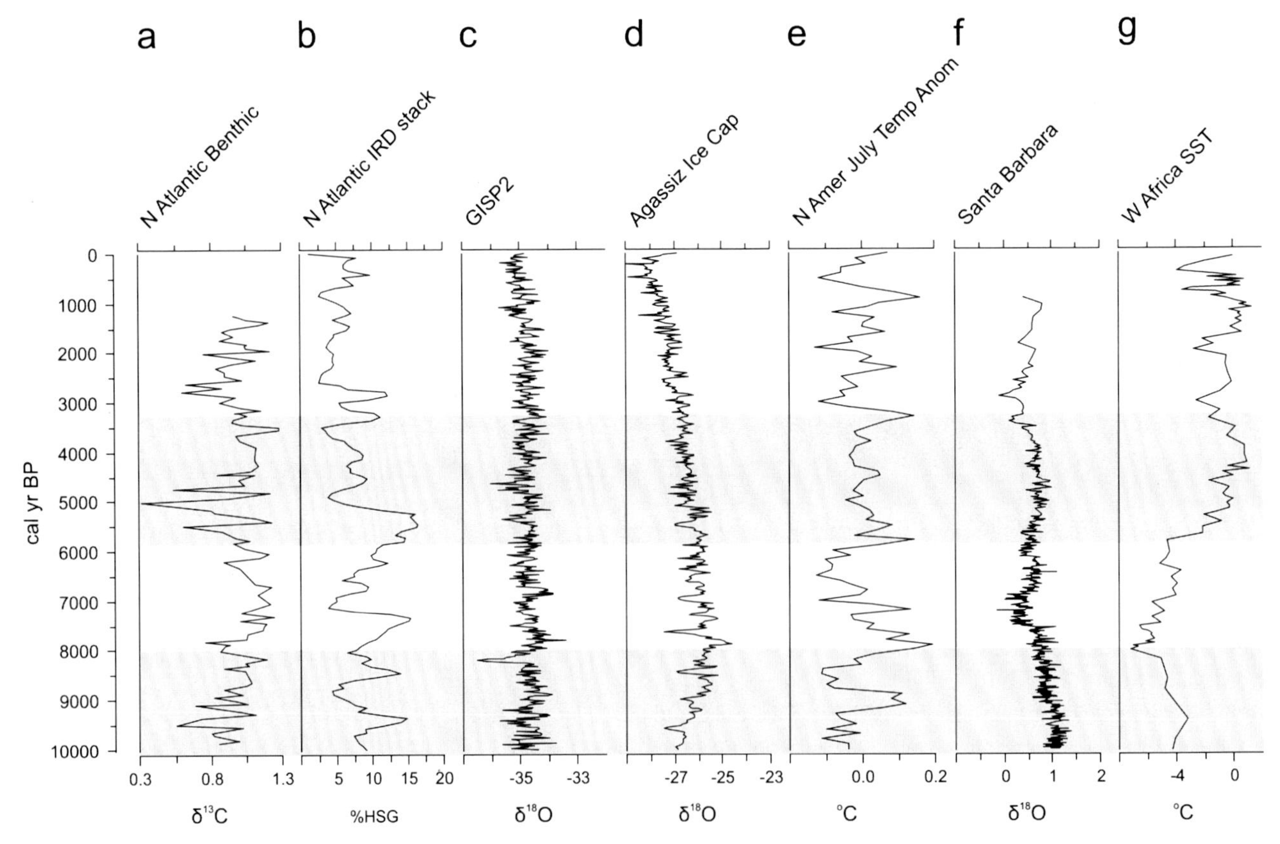

Figure 4. Several high-resolution paleoclimate series from North America and adjacent regions: (a) North Atlantic benthic $\delta^{13}C$ [*Oppo et al.*, 2003]; (b) "stacked" record of hematite-stained glass, an ice-rafted detritus record from *Bond et al.* [2001]; (c) $\delta^{18}O$ record from the GISP2 ice core, Greenland [*Alley et al.*, 1997]; (d) $\delta^{18}O$ record from the Agassiz Ice Cap, Ellesmere Island [*Fisher et al.*, 1995]; (e) mean July temperature for North America, reconstructed using pollen data [*Viau et al.*, 2006]; (f) $\delta^{18}O$ for the Santa Barbara Basin [*Fridell et al.*, 2003]; and (g) warm season sea surface temperatures off the West African coast [*deMenocal et al.*, 2000]. Shading delimits the four major periods in the North American pollen-based July temperature anomaly reconstruction.

4. DISCUSSION

Results such as those presented by *Viau et al.* [2006] show that the climate was continually changing during the Holocene, at many timescales. Many of the "events" identified in some records are part of a continuous series of rapid climate changes, including variations such as the Younger Dryas and the latest of which are the MWP and LIA. For example, the Holocene in North America can be divided into four general periods, with abrupt transitions at ~8, 6, and 3 ka (Figure 4). The millennial-scale variability, with a prominent spectral peak at ~1000 ± 100 years (Table 1), varies as a function of these four periods of time and shows increased variability during the early Holocene and the past 3 kyr. Transitions between these periods are abrupt in many cases. The LIA, for example, is part of a continual series of events during the Holocene and is especially prominent due to the scale interaction with the slower Milankovitch-forced variability. Rapid events identified in some series, such as the 4.2 or 5.2 "events" are part of this continual variability.

These results suggest that the "relative climate stability" supposedly seen in records such as the Greenland ice cores during the Holocene is overstated, and although the variability during interglacial periods is less than that during glacial regimes, it remains significant and large enough to be of relevance to human activities. The fundamental causes of this variability are not entirely explained; however, the evidence suggests that rapid climate change may be related to the interactions of changes in incoming solar radiation and other forcing and may be amplified through internal feedback mechanisms.

Acknowledgments. This research is funded by the Natural Sciences and Engineering Research Council of Canada. We acknowledge the contributors to the NAPD and other scientists for contributing their data online.

REFERENCES

Alley, R. B. (2004), GISP2 Ice Core Temperature and Accumulation Data, IGBP PAGES/World Data Center for Paleoclimatology *Data Contrib. Ser. 2004-013*, http:\\www.ncdc.noaa.gov/paleo, NOAA/NCDC Paleoclimatol. Program, Boulder, Colo.

Alley, R. B., P. A. Mayewski, T. Sowers, M. Stuiver, K. C. Taylor, and P. U. Clark (1997), Holocene climatic instability: A prominent widespread event 8200 yr ago, *Geology*, *25*, 483–486.

Battarbee, R. W. (2000), Palaeolimnological approaches to climate change, with special regard to the biological record, *Quat. Sci. Rev.*, *19*, 107–124.

Bernabo, J. C., and T. Webb III (1977), Changing patterns in the Holocene pollen record of northeastern North America: A mapped summary, *Quat. Res.*, *8*, 64–96.

Bianchi, G. G., and I. N. McCave (1999), Holocene periodicity in North Atlantic climate and deep-ocean flow south of Iceland, *Nature*, *397*, 515–517.

Birks, H. J. B., and H. H. Birks (1980), *Quaternary Palaeoecology*, Blackburn, Caldwell, N. J.

Bond, G., W. Showers, M. Cheseby, R. Lotti, P. Almasi, P. deMenocal, P. Priore, H. Cullen, I. Hajdas, and G. Bonani (1997), A pervasive millennial-scale cycle in North Atlantic Holocene and glacial climates, *Science*, *278*, 1257–1265.

Bond, G., B. Kromer, J. Beer, R. Muscheler, N. E. Evans, W. Showers, S. Hoffman, R. Lotti-Bond, I. Hajdas, and G. Bonani (2001), Persistent solar influence on North Atlantic climate during the Holocene, *Science*, *294*, 2130–2136.

Bouchard, G., K. Gajewski, and P. Hamilton (2004), Freshwater diatom biogeography in the Canadian Arctic Archipelago, *J. Biogeogr.*, *31*, 1955–1973.

Bourgeois, J. (2001), Seasonal and annual variation of pollen variability in snow layers of Arctic ice caps, *Rev. Palaeobot. Palynol.*, *108*, 17–36.

Bourgeois, J., R. Koerner, K. Gajewski, and D. Fisher (2000), A Holocene ice core pollen record from Ellesmere Island, Nunavut, Canada, *Quat. Res.*, *54*, 275–283.

Bryant, V. M., and R. G. Holloway (1985), *Pollen Records of Late-Quaternary North American Sediments*, Am. Assoc. of Stratigr. Palynol., Dallas, Tex.

Chapman, M. R., and N. J. Shackleton (2000), Evidence of 550-year and 1000-year cyclicities in North Atlantic circulation patterns during the Holocene, *Holocene*, *10*, 287–291.

COHMAP Members (1988), Climatic changes of the last 18,000 years: Observations and model simulations, *Science*, *241*, 1043–1052.

Cowling, S. A., and M. T. Sykes (1999), Physiological significance of low atmospheric CO_2 for plant-climate interactions, *Quat. Res.*, *52*, 237–242.

deMenocal, P. B., J. Ortiz, T. Guilderson, and M. Sarnthein (2000), Coherent high- and low-latitude climate variability during the Holocene warm period, *Science*, *288*, 2198–2202.

Dansgaard, W., et al. (1993), Evidence for general instability of past climate from a 250 kyr ice core, *Nature*, *364*, 218–219.

Davis, B. A. S., S. Brewer, A. Stevenson, and J. Guiot (2003), The temperature of Europe during the Holocene reconstructed from pollen data, *Quat. Sci. Rev.*, *22*, 1701–1716.

Davis, M. B. (1986), Climatic instability, time lags, and community disequilibrium, in *Community Ecology*, edited by J. Diamond and T. Case, pp. 269–284, Harper and Row, New York.

EPICA Community Members (2006), One-to-one coupling of glacial climate variability in Greenland and Antarctica, *Nature*, *444*, 195–198.

Fisher, D. A. (1982), Carbon 14 production compared to oxygen isotope records from Camp Century, Greenland and Devon Island, *Clim. Change*, *4*, 419–426.

Fisher, D. A., R. M. Koerner, and N. Reeh (1995), Holocene climatic records from Agassiz Ice Cap, Ellesmere Island, NWT, Canada, *Holocene*, *5*, 19–24.

Fortin, M.-C., and K. Gajewski (2009), Assessing sediment production indices in Arctic lakes, *Polar Biol.*, *32*, 985–998.

Fortin, M.-C., and K. Gajewski (2010a), Holocene climate change and its effect on lake ecosystem production in northern Victoria Island, Canadian Arctic, *J. Paleolimnol.*, *43*, 219–234.

Fortin, M.-C., and K. Gajewski (2010b), Postglacial environmental history of western Victoria Island, Canadian Arctic, *Quat. Sci. Rev.*, *29*, 2099–2110.

Fridell, J. E., R. C. Thunell, T. P. Guilderson, and M. Kashgarian (2003), Increased northeast Pacific climatic variability during the warm middle Holocene, *Geophys. Res. Lett.*, *30*(11), 1560, doi:10.1029/2002GL016834.

Gajewski, K. (1983), On the interpretation of climatic change from the fossil record: Climatic change in central and eastern United States for the past 2000 years estimated from pollen data, Ph.D. thesis, Univ. of Wisconsin-Madison, Madison.

Gajewski, K. (1987), Climatic impacts on the vegetation of eastern North America for the past 2000 years, *Vegetatio*, *68*, 179–190.

Gajewski, K. (1993), The role of paleoecology in the study of global climatic change, *Rev. Palaeobot. Palynol.*, *79*, 141–151.

Gajewski, K. (2006), Is Arctic palynology a "blunt instrument"?, *Geogr. Phys. Quat.*, *60*, 95–102.

Gajewski, K. (2008), The global pollen database in biogeographical and paleoclimatic studies, *Prog. Phys. Geogr.*, *32*, 379–402.

Gajewski, K., and D. Atkinson (2003), Climatic change in northern Canada, *Environ. Rev.*, *11*, 69–102.

Gajewski, K., et al. (2000), The climate of North America and adjacent ocean waters ca. 6ka, *Can. J. Earth Sci.*, *37*, 661–681.

Gajewski, K., A. E. Viau, M. C. Sawada, D. Atkinson, and P. Fines (2006), Synchronicity in climate and vegetation transitions between Europe and North America during the Holocene, *Clim. Change*, *78*, 341–361.

Gajewski, K., A. E. Viau, and M. C. Sawada (2007), Millennial-scale climate variations in the Holocene – the terrestrial record, in *Climate Variability and Change: Past, Present and Future—John E Kutzbach Symposium*, edited by G. Kutzbach, pp. 133–154, Cent. for Clim. Res., Univ. of Wisconsin, Madison.

Grimm, E. (2010), Global Pollen Database, www.ncdc.noaa.gov/paleo, Natl. Clim. Data Cent., Asheville, N. C.

Grimm, E. C., and G. L. Jacobson Jr. (1991), Fossil-pollen evidence for abrupt climate changes during the past 18,000 years in eastern North America, *Clim. Dyn.*, *6*, 179–184.

Heinrich, H. (1988), Origin and consequences of cyclic ice rafting in the northeast Atlantic Ocean during the past 130000 years, *Quat. Res.*, *29*, 142–152.

Heusser, L. E., and J. J. Morley (1985), Correlative 90 kyr northeast Asia–northwest Pacific climate records, *Nature*, *313*, 470–472.

Hu, F. S., D. Kaufman, S. Yoneji, D. Nelson, A. Shemesh, Y. Huang, J. Tian, G. Bond, B. Clegg, and T. Brown (2003), Cyclic variation and solar forcing of Holocene climate in the Alaskan subarctic, *Science*, *301*, 1890–1893.

Huntley, B., and T. Webb III (1989), Migration: Species' response to climatic variations caused by changes in the earth's orbit, *J. Biogeogr.*, *16*, 5–19.

Imbrie, J., A. McIntyre, and A. C. Mix (1989), Oceanic response to orbital forcing in the Late Quaternary: Observational and experimental strategies, in *Climate and Geosciences, A Challenge for Science and Society in the 21st Century*, edited by A. Berger, S. H. Schneider, and J.-C. Duplessy, pp. 121–164, D. Reidel, Dordrecht, Netherlands.

Intergovernmental Panel on Climate Change (2007), *Climate Change 2007: The Physical Science Basis: Working Group I Contribution to the Fourth Assessment Report of the IPCC*, edited by S. Solomon et al., 996 pp., Cambridge Univ. Press, Cambridge, U. K.

Jackson, S. T., and J. W. Williams (2004), Modern analogs in Quaternary paleoecology: Here today, gone yesterday, gone tomorrow?, *Annu. Rev. Earth Planet. Sci.*, *32*, 495–537.

Johnsen, S. J., D. Dahl-Jensen, N. Gundestrup, J. P. Steffensen, H. B. Clausen, H. Miller, V. Mason-Delmotte, A. E. Sveinbjörnrndottir, and J. White (2001), Oxygen isotope and palaeotemperatures records from six Greenland ice-core stations: Camp Century, Dye-3, GRIP, GISP2, Renland and NorthGRIP, *J. Quat. Sci.*, *16*, 299–307.

Kaufman, D. (2009), An overview of late Holocene climate and environmental change inferred from Arctic lake sediment, *J. Paleolimnol.*, *41*, 1–6.

Kauffman, D., et al. (2004), Holocene thermal maximum in the western Arctic (0° to 180°W), *Quat. Sci. Rev.*, *23*, 529–560.

Langdon, P. G., K. E. Barber, and P. D. M. Hughes (2003), A 7500-year peat-based palaeoclimatic reconstruction and evidence for an 1100-year cyclicity in bog surface wetness from Temple Hill Moss, Pentland Hills, southeast Scotland, *Quat. Sci. Rev.*, *22*, 259–274.

Lezine, A.-M., and M. Denefle (1997), Enhanced anticyclonic circulation in the eastern North Atlantic during cold intervals of the last deglaciation inferred from deep-sea pollen records, *Geology*, *25*, 119–122.

MacManus, J. F., D. W. Oppo, and J. L. Cullen (1999), A 0.5 million-year record of millennial-scale climate variability in the North Atlantic, *Science*, *283*, 971–974.

Munoz, S., and K. Gajewski (2010), Distinguishing prehistoric human influence on late Holocene forests in southern Ontario, Canada, *Holocene*, *20*, 967–981.

O'Brien, S. R., A. Mayewski, L. D. Meeker, D. A. Meese, M. S. Twickler, and S. I. Whitlow (1995), Complexity of Holocene climate as reconstructed from a Greenland ice core, *Science*, *270*, 1962–1964.

Oppo, D. W., J. E. McManus, and J. L. Cullen (1998), Abrupt climate events 500,000 to 340,000 years ago: Evidence from subpolar North Atlantic sediments, *Science*, *279*, 1335–1338.

Oppo, D. W., J. F. McManus, and J. L. Cullen (2003), Deepwater variability in the Holocene epoch, *Nature*, *422*, 277–278.

Overpeck, J. T. (1987), Pollen time series and Holocene climate variability of the Midwest United States, in *Abrupt Climatic*

Change, edited by W. H. Berger and L. D. Labeyrie, pp. 137–143, D. Reidel, Norwell, Mass.

Overpeck, J. T., T. Webb III, and I. C. Prentice (1985), Quantitative interpretation of fossil pollen spectra: Dissimilarity coefficients and the method of modern analogs, *Quat. Res.*, *23*, 87–108.

Peros, M., and K. Gajewski (2008), Holocene climate and vegetation change on Victoria Island, western Canadian Arctic, *Quat. Sci. Rev.*, *27*, 235–249.

Peros, M., K. Gajewski, and A. E. Viau (2008), Continental-scale tree population response to rapid climate change, competition and disturbance, *Global. Ecol. Biogeogr.*, *17*, 658–669.

Peros, M., K. Gajewski, T. Paull, R. Ravindra, and B. Podritske (2010), Multi-proxy record of postglacial environmental change of south-central Melville Island, Northwest Territories, Canada, *Quat. Res.*, *73*, 247–258.

Podritske, B., and K. Gajewski (2007), Diatom community response to multiple scales of Holocene climate variability in a small lake on Victoria Island, NWT, Canada, *Quat. Sci. Rev.*, *26*, 3179–3196.

Rahmstorf, S. (2003), Timing of abrupt climate change: A precise clock, *Geophys. Res. Lett.*, *30*(10), 1510, doi:10.1029/2003GL017115.

Raymo, M. E., K. Ganley, S. Carter, D. W. Oppo, and J. MacManus (1998), Millennial-scale climate instability during the early Pleistocene epoch, *Nature*, *392*, 699–702.

Ruddiman, W. (2003), The anthropogenic greenhouse era began thousands of years ago, *Clim. Change*, *61*, 261–293.

Sawada, M., K. Gajewski, A. deVernal, and P. Richard (1999), Comparison of marine and terrestrial Holocene climates in eastern North America, *Holocene*, *9*, 267–278.

Sawada, M., A. E. Viau, G. Vettoretti, W. R. Peltier, and K. Gajewski (2004), Comparison of North-American pollen-based temperature and global lake-status with CCCma AGCM2 output at 6 ka, *Quat. Sci. Rev.*, *23*, 225–244.

Schulz, M., and A. Paul (2002), Holocene climate variability on centennial to millennial time scales: 1. Climate records from the North-Atlantic realm, in *Climate Development and History of the North Atlantic Realm*, edited by G. Wefer et al., pp. 41–54, Springer, New York.

Shuman, B., P. Bartlein, N. Logar, P. Newby, and T. Webb III (2002a), Parallel climate and vegetation responses to the early-Holocene collapse of the Laurentide Ice Sheet, *Quat. Sci. Rev.*, *21*, 1793–1805.

Shuman, B., T. Webb III, P. J. Bartlein, and J. W. Williams (2002b), The anatomy of a climatic oscillation: Vegetation change in eastern North America during the Younger Dryas chronozone, *Quat. Sci. Rev.*, *21*, 1777–1791.

Tape, K., M. Sturm, and C. Racine (2006), The evidence for shrub expansion in Northern Alaska and the Pan-Arctic, *Global Change Biol.*, *12*, 686–702.

Viau, A. E., and K. Gajewski (2009), Reconstructing millennial-scale, regional paleoclimates of boreal Canada during the Holocene, *J. Clim.*, *22*, 316–330.

Viau, A. E., K. Gajewski, P. Fines, D. Atkinson, and M. C. Sawada (2002), Widespread evidence of 1500 yr climate variability in North America during the past 14,000 yr, *Geology*, *30*, 455–458.

Viau, A. E., K. Gajewski, M. C. Sawada, and P. Fines (2006), Millennial-scale temperature variations in North America during the Holocene, *J. Geophys. Res.*, *111*, D09102, doi:10.1029/2005JD006031.

Viau, A. E., K. Gajewski, M. C. Sawada, and J. Bunbury (2008), Low- and High-frequency climate variability in Beringia during the past 25,000 years, *Can. J. Earth Sci.*, *45*, 1435–1453.

Walker, I. R., and L. C. Cwynar (2006), Midges and palaeotemperature reconstruction—The North American experience, *Quat. Sci. Rev.*, *25*, 1911–1925.

Webb, T., III (1986), Is vegetation in equilibrium with climate? How to interpret Late-Quaternary pollen data, *Vegetatio*, *67*, 75–91.

Webb, T., III, and R. A. Bryson (1972), Late- and postglacial climatic change in the northern midwest USA: Quantitative estimates derived from fossil pollen spectra by multivariate statistical analysis, *Quat. Res.*, *2*, 70–115.

Wendland, W., and R. A. Bryson (1974), Dating climatic episodes of the Holocene, *Quat. Res.*, *4*, 9–24.

Whitmore, J., et al. (2005), Modern Pollen Data from North America and Greenland for Multi-scale Paleoenvironmental Applications, *Quat. Sci. Rev.*, *24*, 1828–1848.

Willemse, N. W., and T. E. Törnqvist (1999), Holocene century-scale temperature variability from West Greenland lake records, *Geology*, *27*, 580–584.

Williams, J. W., and S. T. Jackson (2007), Novel climates, no-analog plant communities, and ecological surprises: Past and future, *Front. Ecol. Evol.*, *5*, 475–482.

Williams, J. W., T. Webb III, B. N. Shuman, and P. J. Bartlein (2000), Do low CO_2 concentrations affect pollen-based reconstructions of LGM climates? A response to 'Physiological significance of low atmospheric CO_2 for plant-climate interactions' by Cowling and Sykes, *Quat. Res.*, *53*, 402–404.

Williams, J. W., D. M. Post, L. C. Cwynar, A. F. Lotter, and A. J. Levesque (2002), Rapid vegetation responses to past climate change, *Geology*, *30*, 971–974.

Williams, J. W., B. N. Shuman, T. Webb III, P. J. Bartlein, and P. Leduc (2004), Quaternary vegetation dynamics in North America: Scaling from taxa to biomes, *Ecol. Monogr.*, *74*, 309–334.

Williams, J. W., et al. (2006), *An Atlas of Pollen-Vegetation-Climate Relationships for the United States and Canada*, Am. Assoc. of Stratigr. Palynol., Dallas, Tex.

Wright, H. E. (1976), The dynamic nature of Holocene vegetation, *Quat. Res.*, *6*, 581–596.

Wright, H. E., J. E. Kutzbach, T. Webb III, W. F. Ruddiman, F. A. Street-Perrott, and P. J. Bartlein (1993), *Global Climates Since the Last Glacial Maximum*, Univ. of Minn. Press, Minneapolis.

K. Gajewski and A. E. Viau, Laboratory for Paleoclimatology and Climatology, Department of Geography, University of Ottawa, Ottawa, ON K1N 6N5, Canada. (gajewski@uottawa.ca; aviau@uottawa.ca)

Abrupt Holocene Climatic Change in Northwestern India: Disappearance of the Sarasvati River and the End of Vedic Civilization

B. S. Paliwal[1]

Department of Geology, Faculty of Science, Jai Narain Vyas University, Jodhpur, India

Abrupt climatic changes during the Holocene period accompanied by neotectonic disturbances in the northwestern part of the Indian Peninsular Shield brought about major changes in the region, popularly known as the Thar Desert. These changes not only disorganized the drainage pattern of the region but caused the disappearance of the Vedic Sarasvati River and several other tributaries of the Indus River. Many new tributaries like Jhelam, Chinab, Ravi, Satluj, and Vyas were added to the main course of the Indus River. Ancient mythological literature of India, particularly, Vedas, Puranas, Manusmriti, and Mahabharata of Hinduism, provides evidence showing that the Sarasvati and Drishadvati rivers were flowing in this region during the Vedic era. The famous Vedic civilization flourished along the banks of this mighty river. Because of these abrupt climatic changes and neotectonic activities the Vedic civilization, a civilization much older than those of the Indus Valley and the Harappan civilization, met its end, and several saline lakes were formed in the green fertile region.

1. INTRODUCTION

Abrupt climate change and global warming are important issues that are threatening our civilization and need to be addressed seriously. To find a clue for the future, we will have to look into the changes in the climate pattern of the recent past. This chapter describes abrupt climatic changes during the Holocene period and the effects in the region presently occupied by the great Thar Desert. This region to the west of the Aravalli Mountain Range (Figure 1), extending up to the plain of the Indus River in the northwestern part of the Indian Peninsula, has witnessed several climatic changes and neotectonic disturbances [*Sinha-Roy*, 1986; *Kenoyer*, 1998; *Sridhar et al.*, 1999; *Roy*, 1999; *Radhakrishna and Merh*, 1999; *Possehl*, 2002; *Madella and Fuller*, 2006] resulting in stabilizing and destabilizing a number of civilizations during the Quaternary period. Evidence of some has been excavated, and many more are yet to be discovered. The present trend of climate change is a small arch of a big circle. Abrupt climate changes during the Pleistocene and the Holocene have left signatures in the region, in addition to the geology ranging from the Precambrian to the Recent concealed below the sands of the great Thar Desert [*Heron*, 1953; *Paliwal*, 1993; *Roy and Jakhar*, 2002]. The drainage pattern of the tributaries has been severally affected; the main stream of the Indus River (Figure 2) has not. It follows the Indus Suture that was almost constant for the last 5 million years [*Clift and Blusztajn*, 2005; *Clift et al.*, 2008; *Paliwal*, 2008]. Therefore, whether it is the beginning or end of different civilizations in the region, the disappearance of Vedic Sarasvati River, formation of saline lakes, or desertification, just how all the events have been governed by the abrupt climate change in the region during the recent past is not well understood.

[1]Deceased.

Abrupt Climate Change: Mechanisms, Patterns, and Impacts
Geophysical Monograph Series 193

10.1029/2010GM001028

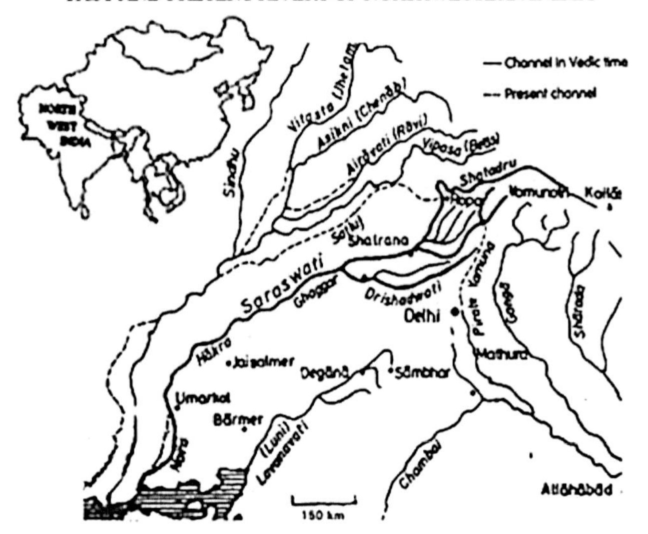

Figure 1. Distribution of the past and present rivers in northwestern India, 65.8°–80.2°E; 24.3°–36°N. From *Oldham* [1893].

2. PHYSIOGRAPHIC CONDITIONS

The region to the west of the Aravalli Mountain Range and southwest of the Himalaya is occupied by a series of rock formations ranging from the Precambrian to the Recent covered by a veneer of the fluvial and lacustrine sediments of the Quaternary period and windblown sands of the great Thar Desert.

Landforms in the area have been geomorphologically classified into three types based on their origin: aeolian (aeolian plain), alluvial (alluvial plain and floodplain), and denudational (structural hills, denudational hills, piedmont zone, and pediplain). The area of the aeolian plain covers a major part of the region. The pediplain is the second largest category that occupies the central and southern areas. Denudational hills occur near the towns of Jodhpur and Jaisalmer. Structural hills occur near Siwana and Jodhpur. Alluvial plains occupy the eastern part [*Singhvi and Kar*, 1992; *Roy and Jakhar*, 2002]. The region is almost sandy, dry and ill watered, forming the arid and semiarid regions of India. However, in the neighborhood of Jaisalmer, the terrain is stony containing numerous rocky ridges. With these exceptions, the region looks like a typical desert with numerous dunes, including the biggest ones at Ramgarh and Sam near Jaisalmer. The landscape of the region has evolved primarily through the aeolian processes [*Roy and Jakhar*, 2002; *Valdiya*, 2010]. Fluvial processes and the paleodrainage system have contributed significantly to the region. At the base of these hills, rocky/gravelly and buried pediments are located, which are mainly concentrated in the central, eastern, and northern parts of the region. Sand dunes and interdunal plains are located in the northwestern parts. Flat and sandy undulating plains of various length and width are located in between these dunes.

The Indus River and its tributaries (Jhelam, Chinab, Ravi, Satluj, and Beas and ephemeral river Luni) are the only rivers of this region that join the Arabian Sea (Figure 1). In addition to the Ghaggar River, which ends in the sands of the Thar Desert, paleochannels have also been identified near Agolai village and west of Jaisalmer constituting the remnants of the lost Sarasvati River [*Radhakrishna and Merh*, 1999; *Roy*, 1999; *Valdiya*, 2002, 2010].

3. FOSSIL GROUNDWATER IN THE REGION

To meet the increased demand for water for drinking and for agriculture in this water- scarce region where there is no surface water source, both Indian central and Rajasthan

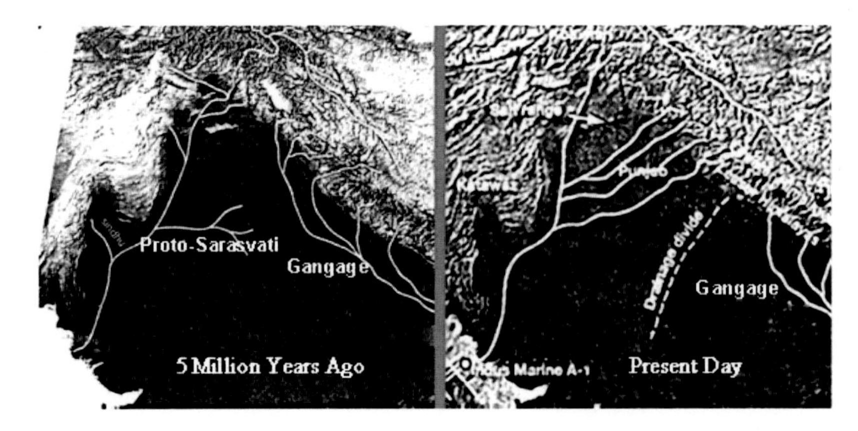

Figure 2. Topographic map of the region, 64.2°–83°E; 23.8°–36°N, showing changes in drainage pattern during the last 5 million years. After *Clift and Blusztajn* [2005] and *Paliwal* [2008]. Reprinted by permission from Macmillan Publishers Ltd: *Nature*, copyright 2005.

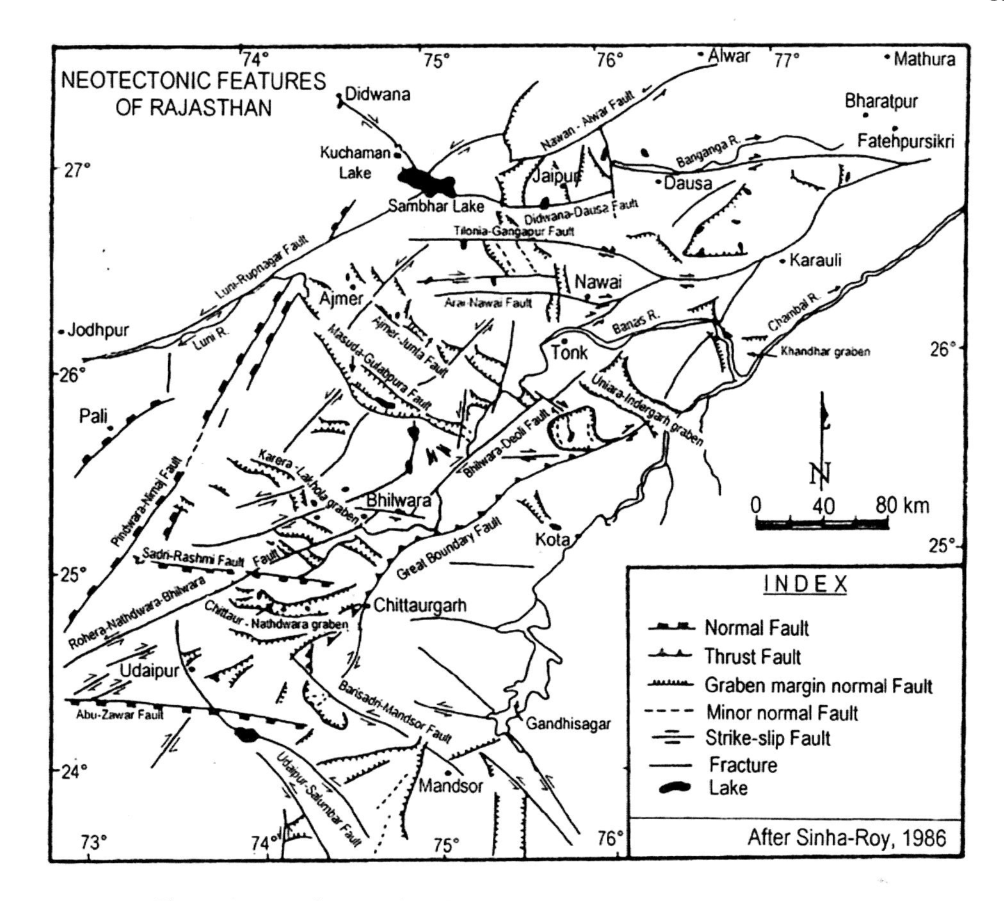

Figure 3. Tectonic map of northwestern Rajasthan. After *Sinha-Roy* [1986].

groundwater agencies have been involved in the exploration and extraction of the groundwater in this part of the country during the last half century. Several deep as well as shallow tube wells were drilled throughout region in different geological formations. Some of the groundwater samples were sent to the Bhabha Atomic Research Centre for isotopic dating [*Nair et al.*, 1999]. It was noted that there are isolated pockets of deep fresh groundwater. Tube wells dug in these pockets show no indication of a lowering of water level despite intense exploitation for the last 35–40 years. Most of these pockets are located along the paleochannels of the Sarasvati River system. In the Jaisalmer- Hanumangarh region, fresh groundwater has been located at a depth of 30–50 m and given a date of 1800 to 5000 years B.P. The fresh groundwater located at a depth of 60–250 m gives a date of 6000 to 22,000 years B.P. [*Nair et al.*, 1999]. In the area of the Hakra floodplain a fresh groundwater pocket with a 100 m thick aquifer has yielded a radiocarbon date of 4700 to 12,900 years B.P. [*Geyh and Ploethner*, 1995]. These dates indicate the existence of a large river system between 22,000 and 4700 years B.P.

4. NEOTECTONIC DISTURBANCES

Neotectonic disturbances in the region (Figure 3) have been taking place for the last 5 million years. More conspicuous signatures of these disturbances have been recorded during the Holocene period at (1) around 10,000 years B.P., (2) around 3000–3500 years B.P., and (3) around the eleventh/thirteenth centuries [*Sridhar et al.*, 1999]. The transform fault, a plate boundary of the Indo-Australian Plate along the northwestern margin, gradually turned into a suture zone. The Indus River has been flowing along the Indus Suture, and maintaining almost the same path, for at least last 5 million years. The geomorphologic map of the region depicting the drainage pattern that existed about 5 million years ago [*Clift and Blusztajn*, 2005], when compared with the present geomorphologic setup (Figure 2), reveals that there is no major change in the course of the Indus River for the last 5 million years. However, there are some changes in the pattern of its tributaries. The lowermost eastern tributary changes its course remarkably and subsequently disappears completely during this period. There is a possibility that this tributary represents the course of the proto-Vedic Sarasvati River in the region [*Paliwal*, 2008]. Similarly, the western lowermost tributary, which may represent the original Sindhu River, has also disappeared. The seven tributaries of the Indus River shown in the 5 million years ago geomorphologic map of the region represent "Sapt Sindhu," i.e., the seven tributaries of the Indus River described in the Vedic literature. It is possible that during this interval India moved farther north [*Paliwal*, 2008].

5. INDIAN MYTHOLOGY AND THE SARASVATI RIVER

India that is known as Bharat (Bharatvarsha/Aryavrita/Jamboo Dweep) hosts the world's most ancient civilization; the Dravidians and Aryans were ab intitio inhabitants. These people followed Hinduism in three different ways: (1) as Vaishnava, (2) as Shaiva, and (3) as Shakta and enjoyed the Vedic culture in an orderly and organized manner. It is believed that before the Vedas were written, Hinduism was transferred from one generation to the other through oral recitation. The Vedas were followed by the Upanishada, Epics, Puranas, Ramayana, Mahabharata, and Geeta texts. Hinduism has persisted since the beginning of cultured civilization and is followed by the majority of the people of India. Hinduism is more a way of life rather than a religion; Buddhism and Jainism followed later. Buddhism became more popular in the neighboring countries like China, Japan, Myanmar, Thailand, Sri Lanka, etc. Through Mughal and British invasions, Islam and Christianity crept into India.

The ancient mythological literature of India provides the evidence of a mighty river known as the Sarasvati River flowing in the region between the Indus River in the west and Aravalli Mountain Range in India [*Singhvi and Kar*, 1992; *Chauhan*, 1999]. A well-developed civilization older than the Harappan civilization and the Indus Valley civilization, popularly cited as the Vedic civilization, flourished along its banks, and Vedas and Epics were written by ancient scholars [*Radhakrishna and Merh*, 1999]. According to Hinduism the Holocene period has been divided into the (1) Vedic era (10,000 years B.P.), (2) Epic era (8000 to 5000 years B.P.), (3) Puranic era (5000 to 2500 years B.P.), and (4) post-Buddha era (2500 years B.P. to present). There are numerous citations in the Vedas, Epics, Puranas, and Mahabharat about the Sarasvati River flowing during these periods [*Chauhan*, 1999]. Out of the four Vedas (Rigaveda, Yajurvaveda, Atharvaveda, and Samaveda), there are eleven references to the Sarasvati River in the Rigaveda (1:32:1, 2:41:16, 6:61:2, 6:61:8, 6:61:10, 6:61:13, 6:61:14, 7:36:6, 7:95:2, 10:64:9, and 10:75:5 versions). There are two references to the holy river in the Yajurvaveda (34:11, 18:03), one reference in the Atharvaveda (6:30:1), and one reference in the Manusmriti (3.17). In the Puranas, there are several references to this ancient river, which depict the Sarasvati River flowing in the region in the Vamana Purana (32:1-4), Shrimad Bhagawata Purana, Vayu Purana, Bhramana Purana, Skanda Purana, Markandeya Purana, and Mahabharata (3:88:2, 3:130:4, 9:36:1, 9:53:11; Shantiparva: 1:29:20; and Wazzu: 3.80.118).

6. ARCHEOLOGICAL FINDINGS IN THE REGION

In the history of early human migration [*Singh*, 2009], India was the first stopover (70,000–50,000 years ago) after the first settlers moved out of Africa in about 2 Ma. Australia (50,000 years ago), Europe (40,000–30,000 years ago), and north central Asia (40,000 years ago) were the second, the third, and the fourth stopovers, respectively, in the journey. Similarly, Nordic countries (20,000–15,000 years ago) and South Africa (15,000–12,000 years ago) were the fifth and the sixth stopovers, respectively. During the Last Glacial Maximum (18,000 years ago), all these places were connected with land bridges. It is interesting to note that the region of the second stopover (70,000–50,000 years ago) around Greater India exhibits the maximum number of climatic changes and most human racial diversity. In the region under consideration, sites of Stone Age (700,000–7000 years B.P.), pre-Harappan (5500–4600 years B.P.), Harappan (4600–3300 years B.P.), and post-Harappan (3500–2100 years B.P.) civilizations have been excavated [*Kenoyer*, 1998; *Possehl*, 2002; *Madella and Fuller*, 2006]. Archaeological excavations (Figure 4) proved the existence of Paleolithic (lower 700,000–100,000 years B.P., middle 102,000–32,000 years B.P., and upper 32,000–12,000 years B.P.), Mesolithic (12,000–8500 years B.P.), and Neolithic or Chalcolithic (8500–7000 years B.P.) sites in the region [*Kenoyer*, 1998; *Possehl*, 2002; *Madella and Fuller*, 2006; *Valdiya*, 2002, 2010]. Paleolithic sites have been found in the floodplain of the Luni River, near Didwana and the hills of the Aravalli Mountain Range. Pre-Harappan (5500–4600 years B.P.) settlements indicate life based on agriculture. The pre-Harappan sites have been located in the Baluchistan and Ghaggar-Hakra reaches of the Sarasvati River up to Jind and Hisar. The settlement discovered at Mehrgarh in Pakistan [*Kenoyer*, 1998] dates back to 9000–5500 years B.P. [*Valdiya*, 2002].

Fossils of elephants (Figure 5), discovered [*Paliwal*, 2003] in the Quaternary gypsum deposit forming the upper layers (Figure 6) at Bhadwasi north of Nagaur in the Thar Desert of India, have been dated at 6808 years B.P. by the thermoluminescence method. Recently, fossilized Bamboo Curtain (Figures 7 and 8) has been discovered in the gypsum deposit

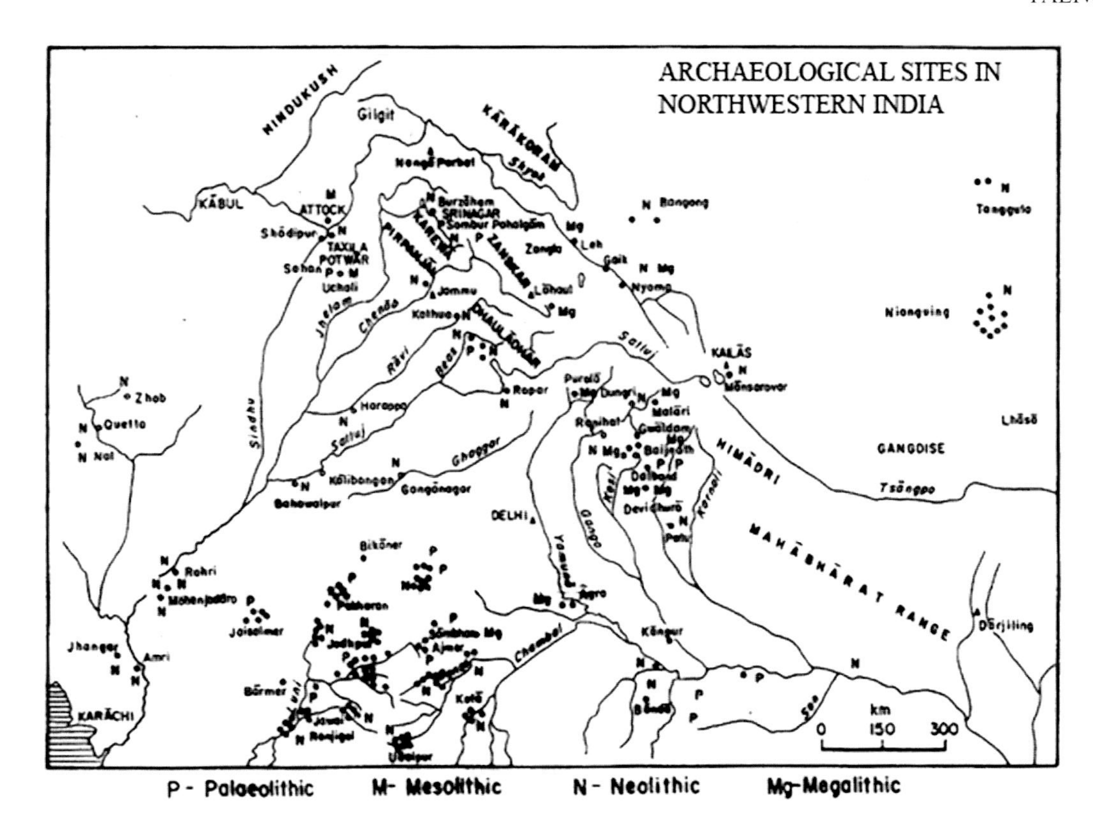

Figure 4. Archaeological sites in northwestern India, 65°–82°E; 25.1°–34°N.

forming the lower part of the Quaternary succession (Figure 6) at Jamsar near Bikaner. The artifact is, in fact, a roll-down door or window blind or curtain made from chopped sticks of bamboo, neatly woven with thread or string (Figure 8). The bamboo sticks were held together with the help of threads/string in a systematic manner. The thread or string appears to

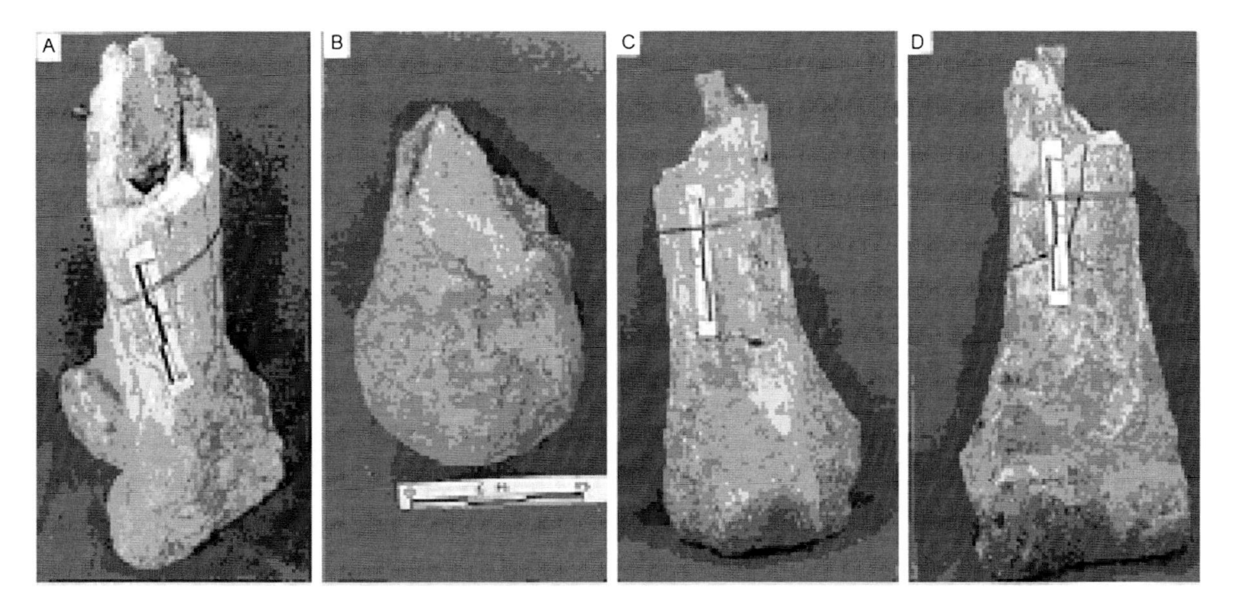

Figure 5. Elephant fossils in the gypsum deposit in the Thar Desert. After *Paliwal* [2003].

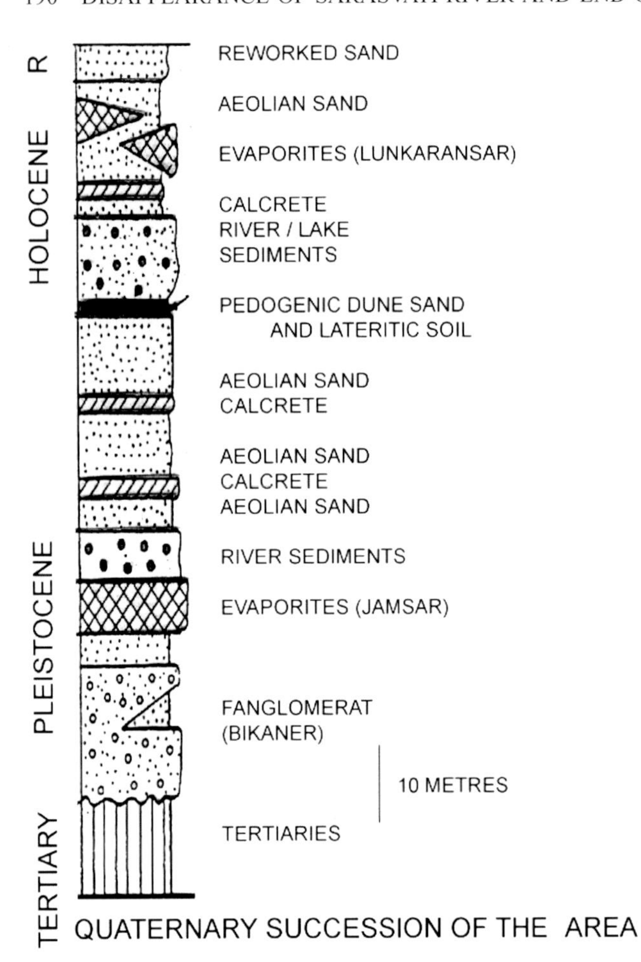

Figure 6. Quaternary succession of the Thar Desert. After *Sinha-Roy et al.* [1998], reprinted with permission of Geological Society of India.

be made of cotton. The man-made artifact was deposited along with gypsum in the lake and was later fossilized. The sample is 37 × 26 cm in size. The partly preserved molds and casts of the bamboo sticks in the sample are about 26 cm long but originally might have been much longer. The impressions of the casts of the bamboo sticks are not perfectly rounded but are slightly angular in cross section. Their diameter ranges from 4 to 5 mm. The distance between two thread string lines varies from 3.5 to 4.5 cm. The diameter of the thread/string is about 2 mm. The two supporting sticks at the top and bottom of the structure are slightly flat and are about 5 mm in thickness.

The Younger Dryas to early Holocene (12.9–11.5 ka), marking a warming period in the northwestern part of India, is quite interesting from a climate variation point of view. During the end of the Pleistocene to the Holocene, there was a sudden warming event that took place in the region. Sea level, low at 100 m depth at 14.5 ka, rose to 80 m at 12.5 ka

Figure 7. Fossilized bamboo blind or curtain (sample size 37 × 26 cm). After *Paliwal* [2011].

and continued to rise with a very high rate during 10.0–7.5 ka [*Hashimi et al.*, 1999]. Hypersaline conditions prevailed in the region during 20–13 ka. It is also believed that Bap-Malar (Jodhpur) and Kanod (Jaisalmer) playas originated during the Last Glacial Maximum [*Deotare et al.*, 2004], and both of these playas host thick-bedded gypsum deposits at the lower levels. At Kanod, sediments from the depth of 180 cm have been dated at 8701 ± 198 years B.P., and another from a depth of 250 cm has been dated at 9567 ± 159 years B.P. Here gypsum occurs from 176 to 250 cm. Evaporite sequences have been reported from the Holocene period at Lunkaransar (H2) and from the Pleistocene period at Jamsar [*Sinha-Roy et al.*, 1998]. If the gypsum deposit of Jamsar is coeval with the Kanod gypsum, the artifact should be of this age.

Figure 8. Enlarged view of the fossilized bamboo curtain showing molds of bamboo sticks neatly woven with the thread (coin size 2.6 cm). After *Paliwal* [2011].

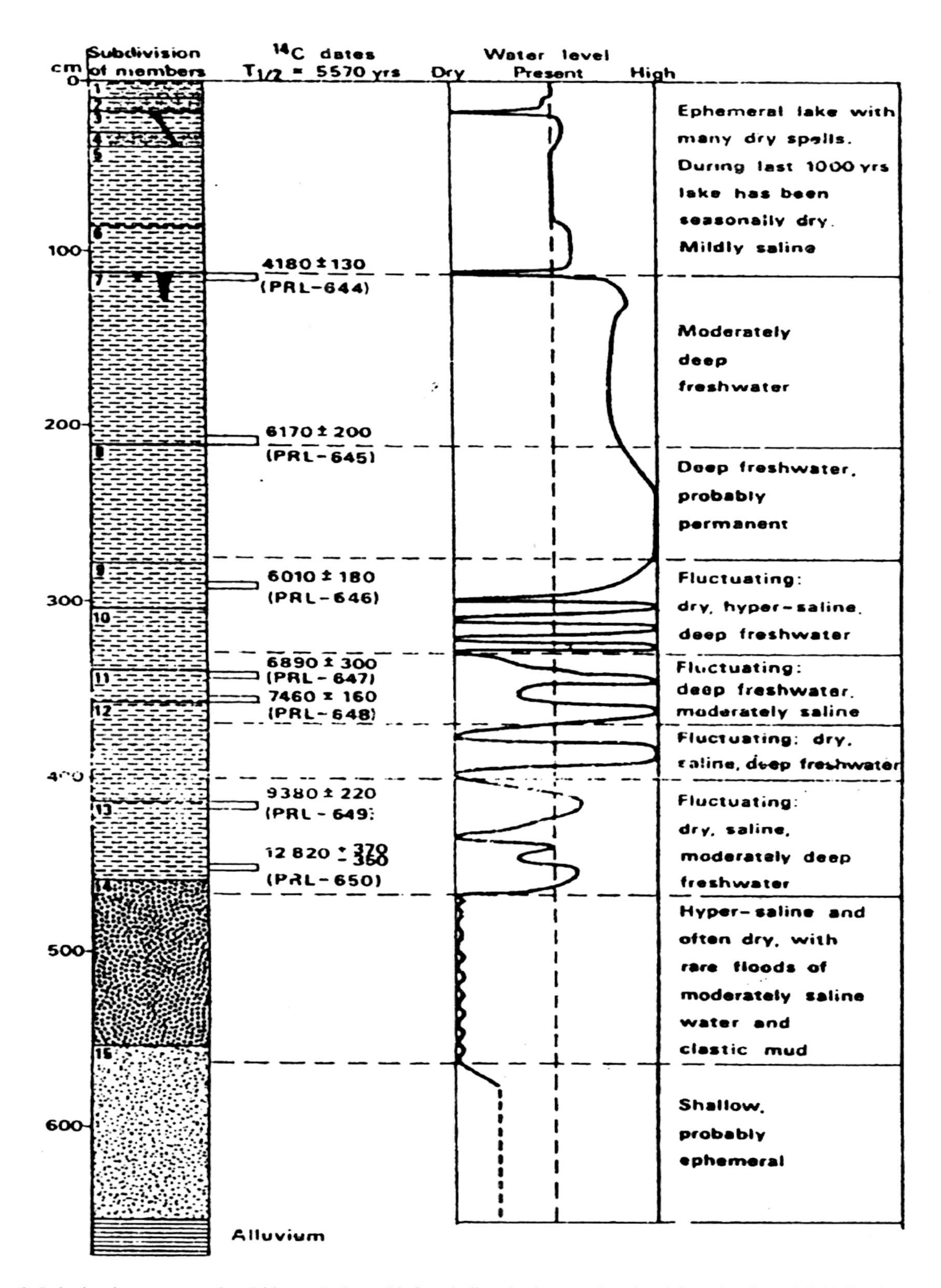

Figure 9. Lake level reconstructed at Didwana Lake and inferred climatic changes. Reprinted from *Singh et al.* [1990], with permission from Elsevier.

There is a possibility that extreme aridity during the terminal Pleistocene and development of hypersaline conditions in the playas, formed by tectonic activity in the region, have caused the deposition of older gypsum deposits like that of Jamsar in the region. The findings point toward the existence of an advanced civilization, possibly the Vedic civilization, which is much older than the Harappan and the Indus Valley civilizations. Dating of the sample will open up new avenues to search for a civilization in the region even older than the Mesolithic sites of the early Holocene period with the ceramic tradition in the Thar Desert of India and Pakistan or that of Mehrgarh settlements at the foot of the Kirther Hills, dating back to 9.0–5.5 ka [*Valdiya*, 2002].

The early Harappan sites [*Kenoyer*, 1998] have been excavated at Balakot, Amri, and Kot Diji in the Sindhu floodplain; Saraikhola Gulma near Taxila in Peshawar plain; Hakra Valley in Cholistan; Kunal, Kalibangan, Banawali, and Rakhigarhi in the Sarasvati plain; and Sothi, Siswal, and Mitathal in the Drishadvati plain. The Harappan civilization [*Kenoyer*, 1998], named after the type of area near the Ravi River, had an agrarian base. Important sites are Kalibangan and Baniwal on the bank of the Sarasvati River; Mohenjodaro in the Sindhu Valley; Harappa on the bank of the Ravi River; Dholavira in Kutch; and Lothal at the Khambhat coast [*Kenoyer*, 1998; *Cullen et al.*, 2000; *Possehl*, 2002; *Madella and Fuller*, 2006].

7. HOLOCENE CLIMATIC CHANGES

The region to the west of the Aravalli Mountain Range and south of the western Himalaya experienced desert conditions as long ago as 200,000 years B.P. with several intermittent glacial and interglacial phases. A moist and humid climate most suitable for habitation prevailed between 40,000 and 20,000 years B.P. The Last Glacial Maximum (a period between 18 and 25 ka) left its signature in the region. Major deglaciation transition occurred between 12,000 and 13,000 years B.P. Deglaciation intensification of the Southwest Monsoon occurred at 12,000 years B.P., peaking between 9000 and 6000 years B.P. [*Swain et al.*, 1983; *Fleitmann et al.*, 2003; *Phadtare*, 2000; *Sinha et al.*, 2005; *Rashid et al.*, 2007; *Valdiya*, 2010]. Ultimately, the Southwest Monsoon became weak at 5000 years B.P. Onset of an arid climate in the region began approximately 3500 years B.P., which began the decline of the Indus Valley civilization. Since then the present arid conditions have prevailed. Figure 9 depicts the lake level reconstruction of Didwana Lake and inferred climatic changes [*Singh et al.*, 1990]. Sediments of all the saline lakes in the region depict identical climatic conditions. After the Last Glacial Maximum, there were abrupt climatic fluctuations in the region [*Swain et al.*, 1983; *Fleitmann et al.*, 2003; *Phadtare*, 2000; *Sinha et al.*, 2005].

The area west of the Aravalli Mountain Range is occupied by several saline lakes and evaporite deposits of the Quaternary period producing a large quantity of different salts. Different opinions have been put forward to explain the origin of salt and saline lakes in the region [*Paliwal*, 1982], including neotectonic disturbances resulting in drainage disorganization, reversal, and at places blocking, causing the formation of saline lakes [*Paliwal*, 1986; *Roy*, 1999]. Considering the enormous quantity of salt produced from these lakes, we will have to account for its source. The "Na^+" component of "NaCl" might have been derived from the surrounding rocks, but the atmosphere is the only source for "Cl^-," and it is bit difficult to account for a huge deposit of salt in these lakes in a short time [*Paliwal*, 2008]. The oldest sediments in these lakes have been dated at more than 12,820 years B.P. [*Wasson et al.*, 1983]. There is a possibility that these inland basins had connections with the seawater through trenches developed during neotectonic activity [*Paliwal*, 2008]. Alternatively, these saline lakes located at the junctions of deep-seated subvertical faults (Figure 3) have groundwater connections with the older evaporates (Hanseran Group of the Marwar Supergroup and Tertiary) situated below the region. Salt moves through these faults when the lakes are filled with the rainwater and gets evaporated, precipitating salt on desiccation [*Paliwal*, 2008].

8. CONCLUSIONS

The Vedic civilization flourished along the banks of the Sarasvati River, and the saline lakes were formed by neotectonic disturbances resulting in drainage disorganization and blocking. In these circumstances the Sarasvati River and the Vedic civilization that flourished along its banks must date back to more than 12,820 years B.P. The region suffered abrupt climatic changes, reflected in lowering of sea level at the west coast of India by 100 m around 14,500 years B.P. and rising by 80 m by 12,500 years B.P. Neotectonic disturbances in the region that took place at 10,000 years B.P. were accompanied by an intense southwest monsoon. The Sarasvati River system was being repeatedly flooded, and this caused the end of the Vedic civilization, a civilization much older than the Harappan and the Indus Valley civilizations. The Sarasvati River, mentioned in the Vedas written by Indian scholars, continued until the Southwest Monsoon became weak at the beginning of the Puranic era (5000 to 2500 years B.P.). This was the time of Mahabharata (about 5238 years B.P.), at the end of which the Sarasvati River also disappeared from the scene.

Acknowledgment. I am deeply thankful to Harunur Rashid for inviting me to attend the Chapman Conference on Abrupt Climate Change held at the Ohio State University, Columbus. Thanks are also due to X. Shubham, X. Shakuntala, X. Bhuwan, and X. Gagan for extending all possible help in preparing the text.

REFERENCES

Chauhan, D. S. (1999), Mythological observations and scientific evaluation of the lost Sarasvati River, in *Vedic Sarasvati Evolutionary History of a Lost River of Northwestern India*, edited by B. P. Radhakrishna and S. S. Merh, *Mem. Geol. Soc. India*, *42*, 187–204.

Clift, P. D., and J. Blusztajn (2005), Reorganization of the western Himalayan river system after five million years, *Nature*, *438*, 1001–1003.

Clift, P. D., K. V. Hodges, D. Heslop, R. Hannigan, H. Van Long, and G. Calves (2008), Correlation of Himalayan exhumation rates and Asian monsoon intensity, *Nat. Geosci.*, *1*, 875–880, doi:10.1038/ngeo351.

Cullen, H. M., P. B. deMenocal, S. Hemming, G. Hemming, F. H. Brown, T. Guilderson, and F. Sirocko (2000), Climate change and the collapse of the Akkadian empire: Evidence from the deep sea, *Geology*, *28*, 379–382.

Deotare, B. C., M. D. Kajale, S. N. Rajaguru, and N. Basavaiah (2004), Late Quaternary geomorphology, palynology and magnetic susceptibility of playas in western margin of the Indian Thar Desert, *J. Ind. Geophys. Union*, *8*, 15–25.

Fleitmann, D., S. J. Burns, M. Mudelsee, U. Neff, J. Kramers, A. Mangini, and A. Matter (2003), Holocene forcing of the Indian monsoon recorded in a stalagmite from southern Oman, *Science*, *300*, 1737–1739.

Geyh, M., and D. Ploethner (1995), An applied palaeohydrological study in Cholistan, Thar Desert, Pakistan, in *Applications of Tracers in Arid Zone Hydrology*, edited by E. M. Adar and C. Leibundgut, *IAHS Publ.*, *232*, 119–127.

Hashimi, N. H., R. Nigam, R. R. Nair, and G. Rajagopalan (1999), Holocene sea level fluctuation in western Indian continental margin, *Mem. Geol. Soc. India*, *42*, 297–302.

Heron, A. M. (1953), Geology of central Rajputana, *Mem. Geol. Surv. India*, *79*, 339 pp.

Kenoyer, J. M. (1998), *Ancient Cities of the Indus Valley Civilization*, Oxford Univ. Press, Karachi, Pakistan.

Madella, M., and D. Q. Fuller (2006), Palaeoecology and the Harappan civilisation of South Asia: A reconsideration, *Quat. Sci. Rev.*, *25*, 1283–1301.

Nair, A. R., S. V. Navada, and S. M. Rao (1999), Isotope study to investigate the origin and age of groundwater along palaeochannels on Jaisalmer and Ganganagar districts of Rajasthan, in *Vedic Sarasvati Evolutionary History of a Lost River of Northwestern India*, edited by B. P. Radhakrishna and S. S. Merh, *Mem. Geol. Soc. India*, *42*, 187–204.

Oldham, R. D. (1893), The Sarsawati and the Lost River of the Indian Desert, *J. R. Asiatic Soc.*, *25*, 49–76.

Paliwal, B. S. (1982), The source of salt in Rajasthan-An investigation of the Salt Lake of Didwana, in *Proceedings of the Workshop on the Problems of Deserts in India*, *Misc. Publ. Geol. Surv. India*, *49*, 99–104.

Paliwal, B. S. (1986), Source of salinity in soil, and surface and subsurface waters and its bearing on the evolution of the Thar Desert, India, *Transactions of XIII Congress of the International Society of Soil Science*, p. 69, Int. Congress of Soil Sci., Hamburg, Germany.

Paliwal, B. S. (1993), Geology of Rajasthan - A summarium, in *Natural and Human Resources of Rajasthan*, edited by T. S. Chauhan, pp. 1–22, Sci. Publ., Jodhpur, India.

Paliwal, B. S. (2003), Fossilized elephant bones in the Quaternary gypsum deposits at Bhadawasi, Nagaur district, Rajasthan, *Curr. Sci.*, *84*, 1188–1191.

Paliwal, B. S. (2008), Neotectonic disturbances in the region to the west of the Aravalli Mountain Range and its effect: Disappearance of Sarasvati River, origin of saline lakes and development of the Vedic civilization in the region, *J. Int. Assoc. Gondwana Res. Conf. Ser.*, *5*, 95–97.

Paliwal, B. S. (2011), Fossilized bamboo curtain from the Quaternary gypsum deposit of Thar Desert at Jamsar mine near Bikaner, Rajasthan, India, *Curr. Sci.*, *100*, 471–472.

Phadtare, N. R. (2000), Sharp decrease in summer monsoon strength 4000–3500 cal yr B.P. in the central Higher Himalaya of India based on pollen evidence from alpine peat, *Quat. Res.*, *53*, 122–129.

Possehl, G. L. (2002), *The Indus Civilization: A Contemporary Perspective*, AtlaMira, Lanham, Md.

Radhakrishna, B. P., and S. S. Merh (1999), Vedic Sarasvati evolutionary history of a lost river of northwestern India, *Mem. Geol. Soc. India*, *42*, 329 pp.

Rashid, H., B. P. Flower, R. Z. Poore, and T. M. Quinn (2007), A ~25 ka Indian Ocean monsoon variability record from the Andaman Sea, *Quat. Sci. Rev.*, *26*, 2586–2597.

Roy, A. B. (1999), Evolution of saline lakes of Rajasthan, *Curr. Sci.*, *76*, 290–295.

Roy, A. B., and S. R. Jakhar (2002), *Geology of Rajasthan: Precambrian to Recent*, 418 pp., Sci. Publ., Jodhpur, India.

Singh, L. (2009), Who is an Indian?, *India Today*, 21 Sept., 22–26.

Singh, G., R. J. Wasson, and D. P. Agarwal (1990), Vegetation and seasonal climatic changes since the last full glacial in the Thar Desert, NW India, *Rev. Palaeobot. Palynol.*, *64*, 351–358.

Singhvi, A. K., and A. Kar (Eds.) (1992), *Thar Desert in Rajasthan: Land, Man and Environment*, Geol. Soc. of India, Bangalore.

Sinha, A., K. Cannariato, L. Stott, H.-C. Li, C. F. You, H. Cheng, R. L. Edwards, and I. B. Singh (2005), Variability of southwest Indian summer monsoon precipitation during the Bølling-Allerød, *Geology*, *33*, 813–816.

Sinha-Roy, S. (1986), Himalayan collision and indication of Aravalli orogen by Bundelkhand wedge: Implications in Rajasthan, paper presented at International Symposium on Neotectonics in South Asia, Dehra Dun, India.

Sinha-Roy, S., G. Malhotra, and M. Mohanty (1998), *Geology of Rajasthan*, 278 pp., Geol. Soc. of India, Bangalore.

Sridhar, V., S. S. Merh, and J. N. Malik (1999), Late Quaternary drainage distribution in northwestern India: A geochronological enigma, in *Vedic Sarasvati Evolutionary History of a Lost River of Northwestern India*, edited by B. P. Radhakrishna and S. S. Merh, *Mem. Geol. Soc. India*, *42*, 187–204.

Swain, A. M., J. E. Kutzbach, and S. Hastenrath (1983), Estimates of Holocene precipitation for Rajasthan, India, based on pollen and lake-level data, *Quat. Res.*, *19*, 1–17.

Valdiya, K. S. (2002), *Saraswati: The River That Disappeared*, 116 pp., Univ. Press, Hyderabad, India.

Valdiya, K. S. (2010), *The Making of India: Geodynamic Evolution*, Macmillan, New Delhi, India.

Wasson, R. J., S. N. Rajaguru, V. N. Mishra, D. P. Agarwal, R. P. Dhir, A. K. Singhvi, and K. K. Rao (1983), Geomorphology, late Quaternary stratigraphy and palaeoclimatology of the Thar dune field, *Z. Geomorphol. Suppl.*, *45*, 117–151.

Evidence for Climate Teleconnections Between Greenland and the Sierra Nevada of California During the Holocene, Including the 8200 and 5200 Climate Events

Stephen F. Wathen

Department of Plant Sciences, University of California, Davis, California, USA

Abrupt climate change (ACC) has had significant impacts on environments and human cultures in the past. The extent and potential causes of ACC events (ACCEs) are thus important questions to explore. An 8500 year macrocharcoal record was developed from Coburn Lake, Sierra Nevada, and compared with published high-resolution paleoclimate and paleofire records from the Sierra Nevada, eastern Canada, and Greenland arriving at the following conclusions. Temperatures were generally hot or rising at the beginning of long-term severe droughts (mega-droughts) in the Sierra Nevada. For the most part, severe fires occurred at Coburn Lake at the beginning of severe droughts, suggesting a response by vegetation and slopes to ACC. There was also strong synchronicity between the beginnings of droughts and fires at Coburn Lake and climate and fire events in eastern Canada and Greenland. Evidence is presented that (1) both the 8200 and 5200 ACCEs occurred in the Sierra Nevada and (2) as severe droughts were beginning in the Sierra Nevada, severe droughts were ending in Greenland. Coburn Lake charcoal peaks were also synchronous with the general pattern of soot deposition onto Greenland, reportedly from fires in eastern Canada, from 8500 to 3000 years ago. The beginnings of severe droughts in the Sierra Nevada were synchronous with charcoal peaks in Lac Francis, Quebec, over the past 6750 years. This synchronicity in climate and fire events suggests the possibility that abrupt, large-scale northward and southward shifts in the locations of the northern Polar Front and subtropical high-pressure zone accompanied the onset of ACCEs in the past.

1. INTRODUCTION

1.1. Abrupt Climate Change

Abrupt climate change (ACC) has had significant impacts on natural environments and humans in the past. Thus, the timing, extent, and potential causes of ACC are important questions to investigate. ACC events (ACCEs) occur over a matter of years or a few decades, such that societies or ecosystems have little time to adapt and thus may undergo significant alteration or even collapse [*Committee on Abrupt Climate Change, National Research Council*, 2002]. ACC has been implicated in the collapse of many early complex human societies [*Douglas*, 1988; *Weiss*, 2003; *Leroya et al.*, 2006]. Less is known, however, about the effect of past ACCEs on natural ecosystems. The two most prominent ACCEs within the Holocene period occurred at 8200 and 5200 cal years B.P. (calibrated years before present).

1.1.1. The 8200 year abrupt climate change event. The 8200 year ACCE was probably the most severe ACCE

Abrupt Climate Change: Mechanisms, Patterns, and Impacts
Geophysical Monograph Series 193

10.1029/2010GM001022

during the Holocene. *Alley and Agustsdottir* [2005] determined that extreme cold and drought at 8200 years ago caused rapid vegetation change in many parts of the world. *Morrill and Jacobsen* [2005] reviewed 52 sites worldwide and concluded that the 8200 ACCE was detected in many areas of the Northern Hemisphere extratropics and tropics. They noted, however, that many more records were needed from the Southern Hemisphere. For example, *Tinner and Lotter* [2001] reported the sudden collapse of a depauperate hazel-dominated (*Corylus avellana*) woodland in southern Germany and Switzerland and the woodland's replacement by a more diverse closed-forest type of vegetation 8200 years ago. *Clark et al.* [2002] reported a permanent change from C4- to C3-dominated grasslands on the United States Great Plains as a result of a severe drought at 8200 years ago.

The term "severe drought," as used here, refers to megadroughts lasting decades or centuries. The 8200 year ACCE initiated hundreds of years of drought worldwide. In the western United States, this period from 8200 to 6300 years ago has been referred to variously as the Holocene Altithermal, thermal maximum, etc. *Barron et al.* [2003] reported that sea surface temperatures along the Northern California coast were 1°C–2°C colder at 8200 years ago than at any other time in the Holocene, resulting in drier conditions inland in the Sierra Nevada [*Moser et al.*, 2004]. Cold temperatures off the coast of Northern California and dry conditions inland lasted until 3200 years ago [*Barron et al.*, 2003]. This time interval coincided with a period of widespread aridity in the western interior of North America [*Fritz et al.*, 2001]. In the northern Sierra Nevada, three severe droughts in a row occurred from 7300 years ago (near the beginning of the *Mensing et al.* [2004] precipitation record) until the arrival of moister conditions approximately 6300 years ago [*Mensing et al.*, 2004].

1.1.2. The 5200 year abrupt climate change event. The 5200 year ACCE represented both the driest climate interval and one of the three largest reductions in meridional overturning circulation (MOC), sometimes inaccurately referred to as thermohaline circulation, in the North Atlantic during the Holocene [*Oppo et al.*, 2003]. The other two large reductions in MOC in the North Atlantic occurred at 8200 and 2800–3000 years ago [*Oppo et al.*, 2003]. The 5200 year ACCE ended the moister conditions that had begun approximately 6300 years ago, which changed vegetation in locations as diverse as British Columbia [*Spooner et al.*, 1997], Minnesota [*Smith et al.*, 2002], eastern Greenland [*Cremer et al.*, 2001], Sweden [*Regnell et al.*, 1995], and Australia [*McKenzie and Kershaw*, 1997]. *Clark et al.* [2002] noted that a major vegetation and environmental shift occurred at 5000 years ago in the northern Great Plains.

1.2. In This Chapter

This study seeks to explain the unusually peaked Coburn Lake macrocharcoal record and its relationship to climate and fire events in the Sierra Nevada, eastern Canada, and Greenland. Here I present a high-resolution climate and disturbance history for the Sierra Nevada based upon comparisons between my 8500 year long charcoal record from Coburn Lake and six published high-resolution paleorecords: (1) a drought record from Pyramid Lake, Nevada [*Mensing et al.*, 2004]; (2) a temperature record from Cirque Peak, California [*Scuderi*, 1993]; (3) a charcoal record from eastern Canada [*Carcaillet et al.*, 2001]; and (4) precipitation, (5) temperature, and (6) soot records from Greenland [*National Snow and Ice Data Center (NSIDC) and World Data Center-A for Paleoclimatology (WDC-A)*, 1997].

The results of these comparisons provide evidence for synchronicity between the timing of the following climate and disturbance events: (1) between the timing of severe fires and erosion at Coburn Lake and the timing of the beginnings of severe droughts and high temperatures in the Sierra Nevada, (2) between the timing of severe droughts and high temperatures in the Sierra Nevada and the timing of increasing precipitation and warming temperatures in Greenland, and (3) between the timing of severe droughts in the northern Sierra Nevada and associated fires at Coburn Lake and the timing of severe fires in eastern Canada. This study also provides new evidence that the 8200 and 5200 year ACCEs affected California and that the magnitude of changes in temperature and precipitation increased in Greenland at approximately 3700 and 3100 years ago, respectively.

Based on the above findings, I propose the following hypotheses: (1) that most instances of severe fire and erosion at Coburn Lake occurred in response to stresses that ACCEs put on vegetation adapted to cooler and wetter conditions; (2) that the synchronous beginnings and endings of drought conditions in the Sierra Nevada and Greenland were caused by abrupt, large-scale northward and southward shifts in the locations of the Earth's major precipitation belts (the Intertropical Convergence Zone (ITCZ), the Subtropical Desert Zone, and the Polar Front) during ACCEs; and (3) that increased climate variability increased the number of ACCEs during the late Holocene. Using the Coburn Lake Charcoal record and published data, I wish to provide evidence supporting these hypotheses.

2. STUDY AREA

The Coburn Lake study area (39°32′30″N, 120°27′W) is located just below and to the west of the crest of the northern Sierra Nevada and is also adjacent to the Sierra Valley portion of the Great Basin (Figure 1). I chose Coburn Lake for the study because the lake is deep (13 m) relative to its surface area (1.7 ha). This relationship suggested that Coburn Lake might contain relatively undisturbed lake sediments [*Larsen et al.*, 1998]. Coburn Lake was also chosen for study because it is located at the bottom of a small (44 ha), steep watershed. This small watershed limits the spatial area from which charcoal, that is deposited into Coburn Lake, can come. The Coburn Lake watershed extends from 2280 to 2460 m (7480–8060 ft) in elevation.

The Coburn Lake watershed currently experiences an upper montane Mediterranean climate with cold, snowy winters and cool, dry summers. Located midway between the Pacific northwest and southwest regions, Coburn Lake does not appear to share the typical El Niño–Southern Oscillation (ENSO) response that either region experiences today.

Old-growth red fir (*Abies magnifica* A. Murr.) currently dominates vegetation within the heavily forested Coburn Lake watershed. Subalpine mountain hemlock (*Tsuga mertensiana* (Bong.) Carr.) codominates with red fir on cooler and moister north facing slopes, while western white pine (*Pinus monticola* Dougl. ex D. Don) codominates with red fir on warmer and drier south facing slopes and on dry ridge tops.

Emergent and riparian vegetation sometimes serves to filter out macrocharcoal that might otherwise reach the lake. Emergent and riparian vegetation are present today on Coburn Lake's flatter and partially filled northern and eastern margins. There is little emergent vegetation below steep slopes on the lake's southern and western margins. However, relatively little of the lake's watershed lies above the southern and western margins of the lake.

3. METHODS

3.1. Methods in the Field

My colleagues and I collected two sets of cores from approximately the center of Coburn Lake. These two sets of cores were located >26 m apart to avoid resampling the same location in the lake. Each set of cores included samples of the water-sediment interface (top core sediments) and of deeper sediments (deep core sediments). We used a gravity-activated top corer to sample the water-sediment interface and a modified 5 cm diameter, 1 m length Livingstone piston corer [*Wright*, 1991] to collect a total length of 4.4 m of deep core sediments.

While still in the field, top core sediments were photographed and then subsampled at 1 cm intervals, unless smaller intervals were required to determine the locations of stratigraphic boundaries or charcoal visible in the core. Deep core sediments were described in the field, wrapped in

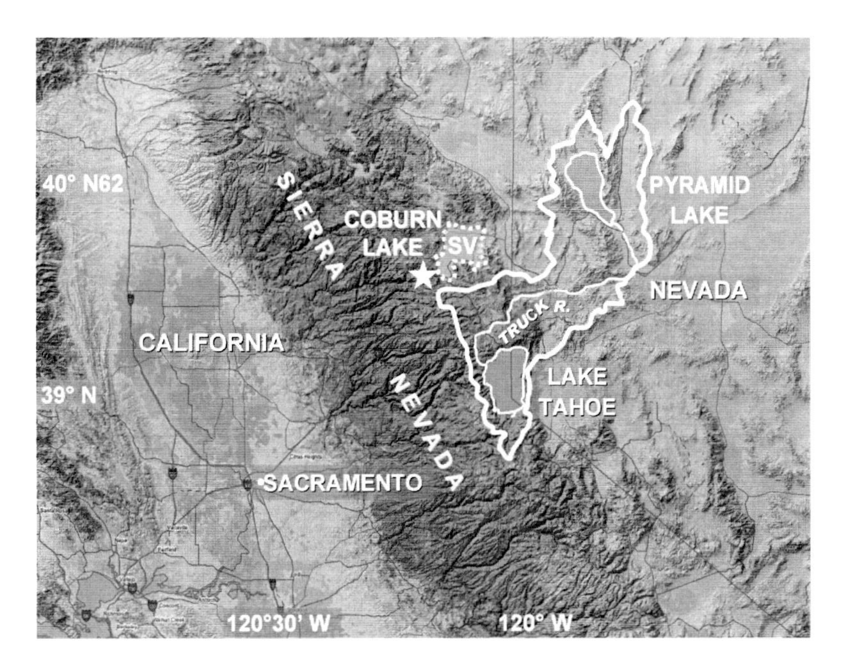

Figure 1. Coburn Lake (θ) is located north of Lake Tahoe, California, just below the east side of the Sierra Nevada crest and adjacent to the Sierra Valley (SV) portion of the Great Basin Desert. The Truckee River (thin solid line) begins at the northern end of Lake Tahoe, passes through Reno, Nevada, runs east and north from Reno, and ends at Pyramid Lake. The watershed for Pyramid Lake is enclosed within a thick solid line.

cellophane and aluminum foil, and then placed in wooden boxes for transport to the laboratory. All samples were kept cool in transit and then stored at 4°C in the laboratory.

3.2. Methods in the Laboratory

One set of cores was designated to be the primary set of cores used for analysis. The other set of cores is referred to here as the secondary set of cores. Both sets of deep core sediments were split horizontally, sketched, and photographed. Sediment layers were then visually classified as to composition: gyttja (fine, organically rich lake sediments), macro-sized organic detritus, clay, or ash (Figure 2). Deep core sediments were also subsampled at 1 cm intervals unless, again, smaller intervals were necessary to locate stratigraphic boundaries or charcoal visible in core segments.

A radiocarbon-based chronology was developed for Coburn Lake based upon ^{14}C accelerator mass spectrometry

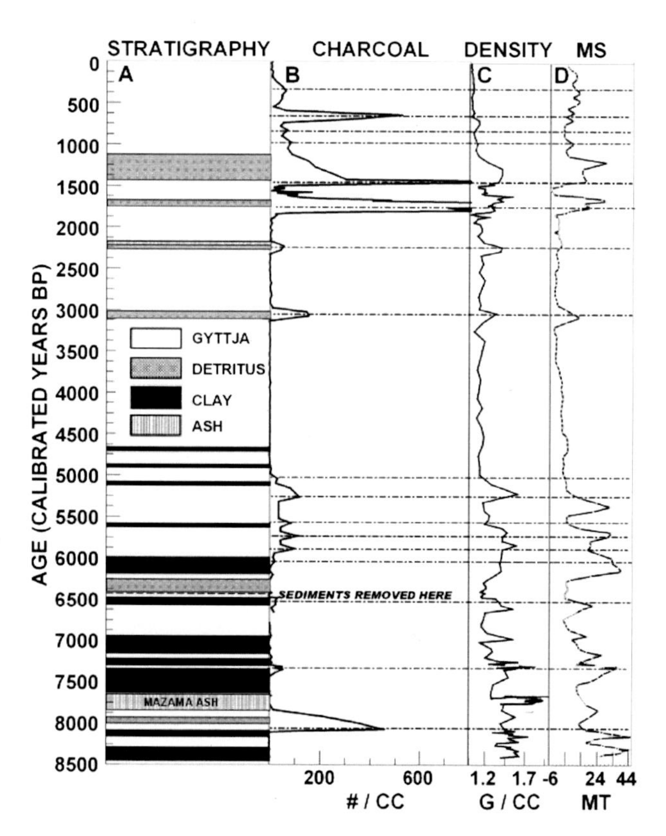

Figure 2. Relationships between stratigraphy layers, charcoal peaks, sediment bulk density, and magnetic susceptibility at Coburn Lake. (a) Stratigraphy of sediments. (b) Number of charcoal particles of >250 μm in size per g^3 found within lake sediments. The two largest charcoal peaks (>1000 charcoal g^{-3}) have been truncated to fit the column size. (c) Bulk density of sediments. (d) Magnetic susceptibility (MS) of sediments.

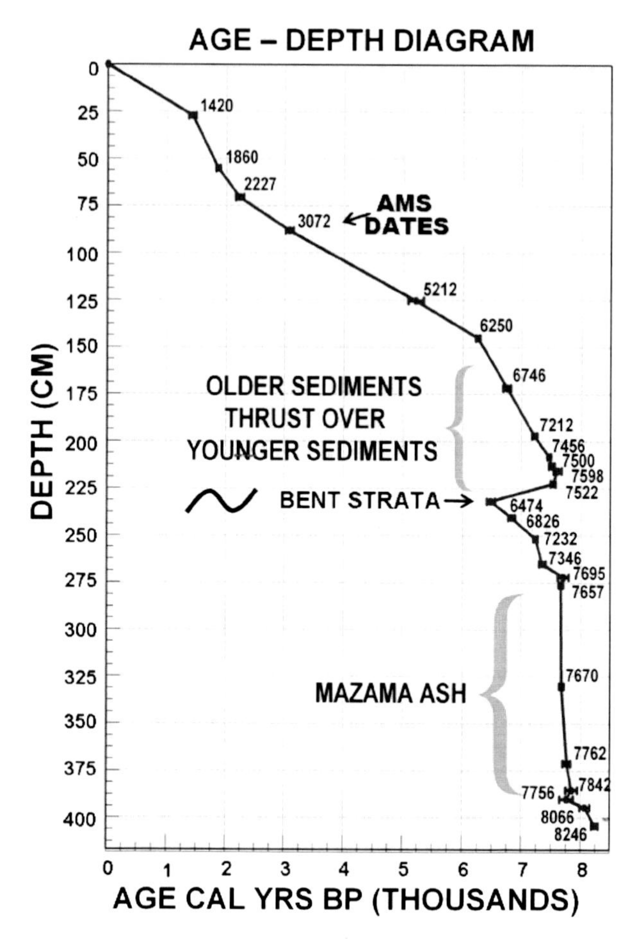

Figure 3. Age-depth diagram for Coburn Lake. Locations of ^{14}C-AMS (accelerator mass spectrometry) dates, a thick layer of Mazama ash, bent strata, and where older sediments (above) were thrust up and over younger lake sediments are highlighted.

(AMS) dating of 24 plant macrofossils taken from the primary set of deep core sediments (Figure 3). Radiocarbon dates (Table 1) were converted to calibrated years before present (cal years B.P., A.D. 1950) using the Cologne Radiocarbon, Calibration, and Paleoclimate Research (CalPal): Online quickcal2004 version 1.2 (B. Weninger et al., 2004, available at http://download.calpal.de/calpal-info/). Layers of organic detritus and clay were not excluded from the age-depth curve for the following reasons: (1) much of the core was made up of nongyttja material, (2) relying on dates in gyttja alone would have produce unrealistic results, and (3) peaks in charcoal, the focus of this study, were often associated with nongyttja layers. Calibrated radiocarbon dates were graphed proportionally between dates without developing a formal age-depth model because (1) the large number of ^{14}C-AMS dates available made such modeling unnecessary, (2) lake deposition appeared to be strongly

Table 1. Calibrated Radiocarbon Dates Used in This Study[a]

Depth (m)	Material/Location	Calculated Livermore 7/2002	Plus/Minus	Calibrated CalPal 7/2005	Plus/Minus
		Core A			
41	cone scale/charcoal lens	1460	60	1366	49
51	needle/clay lens	1920	40	1867	42
148	needle/below charcoal	2210	45	2231	67
212	needle/charcoal lens	4830	100	5540	123
292	wood	6330	80	7245	104
392	leaf	6350	80	7280	91
		Core C			
27	needle/above charcoal	1520	45	1420	62
55	wood/below charcoal	1915	35	1860	35
71	cone scale/charcoal	2200	40	2227	67
88	charred wood	2925	35	3072	67
125	needle/charcoal	4560	70	5211	127
145	rootlet/below clay	5455	40	6250	38
172	rootlet/below clay	5920	60	6746	67
197	rootlet/top detritus lens	6285	40	7214	39
208	rootlet/detritus lens	6540	40	7455	29
213	rootlet/bottom clay lens	6600	35	7500	47
216	rootlet/detritus lens	6750	80	7599	66
222	rootlet/bottom clay lens	6640	40	7523	40
232	seed husk/charcoal lens	5680	60	6474	70
240	wood/dark gyttja lens	6000	45	6826	57
252	forb/charcoal lens	6305	40	7233	36
265	needle/charcoal lens	6420	60	7346	54
272	forb/bottom clay lens	6840	110	7695	92
276	forb	6825	50	7658	37
276	forb (duplicate)	6805	50	7645	32
330	needle/top of unit	6845	40	7670	37
371	forb/bottom of unit	6930	70	7762	65
385	needle/silt lens	7030	110	7842	99
390	needle/brown gyttja lens	6910	130	7757	115
390	needle (duplicate)	7020	230	7868	207
395	needle/charcoal	7250	90	8066	85
404	forb/clay lens	7410	40	8246	57
		Core C Dates Not Used			
133	cone scale/outlier	1850	50	1787	59
266	needle/contaminated	5775	50	6572	62
279	cone scale/outlier	modern date			
294	needle/not needed	6540	35	7453	25

[a]Accelerator mass spectrometry (AMS) dates were determined by the Lawrence Livermore Center for Accelerator Mass Spectrometry from plant macrofossils collected from Coburn Lake, California. AMS dates were transformed to calibrated dates cal years B.P. (present is A.D. 1950) using the Cologne Radiocarbon, Calibration, and Paleoclimate Research Package (CalPal): Online quickcal2004 version 1.2 (B. Weninger et al., 2004, available at http://download.calpal.de/calpal-info/).

episodic rather than uniform, and (3) a coherent pattern of sedimentation resulted from using the methods employed.

The two sets of deep core sediments were compared with each other to confirm the dating and stratigraphy of the primary set of deep core sediments that was to be used for further analysis. The two sets of cores were cross-correlated with each other using (1) the initial 24 ^{14}C-AMS dates; (2) additional AMS dates derived from the secondary set of deep core sediments; and (3) peaks in bulk density, magnetic susceptibility, and charcoal density found in both sets of cores.

One cubic centimeter of material was extracted from each subsample from the primary set of cores (top core sediments and deep core sediments) and processed to determine charcoal abundance, magnetic susceptibility, and bulk density. Drying preweighed sediments in a muffle oven overnight at 40°C, 450°C, and 900°C was used to determine percentages of water, organic carbon, and calcium carbonate within subsamples, respectively, after accounting for the weight of material lost.

Charcoal from Coburn Lake was processed using standard procedures [*Whitlock*, 2001]. Samples were disaggregated in 5% KOH solution overnight and then washed through a series of 250, 125, and 63 μm sized sieves in order to concentrate charcoal particles into progressively smaller sizes. Charcoal particles of >250 μm size were counted within every 1 cm^3 subsample using a 36× dissecting microscope. I also counted 125–250 μm sized charcoal in a portion of both the primary and secondary sets of core sediments to check for differences in charcoal patterns within the two sets of cores.

The Coburn Lake study was designed to develop a record of local fires within the Coburn Lake watershed. Individual charcoal particles of >125 μm in size that can be seen with the naked eye are referred to as macrocharcoal. Macrocharcoal particles are thought to be heavy enough to travel only a few hundred yards through the air before deposition [*Clark*, 1988; *MacDonald et al.*, 1991; *Clark and Royall*, 1996; *Clark et al.*, 2002]. I chose to focus on macrocharcoal particles of >250 μm in size as my proxy for past local fires in order to further strengthen the relationship between particle size and aerial distance traveled. However, stream or sheet flow can potentially carry charcoal of any size into the lake. From this point onward, I will refer to macrocharcoal particles from Coburn Lake simply as charcoal. Charcoal numbers were not converted to charcoal accumulation rates because the purpose of this study was to investigate individual charcoal peaks in relation to individual ACCEs.

4. RESULTS

The accuracy of the dating of the primary set of cores was confirmed by comparing the two sets of cores as described in section 3. Little difference in charcoal patterns resulted from counting 125–250 μm versus >250 μm sized charcoal particles within the same time intervals of both sets of cores.

A thrusting of older sediments over newer sediments was visible in the cores in the field. This overthrusting was confirmed by radiocarbon dating (Figure 3, center right). This overthrusting was potentially caused by a paleoearthquake that occurred sometime between 6474 and 6250 years ago. This paleoearthquake would probably have occurred along a still active fault located just to the east of Coburn Lake within the Sierra Valley.

Visible layers of charcoal, organic detritus, and clay were interspersed within a matrix of lake-derived, organic-rich gyttja (Figure 2). The number of charcoal particles usually peaked in association with peaks in organic detritus, inorganic sediments (clay), sediment bulk density values, and magnetic susceptibility values: all proxies for severe erosion.

As a result, charcoal from Coburn Lake had a very peaked distribution suggesting episodic deposition over time. The average lake sediment subsample over the entire 8500 year span of the Coburn Lake charcoal record contained <10 macrocharcoal particles per cm^3. However, long intervals of time produced little or no charcoal, while relatively short intervals of time produced very large amounts of charcoal (Figure 2).

From 8400 to 6900 years ago, deposition was almost all clay. A large and a small peak of charcoal were deposited into the lake at 8120 and 7300 years ago, respectively. From 6900 to 6560 years ago, a lot of gyttja was deposited into Coburn Lake. However, high rates of clay deposition occurred from 6560 to 6000 years ago in association with a small charcoal peak.

Clay deposition declined after 6000 years ago with deposition being primarily gyttja from that time to the present. Only four thin lenses of clay were visible between 6000 and 4600 years ago, two being associated with charcoal peaks. No clay lenses were visible after 4600 years ago.

Charcoal was deposited continuously into Coburn Lake from approximately 6000 to 5000 years ago, including the deposition of four large charcoal peaks. Conversely, almost no charcoal was deposited into Coburn Lake during the time period from 5000 to 3100 years ago. Charcoal deposition into Coburn Lake recommenced approximately 3100 years ago, becoming massive over the past 1800 years. Four large charcoal peaks from 3100 to 1100 years ago were associated with visible lenses of detritus.

5. DISCUSSION

5.1. Sierra Nevada Climate Records

5.1.1. Pyramid Lake drought record. The Coburn Lake charcoal record was compared with the 7400 year long Pyramid Lake drought record for the northern Sierra Nevada [*Mensing et al.*, 2004] (Figure 4). There was good agreement between the timing of Coburn Lake charcoal peaks and the beginnings of severe droughts in the Northern Sierra Nevada. Charcoal deposition increased at the beginning of most of the

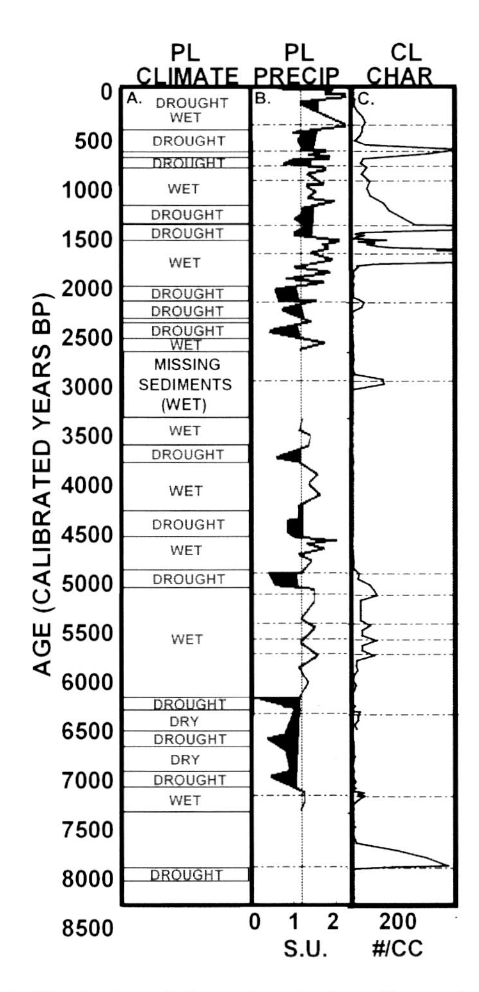

Figure 4. Distribution of charcoal peaks from Coburn Lake (CL) over the past 8500 years compared with the 7400 year long Pyramid Lake (PL) drought record. (a) Drought record in the northern Sierra Nevada [*Mensing et al.*, 2004] with the addition of the of the 8200 year climate event from Coburn Lake. (b) Northern Sierra Nevada drought record from Pyramid Lake, Nevada, based upon the ratio of *Artemisia* (sagebrush) to Chenopodiaceae (shadscale) pollen types (A/C) in Pyramid Lake sediments [*Mensing et al.*, 2004]. S.U. is standard units. (c) Number of charcoal particles of >250 μm in size per g^3 found within lake sediments. The two largest Coburn Lake charcoal peaks (>1000 charcoal g^{-3}) were truncated to fit the column size. The dashed horizontal lines represent the centers of major charcoal peaks.

severe droughts that occurred in the northern Sierra Nevada over the past 7400 years [*Mensing et al.*, 2004]. However, little or no charcoal was deposited into Coburn Lake between 5000 and 3100 years ago (see below). Also, the Pyramid Lake record is missing information for the time period between 3400 and 2750 years ago.

The Pyramid Lake record does not cover the climate record as far back as 8200 years ago. However, a large amount of charcoal was deposited into Coburn Lake beginning 8120 years ago. This date corresponds closely with the 8150 date given by *Muscheler et al.* [2004] for the 8200 year ACCE. As described above, other evidence exists for the occurrence of the 8200 year ACCE in North America [*Alley and Agustsdottir*, 2005; *Morrill and Jacobsen*, 2005] and California [*Moser et al.*, 2004]. *Benson et al.* [1997] reported that Owens Lake, located east of the southern Sierra Nevada, was wet from 10,000 to 8000 years ago. *Meyer et al.* [1995] reported a peak in debris flows at approximately 8200 years ago in Yellowstone, Wyoming. *Dean et al.* [2006] found evidence for the 8200 year ACCE at Bear Lake, located in Utah and Idaho. *Potito et al.* [2006] found that summer lake surface water temperatures in the Sierra Nevada remained steady from approximately 11000 years ago until just before approximately 8000 years ago, at which point summer surface water temperatures became abruptly colder. After this abrupt drop in temperature, temperatures rose steeply, reaching their Holocene maximum at approximately 6500 years ago.

Two smaller peaks of charcoal were deposited into Coburn Lake at 7300 and 6500 years ago at the beginning of two of the three severe droughts that occurred in the northern Sierra Nevada following the 8200 year ACCE [*Mensing et al.*, 2004]. Three smaller charcoal peaks were deposited into Coburn Lake at the beginning of all three of the severe droughts that occurred between 6400 and 5200 years ago, during a relatively wet period in the northern Sierra Nevada [*Mensing et al.*, 2004].

Mensing et al. [2004] reported that a severe drought occurred in the northern Sierra Nevada from approximately 5200 to 5000 years ago. The beginning of this drought corresponds closely with the timing of the 5200 year ACCE. A large amount of charcoal was deposited into Coburn Lake at the beginning of this drought, also synchronous with the 5200 year ACCE.

As described in section 1.1.2, evidence exists for the occurrence of the 5200 year ACCE in locations as close as British Columbia [*Spooner et al.*, 1997] and the Great Plains [*Clark et al.*, 2002]. *Meyer et al.* [1995] reported a peak in debris flows in Yellowstone, Wyoming, at approximately 5000 years ago. *Klem* [2010] found evidence for a glacial advance in the Uinta Mountains of Utah between 5300 and 5200 years ago. *Barron et al.* [2003] reported evidence for increased ocean upwelling, warmer winters, and increased fog off the coast of northern California at 5200 years ago.

In Greenland, the magnitude of variations in precipitation decreased abruptly after the 5200 year ACCE (Figure 5). Total precipitation in Greenland also began a general decline after the 5200 year ACCE. Precipitation continued to decline until approximately 3000 years ago.

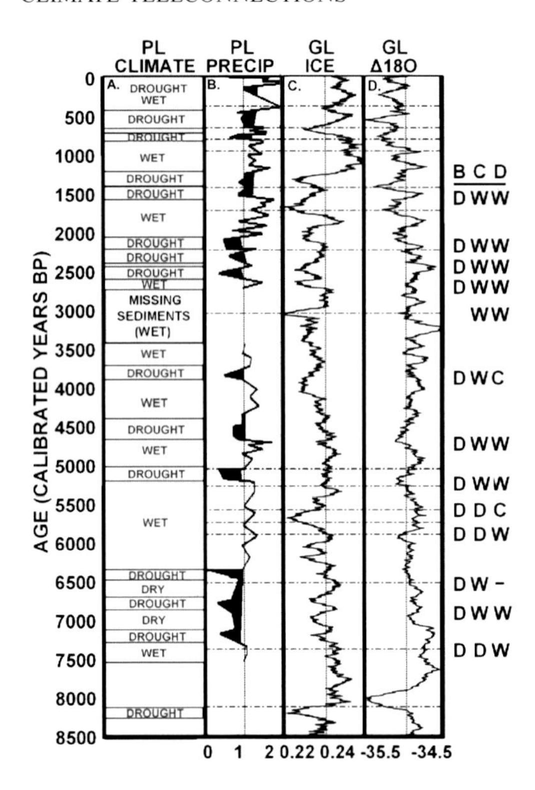

Figure 5. Comparison between the northern Sierra Nevada drought record and Greenland (GL) ice accumulation and temperature records for the past 8500 years. (a) Drought record in the northern Sierra Nevada [*Mensing et al.*, 2004] with the addition of the 8200 year climate event from Coburn Lake. (b) Northern Sierra Nevada drought record from Pyramid Lake (PL), Nevada, based upon the ratio of *Artemisia* (sagebrush) to Chenopodiaceae (shadscale) pollen types (A/C) in Pyramid Lake sediments. Black areas denote periods of drought. (c) Accumulation of ice (m yr^{-1}) through time in central Greenland Greenland Ice Sheet Project 2 (GISP2) ice cores [*Meese et al.*, 1994; *NSIDC and WDC-A*, 1997]. (d) Greenland paleotemperatures derived from delta ^{18}O/mil [*Johnsen et al.*, 1995; *NSIDC and WDC-A*, 1997]. Vertical dashed lines denote median values for Figures 5b–5d. Horizontal lines denote the times of deposition of the largest charcoal peaks from Coburn Lake. Values to the right denote moisture or temperature trajectories in Figures 5b–5d at the beginning of severe droughts in the northern Sierra Nevada. In Figure 5b, D is beginning of severe drought conditions in Sierra Nevada. In Figure 5c, D is toward drier conditions in Greenland; W is toward wetter conditions in Greenland. In Figure 5d, W is toward warmer conditions in Greenland; C is toward colder conditions in Greenland.

At Pyramid Lake, on the other hand, precipitation was generally relatively high after the 5200 year ACCE. Though two severe droughts occurred in the northern Sierra Nevada following the 5200 year ACCE, at approximately 4600 and 3900 years ago, little or no charcoal was deposited into Coburn Lake at the beginning of those ACCEs or at any other time from 5000 to 3000 years ago (Figure 4). It is possible that the vegetation that was growing around Coburn Lake during this 1800 year period did not burn severely at the onset of these severe droughts or, that if the vegetation did burn severely, that the vegetation did not produce much macro-sized charcoal (e.g., a vegetation type dominated by grass or small shrubs).

Temperatures, at least, changed dramatically in the Sierra Nevada beginning approximately 3500 years ago. Following millennia of wide temperature fluctuations, temperatures became cooler and relatively stable in the Sierra Nevada after approximately 3500 years ago [*Potito et al.*, 2006].

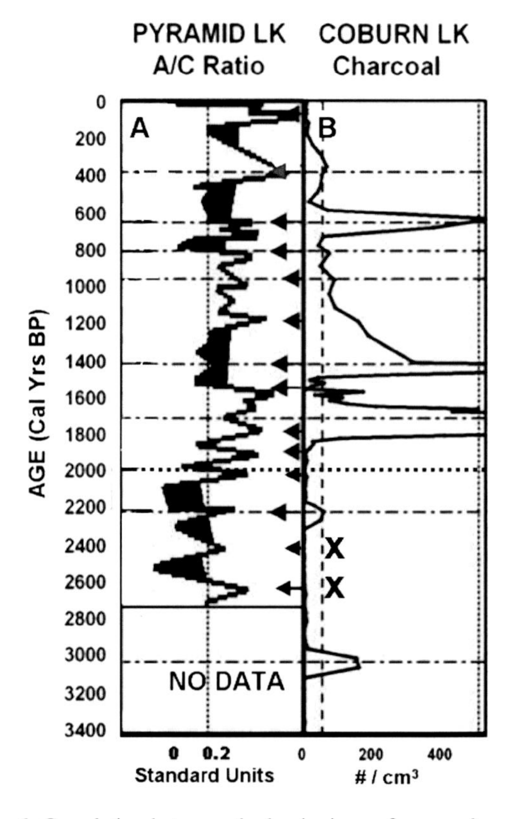

Figure 6. Correlation between the beginnings of severe droughts in the northern Sierra Nevada and Coburn Lake charcoal increases over the past 3400 years. (a) Record of droughts in northern Sierra Nevada from Pyramid Lake, Nevada [*Mensing et al.*, 2004] based upon the ratio of *Artemisia* (sagebrush) to Chenopodiaceae (shadscale) pollen types (A/C) derived from Pyramid Lake sediments. The dotted vertical line represents the mean value for the record. (b) Number of charcoal particles of >250 μm in size per g^3 found within lake sediments. The two largest Coburn Lake charcoal peaks with >1000 charcoal particles g^{-3} were truncated to fit the column size. Dashed horizontal lines represent the centers of charcoal peaks. Dotted vertical line denotes where samples contained >50 particles cm^{-3} of lake sediment. Arrows denote locations of charcoal increases.

A large amount of charcoal was deposited into Coburn Lake approximately 3000 years ago (Figure 4). Though charcoal deposition into Coburn Lake was often associated with the beginning of severe droughts in the Sierra Nevada, I am not aware of evidence for a severe drought in the Sierra Nevada region beginning at approximately 3000 years ago [*Benson et al.*, 2002; *Barron et al.*, 2003]. Unfortunately, the Pyramid Lake record [*Mensing et al.*, 2004] is missing information on precipitation in the northern Sierra Nevada covering the period from approximately 3400 to 2750 years ago (Figure 4). Still, 3000 years ago was a time of great climatic change. A severe drought ended in Greenland (see Figure 5 and section 5.1.3, below) then, and one of the three largest reductions in North Atlantic MOC occurred 3000 years ago [*Oppo et al.*, 2003].

On the other hand, the Coburn Lake charcoal record was strongly correlated with the onset of severe droughts in the Sierra Nevada over the past 2200 years. Increases in charcoal into Coburn Lake coincided with the beginnings of 12 of the 12 severe droughts that occurred in the Northern Sierra Nevada over the past approximately 2200 years, and increases in charcoal did not occur except at the beginnings of those 12 severe droughts (Figure 6) [*Wathen*, 2011].

5.1.2. Cirque Peak temperature record. The beginnings of severe droughts in the northern Sierra Nevada and the deposition of charcoal peaks into Coburn Lake coincided with extreme high temperatures, or an abrupt transition to higher temperatures, in the southern Sierra Nevada over the past 2000 years [*Scuderi*, 1993; *Wathen*, 2011] (Figure 7). As a rule, either high temperatures or an abrupt transition to higher temperatures occurred at the beginnings of severe droughts, while cold temperatures or abrupt changes in temperature running in either direction occurred at the beginnings of pluvial periods. Of the beginnings of 13 of the most severe droughts over the past 1800 years, only a drought that began approximately 1600 years ago was not associated with high temperatures. In addition, severe fires appear to have occurred at Coburn Lake at the beginning of every severe drought-hot temperature event over the past 1800 years.

5.1.3. Abrupt climate change-severe fire and erosion hypothesis. High temperatures at the beginning of severe droughts would have put additional stress on vegetation adapted to predrought climatic conditions. As mentioned above in section 4, charcoal peaks from Coburn Lake were usually associated with indicators of severe soil erosion: peaks in organic detritus, inorganic sediments (clay), sediment bulk density values, and magnetic susceptibility (Figure 2). This close association with severe soil erosion suggests that charcoal peaks from Coburn Lake represent severe stand-replacing fires.

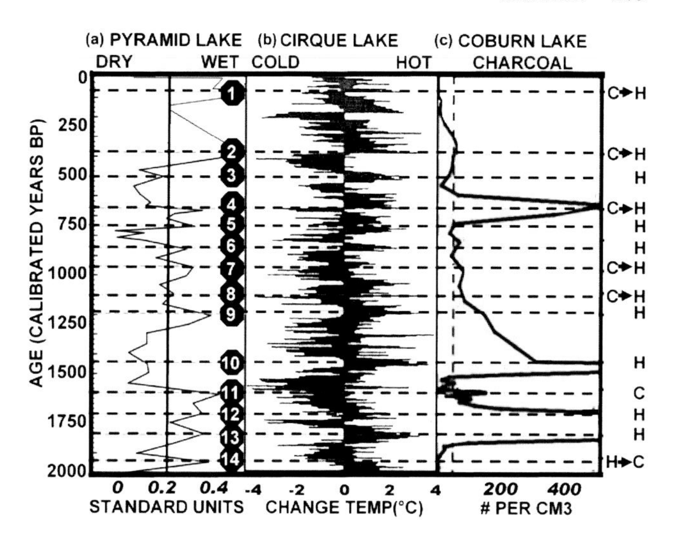

Figure 7. Environmental history of the Sierra Nevada over the past 2000 years based upon the integration of three high-resolution paleorecords. (a) A northern Sierra Nevada drought record from Pyramid Lake, Nevada [*Mensing et al.*, 2004], based upon the ratio of *Artemisia* (sagebrush) to Chenopodiaceae (shadscale) (A/C) pollen types. Horizontal dashed lines denote the beginnings of droughts, which are numbered from top to bottom, from youngest to oldest. The bold vertical solid line denotes the median A/C ratio value for the past 7400 years. (b) A southern Sierra Nevada temperature record based upon the analyses of foxtail pine tree ring chronologies that grew near tree line on Cirque Peak, California. Colder temperature anomalies are to the left, and warmer temperature anomalies are to the right of zero; the mean is from A.D. 1 to 1980 [*Scuderi*, 1993]. (c) The abundance of charcoal particles of >250 μm in size per cm^3 within Coburn Lake sediments. The vertical dashed line designates charcoal abundances of >50 particles cm^{-3}. The two largest Coburn Lake charcoal peaks with >1000 charcoal particles g^{-3} were truncated to fit the column size. Values to the right of Figure 7c denote temperatures, or abrupt changes in temperature, at tree line on Cirque Peak at the beginning of severe droughts in the northern Sierra Nevada. C is cold; H is hot.

Taking into consideration all of these relationships, I propose an "abrupt climate change-severe fire and erosion hypothesis." This hypothesis argues that the abrupt onset of severe droughts and extreme temperatures during the late Holocene caused vegetation and mountain slopes in some ecosystems to be out of balance with new, abruptly changed, climates. This severe imbalance would have resulted in widespread forest die-off, followed by stand-replacing fires and severe soil erosion [*Wathen*, 2011].

5.2. Greenland Precipitation and Temperature Records

Some colleagues and I statistically compared the entire 8500 year long Coburn Lake charcoal record with an ice accumulation-precipitation record from central Greenland [*Meese et al.*, 1994; *NSIDC and WDC-A*, 1997]. *Boucher et al.* [2006] developed a logistics regression model to test for significant correlation between the timing of the deposition of the seven largest charcoal peaks deposited into the Coburn Lake over the past 8500 years and the timing of peak drought conditions in central Greenland (Greenland Ice Sheet Project 2 (GISP2) ice cores) over that same time period.

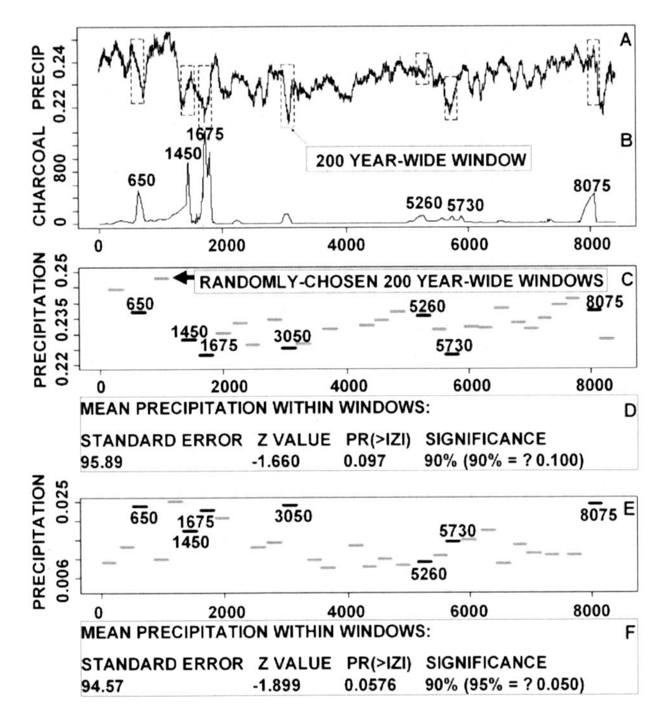

Figure 8. Statistical analysis of the relationship between the timing of the deposition of charcoal peaks from Coburn Lake and peak drought conditions in central Greenland. (a) A reconstructed Holocene North Atlantic snow accumulation record from the central Greenland GISP2 ice core [*Meese et al.*, 1994]. (b) The number of charcoal particles >250 μm in size per sediment sample from Coburn Lake. (c) Mean precipitation values within 200 year-wide windows around the seven largest charcoal peaks from Coburn Lake (black horizontal bars) and around 21 randomly chosen dates (gray horizontal bars). (e) Magnitude of precipitation differences within the same 200 year-wide windows. Bold bars refer to values for the seven largest charcoal peaks from Coburn Lake. Faint bars refer to values for 21 randomly chosen points within the time series. Statistical results from testing the null hypotheses that there are no statistical differences between the sampling windows aligned with Coburn Lake charcoal peaks and randomly placed windows, in terms of either (d) mean precipitation values within windows or (f) the magnitudes of variation within windows, adapted from *Boucher et al.* [2006]. Time is displayed on the *x* axes. The ^{14}C-AMS dates are converted to calibrated years before A.D. 1950 (cal years B.P.).

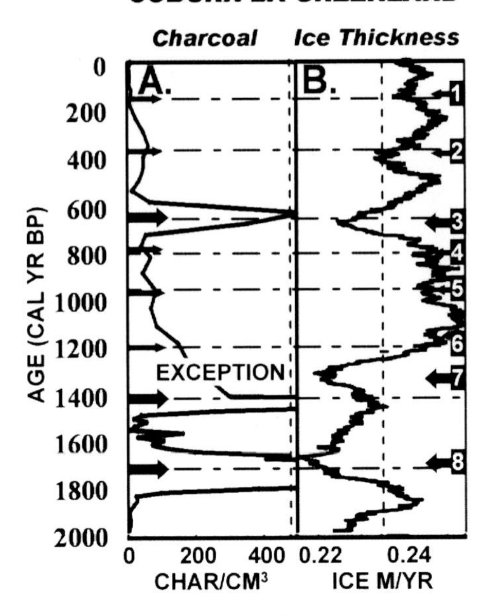

Figure 9. Distribution of charcoal peaks from Coburn Lake compared with a precipitation record from Greenland for the past 2000 years. (a) Abundance of charcoal particles (>250 μm in size) per cm^3 of Coburn Lake sediment. The two largest Coburn Lake charcoal peaks (>1000 charcoal g^{-3}) were truncated to fit the column size. Dashed horizontal lines and arrows in column represent the centers of major charcoal peaks. (b) A reconstructed Holocene North Atlantic snow accumulation record from the central Greenland GISP2 ice core [*Meese et al.*, 1994]. Arrows with numbers represent peak drought conditions in Greenland.

A statistical model was created where (1) x was the predictor variable, ice accumulation; (2) π was the probability that a value of the charcoal time series will be categorized as a peak; and (3) β_o and β_1 were regression coefficients:

$$\log\left(\frac{\pi}{1-\pi}\right) = \beta_0 + \beta_1 x.$$

A 200 year window was centered on each of the dates of the seven charcoal peaks and superimposed onto the Greenland ice accumulation time series, such that the windows would not overlap with other windows (Figure 8). Two hundred year windows allowed for 100 years of variability between the timing of charcoal peaks in Coburn Lake and the timing of peak drought conditions in central Greenland.

Using this model, the seven largest charcoal peaks were compared with 21 randomly chosen noncharcoal peak years. Probabilities were calculated such that the values $Pr(>z$ value) = 0.1 and 0.05 refer to the statistical correlation between charcoal peaks from Coburn Lake and peak drought conditions in central Greenland at the 90% and 95% confidence limits, respectively. Statistical values were calculated separately for both the mean precipitation within each 200 year window and for the magnitude of variance in precipitation within each 200 year window (Figure 8).

The results of the logistics regression model suggest that the largest charcoal peaks from Coburn Lake were deposited during or close to peak drought conditions in the North Atlantic region. The seven largest charcoal peaks from Coburn Lake were significantly correlated at 90% significance levels with peak drought conditions in central Greenland in terms of both their mean values of precipitation within each 200 year window and the magnitude of change in precipitation within each 200 year window.

A closer examination of just the past 1800 years reinforces the idea that charcoal deposition at Coburn Lake was often synchronous with ending of severe drought conditions in the North Atlantic region, particularly during the late Holocene (Figure 9) [*Wathen*, 2011]. The beginnings of severe droughts and the occurrence of severe fires at Coburn Lake in the Sierra Nevada were significantly correlated with the beginning of drought conditions in Greenland (Pearson product moment correlation coefficient, $r = 0.998$) [*Meese et al.*, 1994; *NSIDC and WDC-A*, 1997]. One of the eight charcoal peaks, however, occurred approximately 100 years before the onset of a severe drought, during a time of peak precipitation instead of peak drought conditions in Greenland.

A comparison between the timings of the beginning of severe droughts in the northern Sierra Nevada and a paleotemperature record from Greenland suggests another interesting relationship. The Pyramid Lake drought record was visually compared with a temperature record from Greenland [*Johnsen et al.*, 1995; *NSIDC and WDC-A*, 1997] (Figure 5). Prior to 5200 years ago, the beginnings of droughts in the northern Sierra Nevada were generally associated with the beginnings of droughts in Greenland but not necessarily with high temperatures in Greenland. However, after 5200 years ago, with one exception, the beginnings of droughts in the northern Sierra Nevada were associated with both increasing precipitation as well as increasing temperatures in Greenland.

Warming in Greenland at the beginnings of severe droughts in the northern Sierra Nevada after the 5200 year ACCE agrees with the finding (section 5.1.2, above) that the beginnings of severe droughts in the northern Sierra Nevada were associated with high temperatures in the Sierra Nevada over the past 1800 years [*Wathen*, 2011] (Figure 7).

Taken together, these findings suggest that temperatures and precipitation were rising in Greenland as severe droughts were beginning in the northern Sierra Nevada. These findings agree with other studies that have found that the northern Atlantic Ocean was warm when severe drought conditions were beginning in the western United States [*Enfield et al.*, 2001; *Schubert et al.*, 2004].

This bidirectional synchronicity suggests a hypothesis that large-scale shifts in the locations of the Earth's major precipitation belts may have occurred during periods of ACC in the

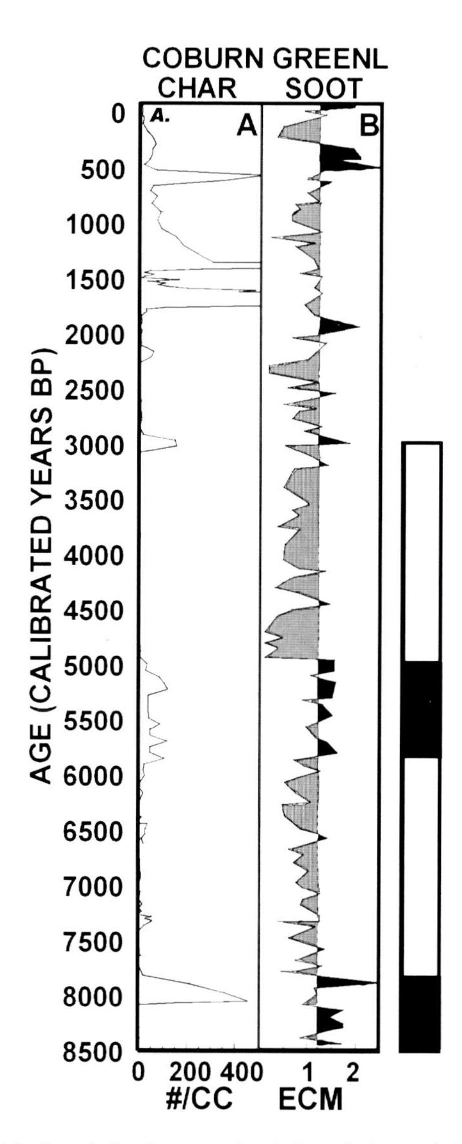

Figure 10. Correlation between the timing of charcoal deposition into Coburn Lake, California, and soot deposition onto central Greenland glaciers over the past 8500 years. (a) Abundance of charcoal particles (>250 μm in size) per cm^3 of Coburn Lake sediment. (b) Peaks in soot accumulation, based on electrical conductivity, from the GISP2 Greenland ice core. Data are from the work of *Taylor et al.* [1996]. The change in shading in Figure 10b denotes the median value. In the vertical box to the right, black shading highlights time periods when Figures 10a and 10b appear to have shared a similar pattern of deposition.

past. This hypothesis suggests that the northern Polar Front and associated storms moved northward over central Greenland at the beginning of moist conditions in Greenland. Simultaneously, the northern subtropical high-pressure zone moved northward over the northern Sierra Nevada bringing severe drought conditions to the northern Sierra Nevada. The opposite precipitation scenario would be true, when the northern Polar Front moved southward away from central Greenland and over the northern Sierra Nevada, the subtropical high-pressure zone would have moved southward away from the northern Sierra Nevada. The shift in these major precipitation belts may have been abrupt.

5.3. Soot and Charcoal Deposition Records From Eastern Canada and Greenland

I compared the 8500 year long Coburn Lake charcoal record with a biomass burning record from the GISP2 ice cores from central Greenland [*Taylor et al.*, 1997; *NSIDC and WDC-A*, 1997]. More than 90% of Coburn Lake charcoal peaks coincided with peaks in soot from Greenland ice cores (Figure 10). Both records also displayed greatest deposition of either charcoal or soot during four time periods: (1) before 7800 years ago, at around the time of the 8200 event; (2) at 5800–5000 years ago, during the wet period that followed the Altithermal, which was ended by the 5200 year event; (3) at 3000 years ago; and (4) after 2000 years ago as Greenland, and presumably the Sierra Nevada, became colder and wetter and precipitation patterns, as described in section 5.4 below, became more unstable.

The soot in the GISP2 ice core is believed to have originated from forest fires that burned in eastern Canada [*Taylor et al.*, 1996]. Lac Francis is located in eastern Canada. A charcoal record from Lac Francis developed by *Carcaillet et al.* [2001] goes back to approximately 6750 years ago. Interestingly, the timing of the beginnings of severe droughts in the Sierra Nevada [*Mensing et al.*, 2004], and thus peak charcoal deposition into Coburn Lake, was significantly correlated (Pearson product moment correlation coefficient is 0.999) with the timing of deposition of the largest charcoal peaks into Lac Francis over the past 2800 years [*Carcaillet et al.*, 2001] (Figure 11). Other large charcoal peaks from Lac Francis appear to correspond to severe droughts in the northern Sierra Nevada that began at approximately 6500, 5200, and 3600 years ago. Though the climates of high elevation subalpine Coburn Lake and high latitude boreal Lac Francis share some similarities today, it is surprising to see such high correlations between two records separated by a distance of almost 5000 km. These results suggest that some ACCEs may have affected locations separated by very large distances.

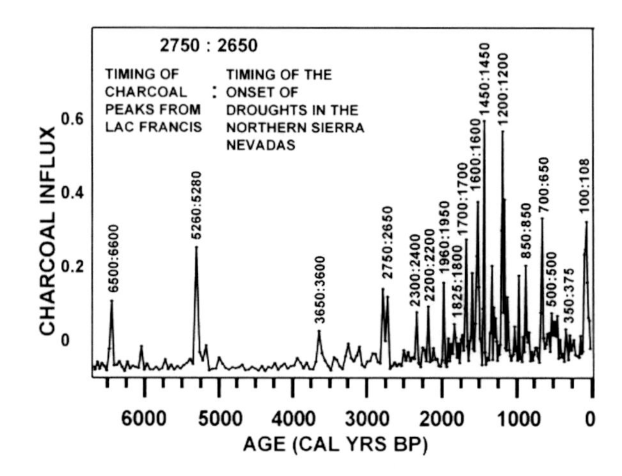

Figure 11. Correlation between the timing of the beginning of droughts in the northern Sierra Nevada and charcoal peaks from Lac Francis, Quebec. The timings for the 14 largest charcoal peaks from Lac Francis over the past 6750 years (adapted from *Carcaillet et al.* [2001]) were compared with the starting dates for 14 severe droughts in northern Sierra Nevada over the same time interval [*Mensing et al.*, 2004]. Vertical numbers above charcoal peaks represent the timing of charcoal peaks from Lac Francis relative to the timing of the onset of droughts in the northern Sierra Nevada.

5.4. Increase in the Frequency of ACCEs During the Late Holocene

In terms of temperature, both Greenland and the Sierra Nevada became cooler during the Late Holocene. In Greenland, the magnitude of fluctuations in temperature increased markedly after approximately 3700 years ago, and long-term cooling began approximately 2400 years ago. Cooling in Greenland accelerated at approximately 1000 years ago [*Meese et al.*, 1994; *NSIDC and WDC-A*, 1997] (Figure 12). In the Sierra Nevada, *Clark et al.* [2002] noted that long-term cooling began by 3400 years ago.

Both Greenland and the Sierra Nevada experienced increased precipitation and more erratic precipitation during the late Holocene. In Greenland, the size or magnitude of fluctuations in precipitation increased substantially after approximately 3000 years ago, and total precipitation increased markedly after approximately 1300 years ago [*Meese et al.*, 1994; *NSIDC and WDC-A*, 1997]. In the Sierra Nevada, the number of abrupt changes in precipitation increased sometime between 3400 and 2700 years ago (Figure 5). Unfortunately, an unintentional gap exists between core sections in the Pyramid Lake record that covers this time period. Therefore, the timing of an increase in the number of abrupt changes in precipitation in the northern Sierra Nevada cannot be further constrained at the moment. Precipitation did

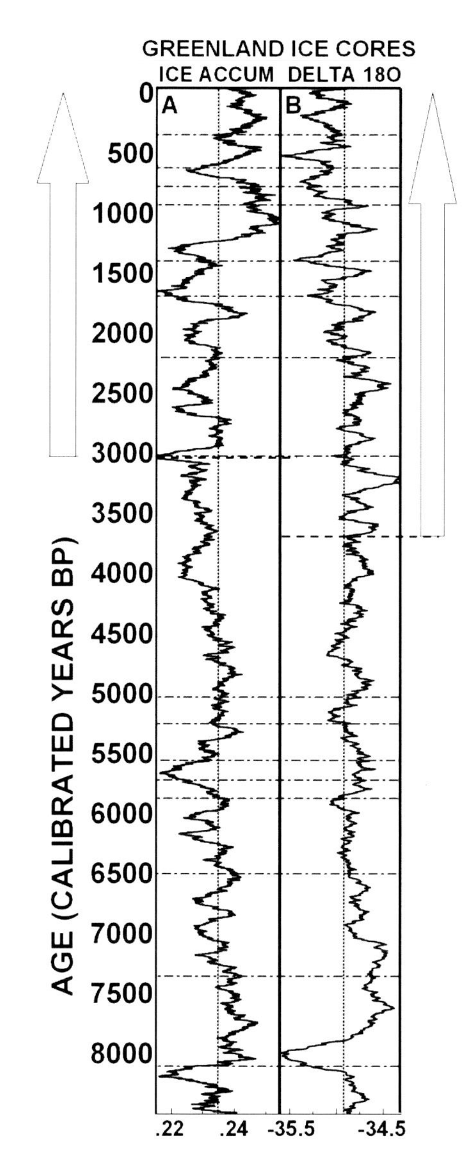

Figure 12. Increase in Greenland climate variability over the past 8500 years. (a) A reconstructed ice accumulation record from the GISP2 ice core serves as a proxy for precipitation [*Meese et al.*, 1994; *NSIDC and WDC-A*, 1997]. (b) A reconstructed $\delta^{18}O$ curve from the GRIP ice core serves as a proxy for air temperature [*Johnsen et al.*, 1995; *NSIDC and WDC-A*, 1997]. The two large vertical arrows highlight the beginning of increases in the magnitude of centennial-scale oscillations of precipitation and temperature at 3000 and 3700 years ago, respectively.

increase in the Sierra Nevada for the long-term beginning approximately 2000 years ago [*Mensing et al.*, 2004].

Increased precipitation, cooler temperatures, and an increase in the frequency of ACCEs in the Northern Hemisphere could explain the increase in severe fires at Coburn Lake, and perhaps at similar environments, during the late Holocene. Increased precipitation and cooler temperatures would have meant higher moisture availability to plants nearly everywhere during the late Holocene. An increase in available soil moisture could have resulted in a change in the dominance of plants toward those better adapted to cooler, moister conditions. Indeed, red fir, the dominant tree growing at Coburn Lake today, grows in the Sierra Nevada where the winter snow pack is the deepest [*Barbour et al.*, 1991].

Increased soil moisture and consequent changes in vegetation would have altered fire regimes at Coburn Lake and in other areas. Light fuels, like grass, may have increased in the generally depauperate understory of red fir stands resulting in more frequent low-severity surface fires. On the other hand, moister and cooler conditions could have resulted in an increase in heavy fuels within red fir stands, along with fewer opportunities for those heavy fuels to dry out. When severe droughts occurred, those heavy fuels would have dried out. This would have resulted in severe stand-replacing fires, of the type I hypothesize are represented in the Coburn Lake charcoal record (section 5.1.3).

5.5. *Controversy*

The results from Coburn Lake were presented at science meetings from August of 2005 [*Wathen*, 2005] through May of 2007 [*Wathen*, 2007]. Attempts to publish results from Coburn Lake began in November of 2005 and have continued since. Opposition to the findings from Coburn Lake have come primarily from three concerns: (1) that I did not use traditional approaches to analyzing the charcoal record from Coburn Lake, (2) that I did not provide enough evidence for climate teleconnections between California and the North Atlantic, and (3) that I did not provide an explanation for those teleconnections.

5.5.1. The study of sediment charcoal. Researchers have used sediment charcoal to study the relationships between past climates and past fire frequencies. The standard methodology for interpreting lake sediment charcoal records was designed to study fire frequencies at the millennial time scale [*Long et al.*, 1998]. This methodology has been used with success to analyze "typical" charcoal records that contain only marginally higher charcoal peaks relative to high levels of background charcoal [*Cwynar*, 1977; *Gajewski et al.*, 1985; *Clark*, 1990; *Whitlock and Anderson*, 2003]. With this standard methodology, a line of demarcation between charcoal peaks and what is considered to be the background level of charcoal for a given record is determined. Peaks that are higher than this predetermined background level of charcoal are assumed to represent individual severe fire events.

In contrast to the usual lake charcoal record, the charcoal pattern derived from Coburn Lake was strongly episodic. Prior to 1500 years ago, there was little background charcoal deposited between peaks (Figure 4). I elected to compare charcoal peaks from Coburn Lake with ACCEs rather than with millennial-scale climate changes, as is most commonly done, because of the episodic nature of the Coburn Lake charcoal record and because of the large number of radiocarbon dates available.

Correlating charcoal peaks with ACCEs also challenges the conventional wisdom for many within the sediment charcoal community that individual charcoal peaks should not be said to represent individual ACCEs. This conventional wisdom is based on the fact that individual charcoal peaks are not well correlated with individual ACCEs in most charcoal records. If a few charcoal peaks are correlated with known ACCEs, most are not. Therefore, those few that are correlated are suspected of being correlated purely by chance.

Exceptions are studies by *Meyer et al.* [1995] and *Pierce et al.* [2004] in which individual charcoal lenses from within hillside debris flow deposits were analyzed in relation to relatively short-term climate changes. As with Coburn Lake, charcoal deposition within these debris flows was episodic. The Coburn Lake record has more in common with the charcoal records from alluvial fans than it does the typical lake charcoal record.

The Coburn Lake charcoal record is unusual in that all charcoal peaks or increases in charcoal occurred at the beginning of severe droughts and only at the beginning of severe droughts over the past 1800 years. Fires from other causes do not seem to be represented. The size and configuration of the Coburn Lake watershed, noted above in section 2, might contribute to understanding why Coburn Lake has been so unusually sensitive to recording ACCE-linked fires while filtering out fires from other causes.

5.5.2. Evidence for interhemispheric teleconnections. Though some remain skeptical, the evidence for teleconnections between climate events in the western United States and the North Atlantic has increased to the point that more researchers have come to acknowledge the evidence. As noted in section 5.1.1, several researchers have published, though not necessarily acknowledged, evidence for the 8200 [*Meyer et al.*, 1995; *Benson et al.*, 1997; *Barron et al.*, 2003; *Potito et al.*, 2006] and 5200 [*Meyer et al.*, 1995; *Spooner et al.*, 1997; *Clark et al.*, 2002; *Barron et al.*, 2003; *Mensing et al.*, 2004; *Klem*, 2010] ACCEs in the western United States or British Columbia.

Numerous studies have recognized climate teleconnections between Greenland and regions surrounding Coburn Lake. In southern California, researchers studying sediments collected off the coast of Santa Barbara have found evidence for correlations with Greenland suggesting global-scale changes in atmospheric circulation [*Kennett and Ingram*, 1995; *Behl and Kennett*, 1996; *Hendy and Kennett*, 1999; *Haug et al.*, 2001; *Hendy et al.*, 2002; *Friddell et al.*, 2002; *Roark et al.*, 2003]. *Wagner et al.* [2005] found isotopic changes in Arizona that closely mirrored the Dansgaard-Oeschger (D-O) events recorded in Greenland ice cores. *Dean* [2007] suggested climate teleconnections between the margins of Alta and Baja California and Greenland covering the past 60,000 years. Farther south, *Peterson et al.* [2000], *Lachniet et al.* [2005], and *Stott et al.* [2002] found close synchronicity between records collected from off the coast of Venezuela, the western coast of Costa Rica, and the western Pacific tropical warm pool, respectively, and D-O events in Greenland.

To the north, *Pisias et al.* [2001] suggested that coastal upwelling, temperatures, and vegetation along the Oregon coast were correlated with the same D-O events. *Hu et al.* [2003] demonstrated synchronicity between climate records in Alaska and Greenland spanning the Holocene. In the Great Basin, *Benson et al.* [1997, 1998], *Lin et al.* [1998], and *Zic et al.* [2002] found evidence for teleconnections with Greenland from Owens and Mono Lakes, California; Searles Lakes, California; and Summer Lake, Oregon, respectively.

The Atlantic-centered Atlantic Multidecadal Oscillation (AMO) has been associated with changes in the frequency of United States droughts, droughts being more frequent or more prolonged when the AMO is in its warm (positive) phase. Two of the most severe droughts of the twentieth century, the droughts of the 1930s and the 1950s, occurred when the AMO was in a warm phase [*Schubert et al.*, 2004]. *Swetnam et al.* [2006] found that fires across the western United States over the past 500 years were most widespread when the AMO was in a warm phase.

McCabe et al. [2004, 2008] found that the Pacific-centered ENSO was also important in understanding drought occurrence in the contiguous United States. McCabe et al. found that the Pacific Decadal Oscillation (PDO) cooling and the AMO each accounted for half of the variance in multidecadal drought frequency. *Kitzberger et al.* [2007] found that both the ENSO and PDO drove the frequency of recent fires in the western United States at interannual to decadal time scales, but that the AMO determined the severity of fires and how widespread fires were on multidecadal time scales.

Seager et al. [2007] suggested that North American megadroughts during the Medieval period (A.D. 800–1300) also occurred during cold La Niña-like conditions in conjunction with warm phases of the AMO. *Feng et al.* [2008] modeled climate conditions during medieval droughts and found that a cool tropical Pacific controlled drought intensity, a warm

North Atlantic controlled the areal extent of droughts, and that the two modes of variability together controlled the severity and longevity of droughts.

Though evidence for the teleconnections between the northern Atlantic Ocean exists, one has to be careful in discussing the causes for megadroughts in the remote past. First, the causes for many of the climatic teleconnections in the remote past, and even today, are not well understood [*Cook et al.*, 2010]. Second, the relationships between the influences of the Atlantic and Pacific oceans in the past were not always consistent [*Conroy et al.*, 2009]. These facts suggest that factors other than the ENSO, PDO, or AMO might also predispose western North America to severe drought.

Synchronicity between climate events in the Western Hemisphere and climate events in Greenland, described here and elsewhere, suggests that there were shared climate teleconnections between the western United States and the North Atlantic in the past. However, it should be noted, with the possible exception of the 8200 years' climate event [*Alley et al.*, 1997; *Barber et al.*, 1999], that shared climate teleconnections between western North America and the North Atlantic does not require that events in one region (e.g., the Sierra Nevada) were directly caused by events in another (e.g., the North Atlantic). Rather, climate teleconnections suggest that both regions were affected by at least some of the same climate drivers at some times in the past.

5.5.3. A possible explanation for climate teleconnections and the implications. Some researchers have been hesitant to acknowledge that climate events in California and Greenland could be correlated with each other unless the mechanisms for those climate teleconnections, which are still unsatisfactory understood at this point, were known. In section 5.2, I noted that the evidence from Coburn Lake suggested that the general locations of the Earth's major precipitation belts might have shifted abruptly during ACCEs in the past. This could be one mechanism to explain climate teleconnections between California and Greenland. Any major relocation of the general location of the polar jet stream, north or south away from today's location, would have meant fewer storms passing over the northern Sierra Nevada and drier conditions during those relocations.

Some evidence suggests that shifting of the Earth's major precipitation belts will occur in the future or may already be occurring. The *Intergovernmental Panel on Climate Change* [2007] suggests that the subtropical high-pressure zones of the world will become drier and will expand poleward in the future due to greenhouse gas-induced warming. *Archer and Caldeira* [2008] suggested that the northern polar jet stream might already have moved northward in recent years as the planet has warmed. *Seager et al.* [2007] noted that recent severe droughts experienced in the southwestern United States may have been associated with a shift of storm tracks northward and an expansion of the subtropical high into the midlatitudes.

What is instructive in terms of ongoing human-caused climate change is how fast climates changed from wetter to drier and from hotter to colder conditions in the Sierra Nevada and Greenland during the relatively recent past. *Seager et al.* [2007] noted that although ongoing warming due to the buildup of greenhouse gasses is a relatively gradual change, the paleoclimate records tell us that the transition to a megadrought can be very abrupt. We know that ACCEs occurred repeatedly in the past without the influence of humans. We do not know, however, if the large amounts of greenhouse gases we are adding to the atmosphere could somehow trigger a future human-caused ACCE.

We do know that past ACCEs have had serious repercussions for people and the world's environments. For instance, a severe drought 8200 years ago reportedly caused the sudden collapse of dry land cereal farming in Mesopotamia and massive loss of human life, leading eventually to the beginning of irrigated agriculture along the Tigris and Euphrates rivers and the rise of the first phase of Mesopotamian civilization, the Sumerian empire, in Iraq [*Weiss*, 2003]. The 5200 ACCE also coincided with a severe drought, leading to the end of the Uruk Phase of Mesopotamia culture [*Weiss*, 2003] and the beginning of the First Dynasty of Ancient Egypt. Therefore, it behooves us to learn about past occurrences of ACCEs and to pay attention to what we are doing to the climate and what the repercussions might be.

6. CONCLUSIONS

This study sought to explain the unusually peaked Coburn Lake macrocharcoal record and its relationship to climate and fire events in the Sierra Nevada, eastern Canada, and Greenland. The entire 8500 year long Coburn Lake charcoal record was compared with a high-resolution 7400 year long precipitation record from Pyramid Lake, Nevada [*Mensing et al.*, 2004]. Severe fires and soil erosion occurred at Coburn Lake at the beginning of six of the seven most severe droughts that occurred in the northern Sierra Nevada over the past 7400 years and at the beginning of every drought that occurred in the northern Sierra Nevada over the past 1800 years.

Two thousand years of both the Coburn Lake charcoal and the Pyramid Lake precipitation records were compared with a 2000 year long Cirque Peak temperature record [*Scuderi*, 1993]. It was found that temperatures at the beginnings of droughts in the Sierra Nevada were either high or times of abrupt transition to higher temperatures. These findings suggest an "abrupt climate change-severe fire and erosion hypothesis" that states that severe fires and erosion occurred in some locations in the past as a result of the stresses put on vegetation and slopes by ACCEs.

The entire Coburn Lake and Pyramid Lake records were also compared with high-resolution temperature and precipitation records from Greenland [*NSIDC and WDC-A*, 1997]. These comparisons provided evidence that the 8200 and 5200 year ACCEs occurred in the northern Sierra Nevada. The largest charcoal peaks from Coburn Lake were deposited as severe droughts were beginning in the northern Sierra Nevada and as severe droughts were ending in Greenland. In addition, five of the seven largest charcoal peaks from Coburn Lake were coincident with peak low, but rising, temperatures in central Greenland.

These findings suggest that temperatures were generally hot or rising in the Northern Hemisphere as severe droughts were beginning in the Sierra Nevada and ending in Greenland. The bidirectional synchronicity of these events suggests a hypothesis that abrupt, large-scale shifts in the general locations of the Earth's major precipitation belts (the ITCZ, the two Subtropical Desert Zones, and the two Polar Fronts) may have accompanied higher temperatures and ACCEs during parts of the Holocene.

The Coburn Lake and Pyramid Lake records were also compared with two records from eastern Canada and Greenland. More than 90% of charcoal peaks from Coburn Lake coincided with peaks in soot deposited onto Greenland glaciers over the past 8500 years [*Taylor et al.*, 1997; *NSIDC and WDC-A*, 1997]. This soot is thought to have originated from fires that burned in eastern Canada [*Taylor et al.*, 1996]. The beginnings of severe droughts in the northern Sierra Nevada [*Mensing et al.*, 2004] and charcoal peaks from Coburn Lake were also highly correlated with the occurrence of severe fires in eastern Canada at Lac Francis over the past 6750 years, including during the 5200 year ACCE [*Carcaillet et al.*, 2001]. These findings suggest the hypothesis that some ACCEs may have affected vegetation at widely dispersed locations.

Climate variability appears to have increased in Greenland and the Sierra Nevada over the past approximately 3700 years. The number of severe fires at Coburn Lake and Lac Francis increased over the past 3000 years, perhaps in response to increased climate variability and increased available moisture. The unusual characteristics of Coburn Lake and its watershed may explain why the Coburn Lake charcoal record appears to contain only fires caused by ACC, while most other lake sediment charcoal records contain fires from various causes. ACCEs have had severe impacts on human and natural systems in the past. Therefore, it is important to pay close attention to what we are currently doing to the climate that might trigger an ACCE.

Acknowledgments. Results from Coburn Lake were first presented in May 2005 in Flagstaff Arizona. Thanks to those individuals who have reviewed versions of this paper since then, including Michael Barbour, Havunur Rashid, and Jennifer Richards; those students who helped me process and count charcoal; the Geological Society of America and the University of California, Davis for funding; Scott Mensing et al. and Christopher Carcaillet et al. and NSIDC and NDCP for generously sharing their data; and a special thanks to Tom Brown at the Livermore CAM Laboratory for his generous radiocarbon dating of my sediments.

REFERENCES

Alley, R. B., and A. M. Agustsdottir (2005), The 8k event: Cause and consequences of a major Holocene abrupt climate change, *Quat. Sci. Rev.*, *24*(10–11), 1123–1149.

Alley, R. B., P. A. Mayewski, T. Sowers, M. Stuiver, K. C. Taylor, and P. U. Clark (1997), Holocene climatic instability: A prominent, widespread event 8,200 years ago, *Geology*, *25*, 483–486.

Archer, C. L., and K. Caldeira (2008), Historical trends in the jet streams, *Geophys. Res. Lett.*, *35*, L08803, doi:10.1029/2008GL033614.

Barber, D. C., et al. (1999), Forcing of the cold event of 8,200 years ago by catastrophic drainage of Laurentide lakes 344, *Nature*, *400*, 344–348.

Barbour, M. G., N. H. Berg, T. G. F. Kittel, and M. E. Kunz (1991), Snowpack and the distribution of a major vegetation ecotone in the Sierra Nevada of California, *J. Biogeogr.*, *18*(2), 141–149.

Barron, J. A., L. Heusser, T. Herbert, and M. Lyle (2003), High-resolution climatic evolution of coastal northern California during the past 16,000 years, *Paleoceanography*, *18*(1), 1020, doi:10.1029/2002PA000768.

Behl, R. J., and J. P. Kennett (1996), Brief interstadial events in the Santa Barbara basin, NE Pacific, during the past 60 kyr, *Nature*, *379*, 243–246.

Benson, L., J. Burdett, S. Lund, M. Kashgarian, and S. Mensing (1997), Nearly synchronous climate changes in the Northern Hemisphere during the last glacial termination, *Nature*, *388*, 263–265.

Benson, L. V., S. P. Lund, J. W. Burdett, M. Kashgarian, T. P. Rose, J. P. Smoot, and M. Schwartz (1998), Correlation of late-Pleistocene lake-level oscillations in Mono Lake, California, with North Atlantic climate events, *Quat. Res.*, *49*(1), 1–10.

Benson, L., M. Kashgarian, R. Rye, S. Lund, F. Paillet, J. Smoot, C. Kester, S. Mensing, D. Meko, and S. Lindström (2002), Holocene multidecadal and multicentennial droughts affecting Northern California and Nevada Pyramid Lake, Nevada, *Quat. Sci. Rev.*, *21*(4–6), 659–682.

Boucher, J., Y. Chen, L. Ziu, and H.-G. Müller (2006), Statistics 401 final consulting report, graduate student consulting report, Univ. of Calif., Davis, Calif.

Carcaillet, C., Y. Bergeron, P. J. H. Richard, B. Fréchette, S. Gauthier, and Y. T. Prairie (2001), Change of fire frequency in the eastern Canadian boreal forests during the Holocene: Does vegetation composition or climate trigger the fire regime?, *J. Ecol.*, *89*(6), 930–946.

Clark, J. S. (1988), Particle motion and the theory of charcoal analysis: Source area, transport, deposition, and sampling, *Quat. Res.*, *30*, 67–80.

Clark, J. S. (1990), Fire and climate change during the last 750 yr in northwestern Minnesota, USA, *Ecol. Monogr.*, *60*, 135–160.

Clark, J. S., and P. D. Royall (1996), Local and regional sediment charcoal evidence for fire regimes in presettlement north-eastern North America, *J. Ecol.*, *84*, 365–382.

Clark, J. S., E. C. Grimm, J. J. Donovan, S. C. Fritz, D. R. Engstrom, and J. E. Almendinger (2002), Drought cycles and landscape responses to past aridity on prairies of the northern Great Plains, USA, *Ecology*, *83*(3), 595–601.

Commitee on Abrupt Climate Change, National Research Council (2002), *Abrupt Climate Change: Inevitable Surprises*, 238 pp., Natl. Acad. Press, Washington, D. C.

Conroy, J. L., J. T. Overpeck, J. E. Cole, and M. Steinitz-Kannan (2009), Variable oceanic influences on western North American drought over the last 1200 years, *Geophys. Res. Lett.*, *36*, L17703, doi:10.1029/2009GL039558.

Cook, E. R., R. Seager, R. R. Heim Jr., R. S. Vose, C. Herweijer, and C. Woodhouse (2010), Megadroughts in North America: placing IPCC projections of hydroclimatic change in a long-term palaeoclimate context, *J. Quat. Sci.*, *25*(1), 48–61.

Cremer, H., B. Wagner, M. Melles, and H.-W. Hubberten (2001), The postglacial environmental development of Raffles So, East Greenland: Inferences from a 10,000 year diatom record, *J. Paleolimnol.*, *26*(1), 67–87.

Cwynar, L. C. (1977), Recent history of fire and vegetation from laminated sediment of Greenleaf Lake, Algonquin Park, Ontario, *Can. J. Bot.*, *56*, 10–21.

Dean, W. E. (2007), Sediment geochemical records of productivity and oxygen depletion along the margin of western North America during the past 60,000 years: teleconnections with Greenland Ice and the Cariaco Basin, *Quat. Sci. Rev.*, *26*(1–2), 98–114.

Dean, W. E., J. Rosenbaum, G. Skipp, S. Colman, R. Forester, A. Liu, K. Simmons, and J. Bischoff (2006), Unusual Holocene and late Pleistocene carbonate sedimentation in Bear Lake, Utah and Idaho, *Sediment. Geol.*, *185*, 93–112.

Douglas, D. J. (1988), A modified light weight piston corer for sampling soft lake sediment, *Ann. Limnol.*, *24*(2), 193–196.

Enfield, D. B., A. M. Mestas-Nuñez, and P. J. Trimble (2001), The Atlantic multidecadal oscillation and its relation to rainfall and river flows in the continental US, *Geophys. Res. Lett.*, *28*(10), 2077–2080.

Feng, S., R. J. Oglesby, C. M. Rowe, D. B. Loope, and Q. Hu (2008), Atlantic and Pacific SST influences on Medieval drought in North America simulated by the Community Atmospheric Model, *J. Geophys. Res.*, *113*, D11101, doi:10.1029/2007JD009347.

Friddell, J. E., R. C. Thunell, and L. E. Heusser (2002), Direct comparison of marine and terrestrial climate variability during marine isotope stages 6 and 5: Results from Santa Barbara Basin ODP Hole 893A, *Paleoceanography*, *17*(1), 1002, doi:10.1029/2000PA000594.

Frits, S. C., S. E. Metcalfe, and W. Dean (2001), Holocene climate patterns in the Americas inferred from paleolimnological records, in *Interhemispheric Climate Linkages*, edited by V. Markgraf, pp. 241–263, Academic, San Diego, Calif.

Gajewski, K., M. G. Winkler, and A. M. Swain (1985), Vegetation and fire history from three lakes with varved sediments in northwestern Wisconsin (USA), *Rev. Palaeobot. Palynol.*, *44*(3–4), 277–292.

Haug, G. H., K. A. Hughen, D. M. Sigman, L. C. Peterson, and U. Röhl (2001), Southward migration of the Intertropical Convergence Zone through the Holocene, *Science*, *292*, 1304–1307.

Hendy, I. L., and J. P. Kennett (1999), Latest Quaternary North Pacific surface-water responses imply atmosphere-driven climate instability, *Geology*, *27*(4), 291–294.

Hendy, I. L., J. P. Kennett, E. B. Roark, and B. L. Ingram (2002), Apparent synchroneity of submillennial scale climate events between Greenland and Santa Barbara Basin, California from 30–10 ka, *Quat. Sci. Rev.*, *21*(10), 1167–1184.

Hu, F. S., D. Kaufman, S. Yoneji, D. Nelson, A. Shemesh, Y. Huang, J. Tian, G. Bond, B. Clegg, and T. Brown (2003), Cyclic variation and solar forcing of Holocene climate in the Alaskan subarctic, *Science*, *301*, 1890–1893.

Intergovernmental Panel on Climate Change (Ed.) (2007), *Climate Change 2007: The Physical Science Basis: Working Group I Contribution to the Fourth Assessment Report of the IPCC*, edited by S. Solomon et al., Cambridge Univ. Press, Cambridge, U. K.

Johnsen, S. J., D. Dahl-Jensen, W. Dansgaard, and N. Gundestrup (1995), Greenland palaeotemperatures derived from GRIP bore hole temperature and ice core isotope profiles, *Tellus, Ser. B*, *47*(5), 624–629.

Kennett, J. P., and B. L. Ingram (1995), A 20,000-year record of ocean circulation and climate change from the Santa Barbara Basin, *Nature*, *377*, 510–514.

Kitzberger, T., P. M. Brown, E. K. Heyerdahl, T. W. Swetnam, and T. T. Veblen (2007), Contingent Pacific-Atlantic Ocean influence on multicentury wildfire synchrony over western North America, *Proc. Natl. Acad. Sci. U. S. A.*, *104*(2), 543–548.

Klem, C. (2010), A sedimentary record of Holocene Neoglaciation from the Uinta Mountains, Utah, B.A. thesis, Middlebury College, Middlebury, Vt.

Lachniet, M. S., Y. Asmerom, S. Burns, V. Polyak, L. Burt, and A. Azouz (2005), Speleothem paleorainfall records from the Isthmus of Panama: Tropical forcing or response to the extratropics?, *Eos Trans. AGU*, *86*(52), Fall Meet. Suppl., Abstract PP11C-02.

Larsen, C. P. S., R. Pienitz, J. P. Smol, K. A. Moser, B. F. Cumming, J. M. Blais, G. M. Macdonald, and R. I. Hall (1998), Relations between lake morphometry and the presence of laminated lake sediments: A re-examination of Larsen and Macdonald (1993), *Quat. Sci. Rev.*, *17*(8), 711–717.

Leroya, S. A. G., H. Jousseb, and M. Cremaschic (2006), Dark nature: Responses of humans and ecosystems to rapid environmental changes, *Quat. Int.*, *151*(1), 1–2.

Lin, J. C., W. S. Broecker, S. R. Hemming, I. Hajdas, R. F. Anderson, G. I. Smith, M. Kelley, and G. Bonani (1998), A reassessment of U-Th and ^{14}C ages for late-glacial high-frequency hydrological events at Searles Lake, California, *Quat. Res.*, *49*(1), 11–23.

Long, C. J., C. Whitlock, P. J. Bartlein, and S. H. Millspaugh (1998), A 9000-year fire history from the Oregon coast range, based on a high-resolution charcoal study, *Can. J. For. Res.*, *28*, 774–787.

MacDonald, G. M., C. P. S. Larsen, J. M. Szeicz, and K. A. Moser (1991), The reconstruction of boreal forest fire history from lake sediments: A comparison of charcoal, pollen, sedimentological, and geochemical indexes, *Quat. Sci. Rev.*, *10*(1), 53–71.

McCabe, G. J., M. A. Palecki, and J. L. Betancourt (2004), Pacific and Atlantic Ocean influences on multidecadal drought frequency in the United States, *Proc. Natl. Acad. Sci. U. S. A.*, *101*(12), 4136–4141.

McCabe, G. J., J. L. Betancourt, S. T. Gray, M. A. Palecki, and H. G. Hidalgo (2008), Associations of multi-decadal sea-surface temperature variability with US drought, *Quat. Int.*, *188*, 31–40, doi:10.1016/j.quaint.2007.07.001.

McKenzie, G. M., and A. P. Kershaw (1997), A vegetation history and quantitative estimate of Holocene climate from Chapple Vale, in the Otway Region of Victoria, Australia, *Aust. J. Bot.*, *45*(3), 565–581.

Meese, D. A., A. J. Gow, P. Grootes, M. Stuiver, P. A. Mayewski, G. A. Zielinski, M. Ram, K. C. Taylor, and E. D. Waddington (1994), The accumulation record from the GISP2 core as an indicator of climate change throughout the Holocene, *Science*, *266*(5191), 1680–1682.

Mensing, S. A., L. V. Benson, M. Kashgarian, and S. Lund (2004), A Holocene pollen record of persistent droughts from Pyramid Lake, Nevada, USA, *Quat. Res.*, *62*(1), 29–38.

Meyer, G. A., S. G. Wells, and A. J. T. Jull (1995), Fire and alluvial chronology in Yellowstone National Park: Climatic and intrinsic controls on Holocene geomorphic processes, *Geol. Soc. Am. Bull.*, *107*(10), 1211–1230.

Morrill, C., and R. M. Jacobsen (2005), How widespread were climate anomalies 8200 years ago?, *Geophys. Res. Lett.*, *32*, L19701, doi:10.1029/2005GL023536.

Moser, K. A., G. M. MacDonald, A. M. Bloom, A. Potito, and D. F. Porinchu (2004), A Holocene record of climate change from the Sierra Nevada, CA, USA: A paleolimnological perspective of California drought, *Eos Trans. AGU*, *85*(47), Fall Meet. Suppl., Abstract PP43A-0600.

Muscheler, R., J. Beer, and M. V. Vonmoos (2004), Causes and timing of the 8200 BP event inferred from the comparison of the GRIP ^{10}BE and the tree ring delta ^{14}C record, *Quat. Sci. Rev.*, *23*, 2101–2111.

National Snow and Ice Data Center (NSIDC) and World Data Center-A for Paleoclimatology (NDCP) (1997), *The Greenland Summit Ice Cores* [CD-ROM], Natl. Snow and Ice Data Cent., Univ. of Colo., Boulder.

Oppo, D. W., J. F. McManus, and J. L. Cullen (2003), Palaeo-oceanography: Deepwater variability in the Holocene epoch, *Nature*, *422*, 277–278.

Peterson, L. C., G. H. Haug, K. A. Hughen, and U. Röhl (2000), Rapid changes in the hydrologic cycle of the tropical Atlantic during the last glacial, *Science*, *290*(5498), 1947–1951.

Pierce, J. L., G. A. Meyer, and A. J. T. Jull (2004), Fire-induced erosion and millennial-scale climate change in northern ponderosa pine forests, *Nature*, *432*, 87–90.

Pisias, N. G., A. C. Mix, and L. Heusser (2001), Millennial scale climate variability of the northeast Pacific Ocean and northwest North America based on radiolaria and pollen, *Quat. Sci. Rev.*, *20*(14), 1561–1576.

Potito, A. P., D. F. Porinchu, G. M. MacDonald, and K. A. Moser (2006), A late Quaternary chironomid-inferred temperature record from the Sierra Nevada, California, with connections to northeast Pacific sea surface temperatures, *Quat. Res.*, *66*(2), 356–363.

Regnell, M., M.-J. Gaillard, T. S. Bartholin, and P. Karsten (1995), Reconstruction of environment and history of plant use during the late Mesolithic (Ertebølle culture) at the inland settlement of Bökeberg III, southern Sweden, *Veg. Hist. Archaeobot.*, *4*(2), 67–91.

Roark, E. B., B. L. Ingram, J. Southon, and J. P. Kennett (2003), Holocene foraminiferal radiocarbon record of paleocirculation in the Santa Barbara Basin, *Geology*, *31*(4), 379–382.

Schubert, S. D., M. J. Suarez, P. J. Pegion, R. D. Koster, and J. T. Bacmeister (2004), On the cause of the 1930s Dust Bowl, *Science*, *303*(5665), 1855–1859.

Scuderi, L. A. (1993), A 2000-year tree ring record of annual temperatures in the Sierra Nevada Mountains, *Science*, *259*(5100), 1433–1437.

Seager, R., N. Graham, C. Herweijer, A. L. Gordon, Y. Kushnir, and E. Cook (2007), Blueprints for Medieval hydroclimate, *Quat. Sci. Rev.*, *26*(19–21), 2322–2336.

Smith, A. J., J. J Donovan, E. Ito, D. R Engstrom, and V. A Panek (2002), Climate-driven hydrologic transients in lake sediment records: Multiproxy record of mid-Holocene drought, *Quat. Sci. Rev.*, *21*(4–6), 625–646.

Spooner, I. S., L. V. Hills, and G. D. Osborn (1997), Reconstruction of Holocene changes in alpine vegetation and climate, Susie Lake, British Columbia, Canada, *Arct. Alp. Res.*, *29*, 156–163.

Stott, L., C. Poulsen, S. Lund, and R. Thunell (2002), Super ENSO and global climate oscillations at millennial time scales, *Science*, *297*(5579), 222–226.

Swetnam, T. W., T. Kitzberger, P. M. Brown, E. K. Heyerdahl, and T. T. Veblen (2006), Multi-century fire synchrony across western North America: Contingent effects of ENSO, PDO, and AMO, paper presented at AMQUA 2006 XIX Biennial Meeting, Am. Quat. Assoc., Boseman, Mont.

Taylor, K. C., P. A. Mayewski, M. S. Twickler, and S. I. Whitlow (1996), Biomass burning recorded in the GISP2 ice core: A record from eastern Canada?, *Holocene*, *6*, 1–6.

Taylor, K. C., R. B. Alley, G. W. Lamorey, and P. Mayewski (1997), Electrical measurements on the Greenland Ice Sheet Project 2 core, *J. Geophys. Res.*, *102*(C12), 26,511–26,517.

Tinner, W., and A. Lotter (2001), Central European vegetation response to abrupt climate change at 8.2 ka, *Geology*, *29*(6), 551–554.

Wagner, J. D., J. E. Cole, J. W. Beck, P. J. Patchett, and G. M. Henderson (2005), Abrupt millennial climate change in the Arizona desert inferred from a speleothem isotopic record, *Eos Trans. AGU*, *86*(52), Fall Meet. Suppl., Abstract PP31A-1499.

Wathen, S. F. (2005), The use of high magnitude spikes in sedimentary charcoal and magnetic susceptibility to detect abrupt climate change and subsequent geomorphological and ecological readjustment, paper presented at Annual Meeting, Ecol. Soc. of Am., Montreal, Ont., Canada, 10 Aug.

Wathen, S. F. (2007), Abrupt climate change and ecosystem response: A 2,000 year history of severe droughts, extreme temperatures, and stand-replacing fires in the Sierra Nevada in response to abrupt climate change, paper presented at 23rd Pacific Climate (PACLIM) Workshop, U.S. For. Serv., Pacific Grove, Calif.

Wathen, S. F. (2011), 1,800 Years of abrupt climate change, severe fire, and accelerated erosion, Sierra Nevada, California, USA, *Clim. Change*, doi: 10.1007/s10584-011-0046-4.

Weiss, H. (2003), Work more? The 8.2 kaBP abrupt climate change event and the origins of irrigation agriculture and surplus agro-production in Mesopotamia, *Eos Trans. AGU*, *84*(46), Fall Meet. Suppl., Abstract PP22C-01.

Whitlock, C. (2001), Doing fieldwork in the mud, *Geogr. Rev.*, *91*(1–2), 19–25.

Whitlock, C., and R. S. Anderson (2003), Fire history reconstructions based on sediment records from lakes and wetlands, in *Fire and Climatic Change in Temperate Forests of the Americas*, edited by T. T. Veblen et al., pp. 3–31, Springer, New York.

Wright, H. E. J. (1991), Coring tips, *J. Paleolimnol.*, *6*(1), 37–50.

Zic, M., R. M. Negrini, and P. E. Wigand (2002), Evidence of synchronous climate change across the Northern Hemisphere between the North Atlantic and the northwestern Great Basin, United States, *Geology*, *30*(7), 635–638.

S. F. Wathen, Palm Springs - South Coast Field Office, Bureau of Land Management, 1201 Bird Center Dr., Palm Springs, CA 92264, USA. (sfwathen@gmail.com)

Abrupt Climate Change: A Paleoclimate Perspective From the World's Highest Mountains

Lonnie G. Thompson

Byrd Polar Research Center, School of Earth Sciences, Ohio State University, Columbus, Ohio, USA

Ice core records recovered from high-elevation, low-latitude ice fields, along with other proxy data, provide two primary lines of evidence for past and present abrupt climate change. First, there is strong evidence from paleoclimate records from within and around these glaciers for two widespread and spatially coherent abrupt events during the Holocene: a major isotopic excursion centered on ~5.2 ka B.P. marks the transition from early Holocene warmth to cooler conditions, and a major dust increase occurs between 4.0 and 4.5 ka B.P. Both of these events were concurrent with structural changes in early civilizations. Second, the continuing retreat of most midlatitude to low-latitude glaciers, many having persisted for millennia, signals a recent and abrupt change in Earth's climate system. High-resolution ice core stratigraphic records of $\delta^{18}O$ (temperature proxy) demonstrate that the current warming at high elevations in the middle to lower latitudes is unprecedented for at least the last two millennia, even though they suggest that the early Holocene was much warmer at many sites. The remarkable similarity between changes in the highland and coastal cultures of Peru and climate variability, especially with regard to precipitation, implies a strong connection between prehistoric human activities and climate in this region. Well-documented ice loss on Quelccaya in the Andes, Naimona'nyi in the Himalayas, and Kilimanjaro in eastern Africa paint a grim future for recovery of tropical glacier histories. The current melting of high-altitude, low-latitude ice fields is consistent with model predictions for a vertical amplification of temperature in the tropics.

1. INTRODUCTION

Fifty percent of Earth's surface area lies between 30°N and 30°S, which is defined as the geographic tropics. This region is of immense societal importance as it is home to 70% of the world's ~7 billion people. It also receives and transmits much of the thermal energy that drives Earth's atmospheric and oceanic circulations. Most of the meteorological and climatic events and perturbations affecting Earth's surface originate in or are amplified by ocean/atmosphere interactions in tropical latitudes. The tropics are Earth's "heat engine," where the warmest atmospheric and sea surface temperatures (SSTs) occur. The energy associated with these warm temperatures gives rise to intense convective precipitation and is crucial for the evolution of phenomena such as the El Niño–Southern Oscillation (ENSO), the monsoonal systems of Asia and Africa, and, on shorter time scales, hurricanes and other tropical disturbances that distribute tropical energy (heat) poleward. ENSO has strong impacts on meteorological phenomena that directly or indirectly affect most regions

Abrupt Climate Change: Mechanisms, Patterns, and Impacts
Geophysical Monograph Series 193

10.1029/2010GM001023

on the planet and their populations. ENSO is one of the modes of climate variability discussed by *Wanner et al.* [2008] along with the North Atlantic Oscillation, Pacific Decadal Oscillation, Atlantic Multidecadal Oscillation, and the Atlantic meridional overturning circulation.

Another very important component of the tropical circulation is the monsoonal system. The Asian monsoon circulation influences Asia, Africa, and Australia, regions in which more than 60% of the Earth's population resides [*Webster et al.*, 1998]. Abrupt variations in monsoon intensity can result in devastating droughts/floods during very weak/strong events. At any given time, water evaporated from the world's oceans stores about a sixth of the solar energy reaching the surface of the Earth. It is the release of part of this energy that is responsible for the power and duration of the monsoon rainy season.

The general warming trend during the twentieth century and into the twenty-first century is now well documented. Scientific evidence verifies that Earth's globally averaged surface temperature is increasing. The *Intergovernmental Panel on Climate Change (IPCC)* [2007] strongly suggests that human activities are contributing significantly to observed changes in the Earth system. Although not all regions have warmed, the globally averaged temperature has increased ~0.7°C since 1900 [*Hansen et al.*, 2006]. The best interpretation of proxy records from borehole temperatures, stable isotopes from ice cores, tree ring data, etc. suggests that the decade of the 1990s was the warmest in the last 1800 years [*Jones and Mann*, 2004; *Mann et al.*, 2009].

On decadal and longer timescales, climate models predict that greenhouse gas-induced warming will cause temperatures to rise faster at higher elevations and that vertical amplification will be greatest in the tropics due to upper tropospheric humidity and water vapor feedback [*IPCC*, 2007, chapter 8, p. 602]. This is not surprising given that 37% of the warm tropical ocean surface water is located between 20°N and 20°S latitude. Results from general circulation models indicate that water vapor provides the largest positive radiative feedback and that it alone roughly doubles the warming in response to forcing such as from greenhouse gas increase. As a result of this forcing, the projected changes in mean annual free-air temperatures between 1990 and 1999 and 2090 to 2099, using CO_2 levels from IPCC scenario A2, shows twice as much warming in the higher elevations (middle to upper troposphere) in the tropics than is predicted at Earth's surface [*Bradley et al.*, 2006]. Moreover, it is expected that some high-elevation tropical glaciers may already be responding to these enhanced temperatures, which may well account for the accelerating rate of glacier loss at some of these sites [*Thompson et al.*, 2006a; *Coudrain et al.*, 2005].

Glacier ice is one of the most versatile recorders of Earth's climate, archiving information on past temperatures and precipitation. Many tropical ice fields have provided continuous, annually resolved proxy records of climatic and environmental variability preserved in measurable parameters, especially oxygen and hydrogen isotopic ratios ($\delta^{18}O$ and δD) and the net mass balance (accumulation). Ice core records from Africa, Alaska, Antarctica, Bolivia, China, Greenland, Peru, Russia, and Indonesia have made it possible to study atmospheric and oceanic processes linking the polar regions to the lower latitudes where human activities are most concentrated. For over 35 years, the Ice Core Paleoclimate Research Group (ICPRG) at the Ohio State University's Byrd Polar Research Center has successfully extracted information on past climatic and environmental changes from ice cores drilled around the world, often under very harsh and logistically challenging conditions. Much of this work has been conducted on tropical mountain glaciers in the Andes, in the Himalayas, on Kilimanjaro, and most recently, in Papua, Indonesia. The details of the glaciological and geophysical studies conducted at each site are available in over 200 publications published by the ICPRG group (http://bprc.osu.edu/Icecore/Abstracts/Publications.html). These low-latitude climate records constitute a critical component of the principal objective of our program, the acquisition of a global array of ice cores that provide high-resolution climatic and environmental histories that will contribute to our understanding of the complex interactions within Earth's climate system.

Because tropical glaciers often produce continuous, high-resolution records of climate, ice cores from these glaciers have been valuable in the study of past variations, particularly those of an abrupt (annual to decadal) nature. Additionally, ice cores drilled through ice sheets and ice caps also provide information on phenomena and events that drive climate change, such as volcanic activity, solar variations, vegetation changes, and greenhouse gas concentrations. Tropical glaciers are currently responding to climate changes that can be defined as abrupt, e.g., the condition when "the climate system is forced to cross some threshold, triggering a transition to a new state at a rate determined by the climate system itself and faster than the cause" and "takes place so rapidly and unexpectedly that human or natural systems have trouble adapting to it" [*National Research Council*, 2002, p. 14]. Evidence for abrupt climate changes, particularly those that have occurred during the Holocene, is often captured by at least some of the physical (e.g., insoluble dust or reconstructed accumulation) or chemical (e.g., stable isotopes of oxygen) time series of these continuous records. Here present and past climate changes are discussed along with their impacts on tropical mountain glaciers, and two major abrupt

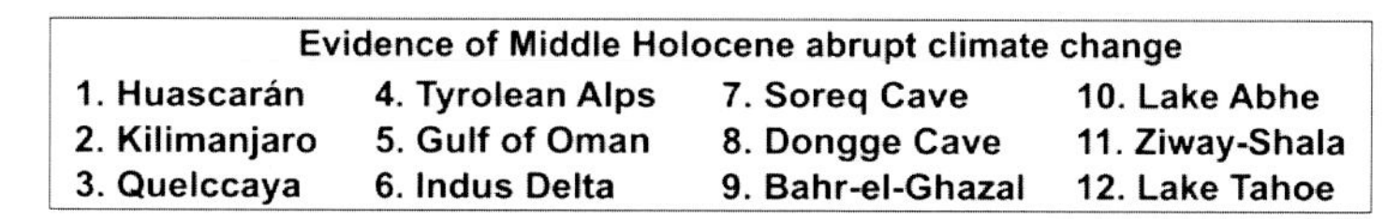

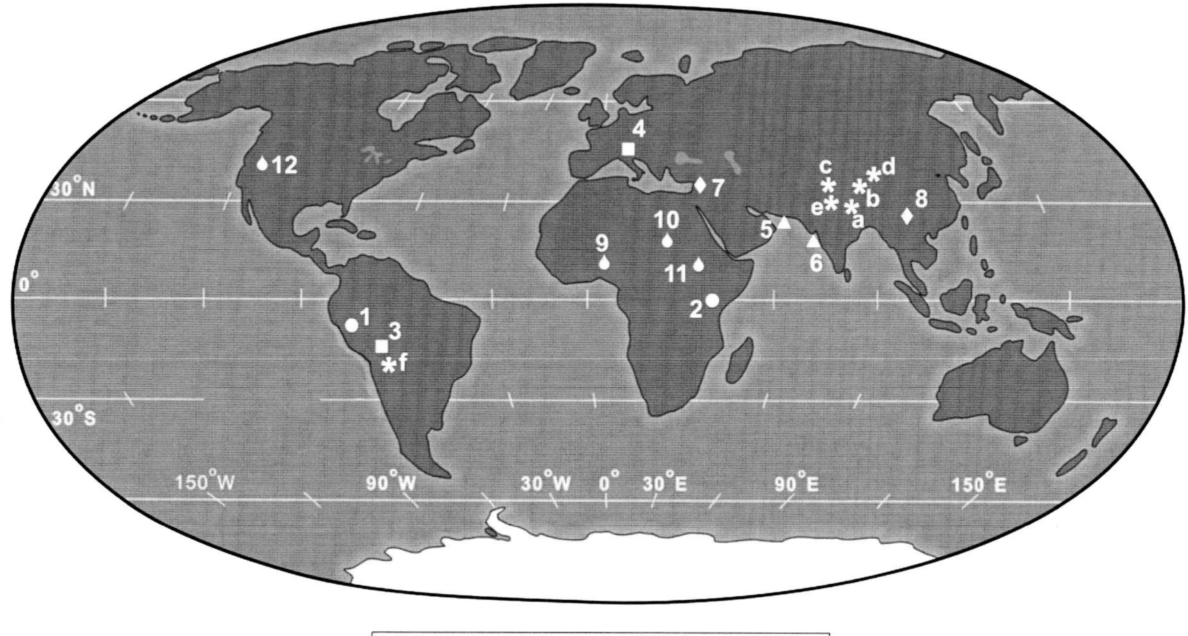

Figure 1. Locations of sites discussed in the text from which evidence of abrupt climate change has been collected listed as 1 through 12. Ice core sites are shown by circles, sites from which biological evidence was obtained are shown as squares, marine cores are shown by triangles, speleothems are shown by diamonds, and lakes are shown by water drops. Other ice core sites mentioned in the text are shown as "a" through "f."

climatic events of the middle Holocene are discussed. These are well documented in several ice core and other proxy records, the locations of which are shown in Figure 1.

2. ABRUPT CLIMATE EVENTS IN THE HOLOCENE

2.1. *The Abrupt Cold Event of 5200 Years B.P.*

Solar insolation in the tropical Northern Hemisphere (NH) increased during the early Holocene as the Earth recovered from the last glacial stage, peaking at ~9 ka B.P., then decreasing toward the present (Figure 2a). The shape of this insolation curve is mimicked by the Holocene records of $\delta^{18}O$ from two tropical ice cores: Huascarán (9°S) in the Cordillera Blanca of northern Peru (Figure 2b) and Kilimanjaro (3°S) in Tanzania, East Africa (Figure 2c). Although Huascarán is located in the Southern Hemisphere (SH), its stable isotope record more closely follows the NH insolation curve because the predominant source of its precipitation is the tropical North Atlantic [*Thompson et al.*, 1995]. The Holocene has been divided into two climatic regimes: the early "Hypsithermal" (also called the "Climatic Optimum"), when insolation and average $\delta^{18}O$ in the ice core records were highest (indicative of the warmer early Holocene), and the later "Neoglacial," by which time insolation had declined to ~60% of the early Holocene high and thereafter decreased to the present levels. A similar increase in the variability of concentrations of several chemical species (Cl^- and SO_4^{2-} in particular) and insoluble dust was noted in the Sajama ice cores starting between 5000 and 6000 years B.P. [*Thompson et al.*, 1998; *Bradley et al.*, 2003]. The division between these two regimes has been placed in Figure 2 at ~5000 ka B.P., which is roughly when the frequency of ENSO underwent a transition to its "modern" mode [*Abarzua and Moreno*, 2008; *Moy et al.*, 2002; *Rodbell et al.*, 1999; *Rowe et al.*, 2002; *Sandweiss*, 2003].

It was also around this time (~5.2 ka B.P.) that one of the most dramatic middle Holocene abrupt climate events occurred. Evidence of this event is obvious in the Kilimanjaro record (Figure 2c) as a sharp but short-lived depletion in ^{18}O, which suggests an abrupt and extreme cooling episode [*Thompson et al.*, 2002]. Evidence for this abrupt change is also seen in numerous climatic, biological, and archeological records from around the world, including several from the tropics [*Staubwasser and Weiss*, 2006]. Many tropical paleoclimate records, especially those located in the African/Asian

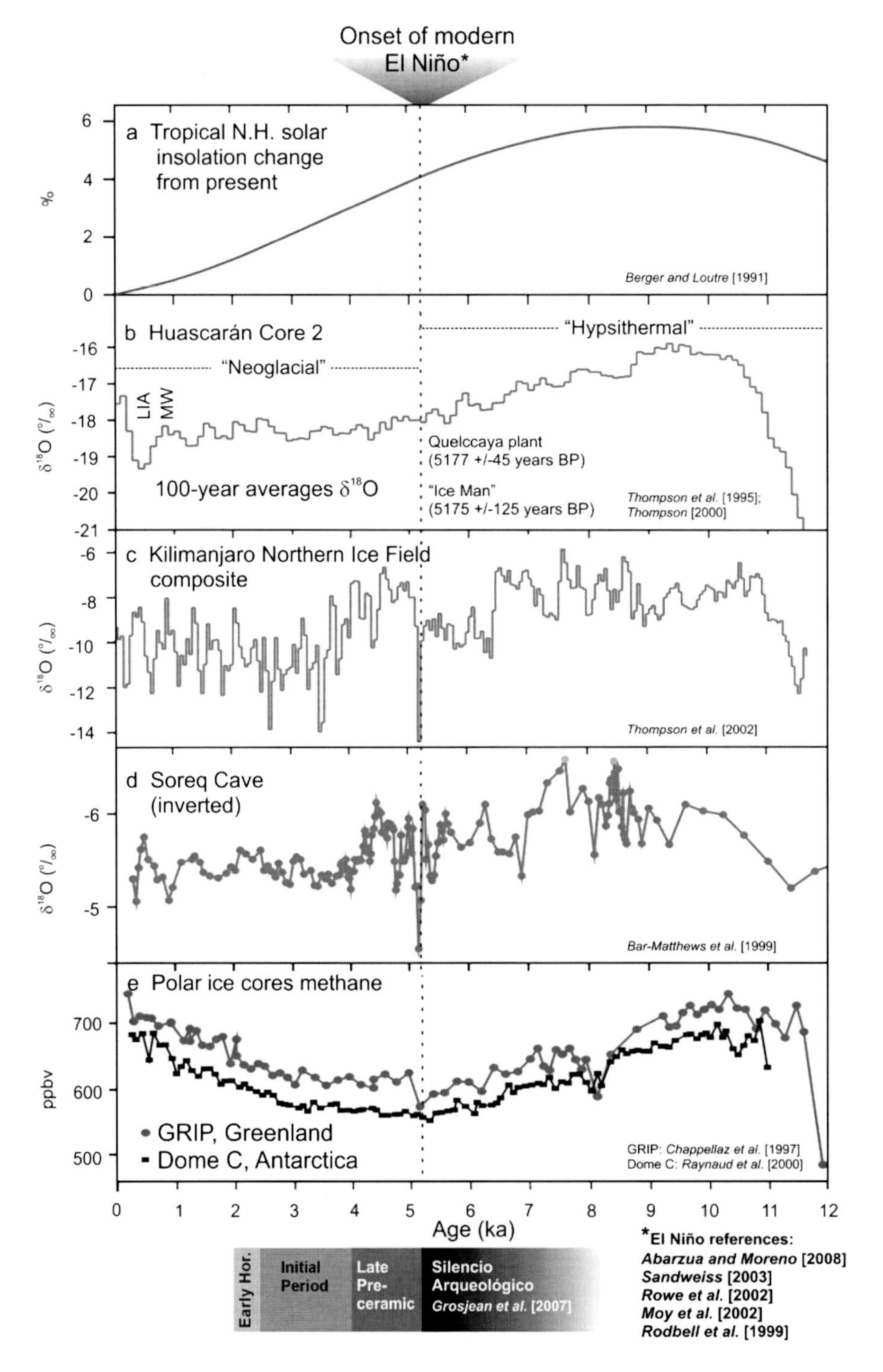

Figure 2

monsoon region, indicate that climate conditions turned either sharply arid or more humid circa 5 ka B.P. [*Magny and Haas*, 2004], and records from East Africa suggest increased aridity around this time [*Chalié and Gasse*, 2002]. The Soreq Cave in Israel contains speleothems, which have provided continuous climate records covering several tens of thousands of years and are shown here (Figure 2d) because of their proximity to the Kilimanjaro ice core records. As in the Kilimanjaro ice core record, the Soreq Cave record (Figure 2d) suggests that an abrupt event also occurred in the Middle East and that it was the most prominent climatic event in the last 13,000 years [*Bar-Matthews et al.*, 1999; *Bar-Matthews and Ayalon*, 2004].

Research at many sites located in extratropical latitudes has yielded information that unusually cold, but humid, conditions dominated in these areas. The most famous of these discoveries was "Otzi" or the "Tyrolean ice man," whose remarkably well-preserved body was discovered in the Alps in 1991 when it melted out of a retreating glacier (location in Figure 1). When his remains were radiocarbon dated, it was discovered that he died (the forensic evidence suggests he was murdered) and was enclosed in a sudden surge of snow and ice around ~5200 years ago [*Baroni and Orombelli*, 1996]. In other locations around the world, vascular plants are being exposed for the first time in 5200 years as glaciers are melting and retreating. The margin of the Quelccaya Ice Cap in southern Peru has retreated substantially since it was first photographed in 1977, and in 2002, a perfectly preserved wetland plant deposit, which contained no woody tissue, was discovered. It was identified as *Distichia muscoides*, which today grows in the valleys below the ice cap and was radiocarbon dated at ~5200 years B.P. [*Thompson et al.*, 2006a]. This is strong evidence that this ice cap has not been smaller than its present size for over five millennia. As the glacier continues to retreat, more plants have been collected and radiocarbon dated, and almost all of these confirm these original findings [*Buffen et al.*, 2009]. Other paleoclimate evidence of this event has been discovered in North America. For example, trees preserved underwater along the southern margin of Lake Tahoe in the Sierra Nevada Range suggest an immersion date of about 5000 years B.P. The preserved condition of the trees indicates that the water in the lake has remained high thereafter [*Lindstrom*, 1990].

Whether effective moisture increased or decreased during this period, there is a consensus among the records that an abrupt and marked cooling episode occurred. This anomalously cold climate lasted a relatively short time, perhaps a few decades to a few centuries, before the climate just as abruptly reverted to its previous state. A possible link to this temperature excursion is seen in the methane records from the Greenland and Antarctic ice cores (Figure 2e), which shows the lowest concentration of this "greenhouse" gas in the Holocene ~5200 years ago [*Chappellaz et al.*, 1997; *Raynaud et al.*, 2000], at the same time as the event in the Kilimanjaro and Soreq Cave records and at a time when NH insolation was rapidly decreasing. More recent measurements of CO_2 concentrations and ^{13}C isotopic compositions in the EPICA Dome C ice core are reported in higher resolution and with an updated chronology [*Elsig et al.*, 2009]. In this record, CO_2 shows a steady rise to preindustrial values (~280 ppm) after 5600 years B.P. before which values were ~260 ppm or slightly lower. Although the declining insolation did not directly drive this abrupt cooling, feedback within the climate system (oceanic, atmospheric, vegetative, and cryospheric) likely amplified the gradual solar variations and the small changes in greenhouse gas abundances [*deMenocal et al.*, 2000]. However, whether a decrease in methane contributed to the forcing of the mid-Holocene abrupt event, or is a consequence of the cooling, is still not known.

The cultural ramifications of the 5200 years B.P. climatic shift are also debatable. *Weiss and Bradley* [2001] argued that from the archaeological perspective, uncertainty about the causal linkage between abrupt climate change and societal collapse is due to chronological imprecision and the uncertain ability of societies to adapt to the abruptness, magnitude, and duration of environmental change. During the Late Uruk period (~5200 years B.P.), societies in southern and northern Mesopotamia collapsed [*Postgate*, 1986; *Weiss*, 2003]. Concomitantly, severe drought was hypothesized to have been responsible for settlement urbanization in southern Mesopotamia around 5200 years B.P. [*Nissen*, 1988]. Coincident or not, this was around the time that some Mesopotamian cultures flourished [*Rothman*, 2004]. Meanwhile, two civilizations on opposite sides of the world created calendars at this time; the Mesoamericans began their long-count calendar in

Figure 2. (opposite) Various Holocene records showing (a) changing tropical Northern Hemisphere summer insolation; (b) the $\delta^{18}O$ record (as 100 year averages) from the Huascarán ice core, northern Peru; (c) the Kilimanjaro $\delta^{18}O$ ice core record (as 50 year averages), in which the time series from 4.2 ka B.P. to the present is a composite of Northern Ice Field isotopic records; (d) the $\delta^{18}O$ record from the Soreq Cave (Israel) speleothem; and (e) the polar ice core methane records. The demarcation at ~5.2 ka B.P. divides the early Holocene "Hypsithermal" as characterized by higher insolation, more enriched ice core $\delta^{18}O$ values, and more depleted $\delta^{18}O$ speleothem values from the late Holocene "Neoglacial" during which insolation and ice core isotopic values were lower and speleothem isotopic ratios were higher. This line also marks an archeological horizon in Peruvian prehistory and the onset of the "modern" El Niño–Southern Oscillation.

3112 before the Common Era (B.C.E.), and two Hindu calendars began in 3102 and 3116 B.C.E. The Mesoamerican (often called Mayan) long-count calendar in particular has received some attention in the popular media lately as the alleged "end date" of the current cycle in the calendar (2012 Common Era (C.E.)) draws near. So far, however, there appears to be little evidence that the development of this calendar had much to do with an anomalous climate event, although there are indications of drought across Mesoamerica ~5800 years B.P. [*Voorhies and Metcalfe*, 2007]. Before the onset of the middle Holocene climatic transition, there is evidence of declining human occupation in the Atacama region, Chile ("Silencio Arqueológico") (Figure 2), which recovered as more humid conditions dominated [*Grosjean et al.*, 1997; *Núñez et al.*, 2002]. Uncertainties about societal collapse and abrupt climate change are directly linked to uncertainties in the dates of past events in both the paleoclimatic and archaeological records. The ability of societies to adapt to abrupt climate change may, more often than not, be linked to a nexus of social, political, and climate events that force the society across a threshold.

2.2. *The Abrupt Arid Event of 4200 Years B.P.*

A substantial collection of proxy climate records from around the tropics (Figure 3) provides evidence for another widespread abrupt climatic reversal that occurred approximately one millennium after the middle Holocene cold event. This particular episode, which appears in the records around 4.0 to 4.5 ka B.P., apparently extended from South America to northern Africa to eastern China and was more a result of extreme decrease in effective moisture than of temperature changes. Evidence appears in a marine core from the Gulf of Oman [*Cullen et al.*, 2000] (Figure 3a), in which an abrupt spike in carbonates has been chemically traced to an archeological site (Tell Leilan) in Syria, the home of the ancient Akkadian culture [*Weiss et al.*, 1993]. Two ice core records from opposite sides of the world, Nevado Huascarán in the Cordillera Blanca of northern Peru and Kilimanjaro in Tanzania (Figure 1), exhibit significant spikes in insoluble dust that occur during this period. The high-resolution Huascarán dust record, which extends to the end of the last glacial stage, contains a large dust peak at ~4.2 to 4.5 ka B.P. (Figure 3b), which is the most prominent feature in the Holocene [*Thompson*, 2000]. The dust that is blown off the African west coast during austral summer (the wet season in the Cordillera Blanca) is entrained by the northeast trade winds and carried, along with moisture, across the tropical North Atlantic and the Amazon Basin. Saharan dust transported during the austral summer has been found in the Amazon Basin [*Swap et al.*, 1992]. The dust that was deposited on Huascarán during the

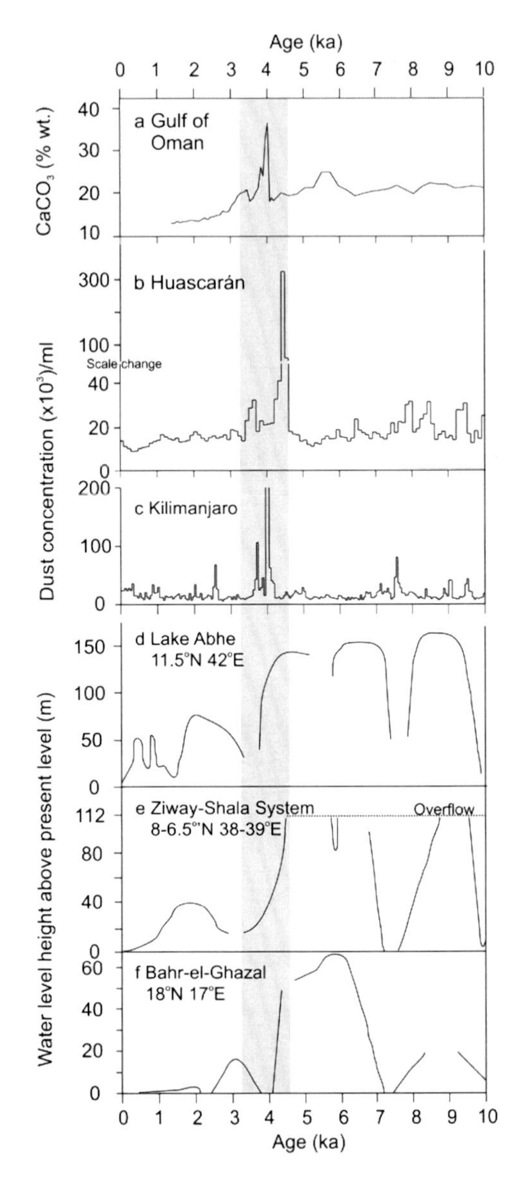

Figure 3. Comparisons of (a) the Gulf of Oman middle Holocene carbonate record on a calibrated ^{14}C time scale (modified from *Cullen et al.* [2000]); (b) the Huascarán Holocene dust record [from *Davis and Thompson*, 2006], reprinted from the *Annals of Glaciology* with permission of the International Glaciological Society; and (c) the dust from the Kilimanjaro ice core, modified from *Thompson et al.* [2002]. Tropical African Holocene lake level records from (d) Lake Abhe [from *Gasse*, 1977], reprinted with permission; (e) Ziway-Shala System [from *Gillespie et al.*, 1983], reprinted with permission; and (f) Bahr-el-Ghazal [from *Servant and Servant-Vildary*, 1980], reprinted with permission. The gray bar marks the middle Holocene arid period in the records.

middle Holocene was most likely transported in this manner during centuries of extreme aridity both in Africa and South America [*Davis and Thompson*, 2006]. The large dust spike in the Kilimanjaro record, which is dated ~4.0 ka B.P. (Figure 3c), was also the product of this "megadrought" that may have lasted a few centuries and during which time the Northern Ice Field, presently the largest on the mountain, dramatically decreased in size [*Thompson et al.*, 2002]. In marine sediments off the coast of Peru, *Rein et al.* [2004, 2005] found that the lithic flux was much lower between ~8.0 and 5.0 ka B. P. than in both the earlier and later Holocene. These records are interpreted to indicate reduced ENSO activity during what was a dry period along coastal Peru.

Lake records from equatorial and North Africa [*Gasse*, 1977; *Gillespie et al.*, 1983; *Servant and Servant-Vildary*, 1980] indicate that water levels had decreased greatly (Figures 3d–3f), as did levels in western Tibet [*Gasse et al.*, 1991, 1996], which are not shown. Other records also contain evidence of a sudden and marked arid climate episode at this time. A speleothem record from the Soreq Cave (location in Figure 1) in Israel [*Bar-Matthews et al.*, 1999; *Bar-Matthews and Ayalon*, 2004] shows an abrupt decrease in $\delta^{13}C$, implying lower precipitation within the 4.2 to 4.5 ka B. P. window. Palynological data from north central China also suggest a cold, dry period from 3.95 to 4.45 ka B.P., which is inferred by a sudden decrease in tree pollen concentration [*Xiao et al.*, 2004]. An oxygen isotope record from a speleothem from Southern China shows evidence for multiple dry periods that coincide with Bond (1.5 kyr) events in the North Atlantic [*Wang et al.*, 2005]. Some of them, including the one at 4.2 ka B.P., occurred at a time of variations in solar activity, which was also observed in the record from a marine core off the Indus Delta [*Staubwasser et al.*, 2003]. The ^{10}Be concentrations measured in the Greenland Ice Core Project ice core [*Vonmoos et al.*, 2006] and in the radiocarbon production rates [*Müller et al.*, 2006] suggest lower solar activity (higher cosmogenic nuclide production) during both the ~5.2 ka B.P. event as well as the ~4–4.5 ka B.P. events. The lack of precise time scales makes direct comparisons difficult. Unfortunately, as yet, there is no final answer on what role, if any, the relatively small changes in solar forcing might have played in these events. This stems from the lack of a clear physically based link between solar activity and solar forcing [*Wanner et al.*, 2008].

It is not only in northern Africa, the Middle East, and Asia that this middle Holocene drought appeared. Stable isotope analyses of planktonic foraminifera from the Amazon fan show that the highest $\delta^{18}O$ values in the Holocene, suggestive of reduced Amazon River flow, occurred at ~4.5 ka B.P. [*Maslin et al.*, 2000], almost contemporaneously with a depletion of ^{18}O in the isotopic record from Lake Junin, Peru [*Seltzer et al.*, 2000]. Evidence for middle Holocene aridity on the Altiplano comes from Lake Titicaca on the border between Peru and Bolivia, where the percent of freshwater plankton reached its lowest levels between ~5 and 2 ka B.P., and the percent of saline diatoms increased at ~6 ka B.P. (^{14}C age), reaching a maximum at ~3.6 to 4.0 ka B.P. (^{14}C age) [*Baker et al.*, 2001]. *Tapia et al.* [2003] offer a more refined time line for climate variation in this interval using the Lake Titicaca record of percent saline planktonic taxa, which confirms that although water levels were low between 3 and 6 ka B.P., the lowest levels were achieved at ~4.5 ka B.P.

In addition to paleoclimate records, archeological and historical records indicate that during the third millennium B.C.E., a sudden and severe drought occurred in many regions that had previously been under the expanded influence of the Asian, Indian, and African monsoon systems in the early Holocene. As with the 5.2 ka B.P. cold reversal, this arid event occurred at a time of widespread cultural disruptions. It was contemporaneous with the decline of the Akkadian Empire in Mesopotamia [*Weiss et al.*, 1993], the failure of Nile River floods that were contemporaneous with severe crises in North African civilizations [*Hassan*, 1997], and the decline of the urban Harappan civilization in the Indus valley [*Staubwasser et al.*, 2003]. In central China, this time also marked the decline of Neolithic culture [*Wu and Liu*, 2004]. Whether the drought was instrumental in the collapse of these societies is controversial among some archeologists [*Butzer*, 1997], but the evidence is compelling that this climatic anomaly occurred at or near the same time as these historical events.

There are several potential mechanisms that could lead to the mid-Holocene abrupt climate change events. For example, several episodes of abrupt aridity in the Indian and Asian monsoon regions have been linked to variations in solar output as was discussed above. Mechanisms discussed by *Wanner et al.* [2008] that were influential in mid- to late Holocene climate change include orbital forcing and shorter-term solar variations and volcanic eruptions, as well as changes in greenhouse gas forcing, land cover, and modes of climate variability. Complex feedback mechanisms among vegetation, the atmosphere, oceans, snow, and ice may also have been critical. While no evidence for volcanic forcing of the mid-Holocene abrupt events discussed above have been found in the ice core records, many of these mechanisms occur gradually over long periods of time and thus would require a nexus of events to push the climate system past some threshold thus leading to rapid, large-scale changes in the climate system.

3. CLIMATE CHANGE IN THE PERUVIAN ANDES AND THE RISE AND FALL OF CIVILIZATIONS

We have the advantage of an increasing abundance of written records from the late Holocene to help reconstruct the timing and extent of abrupt climate changes and their consequences for civilizations of the time. However, the majority of the Mesoamerica and South American writing systems identified before the arrival of the Spanish in 1531 C.E. have not yet been deciphered. In Peru, a succession of preconquest societies such as the Huari (or Wari), the Moche, and the Incas lived in the coastal areas and the highlands, but the only evidence of their rise and demise (other than the Spanish accounts in the case of the Incas) comes from oral tradition and archeological findings. Highland and coastal cultures flourished and waned out of phase with each other from the Early Intermediate period to the Late Horizon [*Paulsen*, 1976], which ended with the destruction of the Inca Empire by the Spanish conquistadors (see Figure 4).

Although written accounts are lacking, the climate record derived from the Quelccaya ice core, Cordillera Vilcanota, southern Peru (location in Figure 1), allows reconstruction of the highland climate and environment in this region of the Andes, especially during critical periods such as the alternating rise and fall of the preconquest agrarian civilizations from the Early Intermediate period to the abrupt disruption of indigenous societies in the mid-sixteenth century C.E. (Figure 4) [*Thompson et al.*, 1994]. The highland cultures (Huari and Inca) existed during two periods, the Middle Horizon from 600 to 1000 C.E. and the Late Horizon from 1450 to 1530 C.E., when the Quelccaya ice core records show above-average ice accumulation (Figure 4) in the southern Peruvian mountains. Alternatively, the lowland cultures appear to have dominated during periods of below-average precipitation in the mountains, when the precipitation was higher in the coastal regions [*Paulsen*, 1976]. Such variations in rainfall would have profound consequences for societies that were highly dependent on agriculture and existed in climatically sensitive environments [*Cardich*, 1985].

Under present climate conditions, coastal Peru experiences higher than normal rainfall during ENSO events, while the highlands, where Quelccaya is located, experience drier than average conditions [*Tapley and Waylen*, 1990]. Based on the geographical and temporal patterns of societal rise and fall in preconquest Peru, it appears as though this spatial pattern of climate variability might have occurred on decadal to centennial time scales in a manner similar to the precipitation distributions governed by the modern ENSO. Archeological evidence shows that during the Late Intermediate period, the southern coastal climate in the region of Peru was relatively humid, then became more arid after ~600 A.D. [*Eitel et al.*, 2005; *Mächtle et al.*, 2006], while the Quelccaya ice core record shows variably above-average accumulation (i.e., precipitation) during this period when the Huari Highland culture dominated. During the twentieth and beginning of the twenty-first century, population migration has been from the highlands to the coastal deserts since at least the 1940s as illustrated in the map of modern population distribution (Figure 4). The longer-term history of the rise and fall of

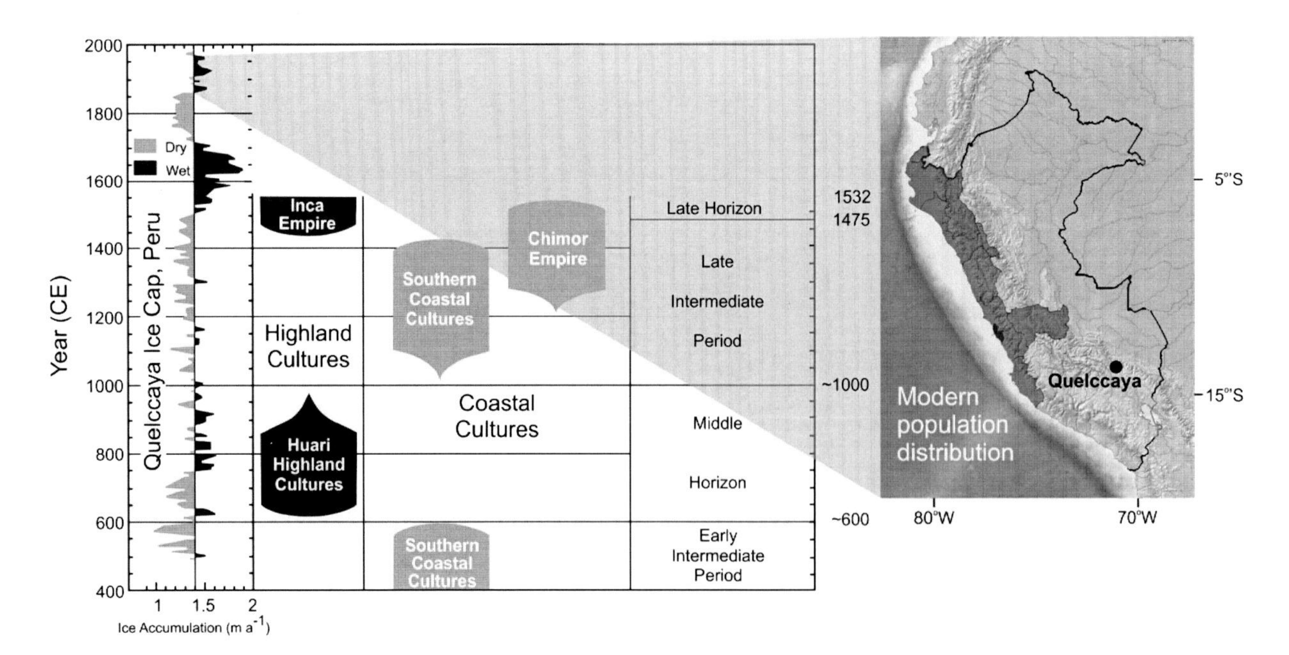

Figure 4. The reconstructed ice accumulation record from the Quelccaya Ice Cap, compared with the Peruvian coastal and highland archeological history from the fifth century C.E. to the Spanish conquest in the 1532 C.E. Map shows modern population distribution which is largely concentrated in the coastal desert areas of Peru.

coastal and highland cultures in Figure 4 argues that the current trends are counter to the patterns of population concentrations over the last 1500 years.

4. CLIMATE CHANGE AND ITS CURRENT EFFECTS ON THE CRYOSPHERE

Figures 5a and 5b show that during the boreal winter (December-January-February) the middle troposphere and SSTs are highest in the tropical latitudes, although tropical atmospheric temperatures tend to remain fairly uniform throughout the year [*Sobel*, 2002]. This thermal energy is instrumental in the production of tropical precipitation (Figure 5c), much of which occurs in areas of highest SSTs and 500 mbar temperatures, especially over the western tropical Pacific. Tropical thermal energy and humidity are the fuel for the latent heat engine that controls so much of the planet's weather systems. A satellite view of the Earth shows the clouds in the intertropical convergence zone just north of the equator, where warm air rises and releases moisture (Figure 6). As the warm, moist air rises over the equator and cools, the maximum latent heat release occurs at the 500 mbar level [*Webster*, 2004]. To the north and south of this band, around the geographic boundaries of the tropics, dry air descends as it cools, giving rise to deserts and arid regions. This major circulation pattern, the Hadley Cell, undergoes

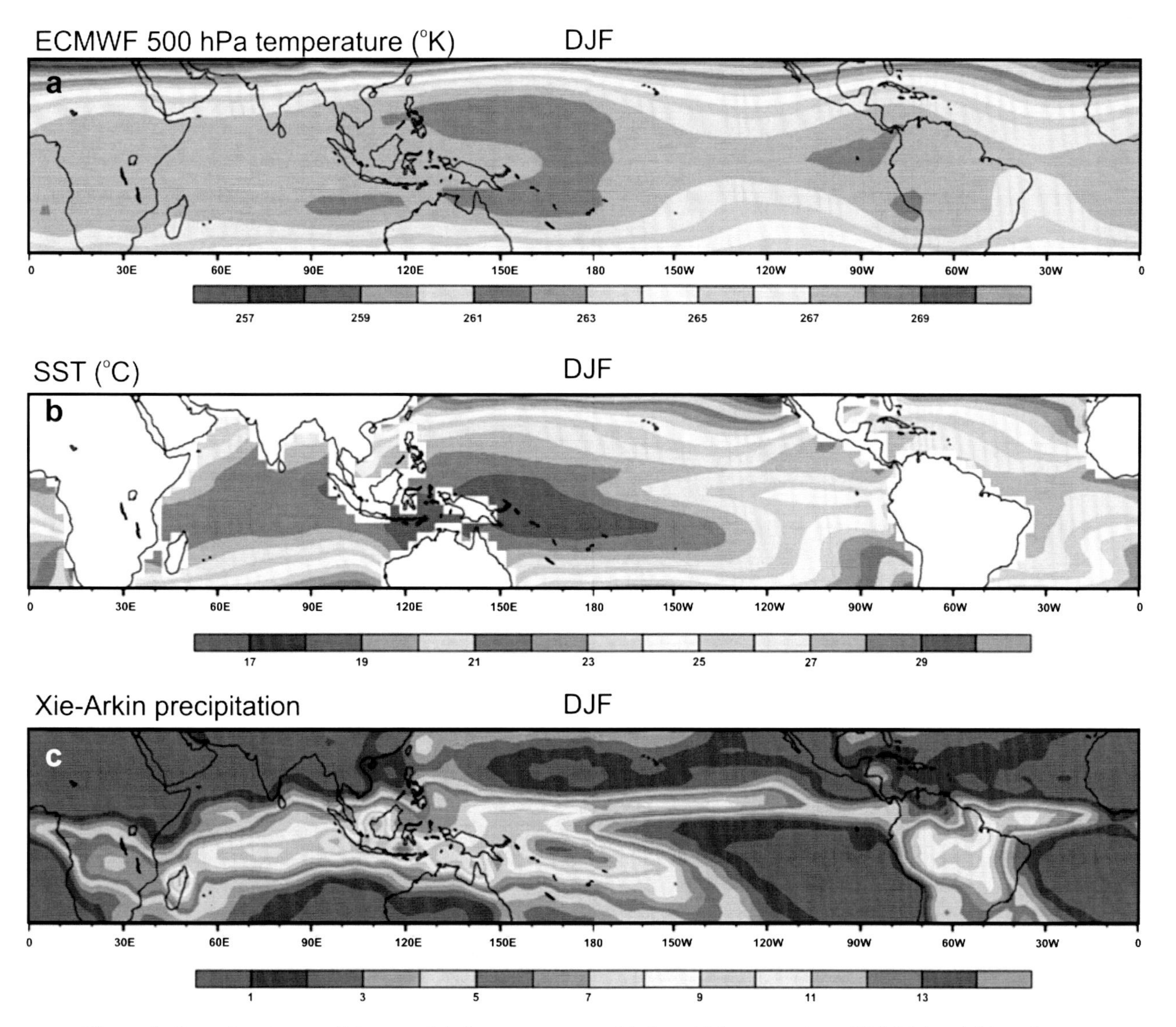

Figure 5. December-January-February global temperatures at (a) the middle troposphere (500 hPa) and (b) at the sea surface. (c) Xie-Arkin global precipitation. Image courtesy of C. Bretherton and A. Sobel.

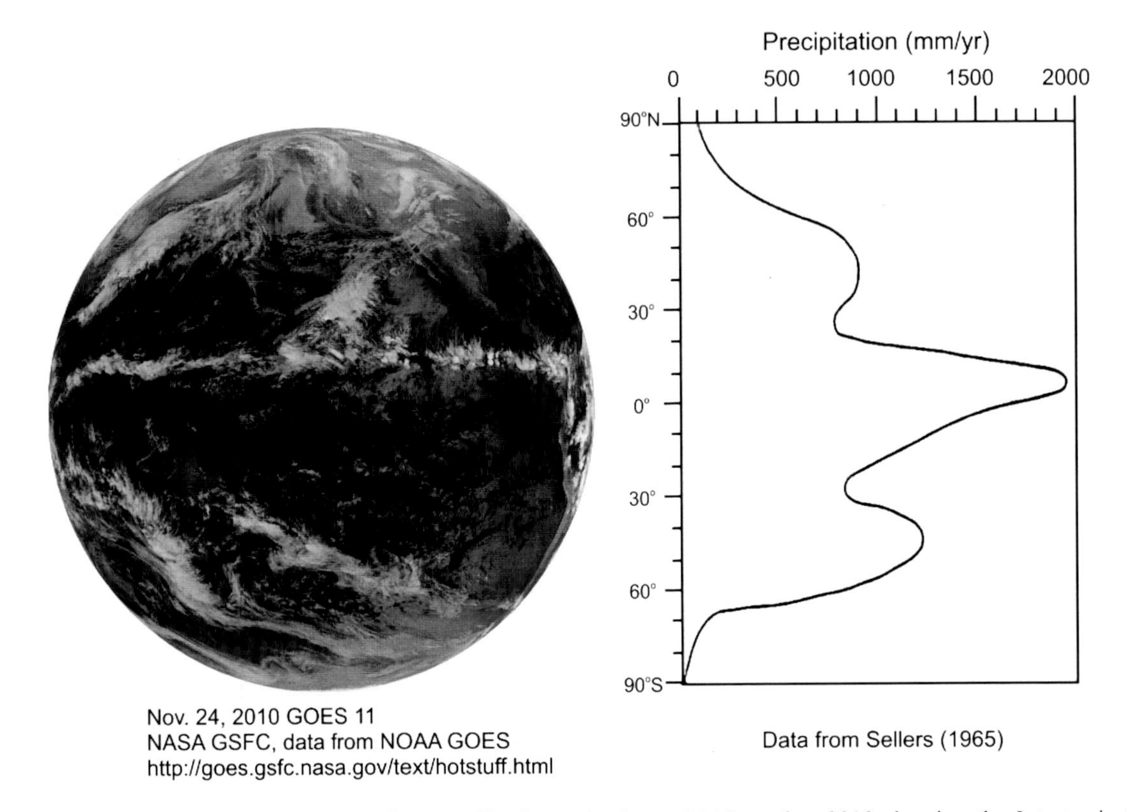

Figure 6. GOES 11 satellite image of the Pacific Ocean basin on 24 November 2010 showing the Intertropical Convergence Zone cloud band in the equatorial Pacific. The deep tropics are also the region of highest precipitation (source of data and time represented) because of the rising of warm, moist air and latent heat release in the region of the 500 mbar level. The zonally averaged precipitation is shown on the right [after *Sellers*, 1965].

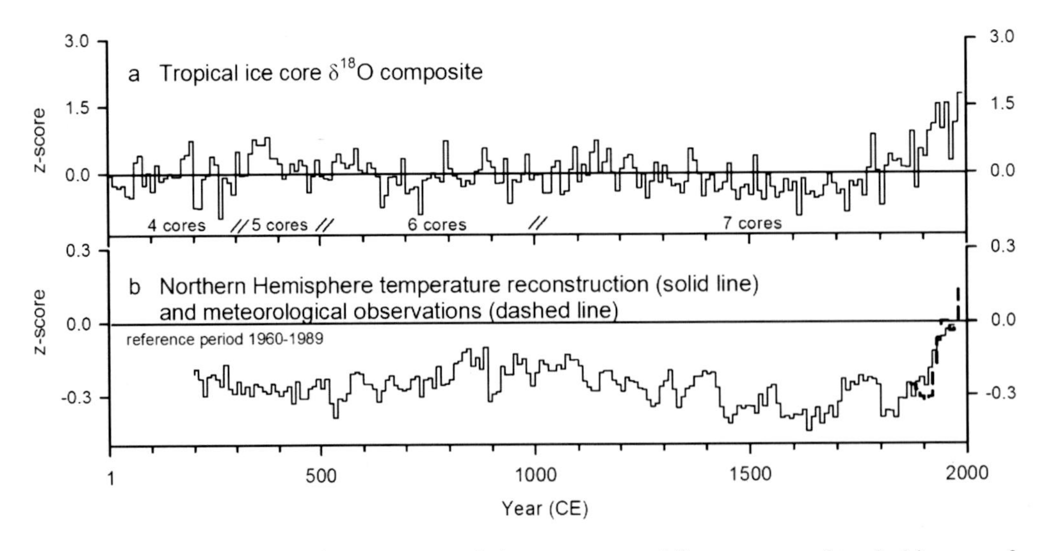

Figure 7. (a) The 2000 year record of temperature variations reconstructed from an array of tropical ice cores from the Andes and the Tibetan Plateau [*Thompson et al.*, 2006a]. (b) Reconstructed temperature records [*Mann and Jones*, 2003] from a variety of proxy data over a comparable time period, overlain by meteorological data [*Jones and Moberg*, 2003] since the midnineteenth century.

changes in its location and intensity through time and is important to tropical and middle-latitude climate variability. Rising temperatures are driven by water vapor feedback in the climate system, which has potential ramifications for Earth's future climate, since it is linked to the circulation of the Hadley cell. There is evidence that over the last 20 years, this cell has expanded north and south by about 2° of latitude, which may broaden the desert zones [*Seidel and Randel*, 2007; *Seidel et al.*, 2008] as the dry descending arms move poleward. Under this scenario, droughts might become more frequent and persistent, not only in the American Southwest, but in the Mediterranean region and SH (Australia and parts of South America and Africa).

The world's ice caps and glaciers are in retreat as tropospheric temperatures warm. Today, ice covers about 10% of the Earth's surface area, compared to 30% coverage during the coldest part of the last ice age. Because this warming is amplified at high altitudes [*Bradley et al.*, 2006], the loss of ice is most drastic in mountainous areas such as the Andes and the Tibetan Plateau. Alpine glaciers at low latitudes are located in the midtroposphere (~5000 to 7000 m above sea level (asl)) where persistent warming is evident in instrumental records over the last few decades [*Liu and Chen*, 2000]. Mountain glaciers are smaller and thinner than the massive polar ice sheets and thus respond more quickly to abrupt temperature and precipitation changes as their surface area to volume ratio is much greater. Although global ice retreat at the beginning of the twenty-first century appears to be driven mainly by increasing temperatures, regional factors such as deforestation and precipitation deficits can also impact individual glaciers. Currently, 60% of the land-based ice loss is occurring on the small glaciers and ice caps. Although alpine glaciers constitute a small percentage of the world's ice cover, their rapid rate of melting in comparison to the large polar ice sheets means they will contribute more to sea-level rise in the short term; the ice loss from mountain glaciers may raise sea level ~0.25 m by 2100 [*Meier et al.*, 2007].

A composite of $\delta^{18}O$ records from high-altitude, low-latitude, and midlatitude ice cores demonstrates that the most isotopically enriched values of the last two millennia occurred during the twentieth century (Figure 7a). This increase in $\delta^{18}O$ is discussed by *Thompson et al.* [2006a] as being representative of temperature, since it is in agreement with NH temperature reconstructions [*Mann and Jones*, 2003] and modern meteorological observations [*Jones and Moberg*, 2003] (Figure 7b). That a similar profile can be determined independently by different climate recorders gives us confidence that the warming trend at the end of the twentieth century and the beginning of the twenty-first century is unprecedented in at least the last two millennia.

There are several well-documented examples of alpine glacier retreat throughout the world from the tropics to the subpolar latitudes, some of which are reviewed briefly here beginning with North American glaciers. A study from the mid-1950s to the mid-1990s of 67 Alaskan glaciers shows that all are thinning, and subsequent measurements on 28 of these in the following years shows that the thinning rate has increased [*Arendt et al.*, 2002]. In the Brooks Range of northern Alaska, 100% of the glaciers are in retreat, while in southeastern Alaska, 98% are shrinking [*Molnia*, 2007]. When Glacier National Park in Montana was established in

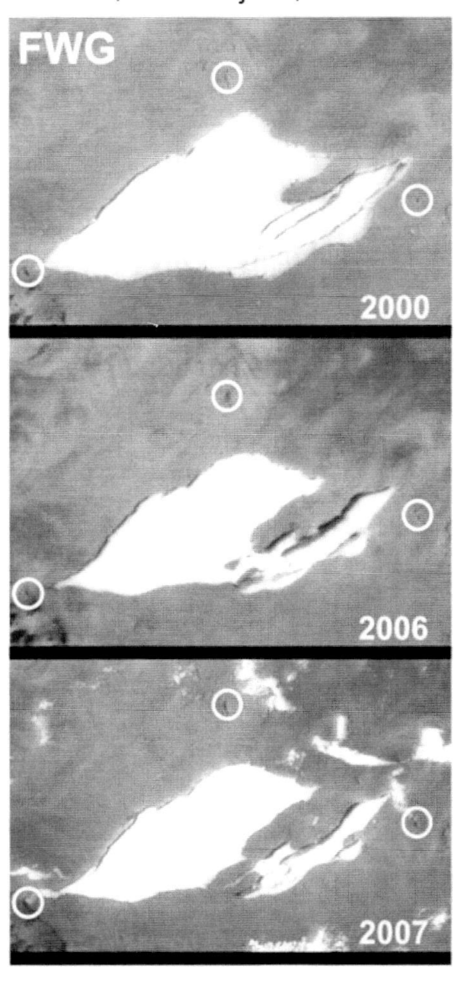

Figure 8. Aerial photographs taken in 2000, 2006, and 2007 of the Furtwängler Glacier in the center of the Kibo Crater, Mount Kilimanjaro, Tanzania, showing the rapid deterioration of this ice field. The open circles indicate identical surface features in each image for reference. Modified after *Thompson et al.* [2009].

1910, there were 150 glaciers within. This number is currently down to 26, and it is estimated that by 2030, at the present rate of decrease, no glaciers will remain in the Park [*Hall and Fagre*, 2003].

The ice on top of Mount Kilimanjaro has been photographed and mapped at irregular intervals since the early twentieth century, but starting in 2000, a systematic study has been underway to record and measure the rate of ice field retreat. Using a combination of terrestrial photogrammetric maps, satellite images, and aerial photographs, it has been determined that the combined surface area of all the ice fields on Kibo, the highest crater on Kilimanjaro, has decreased by 85% since 1912 when the first map was made [*Thompson et al.*, 2009]. At the current rate of retreat, it is estimated that Kibo will be ice free in the next few decades for the first time in 11,700 years. Continued monitoring of these ice fields demonstrates the accelerating rate of their retreat. For example, the deterioration of the Furtwängler Glacier, which lies in the center of the Kilimanjaro crater, is shown in Figure 8 as a series of aerial photographs from 2000 to 2007, when the glacier split into two sections. As it is shrinking in size, it has also thinned rapidly from 9.5 m in 2000 to 4.7 m in 2009, when the latest photographs were taken, and the ablation stake inserted in the borehole to bedrock in 2000 was measured [*Thompson et al.*, 2009, 2010].

The Quelccaya Ice Cap in southern Peru adjacent to the Amazon Basin is the largest tropical ice field on Earth. The Qori Kalis outlet glacier, which flows west from the main ice field, has been measured and photographed since 1963. At the beginning of the study, the end of the glacier extended 1200 m out from the margin, and there was no permanent meltwater at the terminus. By the summer of 2008, 45 years later, Qori Kalis had completely retreated to the margin of the main ice field, and a lake covering 34 ha and 60 m deep had developed in its place (Figure 9). As on Kilimanjaro, continuous monitoring of this glacier has documented an accelerating rate of ice loss: the retreat was calculated as a rate of about 6 m per year from 1963 to 1978, but Quelccaya lost ice 10 times faster on average than this initial rate from 1991 to 2006 [*Thompson et al.*, 2006a]. This loss of ice from Quelccaya is not only occurring on the Qori Kalis Glacier but also along the edge of the ice cap. Since 1978, 25% of this tropical ice cap has disappeared. This reduction in the areal extent of ice cover on both the Quelccaya Ice Cap and on the Qori Kalis Glacier characterizes the larger-scale accelerating glacier retreat throughout the Peruvian Andes (Figure 10).

Figure 11 graphically demonstrates the increasing rate of retreat of three low-latitude alpine ice caps. Qori Kalis and Kilimanjaro, which were discussed above, and Naimona'nyi on the Tibetan side of the southwest Himalaya Mountains are

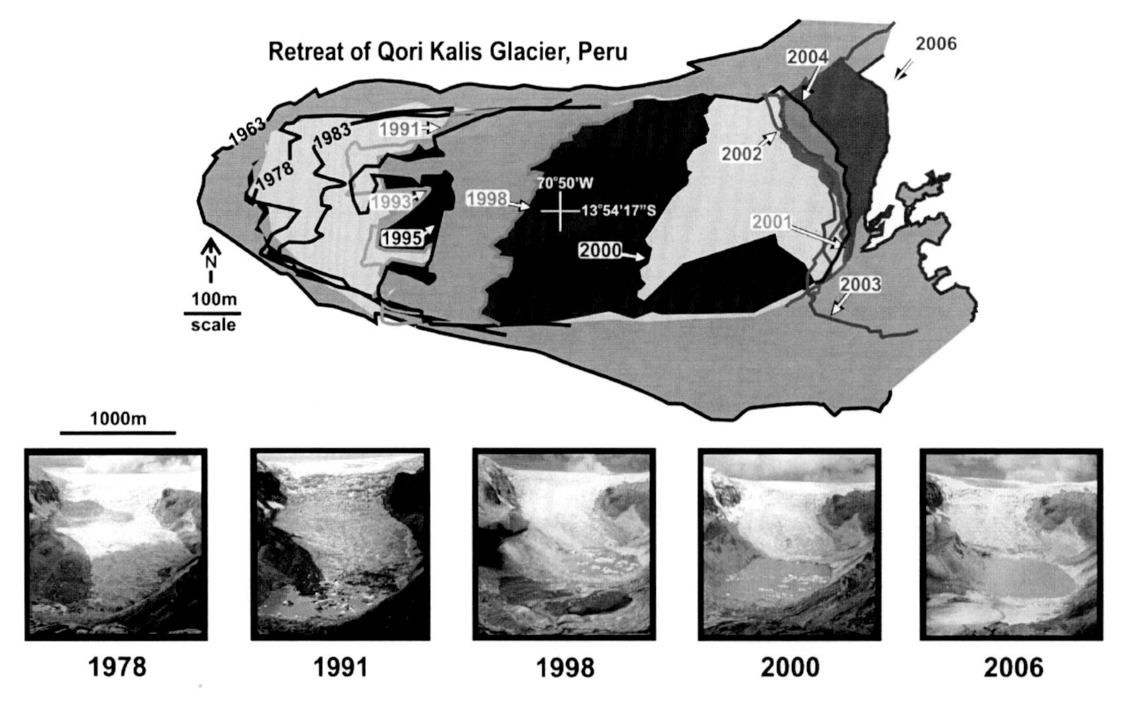

Figure 9. Retreat of the Qori Kalis outlet glacier on the Quelccaya Ice Cap. The shaded areas on the map outline the extent of the glacier through time and the progression of the ice retreat. The photographs along the bottom provide a pictorial history of the melting of this glacier from 1978 to 2006 and the formation of a proglacial lake.

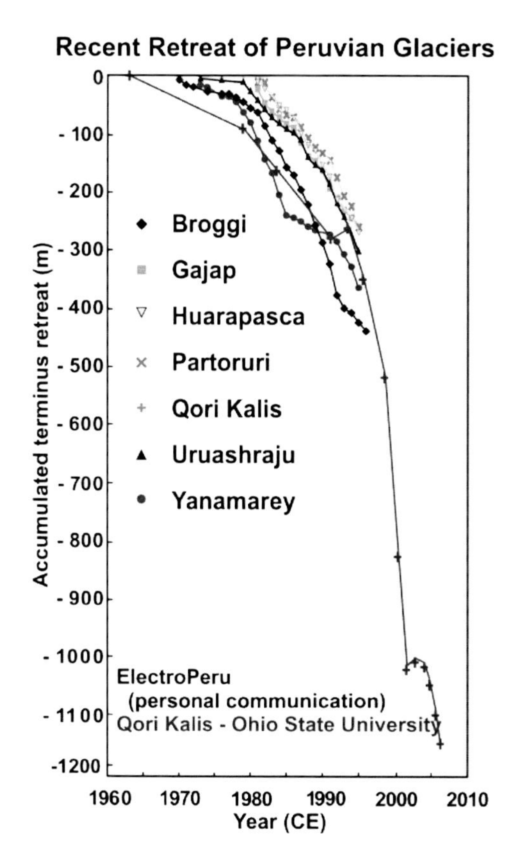

Figure 10. Retreat of the Qori Kalis Glacier, the largest outlet glacier flowing from the Quelccaya Ice Cap. Included are the retreat histories for six other Peruvian glaciers located in the Cordillera Blanca that were monitored by ElectroPeru from the early 1960s to the early 1990s.

all shrinking at rates that have markedly increased during the last few decades. While the percent loss per year on Qori Kalis and Kilimanjaro accelerated beginning in the 1990s, Naimona'nyi has undergone a dramatic decrease in surface area loss over the last decade [*Ye et al.*, 2006]. The Himalaya Mountains are home to more than 15,000 glaciers that constitute an important component of the dry season water supply for India, Nepal, southern China, and Southeast Asia via the major rivers of the region. Unfortunately, only a few of the glaciers have been monitored over an extended period, so the reliable ground observations that are crucial for determining ice retreat rates do not yet exist. However, a recent study on an ice core taken in 2006 from the Naimona'nyi Glacier indicates that ice is disappearing from the top [*Kehrwald et al.*, 2008], as shown by the lack of the radioactive bomb horizons from the 1950s and early 1960s that appear in all Tibetan and Himalayan ice core records [*Thompson et al.*, 1989, 1997, 2000, 2006b]. If the top of the time series from this ice core record is dated to before 1950, then between 30 and 40 m of ice have likely been lost from the surface in recent years. This is estimated by assuming an average regional precipitation rate of 0.80 m of water per year since 1998 as calculated using data from the NASA Goddard Earth Sciences Data and Information Services Center. Figure 11 suggests that much of this loss occurred very recently. Unfortunately, glacier observations are limited around the world but are even more limited in remote mountains such as the Andes and Himalaya, and thus, they do not offer a long-term perspective. The biological material (plants and animals) recently exposed as ice melts and glacier margins retreat can be radiocarbon dated and thereby provide a longer-term perspective. As previously mentioned, plants exposed along the retreating margins of the Quelccaya Ice Cap confirm that Earth's largest tropical ice cap is now smaller than it has been during at least the last ~5000 years.

The glaciologists at the Institute of Tibetan Plateau Research in Beijing have monitored 612 glaciers across the High Asia region since 1980. They found that from 1980 to 1990, 90% of these glaciers were retreating, and from 1990 to 2005, this proportion increased to 95% [*Yao et al.*, 2007]. Meteorological records from the Tibetan Plateau and the Himalayas are scarce and of relatively short duration, most beginning in the mid-1950s to early 1960s; however, those that do exist show that surface temperatures are rising and rising faster at higher elevations [*Liu and Chen*, 2000]. The Tibetan Plateau has been warming at a rate of 0.16°C per decade, with winter temperatures rising 0.32°C per decade. However, while examining 71 meteorological stations in eastern and central Tibetan Plateau, *You et al.* [2008] found a general warming of the plateau from 1961 to 2005 which did not significantly correlate with elevation. In recent decades, but especially since 2000, the Tibetan Plateau has experienced a more rapid warming than surrounding regions [*Qin et al.*, 2009]. Moderate Resolution Imaging Spectroradiometer monthly averaged land surface data for the whole Plateau show that the warming rate increases from 3000 to 4800 m, then becomes stable with a slight decrease near the highest elevations. Using a global database of 1084 high elevation meteorological stations, *Pepin and Lundquist* [2008] found that rising temperature trends through the twentieth century are most rapid near the annual 0°C isotherm due to snow-ice feedback. A 2009 study [*Matsuo and Heki*, 2010] shows that from 2003 to 2009, the average ice loss from the Asian high ice fields, as measured by Gravity Recovery and Climate Experiment satellite observations, had accelerated to twice the rate measured four decades before, but the recent loss was not consistent over space and time. Ice retreat in the Himalayas slowed slightly, while loss in the mountains to the northwest increased markedly over the last few years.

Thus, taking into consideration the surface temperature measurements, satellite studies, ground studies on glaciers,

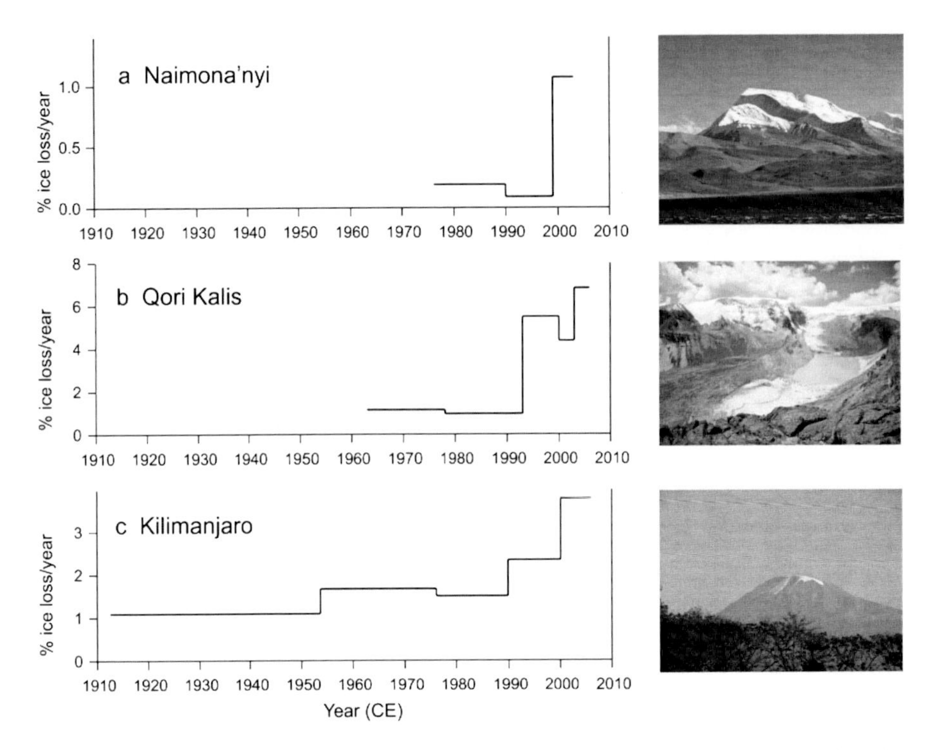

Figure 11. The rate of recent ice retreat, shown as percent ice loss, for three high-altitude tropical glaciers: (a) Naimona'nyi in the Himalayas [*Ye et al.*, 2006], (b) Qori Kalis in the Andes [from *Thompson et al.*, 2006a], and (c) Kilimanjaro in East Africa [from *Thompson et al.*, 2009].

and ice core results, a case can be made that glacier retreat at high elevations is indeed occurring concomitantly with increasing temperatures. This association is consistent with the model results [*Giorgi et al.*, 2001; *IPCC*, 2007] which show not only low-latitude warming but an amplification of that warming at higher elevations where these glaciers are located. A projected planetary-scale rise of 2°C to 4.5°C between 1990–1999 and 2090–2099 translates to a 5°C to 10+°C increase at 18,000 feet asl (inferred from Figure 1 of *Bradley et al.*, [2006]), endangering even the highest alpine glaciers. Figure 12 shows the decadally averaged isotopic ratios for the last thousand years plotted from lower to higher elevation (bottom, Dunde, 5235 m asl; Puruogangri, 6070 m asl; Guliya, 6200 m asl to Dasuopu, 7200 m asl, top). Not only do the mean isotopic values from 1000 to 1987 become more negative with increasing elevation (colder mean temperatures) but averages over their 50 year overlapping period from 1937 to 1987 demonstrate increasing enrichment with higher elevation, consistent with high-elevation amplification of warming trends in the tropics as predicted by the *IPCC* [2007].

Clearly, much more research needs to be conducted on abrupt climate events under warmer climate conditions. In particular, we need higher-resolution Holocene (warmer climate) proxy records to help separate zonal versus meridional shifts. Independently, well-dated records are needed to either establish synchroneity or better quantify leads and lags in the climate system. Rare earth element analysis can potentially fingerprint particulate source regions in paleo archives such as ice cores and marine and lake sediments and thereby better constrain circulation patterns during these abrupt climate events. Finally, an integration of these paleoclimatic records into the latest paleoclimate model reconstructions is urgently needed to reveal the types of forcings that are likely to produce the temporal and spatial patterns recorded in these proxy-based data sets.

5. CONCLUSIONS

Studying the evidence for abrupt climate change in paleoclimate records, especially in conjunction with archeological data on the rise and fall of prehistoric cultures, is an informative interdisciplinary pursuit. We understand that the Holocene, although less climatically variable than the last glacial cycle and deglaciation, has not been as stable and quiescent as originally believed. Abrupt variations in regional and global climate are not confined to the distant past but appear to be underway today. The natural drivers of climate

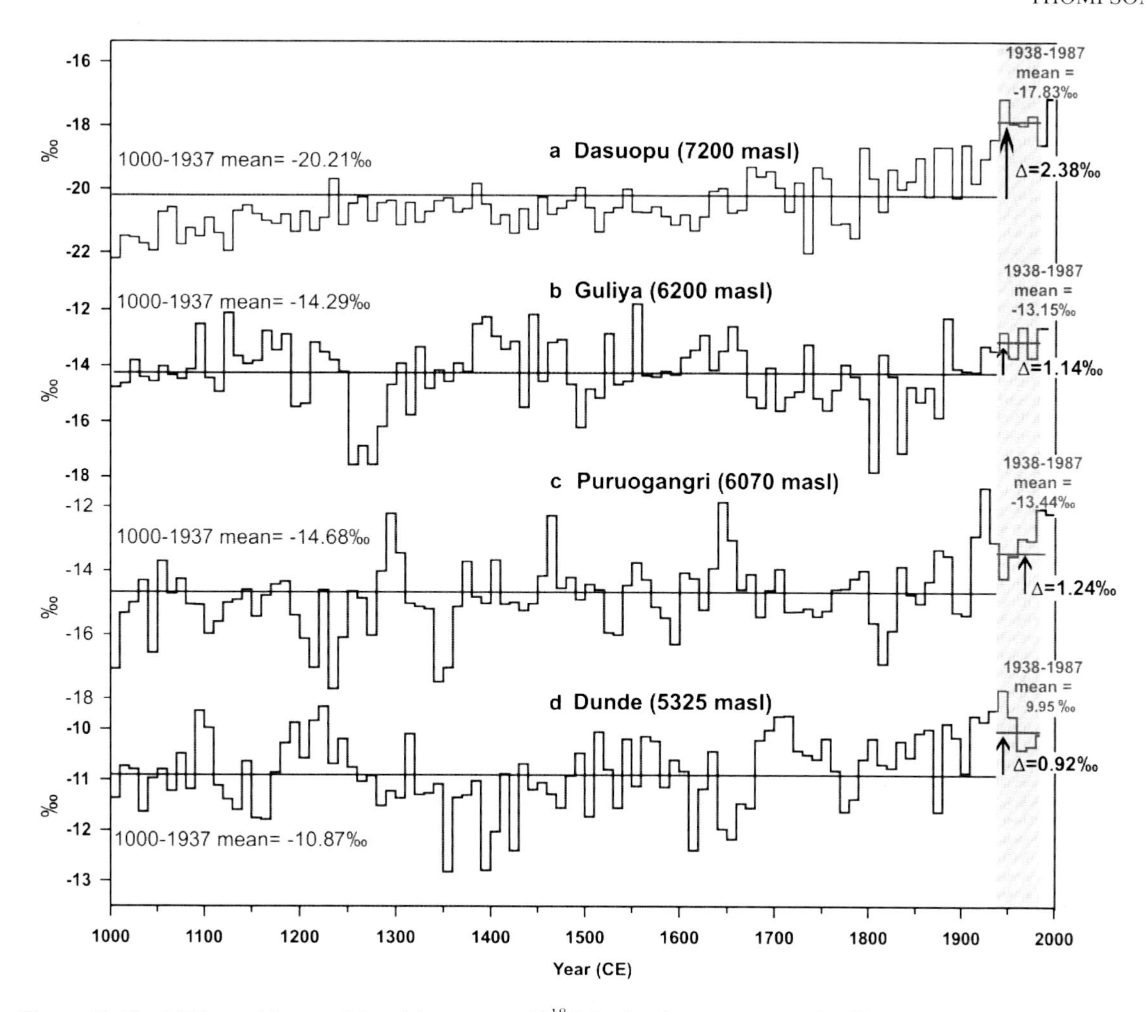

Figure 12. The 1000 year history of decadal averages of $\delta^{18}O$ for four ice cores across the Tibetan Plateau plotted from the top to the bottom with decreasing elevations (a) Dasuopu, 7200 m asl [from *Thompson et al.*, 2000]; (b) Guliya, 6200 m asl [*Thompson et al.*, 2003]; (c) Puruogangri, 6070 m asl [from *Thompson et al.*, 2006b], reprinted from the *Annals of Glaciology* with permission of the International Glaciological Society; and (d) Dunde, 5325 m asl [*Thompson et al.*, 2003]. The isotopic enrichment with increasing elevation in the twentieth century is illustrated by the mean isotopic values for the 50 year period from 1937 to 1987.

change, including insolation variability, volcanic eruptions, and interactions and feedback between the oceans, atmosphere, biosphere, and cryosphere currently are augmented by the radiative gas-driven climate forcing caused by the activities of almost 7 billion people. The increasing global surface and lower to middle tropospheric temperatures and the subsequent rapid disappearance of glaciers and ice caps that have existed for hundreds to tens of thousands of years certainly fits the description of abrupt climate change as stated by the National Research Council. As was the case for the prehistoric Peruvian cultures, many societies in developing countries are still dependent on agriculture for their survival and would be (and some indeed have been) devastated by sudden and prolonged decreases in precipitation. Even highly developed, prosperous nations would be hard-pressed to cope economically with an event like the 5.2 ka B.P. cold reversal or the 4.0–4.5 ka B.P. "megadrought," especially if the onset of such an event were as sudden and extreme as the paleoclimate records indicate and concentrated in the low latitudes where the vast majority of people live. The drivers of these events, and in fact the causes of short-term variability we experience today such as El Niño, are still not fully understood.

The history of rise and fall of past highland and coastal cultures in Peru before the rise of globalization would indicate that population migration such as has occurred since at least the 1940s from the highlands to the coastal plains is inconsistent with the longer-term trends. This may be unsustainable given that over the last 1500 years, highland cultures have flourished when it was wetter in the highlands, as it has been since the turn of the twentieth century. Coastal cultures have declined in the past when the highlands were wetter. It is likely that the abrupt middle Holocene climatic disruptions were triggered by a nexus of possible forcings that include solar variations, greenhouse gases, and changes in dominant modes of climate variability such as ENSO. The abruptness of the mid-Holocene events would argue for complex feedback in the ocean, atmosphere, biosphere, and cryosphere that were unique to that time period and crossed some threshold that induced feedback among Earth's internal systems. These events do show us, however, that the complex and interactive climate system may respond in very dramatic and unexpected ways to slowly changing conditions.

Acknowledgments. The results presented in this chapter reflect the collective efforts of many colleagues who participated in the field programs, the laboratory analyses, and numerous intellectual discussions during the formulation of the ideas presented here. Funding has been provided by the National Science Foundation's Paleoclimate Program Award ATM-0505476 and ATM-08235863 and The Ohio State University's Climate, Water and Carbon Program. This is Byrd Polar Research Center contribution number 1394.

REFERENCES

Abarzua, A. M., and P. I. Moreno (2008), Changing fire regimes in the temperate rainforest region of southern Chile over the last 16,000 yr, *Quat. Res.*, *69*(1), 62–71.

Arendt, A. A., K. A. Echelmeyer, W. D. Harrison, C. S. Lingle, and V. B. Valentine (2002), Rapid wastage of Alas glaciers and their contribution to rising sea level, *Science*, *297*, 382–386.

Baker, P. A., G. O. Seltzer, S. C. Fritz, R. B. Dunbar, M. J. Grove, P. M. Tapia, S. L. Cross, H. D. Rowe, and J. P. Broda (2001), The history of South American tropical precipitation for the past 25,000 years, *Science*, *291*(5504), 640–643.

Bar-Matthews, M., and A. Avalon (2004), Speleothems as paleoclimate indicators, a case study from Soreq Cave located in the Eastern Mediterranean Region, Israel, in *Past Climate Variability Through Europe and Africa*, edited by R. W. Battarbee et al., pp. 363–391, Springer, Dordrecht, Netherlands.

Bar-Matthews, M., A. Ayalon, A. Kaufman, and G. J. Wasserburg (1999), The Eastern Mediterranean paleoclimate as a reflection of regional events: Soreq Cave, Israel, *Earth Planet. Sci. Lett.*, *166*(1–2), 85–95.

Baroni, C., and G. Orombelli (1996), The Alpine "Iceman" and Holocene climatic change, *Quat. Res.*, *46*(1), 78–83.

Bradley, R. S., M. Vuille, D. Hardy, and L. G. Thompson (2003), Low latitude ice cores record Pacific sea surface temperatures, *Geophys. Res. Lett.*, *30*(4), 1174, doi:10.1029/2002GL016546.

Bradley, R. S., M. Vuille, H. F. Diaz, and W. Vergara (2006), Threats to water supplies in the tropical Andes, *Science*, *312*(5781), 1755–1756.

Buffen, A. M., L. G. Thompson, E. Mosley-Thompson, and K. I. Huh (2009), Recently exposed vegetation reveals Holocene changes in the extent of the Quelccaya Ice Cap, Peru, *Quat. Res.*, *72*(2), 157–163.

Butzer, K. W. (1997), Sociopolitical Discontinuity in the Near East C. 2200 B.C.E.: Scenarios from Palestine and Egypt, in *Third Millenium BC Climate Change and Old World Collapse*, edited by H. N. Dalfes, G. Kukla, and H. Weiss, pp. 245–296, Springer, Berlin.

Cardich, A. (1985), The fluctuating upper limits of cultivation in the central Andes and their impact on Peruvian prehistory, *Adv. World Archaeol.*, *4*, 293–333.

Chalié, F., and F. Gasse (2002), Late Glacial-Holocene diatom record of water chemistry and lake level change from the tropical East African Rift Lake Abiyata (Ethiopia), *Palaeogeogr. Palaeoclimatol. Palaeoecol.*, *187*(3–4), 259–283.

Chappellaz, J., T. Blunier, S. Kints, A. Dällenbach, J.-M. Barnola, J. Schwander, D. Raynaud, and B. Stauffer (1997), Changes in the atmospheric CH_4 gradient between Greenland and Antarctica during the Holocene, *J. Geophys. Res.*, *102*(D13), 15,987–15,997.

Coudrain, A., B. Francou, and Z. W. Kundzewicz (2005), Glacier shrinkage in the Andes and consequences for water resources – Editorial, *Hydrol. Sci. J.*, *50*(6), 925–932.

Cullen, H. M., P. B. deMenocal, S. Hemming, G. Hemming, F. H. Brown, T. Guilderson, and F. Sirocko (2000), Climate change and the collapse of the Akdian empire: Evidence from the deep sea, *Geology*, *28*(4), 379–382.

Davis, M. E., and L. G. Thompson (2006), An Andean ice-core record of a middle Holocene mega-drought in north Africa and Asia, *Ann. Glaciol.*, *43*, 34–41.

deMenocal, P., J. Ortiz, T. Guilderson, J. Adkins, M. Sarnthein, L. Baker, and M. Yarusinsky (2000), Abrupt onset and termination of the African Humid Period: Rapid climate responses to gradual insolation forcing, *Quat. Sci. Rev.*, *19*(1–5), 347–361.

Eitel, B., S. Hecht, B. Mächtle, G. Schukraft, A. Kadereit, G. A. Wagner, B. Kromer, I. Unkel, and M. Reindel (2005), Geoarchaeological evidence from desert loess in the Nazca-Palpa region, southern Peru: Palaeoenvironmental changes and their impact on pre-Columbian cultures, *Archaeometry*, *47*, 137–158.

Elsig, J., J. Schmitt, D. Leuenberger, R. Schneider, M. Eyer, M. Leuenberger, F. Joos, H. Fischer, and T. F. Stocker (2009), Stable isotope constraints on Holocene carbon cycle changes from an Antarctic ice core, *Nature*, *461*, 507–510, doi:10.1038/nature08393.

Gasse, F. (1977), Evolution of Lake Abhe (Ethiopia and TFAI), from 70,000 BP, *Nature*, *265*(5589), 42–45.

Gasse, F., et al. (1991), A 13,000-year climate record from western Tibet, *Nature*, *353*(6346), 742–745.

Gasse, F., J. C. Fontes, E. Van Campo, and K. Wei (1996), Holocene environmental changes in Bangong Co basin (western Tibet).

Part 4: Discussion and conclusions, *Palaeogeogr. Palaeoclimatol. Palaeoecol.*, *120*(1–2), 79–92.

Gillespie, R., F. A. Street-Perrott, and R. Switsur (1983), Postglacial arid episodes in Ethiopia have implications for climate prediction, *Nature*, *306*(5944), 680–683.

Giorgi, F., P. H. Whetton, R. G. Jones, J. H. Christensen, L. O. Mearns, B. Hewitson, H. vonStorch, R. Francisco, and C. Jack (2001), Emerging patterns of simulated regional climatic changes for the 21st century due to anthropogenic forcings, *Geophys. Res. Lett.*, *28*(17), 3317–3320.

Grosjean, M., L. Núñez, I. Cartajena, and B. Messerli (1997), Mid-Holocene climate and culture change in the Atacama Desert, northern Chile, *Quat. Res.*, *48*(2), 239–246.

Hall, M. H. P., and D. B. Fagre (2003), Modeled climate-induced glacier change in Glacier National Park, 1850-2100, *Bioscience*, *53*(2), 131–140.

Hansen, J., M. Sato, R. Ruedy, K. Lo, D. W. Lea, and M. Medina-Elizade (2006), Global temperature change, *Proc. Natl. Acad. Sci. U. S. A.*, *103*, 14,288–14,293, doi:10.1073/pnas.0606291103.

Hassan, F. A. (1997), Nile floods and political disorder in early Egypt, in *Third Millennium BC Climate Change and Old World Collapse*, edited by H. N. Dalfes, G. Kukla, and H. Weiss, pp. 1–24, Springer, Berlin.

Intergovernmental Panel on Climate Change (IPCC) (Ed.) (2007), *Climate Change 2007: The Physical Science Basis: Contribution of Working Group I to the Fourth Assessment Report of the IPCC*, edited by S. Solomon et al., Cambridge Univ. Press, Cambridge, New York.

Jones, P. D., and M. E. Mann (2004), Climate over past millennia, *Rev. Geophys.*, *42*, RG2002, doi:10.1029/2003RG000143.

Jones, P. D., and A. Moberg (2003), Hemispheric and large-scale surface air temperature variations: An extensive revision and an update to 2001, *J. Clim.*, *16*(2), 206–223.

Kehrwald, N. M., L. G. Thompson, Y. Tandong, E. Mosley-Thompson, U. Schotterer, V. Alfimov, J. Beer, J. Eikenberg, and M. E. Davis (2008), Mass loss on Himalayan glacier endangers water resources, *Geophys. Res. Lett.*, *35*, L22503, doi:10.1029/2008GL035556.

Lindstrom, S. (1990), Submerged tree stumps as indicators of mid-Holocene aridity in the Lake Tahoe Basin, *J. Calif. Great Basin Anthropol.*, *12*, 146–157.

Liu, X. D., and B. D. Chen (2000), Climatic warming in the Tibetan Plateau during recent decades, *Int. J. Climatol.*, *20*(14), 1729–1742.

Mächtle, B., B. Eitel, A. Kadereit, and I. Unkel (2006), Holocene environmental changes in the northern Atacama desert, southern Peru (14°30′S) and their impact on the rise and fall of Pre-Columbian cultures, *Z. Geomorphol. Suppl.*, *142*, 47–62.

Magny, M., and J. N. Haas (2004), A major widespread climatic change around 5300 cal. yr BP at the time of the Alpine Iceman, *J. Quat. Sci.*, *19*(5), 423–430.

Mann, M. E., and P. D. Jones (2003), Global surface temperatures over the past two millennia, *Geophys. Res. Lett.*, *30*(15), 1820, doi:10.1029/2003GL017814.

Mann, M. E., Z. Zhang, S. Rutherford, R. S. Bradley, M. K. Hughes, D. Shindell, C. Ammann, G. Faluvegi, and F. Ni (2009), Global signatures and dynamical origins of the Little Ice Age and Medieval Climate Anomaly, *Science*, *326*, 1256–1260.

Maslin, M. A., et al. (2000), Palaeoreconstruction of the Amazon River freshwater and sediment discharge using sediments recovered at Site 942 on the Amazon Fan, *J. Quat. Sci.*, *15*(4), 419–434.

Matsuo, K., and K. Heki (2010), Time-variable ice loss in Asian high mountains from satellite gravimetry, *Earth Planet. Sci. Lett.*, *290*(1–2), 30–36.

Meier, M. F., M. B. Dyurgerov, U. K. Rick, S. O'Neel, W. T. Pfeffer, R. S. Anderson, S. P. Anderson, and A. F. Glazovsky (2007), Glaciers dominate Eustatic sea-level rise in the 21st century, *Science*, *317*(5841), 1064–1067.

Molnia, B. F. (2007), Late nineteenth to early twenty-first century behavior of Alaskan glaciers as indicators of changing regional climate, *Global Planet. Change*, *56*(1–2), 23–56.

Moy, C. M., G. O. Seltzer, D. T. Rodbell, and D. M. Anderson (2002), Variability of El Niño/Southern Oscillation activity at millennial timescales during the Holocene epoch, *Nature*, *420*(6912), 162–165.

Müller, S. A., F. Joos, N. R. Edwards, and T. F. Stocker (2006), Water mass distribution and ventilation time scales in a cost-efficient, three-dimensional ocean model, *J. Clim.*, *19*, 5470–5499.

National Research Council (2002), *Abrupt Climate Change Inevitable Surprises*, 244 pp., Natl. Acad. Press, Washington, D. C.

Nissen, H. J. (1988), *The Early History of the Ancient Near East, 9000-2000 B.C*, 224 pp., Univ. of Chicago Press, Chicago, Ill.

Núñez, L., M. Grosjean, and I. Cartajena (2002), Human occupations and climate change in the Puna de Atacama, Chile, *Science*, *298*(5594), 821–824.

Paulsen, A. C. (1976), Environment and empire: Climatic factors in prehistoric Andean culture change, *World Archaeol.*, *8*(2), 121–132.

Pepin, N. C., and J. D. Lundquist (2008), Temperature trends at high elevations: Patterns across the globe, *Geophys. Res. Lett.*, *35*, L14701, doi:10.1029/2008GL034026.

Postgate, N. (1986), The transition from Uruk to Early Dynastic: Continuities and discontinuities in the record of settlement, in *Gemdet Nasr: Period or Regional Style?*, edited by U. Finkbeiner and W. Rollig, pp. 90–106, Ludwig Reichert, Wiesbaden, Germany.

Qin, J., K. Yang, S. Liang, and X. Guo (2009), The altitudinal dependence of recent rapid warming over the Tibetan Plateau, *Clim. Change*, *97*, 321–327, doi:10.1007/s10584-009-9733-9.

Raynaud, D., J.-M. Barnola, J. Chappellaz, T. Blunier, A. Indermühle, and B. Stauffer (2000), The ice record of greenhouse gases: A view in the context of future changes, *Quat. Sci. Rev.*, *19*(1–5), 9–17.

Rein, B., A. Lückge, and F. Sirocko (2004), A major Holocene ENSO anomaly during the Medieval period, *Geophys. Res. Lett.*, *31*, L17211, doi:10.1029/2004GL020161.

Rein, B., A. Lückge, L. Reinhardt, F. Sirocko, A. Wolf, and W.-C. Dullo (2005), El Niño variability off Peru during the last 20,000 years, *Paleoceanography*, *20*, PA4003, doi:10.1029/2004PA001099.

Rodbell, D. T., G. O. Seltzer, D. M. Anderson, M. B. Abbott, D. B. Enfield, and J. H. Newman (1999), An ~15,000-year record of El Niño-driven alluviation in southwestern Ecuador, *Science*, *283*(5410), 516–520.

Rothman, M. S. (2004), Studying the development of complex society: Mesopotamia in the late fifth and fourth millennia BC, *J. Archaeol. Res.*, *12*, 75–119.

Rowe, H. D., R. B. Dunbar, D. A. Mucciarone, G. O. Seltzer, P. A. Baker, and S. Fritz (2002), Insolation, moisture balance and climate change on the South American Altiplano since the Last Glacial Maximum, *Clim. Change*, *52*(1–2), 175–199.

Sandweiss, D. H. (2003), Terminal Pleistocene through mid-Holocene archaeological sites as paleoclimatic archives for the Peruvian coast, *Palaeogeogr. Palaeoclimatol. Palaeoecol.*, *194*(1–3), 23–40.

Seidel, D. J., and W. J. Randel (2007), Recent widening of the tropical belt: Evidence from tropopause observations, *J. Geophys. Res.*, *112*, D20113, doi:10.1029/2007JD008861.

Seidel, D. J., Q. Fu, W. J. Randel, and T. J. Reichler (2008), Widening of the tropical belt in a changing climate, *Nat. Geosci.*, *1*, 21–24.

Sellers, W. D. (1965), *Physical Climatology*, Univ. of Chicago Press, Chicago, Ill.

Seltzer, G., D. Rodbell, and S. Burns (2000), Isotopic evidence for late Quaternary climatic change in tropical South America, *Geology*, *28*(1), 35–38.

Servant, M., and S. Servant-Vildary (1980), L'environnement quaternaire du bassin du Tchad, in *The Sahara and the Nile*, edited by M. A. J. Williams and H. Faure, pp. 133–162, A. A. Balkema, Rotterdam, Netherlands.

Sobel, A. H. (2002), Water vapor as an active scalar in tropical atmospheric dynamics, *Chaos*, *2*(2), 451–459.

Staubwasser, M., and H. Weiss (2006), Holocene climate and cultural evolution in late prehistoric-early historic West Asia, *Quat. Res.*, *66*, 372–387.

Staubwasser, M., F. Sirocko, P. M. Grootes, and M. Segl (2003), Climate change at the 4.2 ka BP termination of the Indus valley civilization and Holocene south Asian monsoon variability, *Geophys. Res. Lett.*, *30*(8), 1425, doi:10.1029/2002GL016822.

Swap, R., M. Garstang, S. Greco, R. Talbot, and P. Kållberg (1992), Saharan Dust in the Amazon Basin, *Tellus, Ser. B*, *44*(2), 133–149.

Tapia, P. M., S. C. Fritz, P. A. Baker, G. O. Seltzer, and R. B. Dunbar (2003), A Late Quaternary diatom record of tropical climatic history from Lake Titicaca (Peru and Bolivia), *Palaeogeogr. Palaeoclimatol. Palaeoecol.*, *194*(1–3), 139–164.

Tapley, T. D., and P. R. Waylen (1990), Spatial variability of annual precipitation and ENSO events in western Peru, *Hydrol. Sci. J.*, *35*(4), 429–446.

Thompson, L. G. (2000), Ice core evidence for climate change in the Tropics: Implications for our future, *Quat. Sci. Rev.*, *19*(1–5), 19–35.

Thompson, L. G., E. Mosley-Thompson, M. E. Davis, J. F. Bolzan, J. Dai, L. Klein, T. Yao, X. Wu, Z. Xie, and N. Gundestrup (1989), Holocene–late Pleistocene climatic ice core records from Qinghai-Tibetan Plateau, *Science*, *246*(4929), 474–477.

Thompson, L. G., M. E. Davis, and E. Mosley-Thompson (1994), Glacial records of global climate: A 1500-year tropical ice core record of climate, *Hum. Ecol.*, *22*(1), 83–95.

Thompson, L. G., E. Mosley-Thompson, M. E. Davis, P.-N. Lin, K. A. Henderson, J. Cole-Dai, J. F. Bolzan, and K.-B. Liu (1995), Late-Glacial Stage and Holocene tropical ice core records from Huascarán, Peru, *Science*, *269*(5220), 46–50.

Thompson, L. G., T. Yao, M. E. Davis, K. A. Henderson, E. Mosley-Thompson, P.-N. Lin, J. Beer, H.-A. Synal, J. Cole-Dai, and J. F. Bolzan (1997), Tropical climate instability: The last glacial cycle from a Qinghai-Tibetan ice core, *Science*, *276*(5320), 1821–1825.

Thompson, L. G., et al. (1998), A 25,000 year tropical climate history from Bolivian ice cores, *Science*, *282*(5295), 1858–1864.

Thompson, L. G., T. Yao, E. Mosley-Thompson, M. E. Davis, K. A. Henderson, and P.-N. Lin (2000), A high-resolution millennial record of the South Asian Monsoon from Himalayan ice cores, *Science*, *289*(5486), 1916–1919.

Thompson, L. G., et al. (2002), Kilimanjaro ice core records: Evidence of Holocene climate change in tropical Africa, *Science*, *298*(5593), 589–593.

Thompson , L. G., E. Mosley-Thompson, M. E. Davis, P.-N. Lin, K. Henderson, and T. A. Mashiotta (2003), Tropical glacier and ice core evidence of climate change on annual to millennial time scales, *Clim. Change*, *59*, 137–155.

Thompson, L. G., E. Mosley-Thompson, H. Brecher, M. Davis, B. León, D. Les, P.-N. Lin, T. Mashiotta, and K. Mountain (2006a), Abrupt tropical climate change: Past and present, *Proc. Natl. Acad. Sci. U. S. A.*, *103*(28), 10,536–10,543.

Thompson, L. G., T. Yao, M. E. Davis, E. Mosley-Thompson, T. A. Mashiotta, P.-N. Lin, V. N. Mikhalenko, and V. S. Zagorodnov (2006b), Holocene climate variability archived in the Puruogangri ice cap on the central Tibetan Plateau, *Ann. Glaciol.*, *43*, 61–69.

Thompson, L. G., H. H. Brecher, E. Mosley-Thompson, D. R. Hardy, and B. G. Mark (2009), Glacier loss on Kilimanjaro continues unabated, *Proc. Natl. Acad. Sci. U. S. A.*, *106*(47), 19,770–19,775.

Thompson, L. G., H. H. Brecher, E. Mosley-Thompson, D. R. Hardy, and B. G. Mark (2010), Response to Mölg et al.: Glacier loss on Kilimanjaro is consistent with widespread ice loss in low latitudes, *Proc. Natl. Acad. Sci. U. S. A.*, *107*(17), E69–E70.

Vonmoos, M., J. Beer, and R. Muscheler (2006), Large variations in Holocene solar activity: Constraints from ^{10}Be in the Greenland Ice Core Project ice core, *J. Geophys. Res.*, *111*, A10105, doi:10.1029/2005JA011500.

Voorhies, B., and S. E. Metcalfe (2007), Culture and climate in Mesoamerica during the middle Holocene, in *Climate Change and Cultural Dynamics: A Global Perspective on Mid-Holocene*

Transitions, edited by D. G. Anderson, K. A. Maasch, and D. H. Sandweiss, pp. 157–188, Elsevier, Amsterdam.

Wanner, H., et al. (2008), Mid- to late Holocene climate change: An overview, *Quat. Sci. Rev.*, *27*, 1791–1828.

Wang, Y. J., H. Cheng, R. L. Edwards, Y. He, X. Kong, Z. An, J. Wu, M. J. Kelly, C. A. Dykoski, and X. Li (2005), The Holocene Asian monsoon: Links to solar changes and North Atlantic climate, *Science*, *308*(5723), 854–857.

Webster, P. J. (2004), The Elementary Hadley Circulation, in *The Hadley Circulation: Past, Present and Future*, edited by H. F. Diaz and R. Bradley, pp. 9–60, Cambridge Univ. Press, Cambridge, U. K.

Webster, P. J., V. O. Magaña, T. N. Palmer, J. Shukla, R. A. Tomas, M. Yanai, and T. Yasunari (1998), Monsoons: Processes, predictability, and the prospects for prediction, *J. Geophys. Res.*, *103*(C7), 14,451–14,510.

Weiss, H. (2003), Ninevite Periods and Processes, in *The Origins of North Mesopotamian Civilization: Ninevite 5 Chronology, Economy, Society*, edited by E. Rova and H. Weiss, pp. 593–624, Brepols, Turnhout, Belgium.

Weiss, H., and R. S. Bradley (2001), What drives societal collapse?, *Science*, *291*(5504), 609–610.

Weiss, H., M.-A. Courty, W. Wetterstrom, F. Guichard, L. Senior, R. Meadow, and A. Curnow (1993), The genesis and collapse of third millennium North Mesopotamian civilization, *Science*, *261*(5124), 995–1004.

Wu, W. X., and T. S. Liu (2004), Possible role of the "Holocene Event 3" on the collapse of Neolithic Cultures around the central Plain of China, *Quat. Int.*, *117*, 153–166.

Xiao, J. L., Q. Xu, T. Nakamura, X. Yang, W. Liang, and Y. Inouchi (2004), Holocene vegetation variation in the Daihai Lake region of north-central China: A direct indication of the Asian monsoon climatic history, *Quat. Sci. Rev.*, *23*(14–15), 1669–1679.

Yao, T., J. Pu, A. Lu, Y. Wang, and W. Yu (2007), Recent glacial retreat and its impact on hydrological processes on the Tibetan Plateau, China, and surrounding regions, *Arct. Antarct. Alp. Res.*, *39*(4), 642–650.

Ye, Q., T. Yao, S. Kang, F. Chen, and J. Wang (2006), Glacier variations in the Naimona'nyi region, western Himalaya, in the last three decades, *Ann. Glaciol.*, *43*, 385–389.

You, Q., S. Kang, N. Pepin, and Y. Yan (2008), Relationship between trends in temperature extremes and elevation in the eastern and central Tibetan Plateau, 1961–2005, *Geophys. Res. Lett.*, *35*, L04704, doi:10.1029/2007GL032669.

L. G. Thompson, Byrd Polar Research Center, School of Earth Sciences, Ohio State University, Columbus, Ohio, USA. (Thompson.3@osu.edu)

AGU Category Index

Index

Note: Page numbers with italicized *f* and *t* refer to figures and tables